COMPUTERIZED ENGINE CONTROLS

Fifth Edition

Dick H. King
Glendale Community College

Delmar Publishers
An International Thomson Publishing Company I(T)P®

Albany • Bonn • Boston • Cincinnati • Detroit • London • Madrid • Melbourne • Mexico City • New York • Pacific Grove • Paris • San Francisco • Singapore • Tokyo • Toronto • Washington

Cover photo courtesy of Mercedes-Benz of North America, Inc.
Cover design by Brucie Rosch

DELMAR STAFF
Publisher: Robert D. Lynch
Acquisitions Editor: Vernon Anthony
Production Coordinator: Karen Smith
Editorial Assistant: Rhonda Kreshover
Art/Design Coordinator: Cheri Plasse

COPYRIGHT © 1997
By Delmar Publishers
an International Thomson Publishing company

The ITP logo is a trademark under license.

Printed in the United States of America

For more information, contact:

Delmar Publishers
3 Columbia Circle, Box 15015
Albany, New York, 12212-5015

International Thomson Publishing Europe
Berkshire House 168-173
High Holborn
London WC1 V 7AA
England

Thomas Nelson Australia
102 Dodds Street
South Melbourne, 3205
Victoria, Australia

Nelson Canada
1120 Birchmount Road
Scarborough, Ontario
Canada M1K 5G4

International Thomson Editores
Campos Eliseos 385, Piso 7
Col Polanco
11560 Mexico D F Mexico

International Thomson Publishing GmbH
Königswinterer Strasse 418
53227 Bonn
Germany

International Thomson Publishing Asia
221 Henderson Road
#05-10 Henderson Building
Singapore 0315

International Thomson Publishing Japan
Hirakawacho Kyowa Building, 3F
2-2-1 Hirakawacho
Chiyoda-ku, Tokyo 102
Japan

Delmar Publishers' Online Services

To access Delmar on the World Wide Web, point your browser to:
http:/www.delmar.com/delmar.html
To access through Gopher: gopher://gopher.delmar.com
(Delmar Online is part of "thomson.com", an internet site with information on more than 30 publishers on the International Thomson Publishing organization.)
For information on our products and services:
email:info@delmar.com
or call 800-347-7707

10 9 8 7 6 5 4   XXX 02 01 00 99 98

Library of Congress Cataloging-in-Publication Data
King, Dick H.
    Computerized engine controls / Dick H. King — 5th ed.
        p.   cm.
    Includes index.
    ISBN 0-8273-7878-5
    1. Automobiles—Motors—Computer control systems.   I. Title.
TL214.C64K56 1997   96–30855
                                                         CIP

Contents

Chapter 18: OBD II
Self-Diagnostics 499

Preface

The application of the microprocessor with its related components and circuits has made automotive technology exciting, fast paced and more complicated. Recent technological developments and those that follow will require entry level automotive service technicians to be well trained in the principles of automotive technology and to be career-long students. Those who respond to this requirement will find the task challenging, but achievable and rewarding.

This text was written in response to a widely recognized need within the industry: to help student-technicians get a commanding grasp of how computerized engine control systems work and how to diagnose problems within them. The student-technician who studies this text will soon come to realize that no single component or circuit within any given computerized engine control system, other than the computer itself, is complicated.

Computerized Engine Controls is written with the assumption that the reader is familiar with the basic principles of traditional engine, electrical system, and fuel system operation. While everything here is within the grasp of a good technician, this is not a beginner's book.

While we recognize that a computerized engine control system does in fact become an integral part of an engine's electrical and fuel system, it is much too significant and complex to be taught as just a unit in an engine performance textbook or class. For purposes of instruction, we have taken this topic out of context and examined it as a stand-alone control-level system. Once the student-technician fully understands the system's purpose, operation, and diagnostic approach, the diagnostic procedures in the service manual will put the system back in its proper perspective as an integral part of the engine's support system.

Computerized Engine Controls presents each popular, multifunction computer control system in a separate chapter. Each system is fully covered, with enough specific information and detail to enable the reader to get a complete and clear picture of how the system works. This text is written with the premise that understanding how the overall system works and what it should be doing not only makes the diagnostic process easier, but also makes the diagnostic literature much easier to understand. Correctly interpreting diagnostic procedure directions is often a problem if the technician is not aware of what the procedure is trying to measure, what normal readings or responses should be, or what conditions will cause abnormal readings or responses.

Frequently, information contained in a manufacturer's service manual is presented in this text to help make a specific concept more clear and to help acquaint the reader with information found in the service manual. Other references are made to the service manual to emphasize the importance of using the service manual in diagnostic procedures. This text is dedicated to helping the student-technician acquire the necessary knowledge to diagnose and repair driveability problems with a computerized engine control system.

Objectives are provided at the beginning of each chapter to help the reader identify the major concepts to be presented. **Review Questions**

are provided at the end of each chapter, which help the readers assess their recall and comprehension concerning the most important concepts. The review questions are also written to reinforce the objectives. Where useful and appropriate, **Real World Problems, Service & Diagnostic Tips,** and **Diagnostic Exercises** are introduced into the text.

Personal safety concerns peculiar to specific computerized engine control systems are highlighted where applicable. The book follows the industry standards for how to use the following terms:

- **Warnings** indicate that failure to observe correct diagnostic or repair procedures could result in personal injury or even death.
- **Cautions** indicate that failure to observe correct diagnostic or repair procedures could result in damage to tools, equipment or the vehicle serviced.
- **Notes** indicate procedures or practices that can make the task at hand easier or that may show another way to solve or understand a particular problem.

Each student should be aware that while working with computerized controls is not inherently dangerous, failure to observe recognized safety practices is. There are unfortunately many more injuries and accidents in the automotive repair business than there should be. Good safety practices if learned early in a student's career literally can be life-saving later on.

Terms unique to computerized engine control systems are highlighted when they are first introduced and their meaning is explained in the related text. A list of such **Key Terms** is provided at the beginning of each chapter. For those chapters where a significant number of abbreviations are used, a list of abbreviations is also provided. Background information that is not of specific concern to understanding a given concept but might be of general interest to the reader is inserted in selected locations throughout the text.

SUGGESTIONS ON HOW TO USE THIS TEXT

Read the **Introduction** first. This section provides background on the operation of many components that are common to most systems. The remaining chapters, which are specific to individual systems, can be read in any order, though most students will find it much easier to read a manufacturer's set of chapters in chronological sequence, the way they appear in the book. Generally systems become more complicated over time, and the later versions are easier to understand with a good background knowledge of the earlier, simpler systems. Many students may not wish to study each specific chapter, or their instructors may not choose to assign the study of each chapter. We suggest, however, that at least three of the specific system chapters be selected for study following the completion of the **Introduction.** Remaining chapters can then be skimmed or serve as a reference for future use.

Acknowledgments

I am grateful to several people who have helped with the many activities involved in the creation of this book:

- to Theresa Murphy who generously gave of her personal time to type the proposal chapter;
- to Margery Rothschild for her hours of work and consistently quick turnaround time in processing illustrations;
- to Tim Turpin for the photographs used in the book;
- to Renault Catalano for his consultation concerning electronics,
- and to the following for their critical reviews and/or answers to questions:

Thomas Baird,
   Butte Community College
Lawrence W. Breeden,
   Albany Technical Institute
Joel Copper,
   Muskingum Area Technical College

John Gahrs,
   Ferris State University
Harold Gubler,
   Columbia-Greene Community College
Dan Hall,
   Kirkwood Community College
Carl Hinkley,
   Central Maine Technical College
Mike Huneke,
   Texas State Technical College
Dale Jaenke,
   John Tyler Community College
Steve Levin
   Columbus State Community College
Randall Pollmeier,
   Indian River Community College
Danny Rakes,
   Danville Community College
Lester Smith,
   Monroe Community College
Patrick Tiekamp,
   University of New Mexico—Gallup

ix

A Review Of Electricity And Electronics

The earliest automobiles had little in the way of electrical systems, but as the automobile has become more complicated and as more accessories have been added, **electrical** or **electronic** systems have become a major feature of today's car. Additional electronic control systems have made and will continue to make the automobile comply with government standards and consumer demands. Automotive manufacturers tell us that before the 1990s have passed, most major automotive systems will be controlled by computers.

This increased use of electrical and electronic systems means two things for the automotive service technician: First, all service technicians need skills in electrical diagnosis and repair to be effective, almost regardless of the technician's service specialty; second, technicians with such skills will command significantly greater financial rewards, and will deserve them.

There are several principles by which electrical systems operate, but they are all fairly simple; learning them is not difficult. As each principle is introduced to you through your reading or in class, ask questions and/or read until you understand it. Review the principles frequently and practice the exercises that your instructor assigns.

ELECTRICAL CIRCUITS VERSUS ELECTRONIC CIRCUITS

The differences between electrical circuits and electronic circuits are not always clear-cut or definite. This has led to some confusion about the use of those terms and how an electronic system is different from an electrical system. Perhaps the comparisons in the table below will help.

Even though the use of solid-state components may often be used as criteria to identify an electronic circuit, solid-state components such as a **power transistor** may also be used in an electrical circuit. A power transistor is a type of transistor designed to carry larger amounts of amperage than are normally found in an electronic circuit. A power transistor is essentially a highly reliable relay.

Electrical Circuits	Electronic Circuits
Do physical work: movement, heat, light, magnetism.	Communicate information: voltages or on-off signals.
Use electromechanical devices: switches, solenoids, relays.	Use solid-state devices (semiconductors) with no moving parts, transistors or diodes.
Operate at relatively high current or amperage.	Operate at relatively low current or amperage.
Circuits have relatively low resistance (ohms).	Circuits have relatively high resistance (ohms).

ELECTRON THEORY

Molecules and Atoms

A study of electricity begins with the smallest pieces of matter. All substances—air, water, wood, steel, stone, and even the various substances that our bodies are made of—are made of the same bits of matter. Every substance is made of units called **molecules.** A molecule is a unit formed by combining two or more atoms and is the smallest unit that a given substance can be broken down to and still exhibit all of the characteristics of that substance. For example, a molecule of water is made up of two atoms of hydrogen and one atom of oxygen (H_2O: H is the chemical symbol for hydrogen and O is the chemical symbol for oxygen). If a molecule of water is broken down into its component atoms, it is no longer water.

As molecules are made up of atoms, atoms are in turn made up of:

- electrons, or negatively charged particles,
- protons, or positively charged particles,
- neutrons, or particles with no charge; at the level of atomic activity concerning us here, these just add mass to the atom.

The smallest and lightest atom is the hydrogen atom. It contains one proton and one electron (it is the only atom that does not have a neutron), Figure I-1. The next smallest and lightest atom is the helium atom. It has two protons, two neutrons, and two electrons, Figure I-2. Since the hydrogen atom is the smallest and lightest, and since it has one electron and one proton, it is given an atomic number of 1. Since helium is the next lightest with 2 electrons, protons, and neutrons, it has an atomic number of 2. Every atom has been given an atomic number that indicates its relative size and weight (or its mass) and the number of electrons, protons and neutrons it contains. An atom usually has the same number of electrons, protons, and neutrons.

Elements

Once the three different bits of matter are united to form an atom, two or more atoms combine to form a molecule. If all of the atoms in the molecule are the same, the molecule is called an **element.** Which element it is depends on how many protons, neutrons and electrons the atoms contain. There are more than 100 different elements. Some examples of elements are hydrogen, oxygen, carbon, iron, lead, gold, silicon, and sodium. An element, then, is a pure substance whose molecules contain only one kind of atom. Substances such as water, which contains hydrogen and oxygen atoms; or steel, which contains iron and carbon atoms, are called **compounds.**

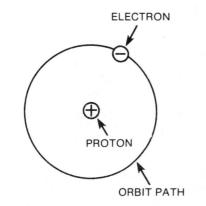

Figure I–1 Hydrogen atom

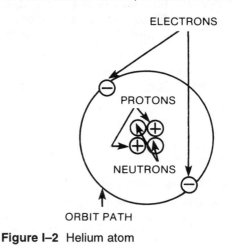

Figure I–2 Helium atom

Atomic Structure and Electricity

Notice in Figures I-1 and I-2 that the protons and neutrons are grouped together in the center of each atom. This is called the **nucleus** of the atom. The electrons travel around the nucleus of the atom in an orbit similar to the way that the earth travels around the sun. But because an atom usually has several electrons orbiting around its nucleus, the electrons form in layers, rather than all of them traveling in the same orbit, Figure I-3. Some, however, will share the same orbit, as seen in Figure I-3. For the purpose of this text, only the electrons in the last layer are of any real importance. This layer is often called the **outer shell** or **valence ring.** The student should realize that we are speaking very loosely here when we describe electrons in shells having orbits. Modern physics actually describes electrons in a shell as probability functions over a general rather than small physical objects spinning in circular orbits. For our purposes, however, the simpler explanation (a model once called the **Rutherford atom**) satisfactorily conveys the nature of the electron.

As mentioned, electrons are negatively charged and protons are positively charged. You have probably heard or know that like charges repel and unlike charges attract. And electrons are always moving; in fact, they are sometimes said to move at nearly the speed of light. These characteristics work together to explain many of the behaviors of an atom that makes current flow. **Current** is defined as having a mass of free electrons moving in the same direction.

NOTE: There are two types of current: **direct current (DC)** and **alternating current (AC).** Direct current always flows in one direction. Current from a battery is the best example. Most of the devices in an automobile use DC. Alternating current circuits continuously switch which end is positive and which end is negative so that current flow (and thus electron movement) reverses direction repeatedly. The power that is available from commercial utility companies is AC and cycles (changes **polarity**) 60 times per second. One cycle occurs when the current switches from forward to backward to forward again. The car's alternator (an alternating current generator) produces AC current, which is converted to DC before it leaves the alternator.

The fast-moving electron wants to move in a straight line, but its attraction to the proton nucleus makes it act like a ball tied to the end of a string twirled around. The repulsive force between the electrons keeps them spread as far apart as their attraction to the nucleus will allow.

The fewer electrons there are in the outer shell of the atom and the more layers of electrons there are under the outer shell, the weaker is the bond between the outer electrons and the nucleus. If one of these outer electrons can somehow be broken free from its orbit, it will travel to a neighboring atom and fall into the outer shell there, resulting in two unbalanced atoms. The first atom is missing an electron. It is now positively charged and is called a **positive ion.** The second atom has an extra electron. It is negatively charged and is called a **negative ion.** Ions

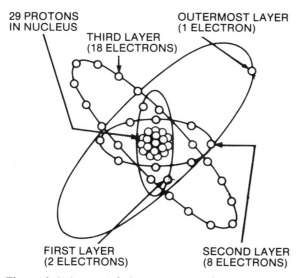

Figure I–3 Layers of electrons around a copper atom nucleus

are unstable. They want to either gain an electron or get rid of one so that they are balanced.

Potential

The first atom in the example has positive **potential.** It has more positive charge than negative charge because it has more protons than electrons. Suppose that this atom is at one end of a circuit, Figure I-4. Further suppose that there is an atom at the other end of the circuit that has an extra electron; it has negative potential. Because of the difference in potential at the two ends of the circuit, an electron at the negatively charged end will start moving toward the positively charged end. The greater the difference in potential (the greater the number of opposite charged ions) at each end of the circuit, the greater the number of electrons that will start to flow.

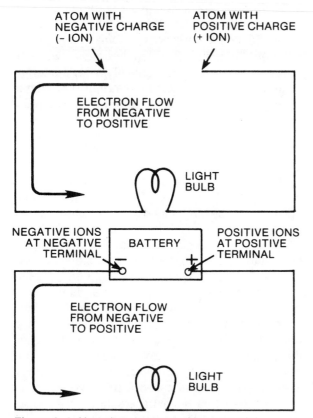

Figure I–4 Negative versus positive potential

This situation can be created by putting something between the two ends of a circuit that will produce positive and negative ions. This is what a battery or generator does in a circuit, Figure I-4. If you connect the ends of the wire to a battery and put some kind of **resistance,** or opposition to a steady electric current, in the wire (why the resistance is needed is explained later), the example will work perfectly. Remember that it is the difference in potential that makes current flow. Actually there are three factors that must be present for an electrical circuit to work, that is, for current to flow through it.

MAGNETISM

Magnetism is closely tied to the generation and use of electricity. In fact, one of the prevailing theories is that magnetism is caused by the movement and group-orientation of electrons. Some materials strongly demonstrate the characteristics associated with magnetism and some do not. Those that strongly demonstrate the characteristics of magnetism such as iron are said to have high **permeability.** Those that do not such as glass, wood, and aluminum are said to have high **reluctance.**

Lines of Force

It is not known whether there actually is such a thing as a **magnetic line of force.** What is known, however, is that magnetism exerts a force which we can understand and manipulate if we assume there are magnetic lines of force. Magnetic force is linear in nature and it can be managed to do many kinds of work. By assigning certain characteristics to these lines of force, the behavior of magnetism can be explained. Magnetic lines of force:

1. have a directional force (north to south outside the magnet).
2. want to take the shortest distance between two poles (just like a stretched rubber band

between the two points from which it is held).

3. form a complete loop.
4. are more permeable to iron than air.
5. resist being close together (especially in air).
6. resist being cut.
7. will not cross each other (they will bend first).

Magnetic lines of force extending from a magnet make up what is commonly called a magnetic field and more correctly called magnetic **flux,** Figure I-5. If a magnet is not near an object made of permeable material, the lines of force will extend from the north pole through the air to the south pole (characteristic 1). The lines of force will continue through the body of the mag-

net to the north pole to form a complete loop (characteristic 3). Every magnet has a north and a south pole. The poles are the two points of a magnet where the magnetic strength is greatest. As the lines of force extend out of the north pole, they begin to spread out. Here you see opposition between characteristics 2 and 5. The lines of force want to take the shortest distance between the poles, but they spread out because of their tendency to repel each other (characteristic 5). The result is a magnetic field that occupies a relatively large area but has greater density near the body of the magnet.

Because the body of the magnet has high permeability, the lines of force are very concentrated in the body of the magnet (characteristic 4). This accounts for the poles of the magnet having the highest magnetic strength.

If there is an object with high permeability near the magnet, the magnetic lines of force will distort from their normal pattern and go out of their way to pass through the object, Figure I-6. The tendency for the lines of force to pass through the permeable object is stronger than their tendency to take the shortest route. The lines of force will, however, try to move the object toward the nearest pole of the magnet.

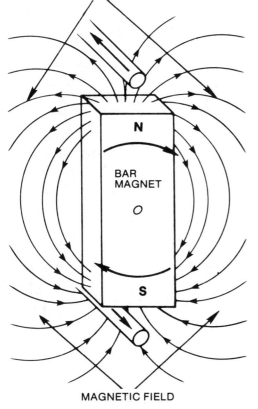

Figure I-5 Magnetic field

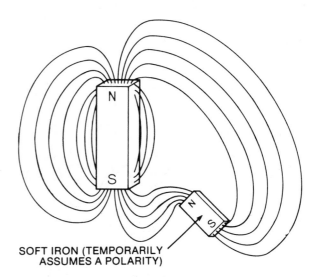

SOFT IRON (TEMPORARILY ASSUMES A POLARITY)

Figure I-6 Magnetic field distortion

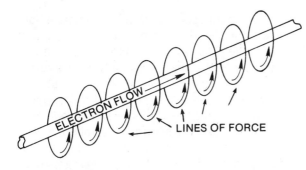

IF YOU IMAGINE PLACING YOUR LEFT HAND
ON THE WIRE WITH YOUR THUMB POINTING
IN THE DIRECTION OF ELECTRON FLOW,
YOUR FINGERS WILL BE POINTING IN THE
DIRECTION OF THE DIRECTIONAL FORCE OF
THE MAGNETIC LINES OF FORCE. WHEN
THINKING IN TERMS OF CONVENTIONAL
CURRENT FLOW, THE SAME WOULD APPLY
FOR THE RIGHT HAND.

Figure I–7 Lines of force forming around a conductor

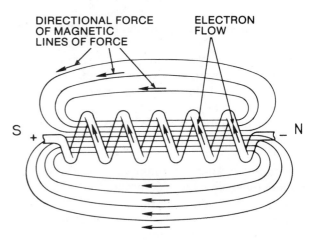

IMAGINE PLACING YOUR LEFT HAND, WITH
THUMB EXTENDED, AROUND THE COIL WITH
YOUR FINGERS POINTING IN THE DIRECTION
OF THE ELECTRON FLOW THROUGH THE
COILS. YOUR THUMB WILL POINT TO THE
NORTH POLE OF THE MAGNETIC FIELD.

Figure I–8 Magnetic field around a coil

Electromagnets

Early researchers discovered that when current passes through a conductor, a magnetic field forms around the conductor, Figure I-7. This principle makes possible the use of electromagnets, electric motors, generators and most of the other components used in electrical circuits.

If a wire is coiled with the coils close together, most of the lines of force wrap around the entire coil rather than go between the coils of wire, Figure I-8. This is because if they do try to wrap around each loop in the coil, they must cross each other, which they will not do (characteristic 7), Figure I-9.

If a high permeability core is placed in the center of the coil, the magnetic field becomes

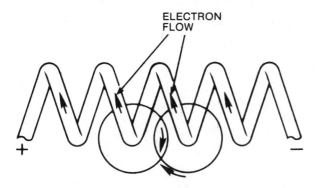

MAGNETIC LINES OF FORCE WILL NOT CROSS.

Figure I–9 Lines of force cannot cross

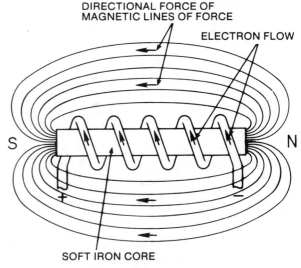

Figure I–10 Electromagnet

much stronger. This is because the high permeability core replaces low permeability air in the center of the coil and more lines of force develop, Figure I-10. If the core is placed toward one end of the coil, Figure I-11, the lines of force exert a strong force on it to move it toward the center so that they can follow a shorter path. If the core is movable, it will move to the center of the coil. A coil around an off-center, movable, permeable core is a **solenoid.** A spring is usually used to hold the core off center.

Motors

In an electric motor, current is passed through a conductor that is looped around the **armature** core, Figure I-12. The conductor loops are placed in grooves along the length of the core. The core is made of laminated discs of permeable, soft iron that are pressed on the arma-

ture shaft. The soft iron core causes the magnetic field that forms around the conductor to be stronger due to its permeability. There are several conductor loops on the armature, but only the loop that is nearest the center of the field poles has current passing through it. The loops are positioned so that when one side of a loop is centered on one field pole, its other side is centered on the other field pole.

The field poles are either permanent magnets or pieces of soft iron that serve as the core of an electromagnet. If electromagnets are used, an additional conductor (not shown in Figure I-12) is wound around each field pole, and current is passed through these field coils to produce a magnetic field between the field poles. The motor frame that the poles are mounted on acts as the magnet body.

Looking at the armature conductor near the

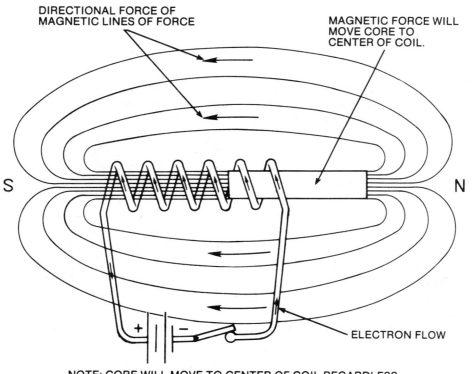

DIRECTIONAL FORCE OF
MAGNETIC LINES OF FORCE

MAGNETIC FORCE WILL
MOVE CORE TO
CENTER OF COIL.

S

N

ELECTRON FLOW

NOTE: CORE WILL MOVE TO CENTER OF COIL REGARDLESS
OF DIRECTIONAL FORCE OF MAGNETIC LINES OF FORCE.

Figure I–11 Solenoid

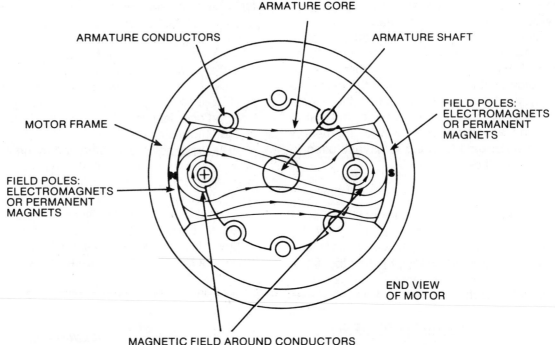

ARMATURE CORE

ARMATURE CONDUCTORS

ARMATURE SHAFT

MOTOR FRAME

FIELD POLES:
ELECTROMAGNETS
OR PERMANENT
MAGNETS

FIELD POLES:
ELECTROMAGNETS
OR PERMANENT
MAGNETS

END VIEW
OF MOTOR

MAGNETIC FIELD AROUND CONDUCTORS

Figure I–12 Electric Motor

north field pole in Figure I-12, you see its magnetic field extending out of the armature core and that it has a clockwise force. The magnetic field between the field poles has a directional force from north pole to south pole. At the top of the armature conductor, the field it has produced has a directional force in the same direction as that of the lines of force between the field poles. The lines of force in this area are compatible, but combining these two fields in the same area produces a high-density field. Remember that magnetic lines of force do not like to be close together.

At the bottom of the armature conductor, the lines of force formed around it have a directional force in the opposite direction from those from the north field pole. The lines of force will not cross each other, so some from the field pole distort and go up and over the conductor into the already dense portion of the field above the conductor, and some just cease to exist. This pro-

duces a high-density field above the conductor and a low density field below it. The difference in density is similar to a difference in pressure. This produces a downward force on the conductor.

The other side of the armature loop, on the other side of the armature, is experiencing the same thing except that the current is now traveling the opposite way. The loop makes a U-shaped bend at the end of the armature. The magnetic field around this part of the conductor has a counterclockwise force. Here, the lines of force around the conductor are compatible with those between the field poles under the conductor but try to cross at the top.

This produces an upward force on this side of the armature loop. The armature rotates counterclockwise. To change the direction in which the armature turns, change either the direction that current flows through the armature conductors, or change the polarity of the field poles.

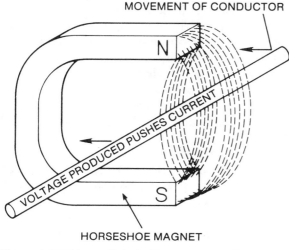

MOVEMENT OF CONDUCTOR

Figure I–13 Magnetic induction

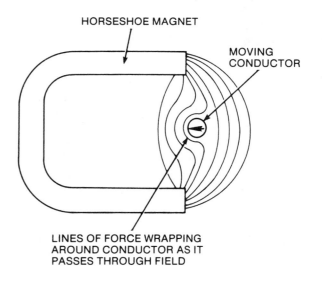

HORSESHOE MAGNET

MOVING CONDUCTOR

LINES OF FORCE WRAPPING AROUND CONDUCTOR AS IT PASSES THROUGH FIELD

Figure I–14 Cutting lines of force

Magnetic Induction

Passing voltage through a wire causes a magnetic field to form around the wire. However, if lines of force can be formed around a conductor, a voltage is produced in the wire and current starts to flow. This assumes, of course, the wire is part of a complete circuit. Lines of force can be made to wrap around a conductor by passing a conductor through a magnetic field, Figure I-13. This phenomenon occurs because of characteristic 6. As the conductor passes through the magnetic field, it cuts each line of force. Because the lines of force resist being cut, they first wrap around the conductor much like a blade of grass would if struck by a stick, Figure I-14. This principle is used in generators to produce voltage and current flow. The principle will work regardless of whether:

- the conductor is moved through a stationary magnetic field, as in a direct current generator.
- a magnetic field is moved past stationary conductors, as in an alternating current generator.
- the lines of force in an electromagnetic field are moved by having the circuit producing the magnetic field turned on and off, as in an ignition coil.

Note that in each case, movement of either the lines of force or the conductor is needed. A magnetic field around a conductor where both are steady state will not produce voltage. The amount of voltage and current that is produced by magnetic induction depends on four factors:

1. The strength of the magnetic field (how many lines of force there are to cut). A tiny amount of voltage is induced in the wire by each line of force that is cut.
2. The number of conductors cutting the line of force. Winding the conductor into a coil and passing one side of the coil through the magnetic field cuts each line of force as many times as there are loops in the coil.
3. How fast and how many times the conductors pass through the magnetic field.
4. The angle between the lines of force and the conductor's approach to them.

Amperage

Amperage is a measure of the amount of current flowing in a circuit. One ampere equals 6,250,000,000,000,000,000 (six and one-quarter billion billion) electrons moving past a given point in a circuit per second. This is often expressed as one **coulomb.**

Voltage

A **volt** is a measure of the force or pressure that caused current to flow; it is often referred to as **voltage.** The difference in potential is voltage. The most common ways of producing voltage are chemically, as in a battery, or by magnetic induction, as in a generator. A more accurate but less used name is **electromotive force.** Note that volts are what drives the electrons through the circuit; voltage is the measurement of that force. Similarly amps are the number of electrons moving; amperage is the measurement of that number.

Resistance

The fact that voltage is required to push current through a circuit suggests that the circuit

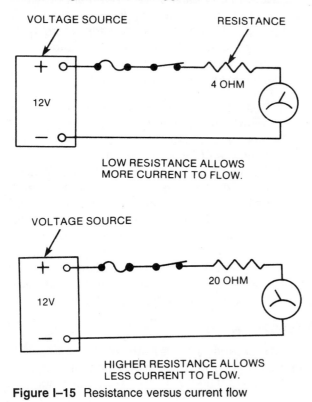

LOW RESISTANCE ALLOWS
MORE CURRENT TO FLOW.

HIGHER RESISTANCE ALLOWS
LESS CURRENT TO FLOW.

Figure I–15   Resistance versus current flow

offers resistance. In other words, you do not have to push something unless it resists moving. Resistance, which is measured in **ohms,** limits the amount of amperage that flows through a circuit, Figure I-15. If a circuit without enough resistance is connected across a reliable voltage source, wires or some other component in the circuit will be damaged by heat because too much current flows.

As mentioned, a bond exists between an electron and the protons in the nucleus of an atom. That bond must be broken for the electron to be freed so that it can move to another atom. Breaking that bond and moving the electron amount to doing work. Doing that work represents a form of resistance to current flow. This resistance varies from one conductive material to another, depending on the atomic structure of the material. For example, lead has more resistance than iron, and iron has more resistance than copper. It also varies with the temperature of the conductor. Loose or dirty connections in a circuit also offer resistance to current flow. Using current flow to do work to create heat, light, a magnetic field, or to move something also amounts to resistance to current flow.

Sometimes students confuse voltage and amperage while doing tests on electrical systems. Review these definitions and consider the influence that voltage and amperage have on a circuit. It might also help to remember that voltage can be present in a circuit without current flowing. However, current cannot flow unless voltage is present.

Voltage Loss

When voltage pushes current through a circuit, voltage is lost by being converted to some other energy form (heat, light, magnetism, or motion). This is usually referred to as **voltage drop.** Every part of a circuit offers some resistance, even the wires, although the resistance in the wires should be very low, Figure I-16. The voltage drop in each part of the circuit is proportional to the resistance in that part of the circuit.

The total voltage dropped in a circuit must equal the source voltage. In other words, all of the voltage applied to a circuit must be converted to another energy form within the circuit.

Ohm's Law

Ohm's law defines a relationship between amperage, voltage, and resistance. Ohm's law says: It takes one volt to push one ampere through one ohm of resistance. Ohm's law can be expressed in one of three simple mathematics equations:

$$E = I \times R$$
$$I = E/R$$
$$R = E/I$$

where: E = electromotive force or voltage
I = "intensity" or amperage
R = resistance or ohms

The simplest application of Ohm's law enables you to find the value of any one of the three factors: amperage, voltage or resistance, if the other two are known. For example, in the following diagram, if the voltage is 12 volts and the resistance

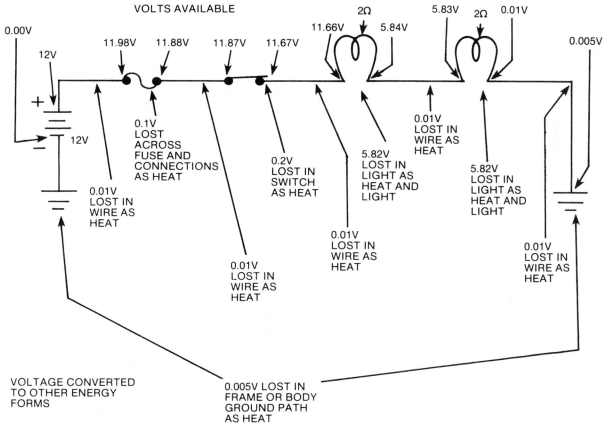

Figure I–16 Voltage drop

is 2 ohms, the current flow can be determined as follows:

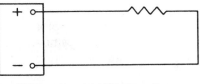

$$I = E/R \text{ or } I = 12 \text{ V}/2 \text{ }\Omega = 6 \text{ amps}$$

(The Greek letter Ω is often used as a symbol or an abbreviation for ohms, and amps is often used as an abbreviation for amperes.)

If the resistance is 4 ohms and the current is 1.5 amps, the voltage applied can be found by:

$$E = I \times R \text{ or}$$
$$E = 1.5 \text{ amps} \times 4 \text{ }\Omega = 6 \text{ V}$$

If the voltage is 12 volts and the current is 3 amps, the resistance can be found by:

$$R = E/I \text{ or}$$
$$R = 12 \text{ V}/3 \text{ amps} = 4 \text{ }\Omega$$

Perhaps the easiest way to remember how to use these equations is to use the following diagram:

To find the value of the unknown factor, cover the unknown factor with your thumb and multiply or divide the other factors as their positions indicate.

There are many other applications of Ohm's law, some of which are quite complex. (A more complicated application is covered later in this chapter.) An automotive technician is rarely required to directly apply Ohm's law in order to find or repair an electrical problem on a vehicle. But knowing and understanding the relationship of the three factors just covered is a must for the technician who wants to effectively diagnose and repair electrical systems.

CONDUCTORS AND INSULATORS

Earlier, in discussing the electrons in the outer shell of an atom, it was said that the fewer electrons there are in the outer shell the easier it is to break them loose from the atom. If an atom has five or more electrons in its outer shell, the electrons become much harder to break away from the atom, and the substance made up of those atoms is a very poor conductor; so poor, it is classed as an insulator. Rubber, most plastics, glass, and ceramics are common examples of insulators. Substances with four electrons in their outer shell are poor conductors but can become good ones under certain conditions. Thus, they are called **semiconductors.** Silicon and germanium are good examples of semiconductors. (Semiconductors are covered in more detail later in this chapter.)

Conductors are substances made up of atoms with three or fewer electrons in their outer shell. These electrons are called free electrons because they can be freed to travel to another atom.

CIRCUITS

Almost all electrical circuits in vehicles will have some type of component to provide each of the following functions:

- *Voltage source,* such as a battery or generator, provides voltage to the power circuit.
- *Circuit protection,* such as a fuse or circuit breaker, serves to open the circuit in the event of excessive current flow.
- *Circuit control,* such as a switch or relay, provides the ability to open and close the circuit.
- *Load,* such as a motor or light, does the func-

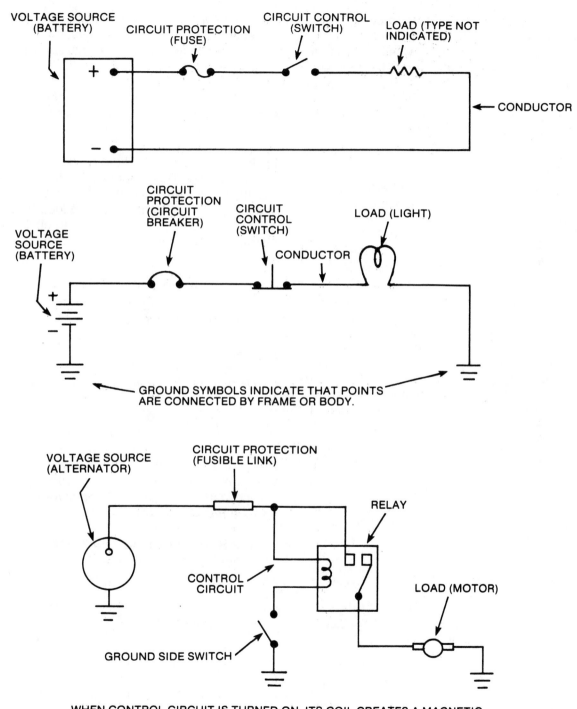

Figure I–17 Circuit components

tion the circuit was designed to do and also provides resistance to limit current.

- *Conduct current,* usually wires or certain parts of the vehicle, such as the frame, body, or engine block.

These components are often represented by symbols as shown in Figure I-17.

Circuit Types

There are two distinct types of electrical circuits, plus combinations of the two.

Series Circuits. In a series circuit, there is only one path for current flow, and all of the current flows through every part of the circuit. Parts A and B of Figure I-17 show simple series circuits. Even though there is only one load in each of these circuits, they qualify as series circuits because there is only one path for current flow. Figure I-18 shows a better example of a series circuit. Not only is there just one path for current flow, there are also two loads in series with each other. When there are two or more loads in series, the current must pass through one before it can pass through the next. The characteristics of a series circuit include the following:

- Current flow is the same at all parts of the circuit.
- Resistance units are added together for total resistance.
- Current flow decreases as resistance units are added.

- A portion of source voltage is dropped at each resistance unit in the circuit.
- An open in any part of the circuit disrupts the entire circuit.

The following problems apply Ohm's law to a series circuit. Refer to Figure I-19, which shows a compound series circuit.

Problem 1. Assume that the resistance of R1 is 2 ohms and that R2 is 4 ohms; find the total current flow. In a series circuit, the resistance value of each unit of resistance can be added together because all of the current passes through each resistor.

$$\begin{aligned}
I &= E/R \text{ or} \\
I &= E/R1 + R2 \text{ or} \\
I &= 12\,V/2\,\Omega + 4\,A \text{ or} \\
I &= 12\,V/6\,\Omega = 2 \text{ amps}
\end{aligned}$$

Problem 2. Assume the resistance values are unknown in Figure I-19, but that the total current flow is 3 amps. To find the total resistance:

$$\begin{aligned}
RT &= E/I \text{ or} \\
RT &= 12\,V/3 \text{ amps} = 4\,\Omega
\end{aligned}$$

Problem 3. Find the volt drop across R1, using the same values from problem 1. Since you are only concerned with the volt drop across R1, use only the resistance of R1.

$$V, = R1. \times I, \text{ or } V1, = 2\,\Omega = 2 \text{ amps} = 4\,V$$

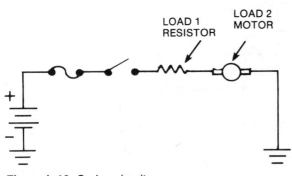

Figure I–18 Series circuit

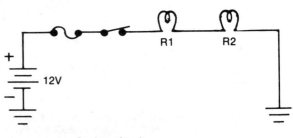

Figure I–19 Series circuit

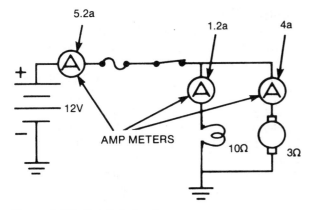

Figure I–20 Parallel circuit

Parallel Circuits.   In a parallel circuit, the conductors split into branches with a load in each branch, Figure I-17 lower and Figure I-20. Some current will flow through each branch with the most current flowing through the branch with the least resistance. Characteristics of a parallel circuit are:

- current varies in each branch (unless resistance in each branch is equal).
- total circuit resistance goes down as more branches are added and will always be less than the lowest single resistance unit in the circuit.
- total circuit current flow increases as more branches are added.
- source voltage is dropped across each branch.
- an open in one branch does not affect other branches.

The following problems apply Ohm's law to a parallel circuit.

Problem 4.   In the following diagram, assume

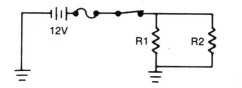

that R1 is 3 ohms and R2 is 4 ohms. Find the current flow through the R1 branch. Note that R1 is not affected by R2, so treat R1 like it is a series circuit.

$$I = E/R \text{ or}$$
$$I = 12 \text{ V}/3\ \Omega = 4 \text{ amps}$$

Find the current flow through R2.

$$I = E/R \text{ or}$$
$$I = 12 \text{ V}/4\ \Omega = 3 \text{ amps}$$

What is the total current flow in the circuit?

$$4 \text{ amps} + 3 \text{ amps} = 7 \text{ amps}.$$

Problem 5.   Now that you know the current flow, you can easily find the total circuit resistance.

$$R1 = E/I \text{ or}$$
$$R1 = 12 \text{ V}/7 \text{ amps} = 1.71\ \Omega$$

Note that to solve for total resistance you have simply solved each branch for current flow, added the flow of each branch, and divided the source voltage by total current. This method will work regardless of how many branches there are in the circuit.

There are other ways to find total resistance in a parallel circuit. If there are only two branches in the circuit, the product over the sum method can be used. In problem 4 there were two resistors in a parallel circuit with values of 3 ohms and 4 ohms. Using those values:

$$\frac{R1 \times R2}{R1 + R2} \text{ or}$$

$$\frac{3\ \Omega \times 4\ \Omega}{3\ \Omega + 4\ \Omega} = \frac{12\ \Omega}{7\ \Omega} = 1.71\ \Omega$$

If there are more than two branches in a circuit, this method can still be used, but it has to be manipulated a bit. Assume there are four branches in a circuit. Select two branches and use the product over the sum method to solve them as just shown. Then take the other two and do the same for them. Now take the two answers and run them through the same formula. The final result is total circuit resistance. There are still other methods, but you have learned all you need to know to solve the problems you are likely to encounter. Many automotive electrical systems texts, and electrical texts and reference books present those methods if you want to pursue them.

Problem 6. Finding voltage drop in a parallel circuit is the easiest of all. Using the values for branch 1 in problem 4 again:

$$E = I \times R1 \text{ or}$$
$$E = 4 \text{ amps} \times 3\ \Omega = 12 \text{ V}$$

Branch 2 looks like this:

$$E = I \times R2 \text{ or}$$
$$E = 3 \text{ amps} \times 4\ \Omega = 12 \text{ V}$$

Voltage drop for the complete circuit:

$$E = I \times RT \text{ or}$$
$$E = 7 \text{ amps} \times 1.71\ \Omega = 12 \text{ V}$$

Or you can just know that the volt drop across any branch of the whole circuit will be source voltage.

Series-Parallel Circuits.   Some circuits have characteristics of a series as well as those of a parallel circuit. There are two basic types of series-parallel circuit. The most common is a parallel circuit with at least one resistance unit in series with all branches, Figure I-21. All of the current flowing through the circuit in Figure I-21 must pass through the indicator light before it divides to go through the two heat elements, which are in parallel with each other.

To solve for current, resistance, or volt drop values in a series-parallel circuit, you must identify how the resistance units relate to each other, then use whichever set of formulas (series circuit or parallel circuit) apply. For example:

Problem 7.   In Figure I-21, assume a resistance of 4 ohms for the indicator light (R3), 20 ohms for R2 and 30 ohms for R1. Find the total resistance for the circuit.

Branches R1 and R2 are in parallel with each other and in series with R3. The battery sees the resistance of R3 plus the resistance of the two parallel branches. Once you find the total resistance of the branch circuits, you can treat that as a single resistance value that is in series with R3. Using the product over the sum method, first solve for the combined resistance of R1 and R2.

$$R = \frac{R1 \times R2}{R1 + R2} \text{ or}$$

$$\frac{20\ \Omega \times 30\ \Omega}{20\ \Omega + 30\ \Omega} = \frac{600\ \Omega}{50\ \Omega} = 12\ \Omega$$

Now add the total resistance of R1 and R2 (12 ohms) to the resistance of R3 (4 ohms).

$$RT = R + R3 \text{ or}$$
$$12\ \Omega + 4\ \Omega = 16\ \Omega$$

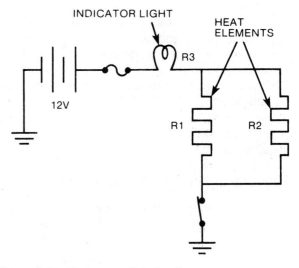

Figure I–21 Series-parallel circuit

Problem 8.   Find the total current flow for the circuit.

$$IT = E/RT \text{ or}$$
$$IT = 12/16 \ \Omega = 0.75 \text{ amp}$$

Before you can find the current flow through RI or R2, you must find the voltage applied to R1 and R2. Remember that R3 is in series with both R1 and R2. Also remember that in a series circuit, a portion of the source voltage is dropped across each resistance unit. Therefore, because there is a volt drop across R3, there is not a full 12 volts applied to R1 and R2.

Problem 9.   Find the volt drop across R3.

$$V3 = IT \times R3 \text{ or}$$
$$V3 = 0.75 \text{ amp} \times 4 \ \Omega = 3 \text{ V}$$

If 3 volts are dropped across R3 from a source voltage of 12 volts, 9 volts are applied to R1 and R2.

Problem 10.   Find the current flow through:

$$R1: I1 = E/R1 \text{ or}$$
$$I1 = 9 \text{ V}/20 \ \Omega = 0.45 \text{ amp}$$

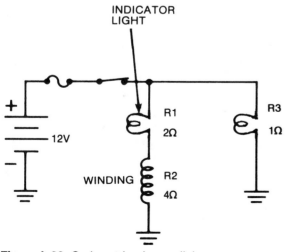

Figure I–22   Series string in parallel

$$R2: I2 = E/R2 \text{ or}$$
$$I2 = 9 \text{ V}/30 \ \Omega = 0.30 \text{ amp}$$

The second type of series-parallel circuit is a parallel circuit in which at least one of the branch circuits contains two or more loads in series, Figure I-22. This type of circuit is often referred to as **series string in parallel.** Finding the total current, resistance, or voltage drop for this circuit is easy. Identify the branch (or branches) that has loads in series and combine the resistances for each branch. Then treat the circuit like any other parallel circuit. When troubleshooting this kind of circuit in a car, failing to recognize the branch of a circuit you are testing as having loads in series could cause confusion.

CAUTION: If you are not absolutely sure about the circuit you are testing, consult the correct wiring diagram for the vehicle.

POLARITY

Just as a magnet has polarity—a north and a south pole that determine the directional force of the lines of force—an electrical circuit has a polarity. Instead of using north and south to identify the polarity, an electrical circuit's polarity is identified by positive and negative. An electrical circuit's polarity is determined by its power source. The best example is a battery, Figure I-23. The side of the circuit that connects to the positive side of the battery is positive and on most vehicles is the insulated side. The voltage on this side is usually near source voltage. Most of the voltage is dropped in the load where most of the work is done. The negative side of the circuit carries the current from the load to the negative side of the battery. There is usually very little resistance in the negative side of the circuit, and the voltage is near zero. Because this side of the circuit has the same potential as the vehicle's frame and sheet metal, this side of the circuit might not be insulated.

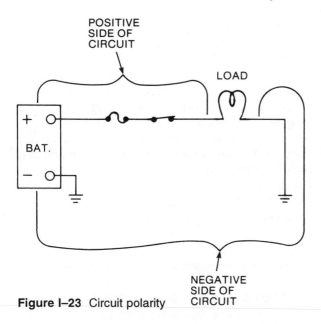

Figure I–23 Circuit polarity

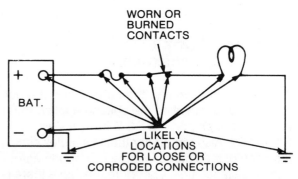

Figure I–24 Possible locations in a circuit for openings or high resistance

CIRCUIT CONTROLS: FAULTS

Three kinds of faults can occur in an electrical or electronic circuit: opens, excessive resistance, and shorts.

Opens

By far the most common fault in electrical and electronic circuits is an open circuit, or **open.** An open circuit means that there is a point in the circuit where resistance is so high that current cannot flow. An open can be the result of a broken wire, loose or dirty connection, blown fuse, or faulty component in the load device. Experienced technicians know that wires rarely break except in applications where the wire experiences a lot of flexing. Most opens occur in connections, switches, and components. A switch in a circuit provides a way to conveniently open the circuit. Of course, when a circuit is opened deliberately, it is not a fault.

Excessive Resistance

A loose or dirty connection or a partially cut wire can cause excessive resistance in a circuit.

Under these conditions, the circuit can still work but not as well as it could because the additional resistance reduces current flow through the circuit, Figure I-24. Excessive resistance can also result from a faulty repair or modification of a circuit in which a wire that is too long or too small in diameter has been installed. The location of excessive resistance in a circuit can be easily found with a series of voltage drop tests that are discussed under **Testing** in this chapter.

Shorts

Most faults in an electrical circuit are called shorts by the uninformed. A short is a fault in a circuit that causes current to bypass a portion or all of the circuit's resistance. The term short as used in electrical terminology means that the current is taking a shortcut rather than following the path it is supposed to take, Figure I-25. Remember Ohm's law: if resistance is reduced, current flow increases.

Short to Positive. In Figure I-25A, electrical contact has developed between two windings of a coil. This causes current to shortcut the loop or loops that exist between the points where the contact occurred, which lowers the coil's resistance. If only a relatively small number of the coil's loops are shorted, the increase in current flow might not be enough to further damage the circuit except possibly to blow a fuse if the circuit is fuse protected. The device might even continue to work but probably at reduced efficiency.

Short to Ground.   If a short occurs up circuit from the load (between the load and the positive source), Figure I-25B, it provides a lower resistance path to ground for current flow. Because of the lower resistance, excessive current will flow, and the fuse element will melt, causing the circuit to open. If the circuit does not have a fuse or some other type of circuit protection, the wire between the positive terminal and the short will overheat, possibly melting the wire or starting a fire.

If a short occurs down circuit (between the load and the negative terminal), Figure I-25C, the switch will not be able to provide the needed resistance to open the circuit. This is because the current has found a shortcut to ground that the switch cannot control. If the circuit in question is controlled by an electronic switching circuit, as you will see in following chapters, this condition threatens the solid-state circuit because the additional current flow might damage it.

TESTING

Basic electrical test procedures are easy to learn. They are essential to effective electrical troubleshooting and can save considerable time in identifying a problem. Effective electrical testing requires the following:

- understanding the principles of electricity.
- understanding the circuit being tested.
- knowing how to use the test equipment.
- knowing the three things above well enough to know what the test results mean.

Testing Voltage

The simplest voltage test is an available volts test. It can be performed with a nonself-powered test light or a voltmeter. The test light usually comes with two leads: one with an alligator clip that hooks to a negative or ground source, and one with a sharp probe that can penetrate the insulation to contact a positive source, Figure I-26. By connecting the clip of a 12-volt test light to a known good ground (nega-

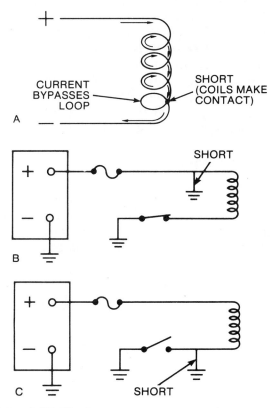

Figure I–25 Shorts

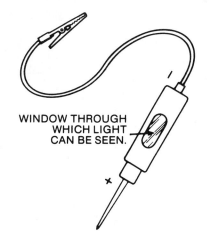

Figure I–26 Non self powered test light

Test Light versus Voltmeter

Most technicians prefer to use a test light for simple available volt tests. The test light is less expensive, requires less careful storage and handling, and does not have to be turned on and off; in short, it is more convenient to use. A voltmeter should be used whenever possible so that:

- you become more comfortable with its use. In some situations you will have to use a voltmeter. Do not make a diagnostic mistake because you make a mistake using or reading the meter.
- you develop a sense of what volt readings to expect on a properly operating circuit and what abnormal readings mean. A reading that is off by a few tenths of a volt can indicate a potential problem to a knowledgeable technician.

tive source) and contacting a positive source with the probe, the light will burn brightly if there are 12 volts available at the point being probed. If there are fewer than 12 volts available, the light will be less bright—as low as about 6 volts. A voltmeter used the same way indicates precisely how much voltage is available, Figure I-27.

The first thing to do when starting an electri-

cal test procedure is to test the tester. With a battery-powered multimeter, test the meter's battery first. Most multimeters have a convenient built-in method to check the battery. A quick way to check a voltmeter or test light is to put it across the vehicle's battery. A fully charged 12-volt battery in good condition should read around 12.6 volts with no load across it.

After verifying that the meter is working properly, select the correct scale on the meter. Most meters have at least two different scales for reading voltage: one for low values and one for high values. Select the scale that goes just a little beyond the voltage value you expect to see. For instance, if you are testing battery voltage on a 12-volt vehicle with the engine running, you should expect to see something between 12 and 15 volts. If your meter has a 3-volt scale and a 20-volt scale, select the 20-volt scale. Some meters are autoranging. That is, they automatically select the correct scale when connected to a circuit. *Do not condemn a component or circuit until you are sure the tester is working properly and you are using it correctly.*

Figure I-28 shows a circuit with several points indicated. Review the following chart and be sure that you can explain why each set of voltage values given indicates the faults listed.

Voltage Drop

Normally, in order for the load device in a circuit to work as well as it was intended, at least 85% of the source voltage should be dropped across it. This assumes, of course, that the source voltage is what the circuit was designed for. When

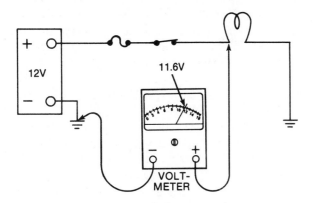

Figure I–27   Available volts test

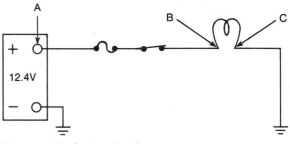

Figure I–28   Series circuit

Points			Fault	Fault Location
A	**B**	**C**		
12.4 v	12.4 v	12.4 v	Open	Down circuit from C
12.4 v	12.4 v	0	Open	At load
12.4 v	0	0	Open	Up circuit from C
12.4 v	12.2 v	.2 v	None	

comparing the volt drop across the load to the source voltage, be sure to take the source voltage reading with the circuit in question turned on. This is because connecting the voltage source's terminals together, which is what the circuit does, lowers the difference in potential between the terminals. The lower the resistance in the circuit, the more it lowers the difference in potential.

Manufacturers usually allow for a maximum of about 15% of the source voltage to be dropped across long runs of wire, circuit protection devices, switches, and connections. If more than 15% of the source voltage is dropped across parts of the circuit other than the load, there is excessive resistance somewhere in the circuit, and current flow is too low. There are exceptions to this, however. Some examples are:

- the current limiting resistors used for the low and intermediate speeds of a blower motor control circuit, Figure I-29A.
- the resistor ("ballast resistor") used to limit current flow through the ignition coil of many ignition systems a few years ago, Figure I-29B.
- the current-limiting resistor used in a temperature-sensing circuit, Figure I-29C.

You have to know or become familiar with the circuit you are testing.

To take a voltage drop reading, place the voltmeter across the portion of the circuit you want to test, Figure I-30. The meter will show the difference in voltage (voltage drop) between the two points being probed. To make this test useful, the technician must know what portion of the circuit is between the probes and how much voltage should be dropped there. Remember that:

- all of the source voltage must be dropped across the circuit.
- the load in the circuit is where the work is being done and that is where you want to drop as much of the voltage as possible.
- the function of the rest of the circuit is to get the voltage and current from the source to the load and back again as efficiently as possible.
- the less voltage drop in the delivery portion of the circuit, the more efficiently that portion is working.
- the higher the volt drop reading, the more resistance there is in the portion of the circuit between the probes.

If the switch in Figure I-30 were open, the voltmeter would show exact source voltage. This is because the positive probe is connected to the positive battery terminal by way of the fuse and insulated circuit, and the negative probe is connected to the battery negative terminal by way of the motor and ground circuit. If the voltmeter were connected across any other part of the circuit while the switch was open (with the exception of the battery), the reading would be zero. This is worth remembering as a good diagnostic tool. If a volt drop reading is zero, there is no current flowing; if there is current flowing, there has to be some amount of volt drop. If the meter shows source voltage, you are across the open. The volt drop is especially useful because it allows a technician to test the efficiency of a connection or component such as a switch that is not easy to get to by probing wires going to and from the connection or component where they are more accessible.

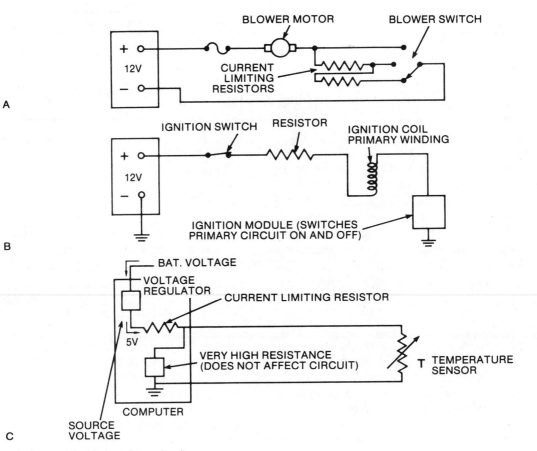

Figure I–29   Current limiting resistor circuits

Ammeter Testing

An ammeter is used to measure the current flowing in a circuit. There are two types: the con-

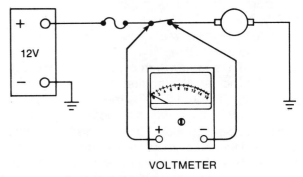

Figure I–30   Volt drop test

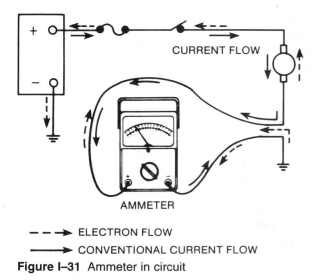

CURRENT FLOW

AMMETER

– – – ▸ ELECTRON FLOW

———▸ CONVENTIONAL CURRENT FLOW

Figure I–31 Ammeter in circuit

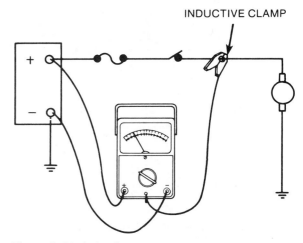

INDUCTIVE CLAMP

Figure I–32 Inductive ammeter

ventional type that must be put in series as part of the circuit, Figure I-31, and the inductive type. The inductive type usually has leads that connect to a power source and one with an inductive clamp that clamps around a conductor in the circuit to be tested, Figure I-32. The clamp contains a **Hall-effect** device that uses the magnetic field around the conductor that it is clamped around to create an electrical signal (Hall-effect devices are discussed in Chapter 2). The strength of the signal depends on the strength of the magnetic field around the conductor, which in turn depends on the amount of current flowing through it. The ammeter displays a reading that corresponds to the strength of the signal. The inductive ammeter is more convenient to use because it eliminates having to open the circuit.

When using a conventional ammeter, it is important to select the correct scale. Passing 5 amps through an ammeter with a 3-amp scale selected can damage the meter. Some meters are fuse protected when on the lowest scale. Become familiar with an ammeter before attempting to use it.

Resistance Testing

Resistance is measured with an **ohmmeter.**

An ohmmeter is self-powered and most have several scales to allow accurate testing of a wide range of resistance values. Most analog ohmmeters (those with a needle) measure from a fraction of 1 ohm to 100,000 ohms. Many digital ohmmeters measure as high as 20 megohms (20,000,000 ohms).

A common mistake when using an ohmmeter is to use a scale that is too low. For example, if you are going to measure the resistance of a circuit that has 1,500 ohms resistance and you use a meter with the scale selector set on R × 1 or R × 10, you will get a reading of infinity; on an analog meter, the needle will not move noticeably from the left side of the scale, and on a digital meter you will get "OL" (open line) or some other indication that the circuit is open. This is because the resistance is higher than the selected scale can measure. To avoid this mistake, start with the highest scale. If you get a low reading (near the right side of the scale), select a lower scale. Continue to select a lower scale until you find the lowest scale that can read the resistance that is present. Some ohmmeters are autoranging. In other words, when the meter is turned on and its leads are contacting a circuit, it automatically selects the correct ohm scale. This type will show an ohm value and display a symbol (K Ω or M Ω)

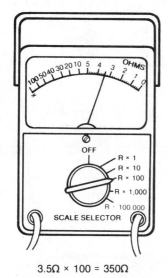

$3.5\Omega \times 100 = 350\Omega$

Figure I–33 Calculating ohmmeter readings

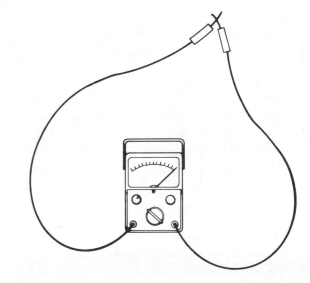

Figure I–34 Calibrating an ohmmeter

to show what scale is used. Be sure to note the symbol so that you can interpret the resistance value correctly.

When reading an ohmmeter on which you have to manually select the scale, multiply the reading on the face of the meter by the value on the scale you have selected. Figure I-33 shows a meter reading of 3.5 ohms, but the scale selector switch is on the 100-ohm scale. So:

$$3.5 \times 100 = 350 \text{ ohms}$$

Most autoranging, digital ohmmeters would show the value shown above as:

$$350 \ \Omega$$

3,500 ohms would be seen as:
$$3.5 \ K \ \Omega \ (K = kilo = 1000)$$

3,500,000 ohms would be seen as:
$$3.5 \ M \ \Omega \ (M = mega = 1,000,000)$$

When using an ohmmeter it is important to remember that *the circuit or component being tested cannot have voltage in it from any source besides the battery in the ohmmeter.* If there is voltage from another source in a circuit when an

ohmmeter is connected to it, the best thing that will happen is an inaccurate reading; the worst is a damaged ohmmeter.

When an ohmmeter is turned on, its battery tries to pass current through the meter coil and the two leads, but unless the leads are electrically connected together outside the meter, the circuit is open. If the two leads are touching or if they contact something that has continuity, the circuit is closed. If the leads are connected directly to each other, enough current should flow to cause the meter to swing full scale to the right, which is zero, Figure I-34. If the leads are across a conductor, component, or part of a circuit, the resistance of that unit will reduce the current flow through the meter's circuit, Figure I-35. The reduced current flow produces a weaker magnetic field, and the meter's needle will not be moved as far to the right. A digital ohmmeter has an electronic circuit that produces a digital readout in response to the amount of current flowing.

Most needle-type (analog) ohmmeters must be calibrated before each use. Calibration adjusts for voltage variation in the meter's battery due to the discharge that occurs during use. This is done by connecting the leads together and

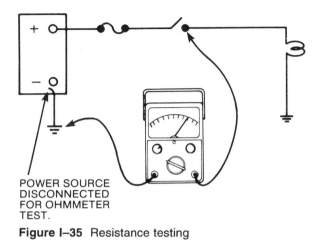

Figure I–35 Resistance testing

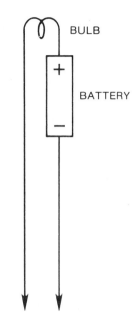

Figure I–36 Self-powered test light

adjusting the zero knob until the needle reads zero, Figure I-34. Doing this establishes that the magnetic field produced by the current flowing through the meter's coil and external leads moves the needle to zero. Any additional resistance placed between the external leads will lower the current flow and will be displayed by the amount the needle lacks in reaching the zero line. Many ohmmeters also must be recalibrated each time a different scale is selected. This is because selecting a different scale changes the meter's internal resistance.

Digital ohmmeters usually do not have a zero adjust knob, but the battery must be checked before each use to see that it meets or exceeds a minimum voltage that should be stated somewhere on the meter. If the battery is low, it must be replaced before the meter can be accurate.

Self-Powered Test Light

A self-powered test light is to an ohmmeter what a nonself-powered test light is to a voltmeter. It contains a battery, a light, and two leads, Figure I-36. If the leads make an electrical circuit with each other, directly or by both touching some other conductive path, the light will come on. This light only tells you if you have continuity or not and is often called a continuity light.

SEMICONDUCTORS

As mentioned earlier in this chapter, a semiconductor is an element with four valence electrons. The two most used semiconductor materials for solid-state components are silicon and germanium. Of these two, silicon is used much more than germanium. Therefore, most of this discussion will apply to silicon.

As previously stated, an atom with three or fewer electrons in its outer shell easily gives them up. If an atom has more than four but fewer than eight electrons in its outer shell, it exhibits a tendency to acquire more until it has eight. If there are seven electrons in the outer shell, the tendency to acquire another one is stronger than if there are only six. Once there are eight, it becomes very stable; in other words, it is hard to get the atom to gain or lose an electron. In a semiconductor material, for example, a silicon crystal, the atoms share valence electrons in what are called covalent bonds, Figure I-37.

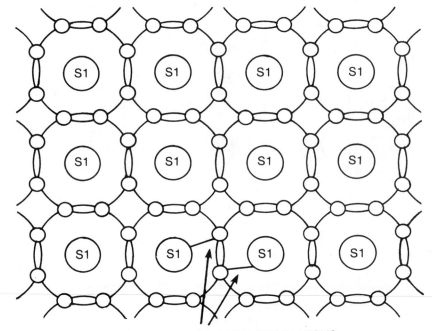

ATTRACTIVE BONDS OF PARENT AT ATOMS.

EACH SILICON ATOM SHARES AN ELECTRON WITH EACH NEIGHBORING ATOM.

Figure I–37 Covalent bonds

Each atom positions itself so that it can share the valence electrons of neighboring atoms giving each atom, in effect, eight valence electrons. This lattice structure is characteristic of a crystal solid and provides two useful characteristics:

1. Impurities can be added to the semiconductor material to increase its conductivity; this is called **doping.**
2. It becomes **negative temperature coefficient,** meaning that its resistance goes down as its temperature goes up. (This principle is put to use in temperature sensors, as we will see in Chapter 2.)

Doping

All of the valence electrons in a pure semiconductor material are in valence rings containing eight electrons as shown in Figure I-37. With this atomic structure there are no electrons that

can be easily freed, and there are no holes to attract an electron even if some were available. The result is this material has a high resistance to current flow. Adding very small amounts of certain other elements can greatly reduce the semiconductor's resistance. Adding trace amounts (about 1 atom of the doping element for every 100 million semiconductor atoms) of an element with either five or three valence electrons can create a flaw in some of the covalent bonds.

Adding atoms with five valence electrons (referred to as **pentavalent** atoms) such as arsenic, antimony, or phosphorus achieves a crystal structure as shown in Figure I-38. In Figure I-38 phosphorus is the doping element and silicon is the base semiconductor element. Four of the phosphorus atom's five valence electrons are shared in the valence rings of neighboring silicon atoms, but the fifth is not included in the covalent bonding with neighboring atoms,

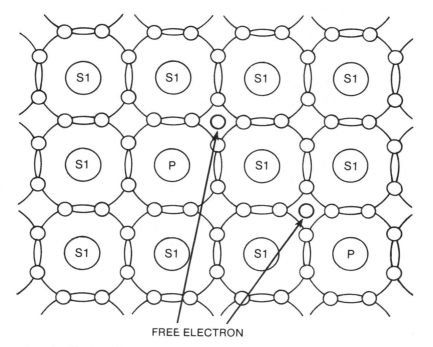

FREE ELECTRON

Figure I–38 Silicon doped with phosphorous

and it is held in place only by its attraction to its parent phosphorus atom – a bond that can be easily broken. This doped semiconductor material is classified as an N–type material. Note that it does not have a negative charge because the material contains the same number of protons as electrons. It does have electrons that can be easily attracted to some other positive potential.

Adding atoms with three valence electrons (referred to as **trivalent** atoms) such as aluminum, gallium, indium or boron achieves an atomic structure as seen in Figure I-39. In Figure I-39, boron is used as the doping material in a silicon crystal. The boron atom's three valence electrons are shared in the valence ring of three of the neighboring silicon atoms, but the valence ring of the fourth neighboring silicon atom is left with a hole instead of a shared electron. Remember that a valence ring of seven electrons aggressively seeks an eighth electron. In fact, the attraction to any nearby free electron is stronger than the free electron's attraction to its compan-

ion proton in the nucleus of its parent atom. Thus this material is classified as a P-type.

Doping Semiconductor Crystals

In actuality, a PN junction is not produced by placing a P- and an N-type semiconductor back to back. Rather, a single semiconductor crystal is doped on one side with the pentavalent atom's opposite sides and on the other with trivalent atoms. The center of the crystal then becomes the junction. The doping is done by first bringing the semiconductor crystal to a molten temperature. In a liquid state, the covalent bonds are broken. The desired amount of doping material is then added. As the semiconductor crystal cools, the covalent bonds redevelop with the doping atoms included.

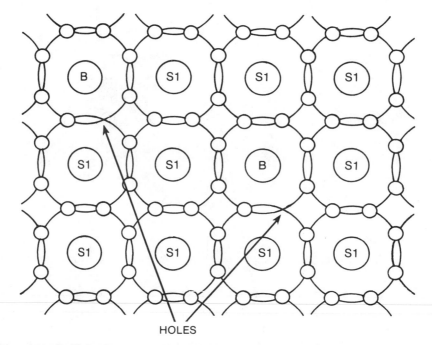

HOLES

Figure I–39 Silicon doped with boron

PN Junction

Figure I-40 shows a P-type crystal and an N-type crystal separated from each other. Only the free electrons in the N-type and the holes in the P-type that result from how the doping atoms bond with the base semiconductor material is shown. If the P-type and the N-type are put in physical contact with each other, the free electrons near the **junction** in the N-type cross the junction and fill the first holes they come to in the seven-electron valence rings near the junction of the P-type, Figure I-41. The junction is the area that joins the P-type and N-type. This action

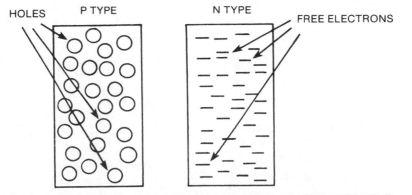

HOLES   P TYPE   N TYPE   FREE ELECTRONS

FOR SIMPLICITY, ONLY THE HOLES AND FREE ELECTRONS THAT RESULT FROM THE PENTAVALENT AND TRIVALENT ATOMS ARE SHOWN; THE STABLE ATOMS ARE LEFT OUT.

Figure I–40 P-type and N-type crystals

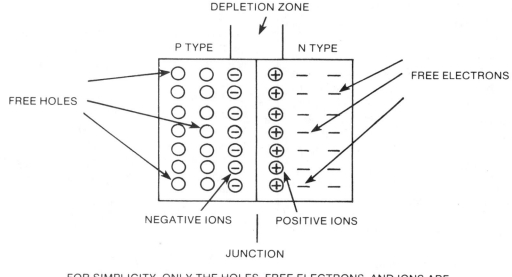

FOR SIMPLICITY, ONLY THE HOLES, FREE ELECTRONS, AND IONS ARE
SHOWN; THE STABLE ATOMS ARE LEFT OUT.

Figure I–41 PN junction

quickly creates a zone around the junction in which:

- there are no more free electrons in that portion of the N-type.
- there are no more free holes in that portion of the P-type.
- the valence rings near the junction in both the N- and P-type have eight shared electrons so they are reluctant to gain or lose any more.

Another way to state the information in the preceding list is to say that there are no longer any current carriers in the zone around the junction. This zone is often referred to as the **depletion zone**. In addition:

- The phosphorus atoms near the junction in the N-type have each lost an electron, which makes them positive ions (the free electron that crossed the junction to fill a hole in the P-type abandoned a proton in the nucleus of the phosphorus atom).
- The boron atoms near the junction in the P-type have each gained an electron, which

makes them negative ions (the electron that dropped into the valence ring around the boron atom is not matched by a proton in the nucleus of the boron atom).

The positive ions near the junction in the N-type are attracted to the free electrons that are farther from the junction, but they are more strongly repulsed by the negative ions just across the junction. Likewise, the negative ions in the P-type are attracted to the holes that are farther from the junction, but those holes are more strongly repulsed by the positive ions on the N-side of the junction. The extra electrons in the negative ion atoms are somewhat attracted to the nearby holes, but they are more strongly bound by the covalent bonding into which they have just dropped.

Keep in mind that a layer of negative ions along the junction in the P-type and a layer of positive ions along the junction of the N-type have been created. The opposing ionic charges on the two sides of the depletion zone create an electrical potential of about 0.6 volt (0.3 volt for germanium). This potential, often referred to as

Conventional Current Flow versus Electron Current Flow

Before electrons were known, Benjamin Franklin surmised that current was a flow of positive charges moving from positive to negative in a circuit. Franklin's belief became so accepted that even after electrons were discovered and scientists learned that current flow consists of electrons moving from negative to positive, the old idea was hard to give up. As a result, the "idea" of positive charges moving from positive to negative is still often used. It is referred to as **conventional current flow** or less formally as **hole flow.**

The conventional current flow theory has gotten a boost in recent years because it helps explain how semiconductors work: positively charged holes move from positive charges to negative charges as electrons move from negative charges to positive charges.

the **barrier potential**, cannot be measured directly, but its polarity prevents current from flowing across the junction unless it is overcome by a greater potential.

Diodes

What was just described in the preceding section is in fact the major part of a diode. A **diode** is basically a one-way electrical check valve; it will allow current to pass in only one direction. The best-known automotive application is probably the diodes used in the alternator to rectify the AC current produced to DC current. There are many other applications of diodes in today's vehicles such as in solid-state voltage regulators, electronic modules, and circuits where one-way control might be needed. Figure I-42 shows a rectifying diode in a circuit. If switch A is on and C is off, light B will get current, but light D will not because the diode will block it. If switch C is on and A is off, both lights will get current.

Here is how the diode works. By applying voltage with correct polarity from an outside source such as a battery, current will flow across the PN junction, Figure I-43. Applying an external voltage of 0.6 volt or more with the positive to the

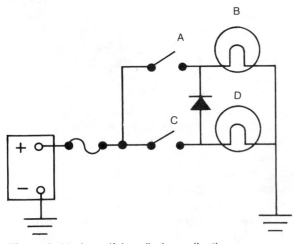

Figure I–42 A rectifying diode application

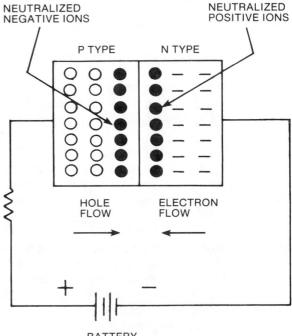

Figure I–43 Forward bias voltage applied to a diode

P-side and the negative to the N-side is called **forward bias voltage.** The higher negative potential introduced by the forward bias voltage at the right side of the crystal repels the free electrons in the N-type. They move toward the junction, canceling the charge of the positive ions. At the same time, the higher positive potential introduced by the forward bias voltage on the left side of the crystal repels the free holes in the P-type. They move toward the junction, canceling the charge of the negative ions. With the barrier potential overcome, current easily flows across the junction, with electrons moving toward the external positive potential, and holes moving toward the external negative potential. When the forward bias voltage is removed, barrier potential redevelops and the diode again presents high resistance to current flow.

NOTE: The amount of current that a diode or any other type of PN junction semiconductor can safely handle is determined by such things as its size, type of semiconductor and doping material used, heat dissipating ability, and sur-

rounding temperature. If the circuit through which forward bias voltage is applied does not have enough resistance to limit current flow to what the semiconductor can tolerate, it will overheat, and the junction will be permanently damaged (open or shorted).

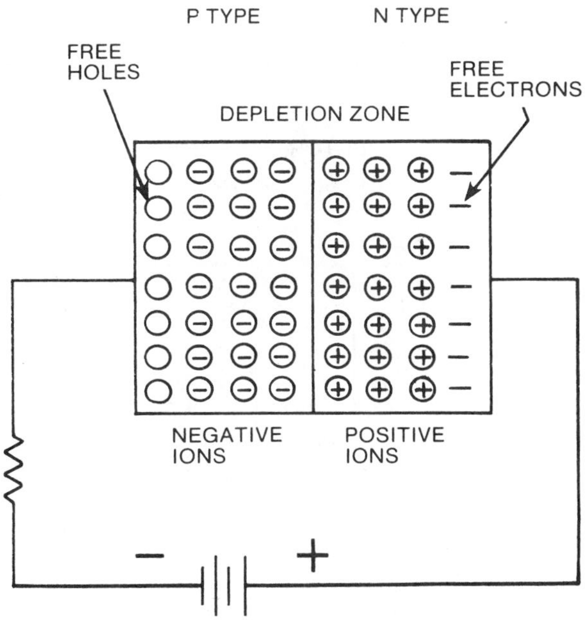

If a reverse bias voltage is applied, with negative external potential to the P-side and positive external potential to the N-side, the positive potential attracts free electrons away from the junction, and the negative potential attracts holes away from the junction, Figure I-44. This causes the depletion zone to be even wider and the resistance across the junction to be even higher.

If the reverse bias voltage goes high enough, that is, above 50 volts for most rectifier diodes (those designed to conduct enough current to do work, and the most common type), current will flow. It will rise quickly, and in most cases the diode will be damaged. This is called the **breakdown voltage.**

Diode Symbols

The most commonly used symbol to represent a diode is an arrow with a bar at the point,

Figure I–44 Reversed bias voltage applied to a diode

Zener Diode

A Zener diode is one in which the crystals are more heavily doped. Because of this the depletion zone is much narrower, and its barrier voltage becomes very intense when a reverse bias voltage is applied. At a given level of intensity the barrier voltage pulls electrons out of normally stable valence rings, creating free electrons. When this occurs, current flows across the diode in a reverse direction without damaging the diode. The breakdown voltage at which this occurs can be controlled by the amount of doping material added in the manufacturing process. Zener diodes are often used in voltage regulating circuits.

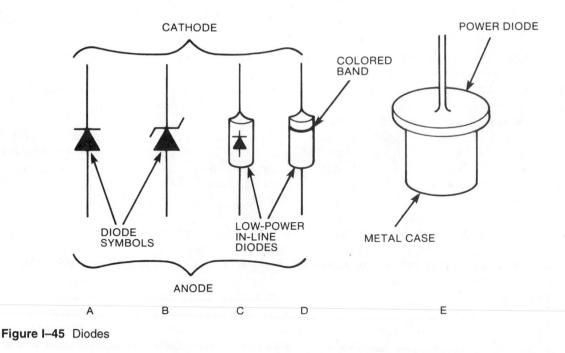

Figure I–45 Diodes

Figure I-45A. The point always indicates the direction of current flow using conventional current. Electron flow would be opposite to the way the arrow is pointing. The arrow side of the symbol also indicates the P-side of the diode, often referred to as the **anode.** The bar at the end of the arrow point represents the N-side and is often called the **cathode.** Figure I-45B shows a modified diode symbol that represents a Zener diode.

On actual diodes, the diode symbol can be printed on the side to indicate the anode and cathode ends, or a colored band can be used instead of the symbol. In this case the colored band will be nearer to the cathode end, Figure I-45C and D. Figure I-45E shows a power diode. A power diode is one large enough to conduct larger amounts of current to power a working device. It will be housed in a metal case, which can serve as either the anode or the cathode connector and will also dissipate heat away from the semiconductor crystals inside. The polarity of a power diode can be indicated by markings or, in the case of specific part number applications such as

in an alternator, may be sized or shaped so that it can be installed only one way.

Transistors

The transistor, probably more than any other single component, has made possible the world

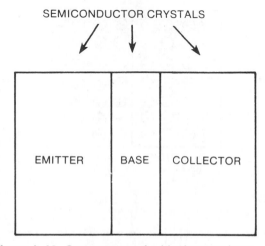

Figure I–46 Components of a bipolar transistor

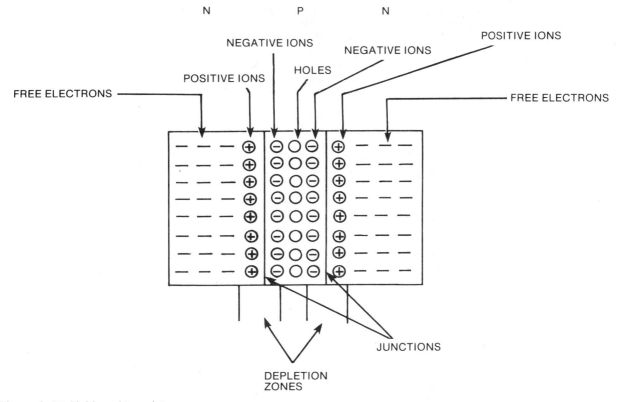

Figure I–47 Unbiased transistor

of modern electronics. Transistors most commonly used in automotive applications are called **bipolar transistors** because they use two polarities—electrons and holes. Bipolar transistors contain three doped semiconductor crystal layers called the **emitter,** the **base,** and the **collector,** Figure I-46. The base is always sandwiched between the emitter and the collector. The major difference between a diode and transistor is that a transistor has two PN junctions instead of one. They can be arranged to have a P-type emitter and collector with an N-type base (a PNP transistor), or an N-type emitter and collector with a P-type base (an NPN transistor). Since most automotive applications are NPN transistors, they are discussed here.

The emitter is heavily doped with pentavalent atoms, and its function is to emit free electrons into the base. The base is lightly doped with trivalent atoms and is physically much thinner than the other sections. The collector is slightly less doped than the emitter but more than the base.

Recalling the discussion of the PN junction (review it if it is not clear), you know that a barrier potential forms at each junction, Figure I-47. If you forward bias the transistor by applying an external voltage of at least 0.6 volt from emitter to base with negative potential to the emitter and positive potential to the base, then apply an external voltage of at least 0.6 volt collector to the emitter with positive potential to the collector and negative potential to the emitter, you can get some unique results, Figure I-48. A slightly higher voltage is usually used between the collector and emitter, thus two separate voltage sources are shown.

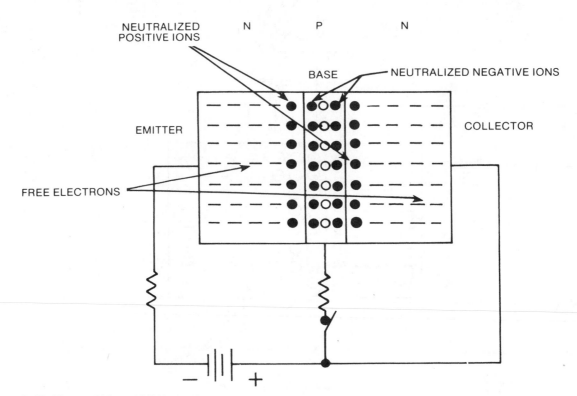

Figure I–48 Forward biased NPN transistor

Because it is lightly doped, the base does not have many holes for the free electrons to drop into. The holes that do exist are quickly filled. Because the base crystal is thin and has so few holes for free electrons to drop into, the majority of the free electrons coming from the emitter cross the base into the collector. The collector readily accepts them because the free electrons in the collector are attracted to the positive potential at the collector electrode and leaves behind a lot of positive ions.

The positive potential that is applied to the base electrode is strong enough to attract some of the valence electrons there. As valence electrons near the base electrode flow from the base toward the emitter-base voltage source, nearby valence electrons move over to take their place.

Because primarily only valence electrons move from the base into the base circuit, and because there are so few holes that will allow free electrons to drop into valence rings in the base (the base is very lightly doped), the current in the base-emitter circuit is quite small compared to the current in the collector-emitter circuit.

If the biasing voltage is removed from the base, the barrier potential is restored at the junctions and both base and collector currents stop flowing. The base current flow caused by the forwarding biasing voltage between the base to the emitter allows current to flow from the emitter to the collector.

Transistors are like switches that can be turned on by applying power to the base circuit. There are many types of transistors, but the most common fall into two categories: power and switching.

Power transistors are larger because the junction areas must be larger to pass more current across them. Passing current across the junctions produces heat; therefore, a power transistor must be mounted on something, ordinarily

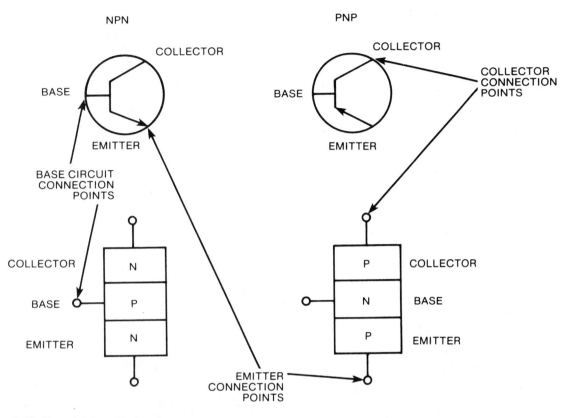

Figure I–49 Transistor symbols

called a **heat sink**, that can draw heat from the transistor and dissipate it into the air. If not it will likely overheat and fail.

A transistor is also much like a relay in that a relatively small current through the base circuit controls a larger current through the collector. The amount of current through the base is determined by the amount of voltage and resistance in the base circuit. In fact, carefully controlling the voltage applied to the base can control how much current flows through the collector. If enough voltage is applied to the base to just start reducing barrier potential, a relatively small amount of current will begin to flow from the emitter to the collector. This is called **partial saturation**. As base voltage is increased, collector current increases. When enough voltage is applied to the base, full

saturation is achieved. There is usually a small voltage spread between minimum and full saturation, and once full saturation is achieved, increasing voltage at the base will not increase collector current. If too much voltage is applied, the transistor will break down. This principle is often used in power transistors. Most power transistors are NPN.

Switching transistors are much smaller. They are often used in information processing or control circuits, and conduct currents ranging from a few milliamps (thousandths of an ampere) down to a few microamps (millionths of an ampere). They are more often designed for extremely fast on and off cycling rates, and the base circuits that control them are designed to turn them on at full saturation, or to turn them off.

Effect of Temperature on a PN Junction

As the temperature of a semiconductor device goes up, the electrons in the valence ring move at a greater speed. This causes some to break out of the valence ring, creating more free electrons and holes. The increase in the number of free electrons and holes causes the depletion layer to become thinner, reducing the barrier potential. Barrier potential voltages of 0.7 for silicon and 0.3 for germanium semiconductors are true at room temperature only. At elevated temperatures, the lower barrier potential lowers the external bias voltage required to cause current to flow across the junction and raises the current flow.

This means that the expected operating temperature range of a semiconductor device must be considered when the circuit is being designed, and that when in operation, the operating temperature must be kept within that temperature limit. If the operating temperature goes higher, unless some kind of compensating resistance is used, current values will go up, and the semiconductor might be damaged.

Transistor Symbols

In electrical schematic illustrations, transistors are usually represented by a symbol. Figure I-49 shows the symbols that are most often used. Of the two types of symbols, the circle containing the lines for emitter, base, and collector are most often used. The arrow is always used to indicate:

- the emitter side of the transistor.
- the direction of conventional current flow.
- whether the transistor is an NPN or PNP. If the arrow points outward, toward the circumference of the circle, it indicates an NPN transistor. If it points inward toward the base, it indicates a PNP transistor.

INTEGRATED CIRCUITS (ICS)

Thanks to scientific research, manufacturers have the capability to produce microscopic transistors, diodes, and resistors. As a result, complete circuits are produced containing thousands of semiconductor devices and connecting conductor paths on a chip as small as two or three millimeters across. These integrated circuits operating with current values as low as a few milliamps or less can process information, make logic decisions, and issue commands to larger transistors. The larger transistors control circuits that operate on larger current values. Personal computers and the computers in today's vehicles became possible because of the development of ICs. Because the components in ICs are so small, they cannot tolerate high voltages. Care must be taken to avoid creating high-voltage spikes such as those produced by disconnecting the battery while the ignition is on. Care must also be taken when handling a component with ICs in it to avoid exposing it to static discharges such as those you sometimes experience when touching something with ground potential after walking across a carpet.

Computers in Cars

KEY TERMS

A/D interface
Analog
Baud
Bit
Byte
Catalyst
Closed Loop
D/A Converter
Digital
Driveability
Engine Calibration
Interface
KAM
Memory
Microprocessor
Modulate
Open Loop
Palladium (Pd)
Platinum (Pt)
PROM
RAM
Rhodium (Rh)
ROM
Signal
Stoichiometric

WHY COMPUTERS?

In 1963, positive crankcase ventilation systems were universally installed on domestic cars as original equipment. People in the service industry felt that dumping all of those crankcase gases in the induction system would plug up the carburetor and be harmful to the engine. Everyone in the service industry knew that all of the crankcase gases in the induction system would plug up the carburetor and ruin the engine. Instead, engine life doubled.

In 1968 exhaust emission devices were universally applied to domestic cars. Compression ratios began to go down; spark control devices denied vacuum advance during certain driving conditions; thermostat temperatures went up; air pumps were installed; heated air intake systems were used during warm-up; and air/fuel ratios began to become leaner. Through the seventies evaporation control systems, exhaust gas recirculation systems, and catalytic converters were added. Although some of the emissions systems actually tended to improve driveability and even fuel mileage, for the most part, in the early days of emission controls, driveability and fuel mileage suffered in order to achieve a dramatic reduction in emissions.

In 1973 and 1974, when domestic car fuel economy was at its worst, we also experienced an oil embargo and an energy shortage. The federal government responded by establishing fuel mileage standards in addition to the already established emissions standards. By the late 1970s, the car manufacturers were hard–pressed to meet the ever more stringent emissions and mileage standards; and the standards set for the eighties looked impossible, Figures 1-1 and 1-2. To make matters worse, the consumer was getting into the picture, too. Not only did the consumer's car get poor fuel mileage, it had poor **driveability** (idled rough, often hesitated or stumbled during acceleration if the engine was not fully warmed up, and had little power). What made the situation so difficult was that the three demands—lower emissions, better mileage, and better driveability—were largely in opposition to each other using the technology available at that time, Figure 1-3.

What was needed was a much more precise way to control engine functions, or **engine calibration**. The automotive industry had already looked to **microprocessors** – processors contained on integrated circuits. In 1968, Volkswagen introduced the first large-scale production, computer-controlled electronic fuel injection system, an early version of the Bosch D-Jetronic system. In 1975, Cadillac had introduced

EMISSION REQUIREMENTS
FOR PASSENGER CARS
(IN GRAMS PER MILE)

MODEL YEAR	HYDROCARBON (HC)		CARBON MONOXIDE (CO)		OXIDES OF NITROGEN (NOx)	
	FEDERAL	CALIFORNIA	FEDERAL	CALIFORNIA	FEDERAL	CALIFORNIA
1978	1.5	0.41	15.0	9.0	2.0	1.5
1979	0.41	0.41	15.0	9.0	2.0	1.5
1980	0.41	0.39	7.0	9.0	2.0	1.0
1981	0.41	0.39	3.4	7.0	1.0	0.7
1982	0.41	0.39	3.4	7.0	1.0	0.4
1983	0.41	0.39	3.4	7.0	1.0	0.4
1984	0.41	0.39	3.4	7.0	1.0	0.4
1985	0.41	0.39	3.4	7.0	1.0	0.7
1986	0.41	0.39	3.4	7.0	1.0	0.7
1987	0.41	0.39	3.4	7.0	1.0	0.7
1988	0.41	0.39	3.4	7.0	1.0	0.7
1989	0.41	0.39	3.4	7.0	1.0	0.4
1990	0.41	0.39	3.4	7.0	1.0	0.4
1960	10.6*		8.4*		4.1*	

*Typical values before controls

Figure 1–1 Federal and California emissions standards compared to the uncontrolled vehicles of 1960

FUEL ECONOMY STANDARDS

MODEL YEAR	MPG	TOTAL IMPROVEMENT OVER THE 1974 MODEL YEAR
1978	18.0	50%
1979	19.0	58%
1980	20.0	67%
1981	22.0	83%
1982	24.0	100%
1983	26.0	116%
1984	27.0	125%
1985	27.5	129%
1986	26.0*	116%
1987	26.0	116%
1988	26.0	116%
1989	26.5	120%
1990	27.5	129%

*Reduced from 27.5   by the federal government in 1986

Figure 1–2 Federally imposed fuel economy standards

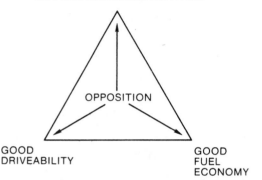

Figure 1–3 Opposition of exhaust emissions, driveability, and fuel economy to each other

a computer-controlled electronic fuel injection (EFI) system. In 1976, Chrysler introduced a computer-controlled electronic spark control system, the Lean-Burn system. And in 1977, Oldsmobile introduced a computer-controlled electronic spark control system that they called MISAR (microprocessed-sensing automatic regulation). All of these systems had three things in common: they all controlled only one engine function, they all used an **analog** computer (a mechanism that continuously varies within a given range with time used for the change to occur), and none of them started any landslide movement toward computer controls.

By the late seventies, the electronics industry had made great strides with **digital** microprocessors. These small computers with their comparatively low cost, compact size and weight, great speed and application flexibility proved to be ideal for the industry. Probably the most amazing feature of the digital computer is its speed. To put it into perspective, one of these computers controlling functions on an eight-cylinder engine running at 3,000 rpm can send the spark timing command to fire a cylinder; reevaluate input information about engine speed, coolant temperature, engine load, barometric pressure, throttle position, air/fuel mixture, spark knock, and vehicle speed; recalculate air/fuel mixture, spark timing, whether or not to turn on the EGR valve, canister purge valve, and the torque converter clutch; and then take a short nap before sending the commands, all before the next cylinder fires. The computer is so much faster than even the fastest engine, it spends most of its time doing nothing but counting time on its internal clock.

At last the automotive manufacturers had the technology to precisely monitor and control, to instantly and automatically adjust, while driving, enough of the engine's calibrations to make the vehicle comply with the government's demands for emissions and fuel economy, and still satisfy the consumer's demands for better driveability. This brought about what is probably the first real revolution in the automotive industry's recent history; other changes have been evolutionary by comparison.

After some experimental applications in 1978 and 1979 and some limited production in 1980, the whole domestic industry was using comprehensive computerized engine controls in 1981. At the same time, General Motors alone produced more computers than anyone else in the world and used half of the world's supply of computer parts.

How Computers Work

Contrary to some "informed" opinions and numerous suggestions from popular movies, computers cannot think for themselves. When properly programmed, however, they can carry out explicit instructions with blinding speed and almost flawless consistency.

Communication Signals. A computer uses voltage values as communication signals, thus voltage is often referred to as a **signal** or a voltage signal. There are two types of voltage signals: analog and digital, Figure 1-4. An analog signal's voltage is continuously variable within a given range and time is used for the voltage to change. An analog signal is generally used to convey information about a condition that changes gradually and continuously within an established range. Temperature-sensing devices usually give off an analog signal. Digital signals also vary but not continuously, and time is not needed for the change to occur. Turning a switch on and off creates a digital signal; voltage is either there or it is not. Digital signals are often referred to as square wave signals.

Binary Code. A computer converts a series of digital signals to a binary number made up of ones and zeros; voltage above a given threshold value converts to 1, and voltage below that converts to 0. Each 1 or 0 represents a **bit** of information. Eight bits equal a **byte** (sometimes referred to as a word). All communication between the microprocessor, the memories, and the interfaces is in binary code, with each infor-

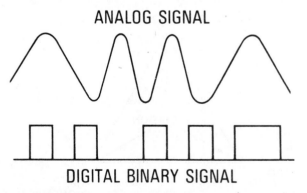

ANALOG SIGNAL

DIGITAL BINARY SIGNAL

Figure 1–4 Graphic illustration comparing analog and digital voltage signals. *(Courtesy of General Motors Corporation, Service Technology Group.)*

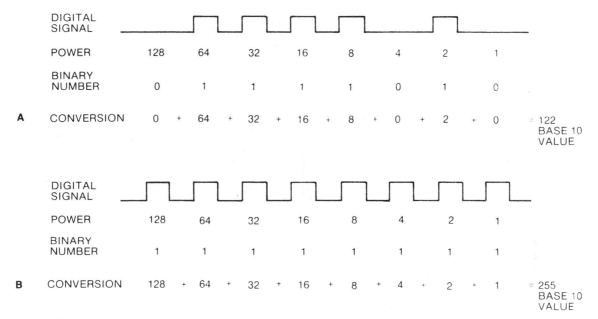

Figure 1–5 Binary numbers

mation exchange being in the form of a byte. If a series of digital signals converts to a binary number of 01111010, the numerical value (in base, 10, which we are most familiar with) can easily be derived, as shown in Figure 1-5A. A power is assigned to each place in the binary number, starting from the right and working to the left. The rightmost place is given a power of 1, with the power doubling with each successive place to the left. Each digit in the binary number is multiplied by its respective power. The products are then added together to yield the base-10 numerical value. As you will see in Figure 1-5B, the largest value that an 8-bit computer can communicate is 255. Most automotive computers have 8-bit microprocessors, but as the computers take on more responsibility and have to communicate more information more rapidly, 16-bit units will become commonplace.

Interface.   The microprocessor is the heart of a computer, but it needs several support functions, one of which is the **interface.** A computer has an input and an output interface circuit, Figure 1-6. The interface has two functions: it

protects the delicate electronics of the microprocessor from the higher voltages of the circuits attached to the computer, and it translates input and output signals. The input interface translates all analog input data to binary code; most sensors produce an analog signal. It is sometimes referred to as A/D, analog to digital. The output interface, D/A, translates digital signals to analog for any controlled functions that need an analog voltage.

Memories.   The microprocessor of a small computer does the calculating and makes all of the decisions, or data processing, but it cannot store information. The computer is therefore equipped with information storage capability called **memory.** The computer actually has three memories: read-only memory (**ROM**), programmable read-only memory (**PROM**), and random access memory (**RAM**), Figure 1-7.

The ROM contains permanently stored information that instructs (programs) the microprocessor on what to do in response to input data. The microprocessor can read information from the ROM but cannot put information into it. The ROM

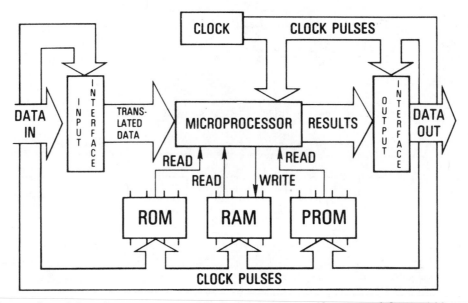

Figure 1–6 Interaction of the microprocessor and its support system. *(Courtesy of General Motors Corporation, Service Technology Group.)*

unit is soldered into the computer and is not easily removed.

The PROM differs from the ROM in that it plugs into the computer and is more easily removed and reprogrammed or replaced with one containing a revised program. It contains program information specific to different vehicle model calibrations.

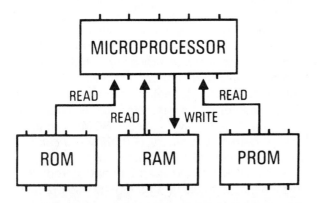

Figure 1–7 The three memories within a computer. *(Courtesy of General Motors Corporation, Service Technology Group.)*

Some of the information stored in both the ROM and the PROM is stored in the form of lookup tables similar to the tables one might find in the back of a chemistry or math textbook. These tables enable the computer to interpret and make decisions in response to sensor data. For example, a piece of information (a voltage value of 2.4 volts, about half throttle) is received from the throttle position sensor. This information plus information from the engine speed sensor is compared to a table for spark advance. This preprogrammed table tells the computer what the spark advance should be for that throttle position and engine speed. The computer will then modify this spark advance value, by consulting other tables, with information concerning engine temperature and atmospheric pressure. These tables are sometimes referred to as maps.

The RAM is where temporary information is stored. It can be both read from and written into. If, for example, data comes in from a sensor and will be used for several different decisions, such as manifold vacuum information, the microprocessor will record it in RAM until it is updated.

In the example presented in the preceding paragraph, the information concerning throttle position and engine speed came from the RAM.

In most automotive computers, the computer tests at least some of the input and output signals to see whether the circuit that sent the signal or received the command is working properly. The computer knows what the voltage will be if the circuit were open and what the voltage will be if it were shorted. It watches to see that the voltage remains between those voltage values. On some circuits, the oxygen sensor circuit for instance, the computer expects to see the voltage go up and down, recrossing a certain value. If during its continuous testing it sees something wrong (a fault), it will record information in RAM concerning the fault. The RAM is also soldered in place and not easily removed.

There are two kinds of RAMS: volatile and nonvolatile. A volatile RAM (sometimes called **KAM**, keep-alive memory) must have a constant source of voltage to continue to exist and is erased when disconnected from its power source. In automotive applications, a volatile RAM is usually connected directly to the battery by way of a fuse or fusible link so that when the ignition is turned off, the RAM is still powered. A nonvolatile RAM does not lose its stored information if its power source is disconnected. Vehicles with digital display odometers usually store mileage information in a nonvolatile RAM.

The terms ROM, PROM, and RAM are fairly standard throughout the computer industry; however, all car manufacturers do not use the same terms in reference to their computer's memory. For instance, instead of using the term PROM, Ford calls their equivalent unit an engine calibration assembly. Another variation of a PROM is an E-PROM. An E-PROM has a housing with a transparent top that will let light through. If ultraviolet light strikes the E-PROM, it reverts back to its unprogrammed state. To prevent deprogramming or memory loss, a cover must protect it from light. The cover is often a piece of tape. An E-PROM is sometimes used where ease of changing stored information is important. General

Motors uses it as a PROM on some applications because it has more memory capacity.

Another variation of RAM is KAM (keep-alive memory). Often used as a subsection of the RAM, KAM is usually powered by a fuse straight from the battery rather than by the ignition or an ignition-controlled circuit. Therefore, when the ignition is turned off, the KAM remains powered and its stored information is retained. Other types of memory that are unique to specific manufacturers will be introduced in appropriate chapters.

Clock Pulses.   In order to maintain an orderly flow of information into, out of, and within the computer, a quartz crystal is used to produce a continuous, consistent time pulse, Figure 1-8. The pulse acts as a kind of rapid cadence by which information is transmitted. During the space or time interval between each pulse, one bit of binary code information is communicated from one part of the computer to another. This method provides an orderly flow of information.

The clock pulse is often referred by the term **baud**. The baud identifies the rate of the pulses. For example, a computer with a baud of 5 000 could transmit 5 000 bits of binary code information per second. Bauds in automotive computers have gone up considerably since those used in the early eighties. The computers used by General Motors in the early eighties had a baud of 160 (some of those computers were still used on some applications in the late eighties), while their P-4 electronic control module introduced in the late eighties has a baud of 8 192 with development underway to make them even faster. Ford introduced a computer in 1988 with a baud rate of 12 500.

Data Links. When computers communicate with other electronic devices such as control panels, modules, some sensors, or other computers in the form of digital signals, they communicate through circuits called data links. Some data links transmit data in only one direction, although others transmit in both directions. What makes a data link different from an ordinary circuit is the manner in which it is controlled. Some control must be used so that the devices at each end of

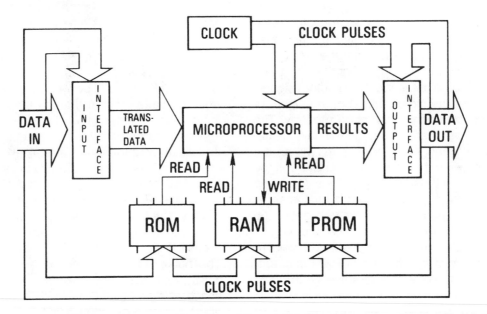

Figure 1–8 Clock pulses within a computer. *(Courtesy of General Motors Corporation, Service Technology Group.)*

the data link know when to transmit and when to receive. Several control methods can be used, but to discuss them here would go beyond the goals of this text and the interest of most automotive students and technicians.

EXHAUST GASES

The computer of a computerized engine control system has a mission: to reduce emissions, improve mileage and improve driveability. The priority varies, however, under some driving conditions. For example, during warm-up, driveability has a higher priority than mileage; and at full throttle, some systems give the performance aspect of driveability a higher priority than emissions or mileage.

There are six major exhaust gases on an automobile: oxygen (O_2), hydrocarbons (HC), water vapor (H_2O), carbon dioxide (CO_2), carbon monoxide (CO) and oxides of nitrogen (NO_x).

1. *Oxygen.* Oxygen forms the basis for all combustion. It constitutes just under 20% of our air and is the chemical that supports all flame in the atmosphere. The burning of oxygen and fuel releases the chemical energy that does all the work of the engine and the vehicle. A properly operating feedback control system will result in a slightly fluctuating amount of residual oxygen in the exhaust. A very rich mixture would have so much fuel that all the oxygen would be consumed; a very lean mixture would leave too much oxygen unused in the exhaust.

2. *HC.* Gasoline is a hydrocarbon compound, as is engine oil. When hydrocarbons burn properly, the hydrogen and carbon atoms separate; each combines with oxygen to form either water (H_2O) or carbon dioxide (CO_2), Figure 1-9. If for any reason the gasoline in the cylinder fails to burn, it is pumped into the exhaust system as a raw HC molecule, a most undesirable emissions result. There can be several reasons for all or part of the HC in the cylinder not to burn:

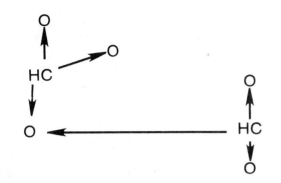

Figure 1–9 Chemical reaction during ideal combustion

- Any electrical malfunction, which prevents the spark plug from igniting the fuel mixture.
- A lack of mechanical integrity, which allows fuel to escape from the cylinder before the spark occurs or to fail to reach a sufficiently high compression pressure to ignite.
- An overly lean mixture, which hampers flame propagation (the flame goes out too soon).
- An exceedingly rich mixture, which uses up the oxygen before all of the HC has been consumed.
- A low cylinder combustion chamber surface temperature, which rapidly draws heat from those HC molecules and the oxygen in contact with it, and thus prevents them from reaching or maintaining ignition temperature. A low cylinder surface temperature can be caused by a low coolant (low temperature or stuck-open thermostat) or an over-advanced ignition timing. When an engine is first started on a cold day, of course, the combustion chamber surfaces will all initially be lower than enough to sustain clean emissions performance.

Any hydrocarbons not burned in the combustion chamber are burned in the catalytic bed of the converter.

3. *H_2O*. Water vapor forms the bulk of the exhaust product of most engines, even those running very poorly. The major source of energy in the fuel is the hydrogen, which readily combines chemically with the atmospheric oxygen drawn in with the intake air, releasing heat energy and forming water vapor. This is the most benign product of the engine's combustion, and is of concern only if the engine is run so little that condensed water is left on the interior surfaces, providing a favorable environment for rust.

4. *CO_2*. Carbon dioxide forms in the burning of the carbon portion of the fuel. Carbon dioxide is a harmless portion of the exhaust. It is the same gas used by plants in their respiration cycle, and it is the same gas that is used to carbonate beverages. While some carbon dioxide is harmless, environmental concerns have focused recently on the sheer amount of the gas produced by internal combustion engines. Some people have expressed the fears that a significant increase in the level of atmospheric carbon dioxide could have a "greenhouse effect" on the earth's atmosphere, eventually changing the climate.

5. *CO*. Carbon monoxide is a very deadly poison gas, odorless and colorless, and lethal in almost undetectably small concentrations. During normal combustion, each carbon atom tries to take on two oxygen atoms (CO_2). If, however, there is a deficiency of oxygen in the combustion chamber, some carbon atoms are only able to combine with one oxygen atom and thus produce CO. The carbon monoxide actively tries to obtain another oxygen atom wherever it can find one, and this oxygen-scavenging property of CO is what makes it so lethal. If breathed into the lungs, it not only provides no oxygen, it removes what oxygen it finds there. An overly rich air/fuel ratio is the only cause of CO production.

Other Exhaust Emissions from Gasoline Engines

Aldehydes
Ammonia
Carboxylic acids
Inorganic solids
   lead (from leaded fuel)
   soot
Sulfur oxides (from fuel impurities)

6. *NO_x.* Air is made up of about 80% nitrogen and 20% oxygen. Under normal conditions nitrogen and oxygen do not chemically unite. When raised to a high enough temperature, however, they unite to form a nitrogen oxide compound, NO. In the presence of atmospheric oxygen, NO rapidly becomes nitrogen dioxide, NO_2. These compounds are grouped into a family of compounds referred to as oxides of nitrogen, NO_x. Oxide of nitrogen compounds start to form at 1,120 degrees centigrade (2,040 degrees Fahrenheit), with production increasing as temperatures go higher.

The reduction beds of the catalytic converter reduce any excessive NO_x gas into harmless constituents.

Catalytic Converter

Controlling the air/fuel ratio and the spark timing, keeping the coolant temperature near 93 degrees centigrade (200 degrees Fahrenheit) or more, and controlling intake air temperature and combustion temperature dramatically reduce exhaust emissions. Even these measures, however, do not eliminate them. The catalytic converter is the single most effective device for controlling exhaust emissions. Placed in the exhaust system, the **catalyst** agents cause low-temperature oxidation of HC and CO and reduction of NO_x, yielding H_2O, CO_2, O_2, and nitrogen. Placing the converter downstream in the exhaust system allows the exhaust gas temperature to drop significantly before the gas enters the con-

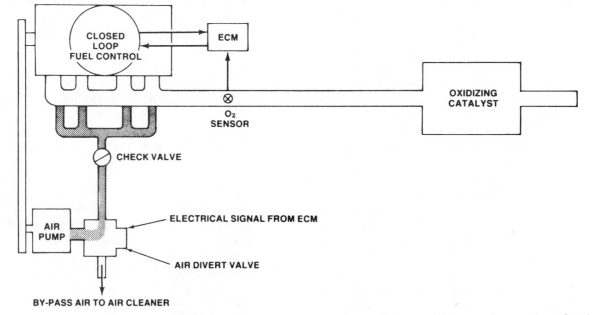

Figure 1–10 Single-bed, three-way catalytic converter. *(Courtesy of General Motors Corporation, Service Technology Group.)*

verter, and thus prevents the production of more NO_x in the oxidation process. The catalysts that are most often used, **platinum** and **palladium** for oxidation and **rhodium** for reduction, are coated over a porous material, usually aluminum oxide in pellet form or aluminum oxide in a honeycomb structure. Aluminum oxide's porosity provides a tremendous amount of surface area on which the catalyst material can be applied to allow maximum exposure of the catalyst to the gases.

There have been three different types of catalytic converters used on automobiles: 1) a single-bed, two-way, introduced on cars in the mid-seventies; 2) a single-bed, three-way; and 3) a dual-bed, three-way. The three-way converters were introduced on cars in the late seventies.

A single-bed converter has all of the catalytic materials in one chamber, Figure 1-10. A dual-bed converter has a reducing chamber for breaking down NO_x and an oxidizing chamber for oxidizing HC and CO. The exhaust passes through the reducing chamber first, Figure 1-11.

Additional oxygen is introduced into the oxidizing chamber to make it more efficient. A two-way converter oxidizes HC and CO; a three-way converter oxidizes HC and CO and reduces NO_x.

The rhodium-reducing catalyst of a catalytic converter causes the NO_x molecules to break down again into nitrogen and oxygen. It can only do so, however, if there is deficiency of oxygen in the exhaust stream. A rich air/fuel ratio is therefore needed; a rich mixture yields only a small amount of leftover oxygen. On the other hand, the platinum- and palladium-oxidizing catalysts need abundant available oxygen to oxidize HC and CO. A lean air/fuel mixture is needed to yield leftover oxygen, Figure 1-12 . A **stoichiometric** air/fuel ratio (approximately 14.7 to 1 at sea level) is essential to make the three-way converter work effectively, Figure 1-13. If the fuel mixture is leaner than 14.7 to 1, the reducing side of the converter is not effective. If the mixture is richer than 14.7 to 1, the additional CO and the oxygen liberated by the reducing catalyst, plus additional

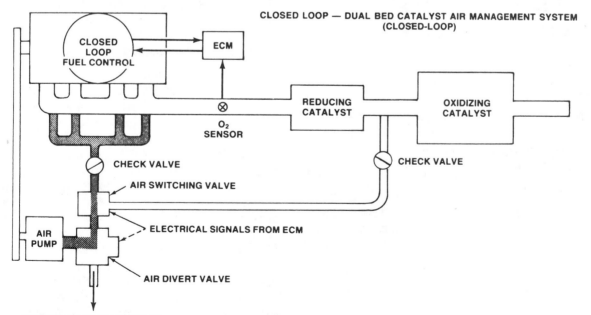

Figure 1–11 Dual-bed, three-way catalytic converter. *(Courtesy of General Motors Corporation, Service Technology Group.)*

3 WAY CONVERTER

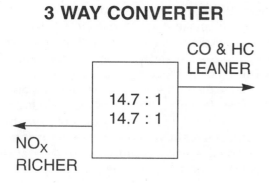

Figure 1–12 Three-way catalytic converter

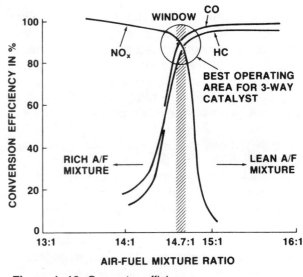

Figure 1–13 Converter efficiency

oxygen pumped into the oxidizing chamber in the case of a dual-bed converter, causes the converter to overheat as a result of the abundant oxidation process. One of the major reasons for creating a computerized engine control system was to have the ability to maintain a 14.7 to 1 air/fuel ratio and therefore enhance the effectiveness of the three-way catalytic converter.

Most manufacturers have experienced some problems with catalytic converters becoming

---

Stoichiometry and 14.7 to 1

Keep in mind that a stoichiometric ratio is the one that most effectively allows all the components of the emissions control system, most notably the catalytic converter, to minimize exhaust emissions. When these systems were first invented, 14.7 to 1 was the ideal ratio for standard gasoline. Since that time there have been many changes to the fuels also, including different blends for different seasons of the year, different altitudes and even different areas depending on their compliance with federal air quality standards. But while *oxygenated* and other special fuels have changed the numbers of the ratio somewhat, the objective remains the same: to maintain the fuel/air ratio so that the resulting exhaust emissions are kept at their optimal state.

---

restricted and have developed methods of testing for this condition. Depending on the severity of the restriction, the symptoms vary from slight loss of power after several minutes of driving to the engine not starting in the first place.

Other Emission-Control Devices

Other emission-control devices currently used on most light-duty vehicles and those that can be included as part of the computerized engine control system, or added to it, are discussed here.

EGR Valve. This valve introduces exhaust gas into the intake manifold to reduce combustion temperature. It is included as a part of and is controlled by the control system on most vehicles, but it is independent of the control system on other vehicles. The EGR valve is not used at all on an increasing number of engines because of the option of using valve overlap to retain a certain amount of exhaust in the cylinder.

Intake Charge Heating. Either electric or exhaust heat is used to heat the air/fuel mixture as it enters the intake manifold. Intake charge heating also is included as one of the controlled

Stoichiometric Air/Fuel Ratios

A stoichiometric air/fuel ratio provides just enough air and fuel for the most complete combustion resulting in the least possible total amount of leftover combustible materials: oxygen, carbon, and hydrogen. This maximizes the production of CO_2 and H_2O, and minimizes the potential for producing CO, HC, and NO_x. It also provides good driveability and economy.

The most power is obtained from an air/fuel ratio of about 12.5 to 1, although the best economy is obtained at a ratio of about 16 to 1. Air/fuel ratios this far from stoichiometric, however, are not compatible with the three-way catalytic converter. The leaner mixtures also increase engine temperature as a result of slower burn rate.

functions on some applications and not on others. It is not used at all on some systems.

Fuel Evaporation Control. This system prevents HC vapors from escaping from the fuel tank and carburetor float bowl. It is included as a controlled function on some control systems and not on others.

Air Injection System (Pump or Pulse Air). This system pumps air into the exhaust ports to oxidize HC and CO coming out of the exhaust valve. Air injection system is controlled by the computer on some applications, while on others it is controlled by conventional methods or not used at all.

Heated Air Inlet. This system heats the air before it enters the air cleaner during engine warm-up. It is usually not controlled by the engine control system and is not used on most port fuel injection systems.

PCV System. This system recirculates crankcase gases through the combustion chambers to burn them. It is a separate system on all engine applications.

ECONOMY AND DRIVEABILITY

Computerized engine control systems contribute to fuel economy and driveability in several ways.

Fuel Economy Controls

Air/Fuel Mixture. Maintaining a 14.7 to 1 air/fuel ratio promotes fuel economy by eliminating the fuel waste that occurs with richer fuel mixtures. Richer mixture usage can easily occur in an uncontrolled engine. It also provides an efficient mixture for combustion and thus enhances economy and driveability. Systems that use electronic fuel injection also contribute to economy with more efficient fuel control such as:

- shutting off fuel during deceleration.
- replacing the mechanical choke with a more articulated, electronically controlled cold engine enrichment.
- replacing the accelerator pump and vacuum-controlled full-power enrichment functions of a carburetor with an electronically controlled acceleration enrichment.

In many cases these features also help to improve driveability.

Electronic Ignition Timing. Ignition timing is well known to be a major factor in both fuel economy and driveability. It is also a critical factor in the control of exhaust emissions. With computer-controlled ignition timing, timing does not have to be dependent on just engine speed and load. It can be adjusted to meet the greatest need during any driving condition. For example, following a cold start, it can be advanced for driveability. During light load operation with a partially warmed-up engine, it can be retarded slightly to hasten engine warmup and reduce exhaust emissions. During acceleration or wide-open throttle operation, it can be adjusted for maximum torque.

Idle Speed Control.   Electronic idle speed control can more nearly maintain the best compromise between a speed at which the engine is prone to stall and one that wastes fuel and produces an annoying lurch as the automatic transmission is pulled into gear. It can keep idle speed from "yo-yoing" up and down as the air-conditioning clutch cycles on and off, and it can be designed to prevent dieseling. It can also adjust idle speed in an effort to compensate for low charging system voltage, low engine temperature, and engine overheating.

Torque Converter Clutch Control.   Many engine control computer systems control a clutch in the torque converter. This function eliminates converter slippage and heat production and thus contributes to fuel economy. On some manual transmission applications, the computer activates an upshift light, which alerts the driver when to upshift to obtain maximum economy.

EGR Valve Control.   The EGR valve has not been the best thing to happen to fuel economy or driveability. Many earlier EGR systems had the EGR valve open anytime the engine was off idle and short of full throttle; any control of it is an improvement. Most engine control computer systems control when the EGR valve is turned on, and leave it off for more of the time when combustion temperatures are not high enough to produce significant amounts of NO_x. Some systems even control how much it opens to **modulate** it, or find a position between two extremes.

Air-conditioning Control.   Some computerized engine control systems feature an air conditioning clutch cutout function that disengages the air-conditioning clutch during heavy throttle operation. This function contributes somewhat to driveability.

Turbocharger Boost Control.   To increase performance, each manufacturer is offering at least one turbocharged engine option. In most cases, the computer controls the amount of boost that the turbocharger can develop. It is worth noting that each of the manufacturers is using an air-to-air intercooler on selected turbo applications.

This lowers manifold air temperature by 120 to 150 degrees Fahrenheit and allows for even more aggressive turbo boost and spark advance.

Closed Loop, Open Loop

Closed Loop.   Made possible by the development and use of the oxygen sensor, a feature of all comprehensive computerized engine control systems is the ability to operate the engine in a closed-loop mode. The term **closed loop** describes an intimate, triangular relationship between the oxygen sensor, the computer, and the fuel-metering control device. The oxygen sensor tells the computer what the delivered air/fuel mixture was. The computer sends a command to the fuel-metering control device to adjust the air/fuel ratio toward a stoichiometric air/fuel ratio (14.7 to 1 at sea level). The adjustment usually causes the air/fuel ratio to cross stoichiometric, and it starts to move away from 14.7 in the opposite direction. The oxygen sensor sees this and reports it to the computer. The computer again issues a command to adjust back toward 14.7 to 1. With the speed at which this cycle occurs, the air/fuel ratio never gets very far from 14.7 to 1 in either direction. This cycle repeats continuously.

Closed loop is the most efficient operating mode, and the computer is programmed to keep the system in closed loop as much as possible. However, the following criteria must be met before any of the systems can go into closed loop:

- The oxygen sensor must reach operating temperature (around 315 degrees Centigrade / 600 degrees Fahrenheit).
- The engine coolant temperature must reach a criterion temperature (varies somewhat but tends to be around 65 degrees Centigrade / 150 degrees Fahrenheit).
- A predetermined amount of time must elapse from the time the engine was started (this varies from a few seconds to one or two minutes).

There are other operating conditions, such as

hard acceleration or, in some cases, prolonged idle, that force the system out of closed loop.

Open Loop. **Open loop** is used during periods of time when a stoichiometric air/fuel ratio is not appropriate such as during engine warm-up or at wide-open throttle (WOT). During this operational mode, the computer uses input information concerning coolant temperature, engine load, barometric pressure, and engine speed to determine what the air/fuel ratio should be.

Once the necessary information is processed, the computer sends the appropriate command to the mixture control device. The command does not change until one of the inputs changes. In this mode, the computer does not use oxygen sensor input and therefore does not know if the command it sent actually achieved the most appropriate air/fuel ratio for the prevailing operating conditions. As a result of this weakness of the open-loop mode, the computer puts the system into closed loop as soon as possible and keeps it there for as long as possible. Certain failures within the system prevent it from going into closed loop or cause it to drop back out of closed loop.

ATTITUDE OF THE TECHNICIAN

Historically, many automotive technicians have had a negative attitude about manufacturers' continuous changes to cars. Changes mean always having to learn something new, having to cope with a new procedure, and so on. There is another way to look at it, though: every time new knowledge or a new skill is required, it represents a new opportunity to get ahead of the pack. Most of us are in this business to make money; the more you know that other people have not bothered to learn yet, the more money you can make. Successful, highly paid automotive technicians *earn* the money they get because of what they *know*. Their knowledge enables them to produce more with the same effort. More important, it enables them to do things that other people cannot.

Computerized automotive control systems represent the newest and probably the most significant and most complicated development ever to occur in the history of automotive service. (It is complicated, not because it is more difficult, but because it involves more processes.) We have seen only the beginning. It is likely that before long, every major component and function on the car will be electronically controlled. Those who appreciate the capability of these systems and learn everything they can about them will be able to service them and do very well while other people scratch their heads and complain.

✔ SYSTEM DIAGNOSIS & SERVICE

Approaching Diagnosis

Diagnosing problems on complicated electronic systems takes a little different approach than many automotive technicians are used to. The flat-rate pay system has encouraged many of us to take shortcuts whenever possible; however, shortcuts on these systems get you into trouble. Use the service manual and follow procedures carefully. It is very useful to spend some time familiarizing yourself with the diagnostic guides and charts that manufacturers present in the service manual or manuals you will be using.

Prediagnostic Inspection. Although most computerized engine control systems feature some degree of self-diagnostic capability, none of them monitors spark plugs, spark plug wires, valves, vacuum hoses, PCV valves, and other engine support and emission-control components that are not a part of the computer system. A seemingly unrelated part such as a spark plug wire can have a direct impact on the performance of the system. For example, a shorted spark plug wire prevents its cylinder from firing. The unburned oxygen coming out of that cylinder, if it is on the same side as the oxygen sensor on a V-type engine, causes the oxygen sensor to mistakenly see a lean condition. The computer responds by enriching the air/fuel mixture. Replacing the oxygen sensor, half or all of the

Replacing the oxygen sensor, half or all of the other sensors, the computer, the headlights, and the front bumper does not solve the problem. Replacing the shorted spark plug wire and possibly cleaning the spark plug *does* solve the problem.

The computer knows the engine operating parameters (throttle position, atmospheric pressure, engine speed, temperature, load, etc.), and with this information it can calculate exactly what the spark timing should be. It continuously makes calculations and sends its spark timing commands. It does not, however, know what base timing is (except on the very latest models); and the command it sends out is added to base timing. If base timing is incorrect, a spark plug will fire at the wrong time in spite of the computer.

Having a closed thermostat stick or a restricted radiator causes the engine to overheat, possibly causing the computer to think that the coolant temperature-sensing device or its connecting circuit is shorted and setting a fault code in its diagnostic memory. The coolant sensing device, however, is only reporting what it sees; the fault is in the cooling system. Do not overlook the possibility that someone has replaced a closed 195 degree thermostat with one that opens at 160 degrees, or that the charcoal canister is saturated with gasoline or that it is being purged when it should not be. Do not overlook the possibility of loose or shorted electrical wires or of cracked or misrouted vacuum hoses.

Any such problem, although it may not be directly related to the computer system, can cause a driveability problem and can in some cases cause the computer to mistakenly report a problem in one of its circuits. The service manual for each vehicle with an engine control computer system should provide instructions for making a prediagnostic inspection, because the system's diagnostic procedures assume that all such unrelated, or perhaps it would be better to say *semirelated,* systems are functioning properly. All of the engine components, especially engine support components, and some nonengine components such as the transmission, are more related in their influence on each other than they have ever been before the development of computer control systems. It is extremely important to remember that for the computer control system to work properly, all other engine-related systems must be operating to manufacturer's specifications. If routine maintenance, such as changing the oil, PCV valve, spark plugs, and spark plug wires, is due, it is advisable to do it before beginning diagnostic procedures.

SUMMARY

In this chapter, we have begun building our conceptual foundation for understanding why carmakers use computers in cars and what the computers are supposed to do. The principal design objectives are to control exhaust emissions to a legally acceptable level, to optimize the vehicle's driveability, and to get the best fuel economy consistent with the driving conditions.

We have learned the major components of a microprocessor, the central element of the engine control computer. It employs a PROM (programmable, read-only memory) or hard-wired memory to store general information about the specific vehicle. It maintains a KAM (keep-alive memory) to store adaptations to its control maps that it has learned in past driving trips. It includes interface connections for all its sensors and outputs as well as power to the computer itself.

We have learned the various elements in a car's exhaust that relate to the emissions laws: hydrocarbons or unburned fuels, carbon monoxide, and carbon dioxide. We have seen how residual oxygen is used as a measure of the computer's mixture control success and how water vapor is produced as a result of the hydrogen in the fuel.

The concept of stoichiometry has been introduced, a concept that will play a role in each successive chapter. A stoichiometric ratio is the exact air/fuel mixture that will allow the catalytic carbon emissions to a minimum. We have seen how this stoichiometric ratio is such an exact mixture there really is no way to maintain it but to use a computer.

We have learned the concepts of open and closed loop. Open loop describes the conditions under which the computer determines air/fuel mixture and ignition timing based on information stored in its memory, about temperature, load, speed and other system parameters. Closed loop, in contrast, is the feedback mode in which the computer principally controls air/fuel mixture by monitoring the output signal from the oxygen sensor.

Finally, we have learned the importance of following precise diagnostic steps if this type of computer system is to be diagnosed. Without such a sequence, the technician can be misled by other problems, or by countermeasures the computer employs to solve some other problem that has affected the system.

▲ DIAGNOSTIC EXERCISE

In an uninformed and socially irresponsible attempt to improve his car's performance, a do-it-yourselfer disconnected, disabled, or removed most of the emissions controls on his car. He slabbed off the EGR, disconnected most of the vacuum lines, and ran the distributor advance vacuum line directly to manifold vacuum.

What will the consequences be for the car?

┌─ Excessive HC Production ─┐

Overadvancing ignition timing by 6 degrees can cause HC production to go up by as much as 25% and NO_x production to increase by as much as 20 to 30%. A thermostat that opens at 160 degrees Fahrenheit can cause HC production to go up as much as 100 to 200 parts per million.

Will its driveability, power or fuel economy improve? What should the repair/diagnostic technician's approach be?

REVIEW QUESTIONS

1. Name two types of voltage signal.
2. The term *square wave* refers to what?
3. Most sensors create what type of signal?
4. How many different characters are in the binary code?
5. Data coming into the automotive-type microcomputer is first processed by _____.
6. What does the abbreviation ROM stand for, and what is its function?
7. Name at least two ways that a PROM is different from a ROM.
8. Name at least two ways that a RAM is different from a ROM.
9. Name two types of RAM.
10. What is the function of the clock pulse in a computer.
11. Name at least three things that cause excessive HC from the exhaust.
12. What causes high CO production?
13. What causes high NO_x production?
14. What air/fuel ratio is necessary for the three-way catalytic converter to work effectively?
15. Name two different types of three-way catalytic converters.
16. Describe the condition(s) that must exist before the computer system can function properly and before a diagnostic procedure should be begun.

ASE-type Questions   (Actual ASE test questions are rarely so product specific.)

17. Technician A says that computerized engine control systems are intended to reduce exhaust emissions. Technician B says that computerized engine control systems are

computerized engine control systems are designed to improve fuel mileage and drive-ability. Who is correct?

a. A only.
b. B only.
c. both A and B.
d. neither A nor B.

18. A car with a driveability complaint is being checked. During the inspection it is found that base timing is off by 4 degrees and one spark plug is faulty. Also the engine control computer's self-diagnostic test shows an oxygen sensor code. Which of the three faults identified are most likely related?

a. the timing and the oxygen sensor code.
b. the oxygen sensor code and the spark plug.
c. all three of them.
d. none of them.

19. Technician A says that a fault code indicating a coolant temperature sensor problem could be the result of a faulty sensor or one of its circuit connections. Technician B says that it could be the result of a faulty thermostat. Who is correct?

a. A only.
b. B only.
c. both A and B.
d. neither A nor B.

20. Technician A says that for a car to use a three-way catalytic converter, it must also have some way to test and control its air/fuel mixture. Technician B says that if a car has a three-way catalytic converter, the converter can be either a single-bed or a dual-bed design. Who is correct?

a. A only
b. B only
c. either A or B
d. neither A nor B

Chapter 2

Common Components for Computerized Engine Control Systems

OBJECTIVES

In this chapter you can learn:
- ❑ the advantages and disadvantages of a multipoint fuel injection system.
- ❑ the concepts of pulse width and duty cycle.
- ❑ how an oxygen sensor works.
- ❑ how a thermistor is used as a temperature-measuring sensor.
- ❑ how a potentiometer is used as position-measuring sensor.
- ❑ the operation of a Hall-effect switch.
- ❑ a sensor that is commonly used to measure air pressure.
- ❑ how a simple switch is used as a sensor
- ❑ what an actuator is and does.

KEY TERMS

Actuator
Barometric Pressure
Clamping Diode
Duty Cycle
E-cell
Gallium Arsenate Crystal
Hall-Effect Switch
Hz (Hertz)
Impedance
Piezoelectric
Piezoresistive
Potentiometer
Pulse Width
Schmitt Trigger
Speed Density Formula
Thermistor
Volumetric Efficiency (VE)
Wide Open Throttle (WOT)
Zirconium Dioxide (ZrO_2)

Among the different car manufacturers, vehicle models, and years, there are many different computerized engine control systems. Although there are significant differences between the various systems, when compared, they are actually more alike than they are different. Some of the more common components and circuits will be discussed in this chapter to avoid needless duplication when discussing the specific systems. As you read the following chapters, you may find it useful to refer to this chapter to clarify how a particular component or circuit works.

COMMON FEATURES

Computers

The comprehensive automotive computer is a special-purpose, small, highly reliable, solid-state digital computer protected inside a metal box. It receives information in the form of voltage signals from several sensors and other input sources. With this information, which the computer rereads several thousand times per second, it can make "decisions" about engine calibration functions such as air/fuel mixture, spark timing, EGR application, and so forth.

These decisions appear as commands sent to the **actuators**—the solenoids, relays, and motors that carry out the output commands of the computer. Commands usually amount simply to turning an actuator on or off. In most cases the ignition switch provides voltage to the actuators

55

either directly or through a relay. The computer controls the actuator by using one of its internal solid-state switches ("power transistors" and "quad drivers" are two types) to ground the actuator's circuit. In this way the transistor only has to deal with the low-voltage side of the circuit (the voltage is dropped, the work is done, and most of the heat is produced in the actuator, away from and outside the computer).

CAUTION: No attempt should be made to open the computer's metal housing. The housing protects the computer from static electricity. Opening it or removing any circuit boards from it outside carefully controlled laboratory conditions will likely result in damage to some of its components. Certain late-model systems use computers with a special kind of memory that can be reprogrammed by exposing that component to light, usually by removing a piece of opaque tape. If you do not intend to do such reprogramming, or do not know what the effect will be, do not open a computer just to see what the internal parts look like. Over time, a technician can expect to find a completely failed computer that can be opened up without doing any additional harm.

Location. On most systems the computer is located inside the passenger compartment. This protects it from the harsh, high-temperature envi-

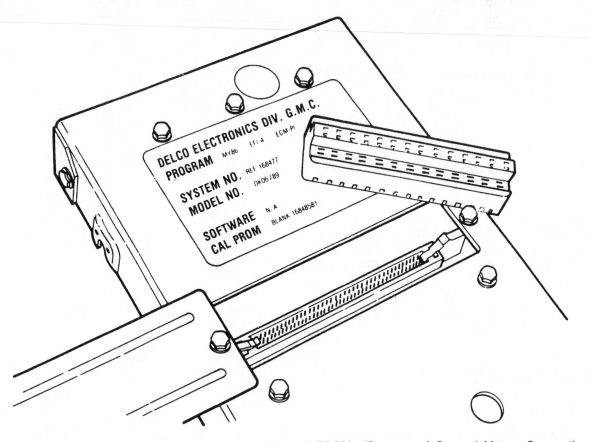

Figure 2–1 General Motors electronic control module and PROM. *(Courtesy of General Motors Corporation, Service Technology Group)*

ronment in the engine compartment. Some early systems positioned the computer in the engine compartment. Indeed, it is because of that experience and the exposure to heat and vibration that they were later moved to the passenger compartment.

Engine Calibration.   Because of the wide range of vehicle sizes and weights, engine and transmission options, axle ratios, and so forth, and because many of the decisions the computer makes must be adjusted for those variables, the manufacturers use some type of an engine calibration unit, Figure 2-1. The calibration unit is a chip that contains information that has to be specific to each vehicle. For example, the vehicle's weight affects the load on the engine, so to optimize ignition timing, ignition timing calibration must be programmed for that vehicle's weight. In a given model year, a manufacturer might have more than a hundred different vehicle models counting different engine and transmission options, but might use less than a dozen different computers. This is made possible by using an engine calibration unit specific to each vehicle.

The engine calibration unit may be referred to as the "PROM," or some other term may be used. Some manufacturers make the calibration unit removable, and some make it a permanent part of the computer.

Five-Volt Reference

With few exceptions, all of the systems use 5 volts to operate their sensors. Within the electronics industry, 5 volts has been almost universally adopted as a standard for information transmitting circuits. This voltage value is high enough to provide reliable transmission and low enough not to damage the tiny circuits on the chips in the computer. Of course, use of a computer-industry standard voltage makes parts specification more economical for the car makers.

Fuel Injection

There are two types of fuel injection systems used with comprehensive computerized engine

Constant Spray and Timed Spray Injector

There are two basic methods of controlling the quantity of fuel introduced by a fuel injection system: the constant spray injection and the timed spray injection. Historically these have been port injection systems (placing an injector in the center of the intake manifold, above the throttle blades, is a comparatively new development). As the term *constant spray* implies, in this system each injector sprays a continuous stream of fuel into each intake port; and the quantity of fuel introduced is controlled by a variable fuel pressure. Fuel pressure is controlled by an airflow meter. The Bosch K-Jetronic is the most common example of this type of system.

In timed spray systems, the difference between fuel pressure and intake manifold pressure is held constant by a pressure regulator, and the injectors are turned on and off. Fuel quantity is controlled by how long the injectors are turned on, and the injectors are controlled by an electronic module (a single-function computer system). The Bosch D-Jetronic and L-Jetronic systems are the best known examples of this type of system.

control systems: single point and multipoint. They each use an intermittent or timed spray to control fuel quantity (There is one exception, the Bosch K-Jet systems, described more fully in the European systems chapter of this book.)

Single-Point Injection.   Single-point injection is often referred to as throttle body injection. Single point means that fuel is introduced into the engine from one location. This system uses an intake manifold similar to what would be used with a carbureted engine, but the carburetor is replaced with a throttle body unit, Figure 2-2. The throttle body unit contains one or two solenoid-operated injectors that spray fuel directly over the throttle blade (or blades). Fuel under pressure is

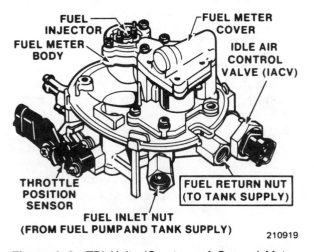

Figure 2–2   TBI Unit. *(Courtesy of General Motors Corporation, Service Technology Group)*

supplied to the injector. The throttle blade is controlled by the throttle linkage just as in a carburetor. The computer controls voltage pulses to the solenoid-operated injector, which opens and sprays fuel into the throttle bore. The amount of fuel introduced is controlled by the length of time the solenoid is energized. This is referred to as **pulse width**. The amount of air introduced is controlled by the opening of the throttle blade.

The EFI system is characterized, especially on smaller engines, by excellent throttle response and good driveability. Experience has shown, however, that the system is best suited for engines with small cross-sectional area manifold runners that at low speeds will keep the fuel mixture moving at a higher velocity. This reduces the tendency for the heavier fuel particles to fall out of the airstream.

Multipoint Injection.   Multipoint injection is often referred to as port injection and means that fuel is introduced into the engine from more than one location. This system uses an injector at each intake port. Fuel is sprayed directly into the port, just on the manifold side of the intake valve.

The multipoint injection system provides the most advanced form of fuel control yet developed. It offers the following advantages:

- Spraying precisely the same amount of fuel directly into the intake port of each cylinder eliminates the unequal fuel distribution inherent when already mixed air and fuel are passed through an intake manifold. There is simply no way to make certain that equal amounts of fuel and air get to each cylinder.
- Because there is no concern about fuel condensing as it passes through the intake manifold, there is less need to heat the air or the manifold.
- Because there is no concern about fuel molecules falling out of the airstream while moving through the manifold at low speeds, the cross-sectional area of the manifold runners can be larger and thus offer better cylinder-filling ability [**volumetric efficiency, (VE)**] at higher speeds. Some manufacturers have even been able to design variable-geometry intake runners, to optimize air ingestion at different engine speeds and loads.
- Most of the manifold-wetting process is avoided, though some wetting still occurs in the port areas and on the valve stems. If fuel is introduced into the intake manifold, some will remain on the manifold floor and walls, especially during cold engine operation and acceleration. Fuel metering has to allow for this fuel in order to avoid an overlean condition in the cylinders. It has to be accounted for again during high-vacuum conditions because it will then begin to evaporate and go into the cylinders.

In general port fuel injection provides better engine performance and excellent driveability while maintaining or lowering exhaust emission levels and increasing fuel economy. The major disadvantages are somewhat greater cost and reduced serviceability because of the larger number of components and their relative inaccessibility.

Fuel Pressure Regulators.   All fuel injection systems that are part of a comprehensive computerized engine control system use pulse width as the primary means of controlling fuel metering. A pressure regulator is used to provide

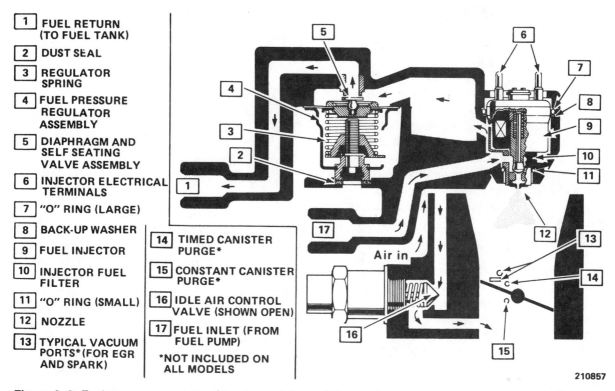

1. FUEL RETURN (TO FUEL TANK)
2. DUST SEAL
3. REGULATOR SPRING
4. FUEL PRESSURE REGULATOR ASSEMBLY
5. DIAPHRAGM AND SELF SEATING VALVE ASSEMBLY
6. INJECTOR ELECTRICAL TERMINALS
7. "O" RING (LARGE)
8. BACK-UP WASHER
9. FUEL INJECTOR
10. INJECTOR FUEL FILTER
11. "O" RING (SMALL)
12. NOZZLE
13. TYPICAL VACUUM PORTS*(FOR EGR AND SPARK)
14. TIMED CANISTER PURGE*
15. CONSTANT CANISTER PURGE*
16. IDLE AIR CONTROL VALVE (SHOWN OPEN)
17. FUEL INLET (FROM FUEL PUMP)

*NOT INCLUDED ON ALL MODELS

210857

Figure 2–3 Fuel pressure regulator. *(Courtesy of General Motors Corporation, Service Technology Group)*

a constant pressure to the injector. What is held constant is the difference between the pressure of the fuel in the injector and the air pressure in the intake manifold. This insures a consistent flow of fuel for a given pulse width; there is no need for the computer to make allowances for flow differences based on pressure differences. Very late model systems include computers that can make these adjustments, and can run at an absolutely constant fuel pressure.

An electric fuel pump supplies fuel under pressure to the regulator. The fuel pressure regulator is located on the fuel rail or manifold and allows excess fuel to return to the tank. It has a diaphragm with fuel pressure on one side and atmospheric pressure on the other side, Figure 2-3. A diaphragm spring pushes the diaphragm against fuel pressure. As the diaphragm moves in the direction the spring is pushing it, a flat disc in

the center of the diaphragm closes off the fuel return passage. This allows less fuel to return to the tank and causes fuel pressure to go up, closer to fuel pump pressure. The increased fuel pressure moves the diaphragm against the spring. This allows more fuel to escape through the return line and thus causes fuel pressure to drop. The diaphragm will always find a balanced position that will keep fuel pressure at the desired value, maintaining an exact pressure differential.

On some single-point injection systems, the spring side of the diaphragm is exposed to atmospheric pressure. Changes in atmospheric pressure produce a slight change in fuel pressure. Some multipoint injection systems connect manifold vacuum to the spring side of the diaphragm.

Changes in atmospheric pressure or in manifold pressure change the total pressure on the

Fuel Injection History

Fuel injection for gasoline engines dates as far back as the middle 1930s. It has captured the interest of automotive engineers and performance enthusiasts for years by promising to be a superior method of fuel induction. Only recently, however, has its technology evolved far enough (see **Measuring Air Mass** in this chapter) to enable fuel injection to surpass the carburetor's ability to deliver performance, driveability, economy, and emissions reduction under all driving conditions.

spring side of the diaphragm and therefore the fuel pressure. The intent, however, is not to increase or decrease fuel delivery but to maintain a constant pressure drop across the injector. For example, when manifold pressure goes up, the injector has to spray fuel into a higher pressure environment. For the correct amount of fuel to be delivered during the pulse width, fuel pressure must be increased in proportion to the manifold pressure increase.

SENSING DEVICES

Exhaust Oxygen Sensors

One of the major components of all comprehensive computerized engine control systems is the oxygen sensor, Figure 2-4. The amount of free oxygen in the exhaust stream is a direct result of the air/fuel ratio. A rich mixture yields little free oxygen; most of it is consumed during combustion. A lean mixture, on the other hand, yields considerable free oxygen because not all of the oxygen is consumed during combustion. With an oxygen sensor in or near the exhaust manifold, measuring the amount of oxygen in the exhaust indirectly reveals the air/fuel ratio burned in the cylinder.

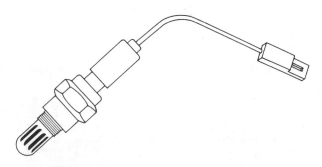

Figure 2–4 Oxygen sensor

The heart of the oxygen sensor is a hollow ceramic body, **zirconium dioxide,** closed at one end, Figure 2-5. The inner and outer surfaces of the ceramic body are coated with separate, superthin, gas-permeable films of platinum. The platinum coatings serve as electrodes. The outer electrode surface is covered with a thin, porous, ceramic layer to protect against contamination from combustion residue. This body is placed in a metal shell similar to a spark plug shell except that the shell has a louvered nose that encloses the exhaust-side end of the ceramic body. When the shell is screwed into the exhaust pipe or manifold, the louvered end extends into the exhaust passage. The outer electrode surface contacts the shell; the inner electrode connects to a wire that goes to the computer. Ambient oxygen is allowed to flow into the hollow ceramic body and to contact the inner electrode.

The oxygen sensor has an operating temperature range of 300 degrees Centigrade to 850 degrees Centigrade (572 degrees Fahrenheit to 1,562 degrees Fahrenheit). When the zirconium dioxide reaches 300 degrees Centigrade it becomes oxygen-ion conductive. If the oxygen in the exhaust stream is less than that in the ambient air, a voltage is generated. The greater the difference, the higher the voltage signal generated, Figure 2-6. Voltage generated by the oxygen sensor will normally range from a minimum of 0.1 volt (100 millivolts) to a maximum of 0.9 volt (900 millivolts). The computer has a preprogrammed value, called a set-point, that it wants to see from

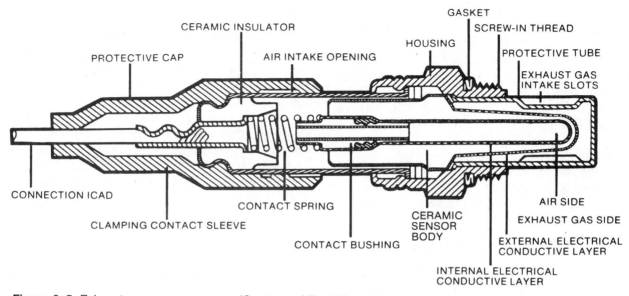

Figure 2–5 Exhaust gas oxygen sensor. *(Courtesy of Ford Motor Company)*

the oxygen sensor during closed-loop operation. The set-point is usually between 0.45 volt and 0.5 volt and equates to the desired air/fuel ratio. Voltage values below the set-point are interpreted as lean (the percent of oxygen in the exhaust

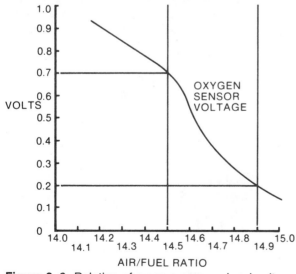

Figure 2–6 Relation of oxygen sensor signal voltage to air/fuel ratio (approximate)

stream and in the ambient air becomes closer), and values above it are interpreted as rich (percentages of oxygen in the two areas move apart).

By the middle to late 1980s, most manufacturers had begun using heated oxygen sensors on some engine applications. The operation of the sensor itself is no different; however, a heat element is placed inside it to heat it quickly and to keep it hot enough during idle and other periods when less heat is produced. This is largely due to a tendency to move the sensor a little lower in the exhaust system so that the exhaust sampled by the sensor is more of an average of what all the cylinders are producing. Placing the oxygen sensor lower in the exhaust pipe means that its wires also attach to the oxygen sensor to power the heat element. In most cases, one powers the heat element when the ignition is turned on, and the other goes to a fixed ground. The circuit is not part of the computer control system, though on some later versions the computer monitors the heater circuit for continuity.

Oxygen sensors are subject to contamination under certain conditions. Contaminating either electrode surface (the outside electrode is more vulnerable) will shield it from oxygen and adversely affect its performance. Prolonged exposure to rich fuel mixture exhaust can cause it to be carbon fouled. This can sometimes be burned off by operating in a lean condition for two or three minutes. The oxygen sensor will also be fouled by exposure to:

- exhaust from leaded fuel.
- the vapors from some silicone-based gasket-sealing compounds.
- the residues of coolant leaking into the combustion chamber.

Fouling by these materials will probably require that the oxygen sensor be replaced.

Diagnostic & Service Tip

Oxygen sensors and mixture. Technicians should remember that, while the oxygen sensor's signal is central to the control of air/fuel mixture, it does not directly sense that mixture. All the oxygen sensor can respond to is the difference in oxygen in the two different gases. Anything other than combustion that allows a change in the oxygen content of the exhaust will thus make the sensor's signal inaccurate. If there is an air leak into the exhaust, perhaps around the exhaust manifold, the sensor will report a lean mixture and the computer will attempt to correct for it by richening the mixture, exactly the wrong tactic. If a spark plug fails to deliver the ignition spark, the air in that cylinder's charge will go into the exhaust unburned. The sensor will again see an overlean condition and the computer will try to richen the mixture. For the computer's sensors, you must remember exactly what each one is measuring before you can know what the possibilities are for corrective measure.

Thermistors

A **thermistor** is a resistor made from a semiconductor material. Its electrical resistance changes greatly and predictably as its temperature changes. At -40 degrees Centigrade (-40 degrees Fahrenheit), a typical thermistor can have a resistance of 100,000 ohms. At 100 degrees Centigrade (212 degrees Fahrenheit) the same thermistor can likely have a resistance between 100 and 200 ohms. Even small changes in temperature can be observed by monitoring the thermistor's resistance. This characteristic makes it an excellent means of measuring the temperature of such things as water (engine coolant) or air.

There are negative temperature coefficient (NTC) and positive temperature coefficient (PTC) thermistors. The resistance of the NTC thermistor goes down as its temperature goes up, while the resistance of the PTC type goes up as its temperature goes up. Most thermistors, including those used as temperature sensors on automotive computer systems, are the NTC type. A good example is an engine coolant temperature sensor, common to just about all computerized engine control systems. The coolant temperature sensor consists of a thermistor in the nose of a metal housing, Figure 2-7. The housing screws

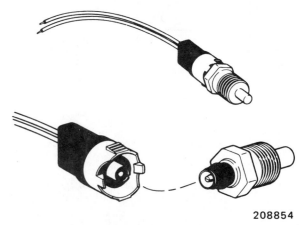

208854

Figure 2–7 Coolant temperature sensor. *(Courtesy of General Motors Corporation, Service Technology Group)*

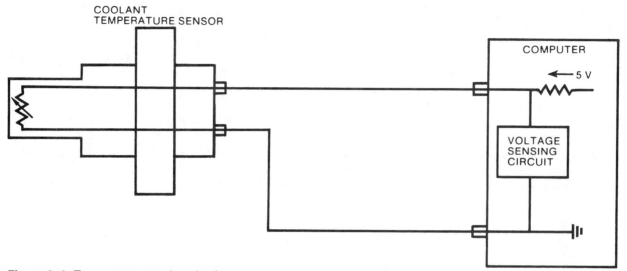

Figure 2–8 Temperature-sensing circuit

into the engine, usually in the head, with its nose extending into the water jacket so the thermistor element will be the same temperature as that of the coolant. The computer sends a regulated voltage signal (reference voltage, usually 5 volts) through a fixed resistance and then on to the sensor, Figure 2-8. A small amount of current flows through the thermistor (usually shown as a resistor symbol with an arrow diagonally across it) and returns to ground. This is a voltage divider circuit (current flows through the first resistance unit and then has an alternate path in parallel with the second resistance unit) and is commonly used as a temperature-sensing circuit. Because the resistance of the thermistor changes with temperature, the voltage drop across it changes also. The computer monitors the voltage drop across the thermistor and, using preprogrammed values, converts the voltage drop to a corresponding temperature value.

Some later systems use a more complicated system which switches a fixed resistance into or out of series with the sensor. The effect of this is to spread the voltage output value of the sensor's signal in the area of normal operating temperature, allowing even more precise control of the mixture and spark advance.

Potentiometers

A **potentiometer** is another application of a voltage divider circuit. It has a movable center contact or wiper that taps or senses the voltage between two resistance units, Figure 2-9. The potentiometer is usually used to measure either linear or rotary motion. If we were going to use a potentiometer to measure throttle position, we would probably attach it to one end of the throttle shaft, as have most car manufacturers. As the throttle is opened, the rotation of the throttle shaft moves the wiper, which slides along the wire-wound resistor.

The computer sends a reference voltage to one terminal of the potentiometer, point A. If the wiper is positioned near point A (wide-open throttle on most applications), there will be a low voltage drop between points A and B (low resistance), and a high voltage drop will exist between points B and C. When the wiper is positioned near point C (idle position on most applications), there is a high voltage drop between points A and B and a low drop between points B and C. The computer monitors the voltage drop between points B and C (C should be the same as D) and interprets a low voltage, 0.5 for example, as idle position. A high voltage, around 4.5, will be interpreted as **wide open throttle (WOT).** Voltages between these values will be interpreted as a proportionate throttle position.

The insert in Figure 2-9 shows a slight variation in the throttle position sensor circuit. Some applications use this variation to enhance diagnostic capability. The high resistance connection between the reference voltage circuit and point B causes near reference voltage to appear at the sensor signal terminal if an open circuit occurs in the harness that connects the sensor to the computer.

Pressure Sensors

Pressure sensors are commonly used to monitor intake manifold pressure and/or **barometric pressure.** Intake manifold pressure is a direct response to throttle position and engine speed, with throttle position being the most significant factor. The greater the throttle opening, the greater the manifold pressure becomes (lower vacuum). At WOT manifold pressure is nearly 100% of atmospheric pressure, reduced only by the slight energy consumed by friction

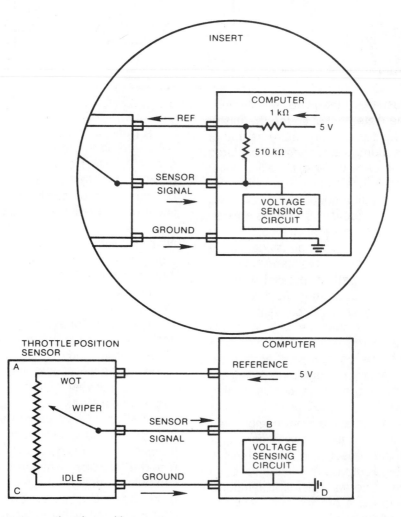

Figure 2–9 Potentiometer as throttle position sensor

with the intake channel walls. Intake manifold pressure can therefore be translated as close to engine load, and it is a critical factor in determining many engine calibrations. Barometric pressure is the actual ambient pressure of the air at the engine. It also impacts on manifold pressure and in most systems is considered when calculating engine calibrations such as air/fuel mixture and ignition timing.

Intake manifold pressure (usually a negative pressure or *vacuum* unless the vehicle is equipped with a turbocharger or supercharger) and/or barometric pressure are, on most systems, measured with a silicon diaphragm that acts as a resistor, Figure 2-10. The resistor/diaphragm (about 3 millimeters wide) separates two chambers. As the pressure on the two sides of the resistor/diaphragm varies, it flexes. The flexing causes the resistance of this semiconductor material to change. The computer applies a reference voltage to one side of the diaphragm. As the current crosses the resistor/diaphragm, the amount of voltage drop that occurs depends on how much the diaphragm

is flexed. The signal is passed through a filtering circuit before it is sent to the computer as a DC analog signal. The computer monitors the voltage drop and, by using a look-up chart of stored pressure values, determines from the returned voltage the exact pressure to which the diaphragm is responding.

There are two slightly different pressure sensor designs that use the **piezoresistive** silicon diaphragm: the absolute pressure and the pressure differential sensors, Figure 2-11. In one design, the chamber under the diaphragm is sealed and contains a fixed reference pressure. The upper chamber is exposed to either intake manifold pressure or to atmospheric pressure. If the upper chamber is connected to the intake manifold, the sensor functions as a manifold absolute pressure sensor. Absolute, as used in MAP (manifold absolute pressure), refers to a sensor that compares a varying pressure to a fixed pressure. The output signal (return voltage) from this sensor increases as pressure on the variable side of the diaphragm increases (wider throttle opening for the engine speed). If the

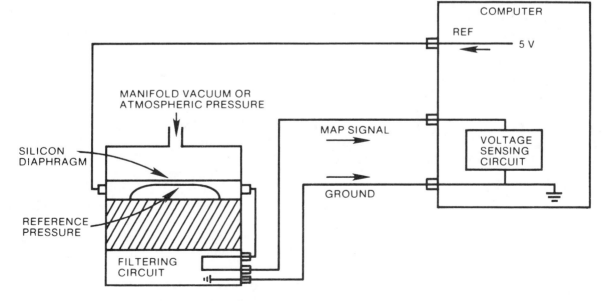

Figure 2–10 Silicon diaphragm pressure sensor

PRESSURE SENSOR DESIGN

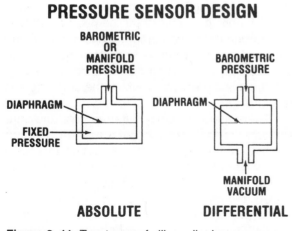

Figure 2–11 Two types of silicon diaphragm pressure sensors. *(Courtesy of General Motors Corporation, Service Technology Group)*

upper chamber is exposed to atmospheric pressure, the sensor functions as a barometric pressure sensor. Many systems take an initial MAP sensor reading as a reference barometric pressure reading for the next driving trip.

The differential pressure sensor combines the functions of both the MAP and the BARO sensors. Instead of using a fixed pressure, one side of the resistor/diaphragm is exposed to barometric pressure and the other side is connected to intake manifold pressure. The output signal is the result of subtracting manifold vacuum from barometric pressure. When the output signal is opposite to the MAP sensor, however, as manifold pressure decreases (higher vacuum), output voltage increases. The pressure differential sensor had limited use on some early General Motors Computer Command Control systems.

Other types of pressure sensors are discussed in the appropriate chapters.

Speed Density Formula. To know how much fuel to meter into the cylinder, the microprocessor must know how much air, as measured by weight (more accurately, by mass), is in the cylinder. Since the microprocessor has no way to actually weigh the air going into a cylinder,

some other method must be used to determine this value. Most systems use a mathematical calculation called the **speed density formula.**

$$\frac{EP \times EGR \times VE \times MAP}{AT} = \text{air density in cylinder}$$

Where:
MAP	=	manifold absolute pressure
VE	=	volumetric efficiency
EP	=	engine parameters
EGR	=	EGR flow
AT	=	air temperature

To make this calculation, all of the factors that influence how much air gets into the cylinders must be considered. Throttle position and engine speed affect air intake, and engine temperature (measured as coolant temperature) affects how much heat the air gains as it passes through the induction system and thus affects the air's density in the cylinder. These factors are grouped together and called engine parameters. The microprocessor gets these pieces of information from some of its sensors that are common to all systems.

The amount of air getting into the cylinders is reduced by the amount of exhaust gas that the EGR valve meters into the induction system. Some systems use a sensor to determine EGR flow, and some use estimates (based on engine parameters) of EGR flow that are stored in computer memory.

Volumetric Efficiency. You know the diameter, length, and shape of the intake manifold runners; the valve size, lift, and timing; the combustion chamber design; and the cylinder size and compression ratio all affect how much air can get into the cylinders. This group of factors is called **volumetric efficiency.** A volumetric efficiency value for any given set of engine parameters is stored in the computer's memory. In principle, volumetric efficiency is the amount of air that gets into the cylinder by the time the intake valve closes, divided by the theoretic amount displaced by the moving piston. Ordinarily this is less than 100%, but the volumetric efficiency varies with engine speed and load. With careful use of valve overlap, volumetric efficiency can sometimes go just above 100% for certain engine speeds. Maximum engine torque, of course, occurs just at the highest volumetric efficiency for a given engine.

Air Temperature.   The air's temperature affects its density and, therefore, how much air can get into the cylinders. Most systems use an air temperature sensor to measure air temperature. Some have estimates of air temperature stored in memory. The estimates of air temperature are based on a fairly predictable relationship between coolant temperature and air temperature.

You also know that barometric pressure pushes air past the throttle valve and manifold pressure pushes that air past the intake valve into the cylinder. The last known values that must be determined are how much air the engine will inhale during any given set of operating conditions, and barometric pressure and manifold pressure. The MAP sensor, as you will see in later chapters, almost always provides both pressure values. A MAP sensor identifies those systems that use the speed density formula.

Measuring Air Mass

Port fuel injection or multipoint injection has become the predominate fuel-metering system. Although a multipoint injection system offers several distinct advantages over other types of fuel-metering systems (discussed earlier in this chapter), it has one shortcoming. A multipoint injection system provides less opportunity (in time duration) for the fuel to evaporate than does a system where the fuel is introduced earlier in the intake channels, into the center of the intake manifold, and only the fuel that evaporates before combustion occurs is usable. In the past, this problem was probably most apparent as a lean stumble when accelerating from idle and was the result of the high manifold pressure that comes from the sudden increase in throttle opening. With the

throttle open, atmospheric pressure forces air into the manifold faster than the engine can use it. The turbulence produced during the intake and compression strokes (and the increased oxygen density on turbocharged applications) helps to atomize the fuel and hasten its evaporation. This condition is only completely overcome, however, by spraying in additional fuel to compensate for the failure of the heavier hydrocarbon molecules to evaporate. The capability of the digital microprocessor provides the ability to calculate just the right amount of fuel to avoid a lean stumble without sacrificing economy or emissions.

For an engine at a given operating temperature and an identified set of atmospheric conditions (air temperature, pressure, and humidity), there is one precise amount of fuel that should be injected into the intake port for each manifold pressure value within the operating range. This amount will provide an evaporated 14.7 to 1 air/fuel ratio in the combustion chamber with the lowest possible leftover unevaporated, unburned HC.

Mass Air Flow (MAF) Sensor.   The MAP sensor used with the speed density formula has been pretty effective in enabling the microprocessor to accurately determine the mass of the air that goes into the cylinder. Keep in mind that the microprocessor quantifies both the air and the fuel by weight when calculating air/fuel ratios. The vane meter has also been fairly effective in providing the microprocessor with the needed information to calculate the air's mass. The vane meter is discussed in the **Inputs** section of the appropriate chapter. Each of these two methods have, however, been unable to measure one factor that affects the air's mass—humidity.

The MAF sensor, introduced in the mid-1980s, provides information to the microprocessor that accounts for the airflow's rate, temperature, and humidity and can actually determine the airflow mass. It does this by measuring the air's ability to cool. Air flowing over an object that is at a higher temperature than the air carries heat away from the object. The amount of heat carried away depends on several factors: the difference

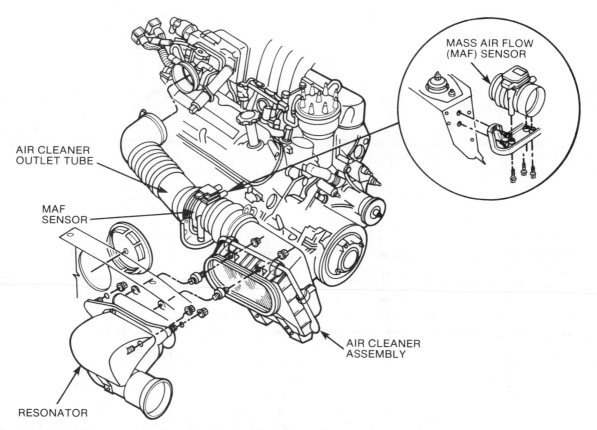

MASS AIR FLOW
(MAF) SENSOR

AIR CLEANER
OUTLET TUBE

MAF
SENSOR

AIR CLEANER
ASSEMBLY

RESONATOR

Figure 2–12 MAF sensor location. *(Courtesy of Ford Motor Company)*

in temperature between the air and the object, the object's ability to conduct heat, the object's surface area, the air's mass, and its flow rate.

The MAF sensor is placed in the duct that connects the air cleaner to the throttle body so that all of the air entering the induction system must pass through it, Figure 2-12. In the MAF sensor's main passageway is a smaller passage that a fixed percentage of the air will pass through. In the smaller passage is a thermistor and a heated wire, Figure 2-13. The thermistor measures the incoming air temperature. The heated wire is maintained at a predetermined temperature above the air's temperature by a small electronic module mounted on the outside of the sensor's body.

The heated wire is actually one of the resis-

tors of a balanced bridge circuit. Figure 2-14 represents a simplified MAF sensor circuit. The module supplies battery voltage to the balanced bridge. Resistors R1 and R3 form a series circuit in parallel to resistors R2 and R4. The voltage at the junction between R1 and R3 is equal to the voltage drop across R3. Because the values of R1 and R3 are fixed, the voltage at this junction is constant; it will vary but only as battery voltage varies. Resistors R1 and R2 have equal values. With no airflow across R4, its resistance is equal to R3, and the voltage at the junction between R2 and R4 is equal to the voltage at the R1/R3 junction.

As airflow across R4 increases, its temperature drops; and because it is a positive temperature coefficient resistor, its resistance goes down with its temperature. As the air's mass increases,

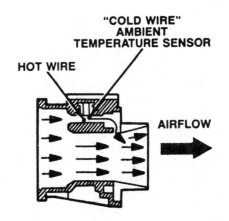

Figure 2–13 MAF sensor. *(Courtesy of Ford Motor Company)*

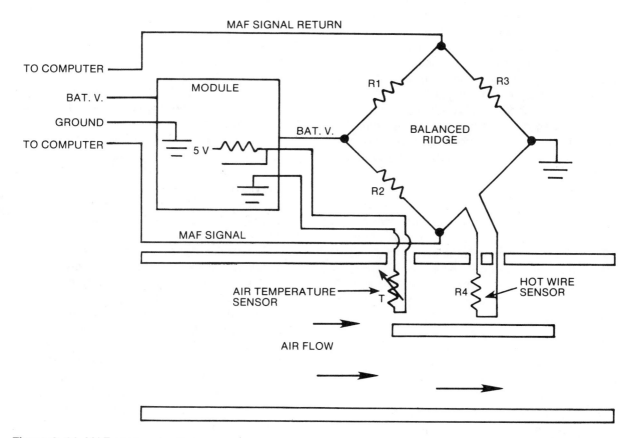

Figure 2–14 MAF sensor circuit

R4's resistance goes down even more. As the resistance of R4 goes down, a smaller portion of the voltage applied to the circuit formed by R2/R4 is dropped across R4. Therefore, a larger portion of the voltage is dropped across R2. So, as the resistance of R4 goes down, the voltage at the R2/R4 junction goes down. The computer reads the voltage difference between the R1/R3 junction and the R2/R4 junction as the MAF sensor value. By looking at a table in the computer's memory, the microprocessor converts the voltage value to a mass airflow rate.

Hall-Effect Switches

A **Hall-effect switch** is frequently used to sense engine speed and crankshaft position. It provides a signal each time a piston reaches top dead center, and its signal serves as the primary input on which ignition timing is calculated. Because of the frequency with which this signal occurs and the critical need for accuracy, the Hall-effect switch is a popular choice because it and its related circuitry provide a digital signal. On some applications, a separate Hall-effect switch is used to monitor camshaft position as well. Systems that do not use a Hall-effect switch use either the magnetic pickup unit in the distributor as the crankshaft speed and position information source, or its equivalent outside the distributor.

A Hall-effect switch consists mainly of a permanent magnet, a **gallium arsenate crystal** with its related circuitry, and a shutter wheel, Figure 2-15. The permanent magnet is mounted so that a small space is between it and the gallium arsenate crystal. The shutter wheel, rotated by a shaft, alternately passes its vanes through the narrow space between the magnet and the crystal, Figure 2-16. When a vane is between the magnet and the crystal, the vane intercepts the magnetic field and thus shields the crystal from it. When the vane moves out, the gallium arsenide crystal is invaded by the magnetic field. A steady current is passed through the crystal from end to end. When the magnetic lines of force from the permanent magnet invade the crystal, the current

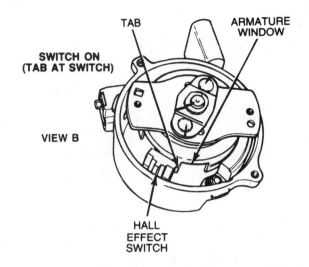

Figure 2–15 Hall-effect switch. *(Courtesy of Ford Motor Company)*

flow across the crystal is distorted. This results in a weak voltage potential being produced at the crystal's top and bottom surfaces—negative at one surface and positive at the other. As the shutter wheel turns, the crystal provides a weak high/low voltage signal.

This signal is usually modified before it is sent to the computer. For example, as shown in Figure 2-16, the signal is sent to an amplifier, which strengthens and inverts it. The inverted signal (high when the signal coming from the Hall-effect switch is low) goes to a **Schmitt trigger** device, which converts the analog signal to a digital signal. The digital signal is fed to the base of a switching transistor. The transistor is switched on and off in response to the signals generated by the Hall-effect switch assembly.

When the transistor is switched on, current flows through it from another circuit to ground. The other circuit can come from the ignition switch or from the computer, as shown in Figure 2-16. In either case it has a resistor between its voltage source and the transistor. A voltage sensing circuit in the computer connects to the circuit between the resistor and the transistor. When the transistor is on and the circuit is complete to ground, a volt drop occurs across the resistor. The voltage signal to the voltage sensing circuit is less than 1 volt. When the transistor is switched off, there is no drop across the resistor; and the voltage signal to the voltage-sensing circuit is near 12 volts. The computer monitors the voltage level and can determine by the frequency at which it rises and falls what engine speed is. Each time it rises, the computer knows that a piston is approaching top dead center.

Detonation Sensors

To optimize performance and fuel economy, ignition timing must be adjusted so that combus-

tion will occur during a specific number of crank-shaft degrees beginning at top dead center of the power stroke. If it occurs any later, less pressure will be produced in the cylinder, and less useful work will be done for the amount of fuel burned. If it occurs any sooner, detonation will occur. Due to variations in fuel quality, cylinder cooling efficiency, machining tolerances and their effect on compression ratios, a preprogrammed spark advance schedule can result in spark knock under certain driving conditions. In order to be able to provide an aggressive spark advance and avoid spark knock, many systems use a detonation sensor, often referred to as a knock sensor. It alerts the computer when spark knock occurs

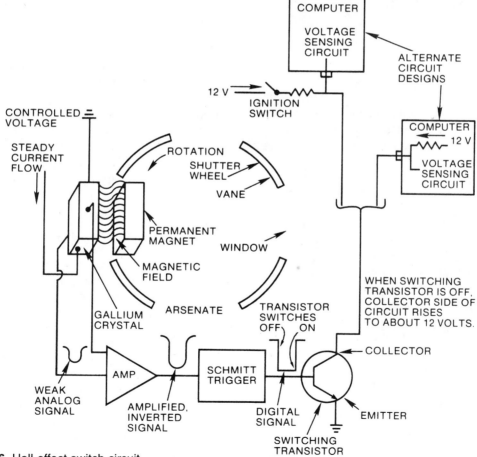

Figure 2–16 Hall-effect switch circuit

Figure 2–17 Knock sensor

so that the timing can be retarded.

The knock sensor, Figure 2-17, screws into some section of the engine where it will be subjected to the high-frequency vibration caused by the spark knock. Most knock sensors contain a **piezoelectric** crystal. Piezoelectric refers to a characteristic of certain materials whereby they produce an electrical signal in response to physical stress, or experience stress in response to an electrical signal. The most commonly used piezoelectric material produces an oscillating voltage signal of about 0.3 volt or more in response to pressure or vibration. The oscillations occur at the same frequency as the spark knock (5,000 to 6,000 **hertz [Hz]**, or cycles per second. Spark knock is essentially the ringing of the engine block when struck by the explosion of detonation). Although most vibrations within the engine and many elsewhere in the vehicle cause the knock sensor to produce a signal, the computer only responds to signals of the correct frequency.

Switches

Most systems have several simple switches that are used to sense and communicate information to the computer. The information includes whether or not the transmission is in gear, the air-conditioning is on, the brakes are applied, and so on. The same type of circuit is usually used for each switch regardless of what condition the switch is measuring.

As an example, let's take a switch used to tell the computer whether or not the automatic transmission is in gear, Figure 2-18. The ignition switch feeds 12 volts through 1 k-ohm of resistance to the P/N switch. Let's assume that when the transmission is in park or neutral, the switch is closed and thus allows current to go to ground. During this condition, the only significant resistance between the voltage source and ground is the 1 k-ohm resistor, and all but a tiny fraction of the 12 volts is dropped across it. The voltage signal to the computer is near zero. When the transmission is shifted into gear, the switch opens. Now the only path to ground is through the 1 k-ohm resistor and then the 10 k-ohm resistor in the computer. As the current goes through the 1 k-ohm resistor, about 1 volt is dropped and the remaining 11 volts is dropped across the 10 k-ohm resistor. The computer recognizes that when voltage is present at the sensing circuit, the transmission is in gear.

E-cells

Some selected systems use an **E-cell** to indicate to the computer when a specified amount of engine operation has occurred. The E-cell actually measures a specific amount of ignition on-time, which equates to an estimated number of vehicle miles. Its purpose is to alert the computer when the predetermined mileage has occurred so that the computer can adjust specific calibrations to compensate for predictable engine wear.

The E-cell contains a silver cathode and a gold anode. Ignition voltage is applied to the E-cell through a resistor. The passage of current through the cell controls a signal to the computer. As current passes through the cell, a chemical reaction takes place; the silver on the cathode is

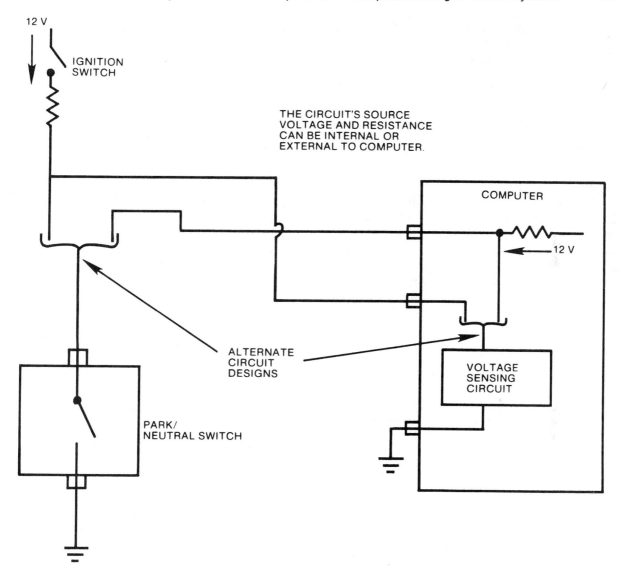

Figure 2–18 Park/neutral switch

attracted to the gold anode. After a certain amount of on-time, the silver cathode is completely depleted and the circuit opens. How long this takes is controlled by the amount of current allowed to flow through the E-cell (how much resistance is in the circuit).

ACTUATORS

Solenoids

The most common actuator is a valve powered by a solenoid. The valve usually controls vacuum, but it can also be a valve designed to control fuel vapor, airflow, oil or water flow, and so forth. A solenoid consists of a coil of wire with a movable iron core, Figure 2-19. When the solid-state switching circuit in the computer grounds the solenoid circuit, the coil becomes energized. The magnetic field developed by the current passing through the coil windings pulls the iron core so that it occupies the entire length of the coil. This movement of the core opens the valve attached to its other end.

Duty-cycled and Pulse Width Solenoids. In some cases the switching circuit that drives the solenoid turns on and off rapidly. If the solenoid circuit is turned on and off (cycled) ten times per second, then the solenoid is duty cycled. To complete a cycle, it must go from off to on, to off again. If it is turned on for 20% of each tenth of a second and off for 80%, it is said to be on a 20% **duty cycle**. In most cases where a solenoid is duty-cycled, the computer has the ability to change the duty-cycle to achieve a desired result. The best example of this is the mixture control device on many computer controlled carburetors, where the duty cycle will frequently change to control the air/fuel mixture.

If a solenoid is turned on and off rapidly but there is no set number of cycles per second at which it is cycled, the on time is referred to as pulse width. For example, let's assume that the computer has decided that the EGR valve should be open, but that it should be open to only 60% of its capacity. The computer can energize a solenoid valve such as the one in Figure 2-19 to allow vacuum to the EGR valve. To control or modulate the amount of vacuum to the EGR valve, the

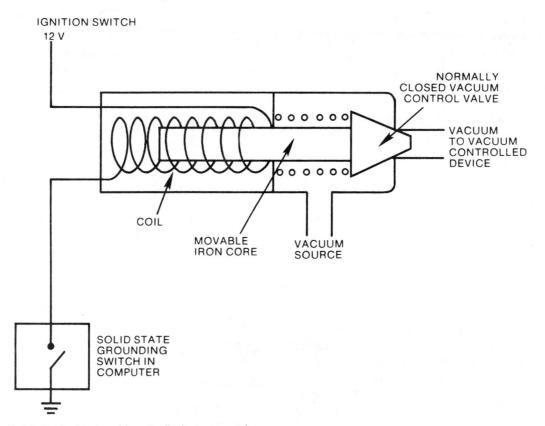

Figure 2–19 Typical solenoid-controlled vacuum valve

computer issues pulse width commands to a solenoid that when energized opens a port that allows atmospheric pressure to bleed into the EGR vacuum line, Figure 2-20. Because the computer wants the EGR valve open to 60% of its capacity, it selects pulse widths from a look-up chart to appropriately weaken the vacuum signal to the EGR valve. Some literature refers to pulse width as dithering.

Relays

A relay is a remote control switch that allows a light-duty switch such as an ignition switch, a blower motor switch, a starter switch or a driver switch in a computer to control a device that draws a relatively heavy current load. The most common example in automotive computer application is a fuel pump relay, Figure 2-21. When the

computer wants the fuel pump on, it turns on its driver switch (transistor). The transistor applies 12 volts to the relay coil, which develops a magnetic field in the core. The magnetic field pulls the armature down and closes the normally open contacts. The contacts complete the circuit from the battery to the fuel pump. On many systems, the fuel pump relay is one of the few examples where the computer supplies the voltage to operate a device instead of the ground connection.

Electric Motors

When electric motors are used as actuators, there are two unique types used.

Stepper Motors. A stepper motor, as used in these applications, is a small electric motor powered by voltage pulses. It contains a permanent magnet armature and usually either two or

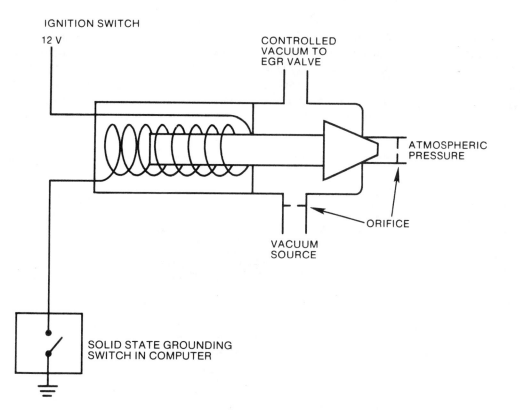

Figure 2–20 Typical pulse-width controlled vacuum valve

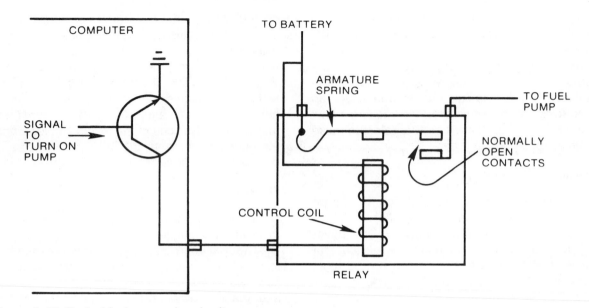

Figure 2–21 Typical fuel pump relay circuit

four field coils, Figure 2-22. A voltage pulse applied to one coil or pair of coils causes the motor to turn a specific number of degrees. The same voltage pulse applied to the other coil or pair of coils makes the motor turn an equal number of degrees in the opposite direction.

The armature shaft usually has a spiral on one end and this spiral connects to whatever the motor is supposed to control. As the motor turns one way, the controlled device, a pintle valve for instance, is extended. As it turns the other way, the valve is retracted. The computer can apply a

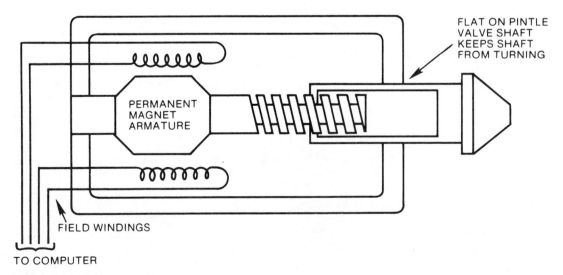

Figure 2–22 Typical stepper motor

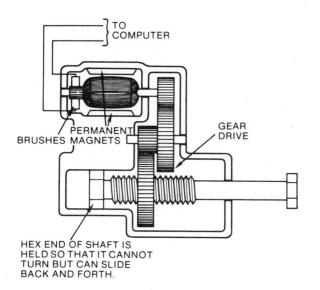

Figure 2–23 Permanent magnet field reversible DC motor

series of pulses to the motor's coil windings in order to move the controlled device to whatever location is desired. The computer can also know exactly what position the valve is in by keeping count of the pulses applied. The stepper motor has been used to control air/fuel mixture and idle speed.

Permanent Magnet Field Motors. Some systems use this type of motor to control idle speed, Figure 2-23. It is a simple, reversible, DC motor with a wire-wound armature and a permanent magnet field. The polarity of the voltage applied to the armature winding determines the direction in which the motor spins. The armature shaft drives a tiny gear drive assembly that either extends or retracts a plunger. The plunger contacts the throttle linkage; and when it is extended, the throttle blade opening is increased. When it is retracted, the throttle blade opening is decreased. The computer has the ability to apply a continuous voltage to the armature until the desired idle speed is reached.

✔ SYSTEM DIAGNOSIS & SERVICE

Special Tools

A good service manual that applies specifically to the vehicle being serviced is essential. It should contain all of the necessary charts and diagnostic procedures and should list the tools necessary for diagnostic operations.

Most systems have a data link used in the factory to test the system's operation as it leaves the assembly line. Several companies have designed and are selling test equipment that can connect to the data link and get service code information from the computer. In most cases other information such as sensor voltage values and the state of position of some actuators can also be obtained. This information is either shown as a digital readout on the test instrument or is printed on a piece of paper. This type of tool is very helpful and in some cases is essential.

High-Impedance Digital Volt-Ohmmeter (DVOM). When diagnosing most systems, a DVOM is required for two reasons:

1. Many of the readings to be taken require a higher degree of accuracy than is obtained by reading a needle position on a scale.
2. Many of the circuits that will be tested have high resistance, and very low voltage and amperage values. Applying a standard meter with its relatively low internal resistance to such a circuit can easily constitute a short to the circuit. In other words the meter's circuit offers an easier electrical path than the circuit or portion of the circuit being tested. A high-**impedance** meter has a minimum of 10 million ohms of internal resistance; a standard meter has only about 10 thousand.

Diagnostic & Service Tips

Oxygen Sensor. When handling or servicing the oxygen sensor, several precautions should be observed:

- A boot is often used to protect the end of the sensor where the wire connects. It is designed to allow air to flow between it and the sensor body, Figure 2-24. The vent in the boot and the passage leading into the chamber in the oxygen-sensing element must be open and free of grease or other contaminants.
- Those with soft boots must not have the boot pushed onto the sensor body so far as to cause it to overheat and melt.
- No attempt should be made to clean the sensor with any type of solvent.
- Do not short the sensor lead to ground, and do not put an ohmmeter across it. Any such attempt can permanently damage the sensor. The single exception is that a digital high-impedance (10 meg ohm minimum) voltmeter can be placed across the sensor lead to ground (the sensor body).

- If an oxygen sensor is reinstalled, its threads should be inspected and if necessary cleaned with an 18 mm spark plug thread chaser. Before installation its threads should be coated with an anti-seize compound, preferably liquid graphite and glass beads. Replacement sensors may already have the compound applied.

WARNING: It is often recommended that the oxygen sensor be removed while the engine is hot to reduce the possibility of thread damage. If doing so, wear leather gloves to prevent burns on the hand.

Vacuum Leaks. Vacuum leaks can have far-reaching effects. Examine Figure 2-25. Assume that the diaphragm ruptured in the primary vacuum break (or it could be a vacuum

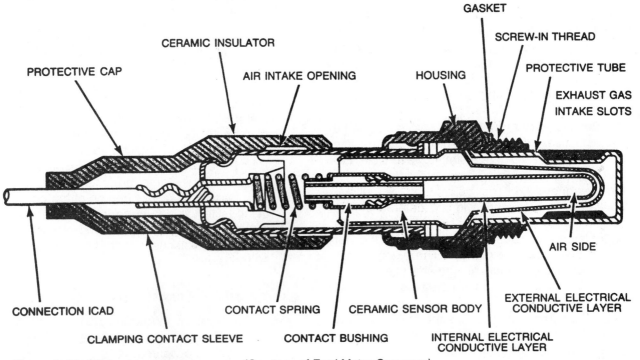

Figure 2–24 Exhaust gas oxygen sensor. *(Courtesy of Ford Motor Company)*

hose cracked or left off anywhere in that circuit). The results would be as follows:

- The primary vacuum break would not work.
- The secondary vacuum break would not work (rich condition, possibly stalling during warm-up).
- The heated air system would not work (poor cold driveability).
- The vacuum sensor would report "no vacuum" to the computer, which would in turn issue erroneous spark and fuel mixture commands.
- A vacuum sensor code would probably be set.

Vacuum leaks downstream from the airflow sensor on a multipoint injection system cause poor idle if the engine idles at all.

Charcoal Canister. Charcoal canister saturation can often be the result of fuel tank overfilling. Slowly filling the tank right to the top of the filler neck causes the antioverfill chamber in the tank to fill and thus leaves no room for fuel expansion. When the fuel does expand, it is pushed up and into the charcoal canister. Canister purging then dumps so much hydrocarbon into the intake manifold that the system may not be able to compensate, even in closed loop. Many driveability problems can be linked to charcoal canister problems; don't overlook it in diagnostic procedures.

Avoid Damaging the Computer.   When servicing a computerized engine control system vehicle, observe the following in order to avoid producing a voltage spike that could damage the computer:

- Avoid starting the vehicle with the aid of a battery charger. Finish charging the battery first, then disconnect the charger before starting.
- Turn the ignition off before connecting jumper cables.
- Turn the ignition off before disconnecting the battery or computer connectors.

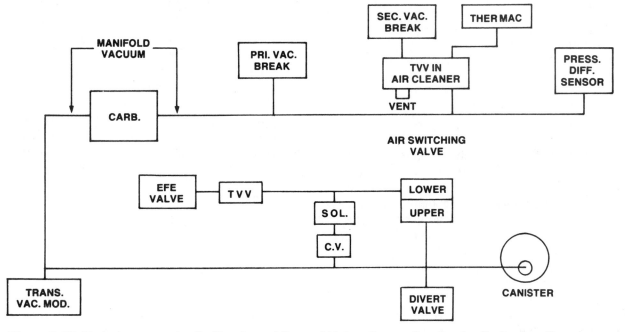

Figure 2–25 Typical vacuum circuit. *(Courtesy of General Motors Corporation, Service Technology Group)*

- Replace any system actuator with less than specified resistance or one that has a shorted clamping diode. A **clamping diode** is a diode placed across an electrical winding to suppress voltage spikes that are produced when the winding is turned on and off.
- Be sure that the winding of any electrical accessory added to the vehicle has sufficient voltage spike containment capability. A resistor of 200- to 300-watt capacity can be used instead of a diode in some cases.
- Use caution when testing the charging system, and never disconnect the battery cables with the engine running.

SUMMARY

This chapter has described single-point and multipoint fuel injection systems. The single-point systems use a single fuel injector, or often a pair of two injectors for a V-form engine, to deliver the fuel upstream of the throttle blade. These systems resemble an electronic carburetor. The multipoint systems use an injector (or sometimes two) for each cylinder, allowing somewhat more precise fuel mixture delivery. Each system varies the amount of fuel delivered by varying the pulse width of the injector, the time the injector solenoid is on, and fuel sprays. Single-point or throttle body systems usually pulse at a fixed frequency and vary the pulse width to fit conditions. Multipoint systems ordinarily spray individually or in cylinder banks in synchronization with the camshaft speed, varying quantity also by pulse width modification.

We have learned how the computer system's most important sensors work, as variable resistors, thermistors varying their internal electrical resistance, or as signal generators. The oxygen sensor produces a voltage corresponding to the amount of oxygen remaining in the exhaust gas; the crankshaft position sensor sends a signal corresponding to the position and speed of the crankshaft; the temperature sensors modify a reference voltage to match their temperature; and the throttle position sensor varies its return voltage signal through a variable potentiometer, so the computer knows exactly where the throttle is and how fast it has moved recently.

We reviewed both pulse-generation sensors and Hall-effect sensors. The pulse generators use a magnet and a coil to produce an alternating current corresponding to the speed of the crankshaft or camshaft rotation monitored. The Hall-effect sensor functions basically as an on-off switch, signaling the same sort of information. We have also seen how simple on-off switches can provide information on subjects like the engagement or disengagement of the air conditioner compressor or when the throttle closes completely, leaving the engine speed under the computer's control. Finally, we have seen how the computer operates the actuators not by powering them directly but by providing them an electric path to ground, thus protecting the computer itself from shorts in the circuit.

▲ DIAGNOSTIC EXERCISE

A car is brought into the shop with various running and driveability problems. Among the things the technicians notice is a great deal of rust throughout the car, including on the connections for the electrical system. What kinds of problems would you anticipate might occur as a result of this kind of excessive rust formation? How would you go about checking circuits to see how much, if at all, they were affected? How would you go about correcting the electrical resistance problems that would occur from the rust?

REVIEW QUESTIONS

1. What does the term *duty-cycle* mean?
2. How is duty-cycle different from pulse width?

3. How does the oxygen sensor know if the air/fuel mixture is rich?
4. What type of thermistor has higher resistance at lower temperatures?
5. Diagram a voltage divider circuit.
6. Diagram a potentiometer circuit.
7. Describe how a Hall-effect switch turns on and off a switching transistor.
8. Describe how the Hall-effect switch's switching transistor signals the computer.
9. What information does the Hall-effect switch provide for the computer?
10. Name a resistance device commonly used to measure air pressure.
11. Why is 5 volts most commonly used as a reference voltage for sensor operation?
12. Describe how a simple switch is used as a sensing device.
13. What is an actuator?
14. Describe the operation of a solenoid.
15. Describe the operation of a relay.
16. Describe how to reverse the direction of rotation of a permanent magnet field motor.
17. For what is a permanent magnet field motor usually used?

ASE-type Questions. (Actual ASE test questions are rarely so product specific.)

18. Technician A says that a rich exhaust stream contains little free oxygen because most of the available oxygen was combined with other elements during combustion. Technician B says that little free oxygen is present in a rich exhaust stream because there was not enough oxygen in the air/fuel mixture from the beginning. Who is correct?
    a. A only.
    b. B only.
    c. both A and B.
    d. neither A nor B.
19. The voltage reading from the oxygen sensor of a car with a computerized engine control system is consistently around 0.7 volt. This indicates that the system is operating in a _____ condition.
    a. rich.
    b. lean.
    c. normal.
    d. abnormal.
20. Technician A says that a potentiometer is usually used to monitor throttle position. Technician B says that a thermistor is usually used to monitor throttle position. Who is correct?
    a. A only.
    b. B only.
    c. both A and B.
    d. neither A nor B.

21. Technician A says that an oxygen sensor can be fouled by fuel additives such as lead and antioxidation agents. Technician B says that an oxygen sensor can be fouled by use of a gasket compound containing silicon and by fuels containing alcohol. Who is correct?
    a. A only.
    b. B only.
    c. both A and B.
    d. neither A nor B.

Chapter 3

General Motors' Computer Command Control

OBJECTIVES

In this chapter, you can learn:
- ❑ how the torque converter lockup clutch operates.
- ❑ how to use a dwell meter to evaluate air/fuel mixture controls.
- ❑ how the mixture control solenoid works in a computer command control system.
- ❑ how idle speed is controlled in the computer command control system.
- ❑ how the air management system directs air to the exhaust manifold, catalytic converter, or atmosphere.
- ❑ what a trouble code is and how to retrieve it from the computer.

KEY TERMS

Aspirator
Bleed
Detonation
Divert Mode
Dualjet
Fail-Safe
Light-Emitting Diode
Maximum Authority
Milliamp
Millivolt
Minimum Authority
Pulsair System
Purge Valve
Vacuum Control Valve

The computer command control (CCC) system was General Motors' first widely used comprehensive computerized engine control system. It was first introduced in mid-1980. Beginning with the 1981 model year, all of General Motors' carbureted passenger cars have used the CCC system. It should be noted that the CCC system varies slightly from one engine application to another and from one model year to another. In 1982 four different CCC systems were used. Most 1982 engines used the full-function system, which was a slightly updated version of the 1981 system. Beginning in 1982, Oldsmobile 5-liter engines were equipped with a limited control system; the T-car (Chevrolet Chevette and Pontiac T-1000) used a minimum-function system. The Chevrolet and GMC S truck (S-10 and S-15, introduced in 1982) with the California emissions

From EFC to CCC

General Motors introduced its first electronic closed-loop fuel control system in 1978. This single function system, called electronic fuel control (EFC), was limited to a few four-cylinder engines. In 1979 GM introduced its first comprehensive engine control system on selected applications: computer-controlled catalytic converter (C-4). The system was improved for model year 1981 and was renamed the computer command control system. Apparently someone at GM noticed the name and decided they did not want their engine control system named after a Ford automatic transmission.

package and a four-cylinder gasoline engine used an imported electronic control module (ECM) slightly different from the other CCC systems. There are also slight variations within each of these systems.

Although all of this seems a bit bewildering, it need not be. All versions of the CCC system are much more alike than different. Once the basic system and function are understood and a good service manual is at hand for the specific directions and specifications of the vehicle, you will find the CCC systems are very manageable.

ELECTRONIC CONTROL MODULE (ECM)

The engine control computer used in the CCC system and in all other General Motors systems is called an electronic control module, Figure 3-1. The inputs it receives and the actuators it controls are shown in Figure 3-2. The ECM

Figure 3–1 ECM. *(Courtesy of General Motors Corporation, Service Technology Group)*

is located in the passenger compartment, usually near the glove compartment or behind the passenger kick panel. The PROM (programmable read-only memory) is located under an access cover, Figure 3-3. It is responsible for fine-tuning

Inputs	Codes	ECM	Codes	Outputs
Coolant Temperature Sensor	14, 15		23	Fuel Mixture
Vacuum Sensor	34		42	Electronic Spark Timing
Barometric Pressure Sensor (if used)	32	ROM and PROM process		Electronic Spark Retard (if used)
Throttle Position Sensor	21	information		Idle Speed
Distributor Reference (crank position—engine speed)	12, 41	(inputs) and issue commands		AIR Management EGR
Oxygen Sensor	13, 44, 45	(outputs).		Canister Purge
Vehicle Speed Sensor	24			Torque Converter Clutch (or shift light on
Ignition On				manual transmission)
Air Conditioner On/Off				Air Conditioner
Park/Neutral Switch		RAM monitors		Early Fuel Evaporation
System Voltage		indicated circuits,		Diagnosis
Transmission Gear Position Switch(es)		sets codes, and reads codes out		(check engine light)
Spark Knock Sensor (if used)	43	when put in		(test terminal)
EGR Vacuum Indicator Switch (only on 1984 and later models)	53	diagnostics.		(serial data)
Idle Speed Control Switch	35			

Figure 3–2 Overview of computer command control. Inputs and outputs with code numbers are monitored for faults. Those without are not.

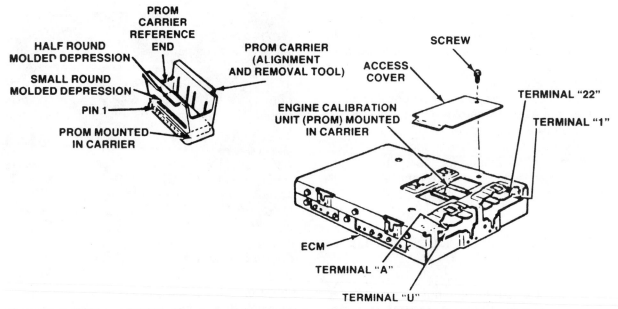

Figure 3–3 ECM and PROM. *(Courtesy of General Motors Corporation, Service Technology Group)*

engine calibrations for each specific vehicle and when plugged in becomes a functioning (in fact, essential) part of the ECM.

Real World Problem

In earlier ECMs, the PROM can easily be installed backwards, which will result in irreversible electrical damage to it. Consult a service manual or the **System Diagnosis and Service** section of this chapter before removing and replacing any GM PROM.

Keep Alive Memory (KAM)

The ECM monitors its most important sensors and selected actuator circuits. If a fault such as an open or short circuit occurs during normal operation, a fault code will be stored in the KAM. This code can be retrieved at a later time to aid the technician locating the problem. This feature is especially useful because it enables the ECM to report a fault that occurred recently but is no longer present, as well as those that are currently in the system. The KAM receives battery power at all times so the codes can be retained in memory. This battery draw is very low, much less than the electrical power loss internal to the battery itself.

OPERATING MODES

There are basically two operating modes, closed loop and open loop.

Closed Loop

Before the system can go into closed loop, the coolant must reach a temperature of approximately 65 degrees Centigrade (150 degrees Fahrenheit), the oxygen sensor must reach at least 300 degrees Centigrade (570 degrees Fahrenheit), and a predetermined amount of time must have passed following engine start-up. This time is programmed into the PROM and varies with the engine application, model year and so on. It can be as little as a few seconds on some engines, or as much as a couple of minutes on others.

The system drops out of closed loop if the vehicle is driven at or near wide-open throttle (WOT), if the coolant temperature drops below the criterion value or if certain crucial system component failures occur.

Real World Problem

1982 and later GM T-cars (Chevrolet Chevette and Pontiac T-1000) used a minimum-function CCC system. These systems use a coolant temperature switch instead of a coolant temperature sensor. As the engine reaches operating temperature, the switch closes and puts the system into closed loop (assuming the oxygen sensor is hot and the proper time has elapsed). If the engine cools down and the switch opens, the system stays in closed loop. Technicians are sometimes confused by the temperature switch, thinking it is shorted or open, when it is functioning normally.

Open Loop

This is an operating mode that includes several suboperational modes.

Start-Up Enrichment. This mode provides a rich mixture command to the mixture control solenoid for a short time after each engine start-up. The duration of the mode depends on engine temperature, as reported by the coolant temperature at the time the engine starts. If the engine is started warm, this mode will be shorter than if it is started cold.

Blended Enrichment. This mode occurs during engine warm-up (engine coolant and/or the oxygen sensor are not up to operating temperatures). In this mode, the ECM controls the air/fuel ratio using information from sensors concerning coolant temperature, throttle position, manifold pressure, barometric pressure, and engine speed. As engine temperature warms, the air/fuel ratio is adjusted leaner by the ECM. In this open-loop mode, the ECM attempts to adjust the air/fuel ratio using the fixed, preprogrammed instructions in its read-only memory. If the air/fuel ratio is not correct because of a carburetor problem, vacuum leak and so forth, the ECM will not be able to recognize the problem and cannot take any countermeasures in response to it. In other words, in open loop the ECM does what it is programmed to do, but the engine does not run correctly unless other engine components function properly according to its expectations.

Power Enrichment. This mode occurs when the vehicle is operated at or near wide-open throttle. The ECM sends a steady power enrichment signal to the mixture control solenoid. This provides the rich mixture required by an engine when manifold vacuum is near zero.

Shutdown. This mode turns off the mixture control solenoid, thus creating a full-lean operating condition, when the reference pulse to the ECM indicates 200 rpm or lower and when battery voltage to the ECM is below 9 volts.

Limp-In. This mode allows the engine to continue operating in spite of most major failures, such as an ECM failure, that can occur in the system. This is a fail-soft mode that provides no fuel mixture control (full-rich or full-metered) and no spark advance. The engine will run, but driveability, fuel economy, and emissions quality performance will be compromised.

INPUTS

Coolant Temperature Sensor (CTS)

With the exception of some T-car applications (Chevrolet Chevette and Pontiac T-1000), all CCC systems use a coolant temperature sensor, the CTS, Figure 3-4, as described in Chapter 2. Most often it is near the thermostat housing. It is the single most important sensor in the system. Its input affects just about every command the computer sends (this statement is generally true of all computerized engine control systems, regardless of make, model, or year).

Some 1982 and later GM T-cars are equipped with a minimum-function system and

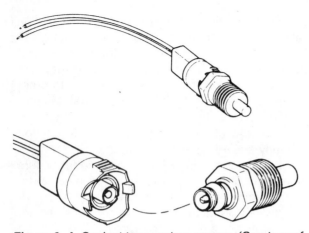

Figure 3–4 Coolant temperature sensor. *(Courtesy of General Motors Corporation, Service Technology Group)*

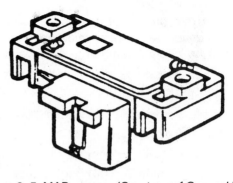

Figure 3–5 MAP sensor. *(Courtesy of General Motors Corporation, Service Technology Group)*

use a coolant temperature switch instead of a CTS. The switch is open while the coolant is below a specified temperature and closes as the coolant approaches the normal operating temperature. In this system, the ECM only knows whether the coolant is below or above a specific temperature.

Pressure Sensors

Three different kinds of pressure sensors were used on various CCC systems. They are all of the piezoresistive silicon diaphragm type and are nearly identical in appearance, Figure 3-5. Each can, however, be distinguished by a colored plastic insert at its electrical harness connector.

Manifold Absolute Pressure (MAP) Sensor. The MAP sensor has either a red or orange connector (orange is for turbocharged engines). The MAP sensor is often used in combination with a BARO sensor.

Barometric Pressure (BARO) Sensor. The BARO sensor uses a blue connector. On some vehicles it is under the hood and near the MAP sensor. On others, it is inside the passenger compartment under the instrument panel.

NOTE: The MAP and BARO sensors for 1980 and early 1981 vehicles were round metal units, both mounted on the same bracket under the hood.

Pressure Differential (VAC—for vacuum) Sensor. The VAC sensor uses a grey or black connector and is used by itself.

Oxygen Sensor

The oxygen sensor used on 1981 and earlier models used two wires with a vented silicone

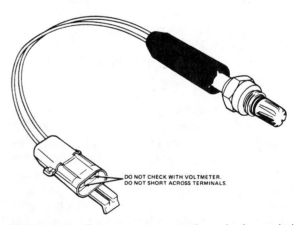

DO NOT CHECK WITH VOLTMETER.
DO NOT SHORT ACROSS TERMINALS.

Figure 3–6 Oxygen sensor, early style (two wire). *(Courtesy of General Motors Corporation, Service Technology Group)*

boot covering its open end, Figure 3-6. The second wire served as a backup ground circuit; the sensor shell is already grounded to the manifold. This backup is used because it is not uncommon for the heat fluctuations of the exhaust manifold to break a ground circuit through it. Beginning in 1982, the ground wire was not used and a metal boot replaced the silicone boot. The silicone boot melted if it was pushed too far over the oxygen sensor toward the exhaust manifold. These systems use an oxygen sensor ground wire running from the ECM to the engine block, not directly connected to the oxygen sensor.

Throttle Position Sensor (TPS)

The TPS is a potentiometer mounted in the bowl section of the carburetor, Figure 3-7. When the throttle is opened, the accelerator pump lever

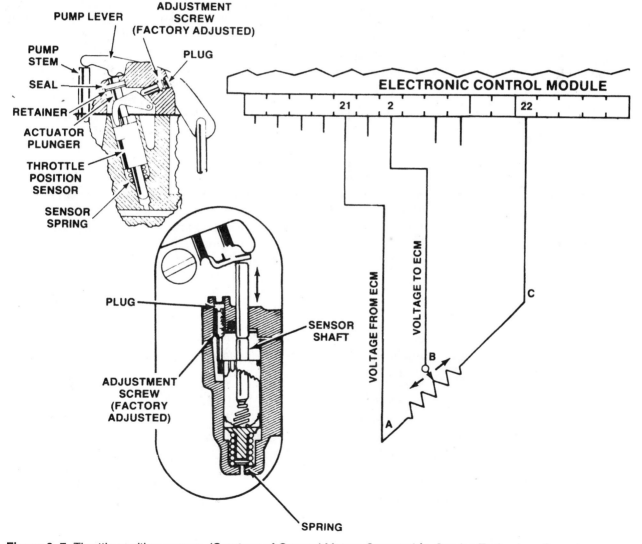

Figure 3–7 Throttle position sensor. *(Courtesy of General Motors Corporation, Service Technology Group)*

Service Interval for Oxygen Sensors

Prior to model year 1981 GM recommended a service interval of 15,000 or 30,000 miles for oxygen sensors. No replacement interval is set for vehicles built after model year 1981; replacement is called for instead only when the sensor is found to be defective.

forces the TPS plunger down. This action moves the potentiometer's wiper. The TPS is the only adjustable sensor in the CCC system. Because engine vacuum and throttle position are so closely related, the ECM expects to see the readings from these two sensors closely synchronized.

Distributor Reference Pulse (REF)

The REF is also abbreviated as DIST or REF pulse. This signal provides the ECM with information about engine speed and crankshaft position. The REF is obtained by tapping into the pickup coil circuit inside the HEI module, Figure 3-8.

The pickup coil signal is generated as the distributor shaft and timer core turn, moving the points of the timer core (or reluctor) past the points of the magnetic pickup unit. Each time the magnetic field strengthens or weakens as a result of the points aligning and misaligning, a voltage signal is produced in the pickup coil. This signal occurs just as the circuit reverses polarity and the points of the reluctor and the pickup cross. In the HEI module a signal converter changes the analog signal produced by the pickup coil to a digital signal. On HEI systems prior to computer controls, the pickup coil signal indicated to the module when to turn off the main switching transistor, which turned off the primary ignition circuit and fired the spark. On CCC systems, the only time the pickup coil signal is used by the module is during startup cranking and in some system failures, in **fail-safe** or limp-home mode. During normal operation the ECM uses the pickup coil signal only as a REF signal, a concept explained further in the **Outputs** section of this chapter.

From 1982 to 1985, a Hall-effect switch, Figure 3-9, was positioned between the pickup coil and the distributor rotor on the Chevrolet 3.8-liter (229 cid) odd-firing V6. This sensor was used to produce greater ignition timing accuracy. The odd-firing engine is especially susceptible to sig-

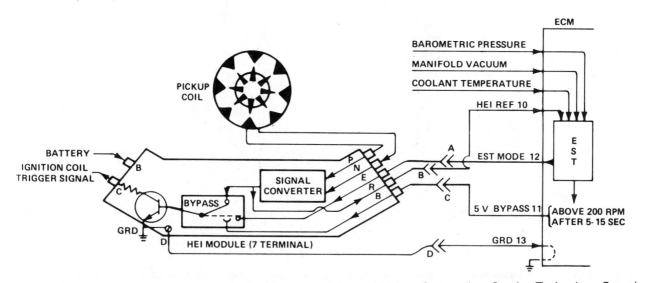

Figure 3–8 Electronic spark control circuit. *(Courtesy of General Motors Corporation, Service Technology Group)*

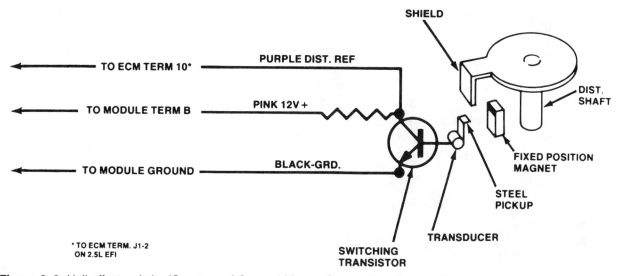

Figure 3–9 Hall-effect switch. *(Courtesy of General Motors Corporation, Service Technology Group)*

nal converter error. Notice in Figure 3-10 the pickup is still there. However, it is used only to start the engine. Notice that terminal R of the HEI module is not used (on applications without a Hall-effect switch, terminal R feeds the REF sig-

nal to the ECM), and that the REF comes directly to the ECM from the Hall-effect switch. Once the engine starts, the solid-state switching device, illustrated in Figure 3-10 as a relay, has disconnected the pickup coil from the main

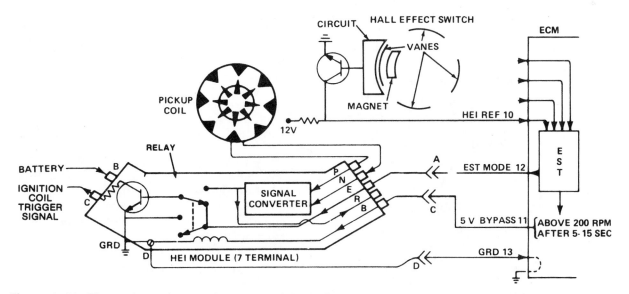

Figure 3–10 Electronic spark control system with Hall-effect switch. *(Courtesy of General Motors Corporation, Service Technology Group)*

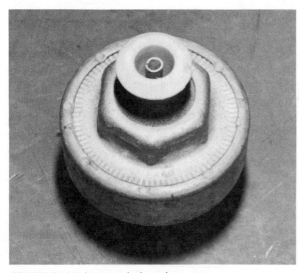

Figure 3–11 Late-style knock sensor

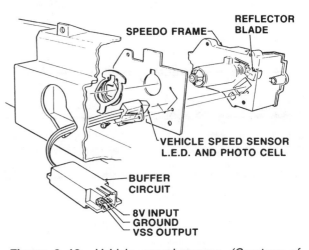

Figure 3–12 Vehicle speed sensor. *(Courtesy of General Motors Corporation, Service Technology Group)*

switching transistor and connected in its place the EST lead from the HEI terminal E. The ECM is now in charge of spark timing.

Detonation Sensor

The **detonation** or knock sensor, Figure 3-11, is part of the electronic spark control (ESC) system, a subsystem of CCC on many applications. A similar spark control system, also called ESC, has been an independent spark retard system for several years on non-CCC vehicles.

On many General Motors engines, the knock sensor is screwed into the block in the crankcase area to reduce its sensitivity. On those applications, it was found that mounting it on the upper part of the engine allowed a clicking valve lifter to trigger the knock signal and erroneously retard the ignition timing.

Vehicle Speed Sensor (VSS)

Two kinds of vehicle speed sensors are used on CCC systems. The earlier type uses a **light-emitting diode (LED)** and a phototransistor. Both are in a plastic connector plugged into the back of the speedometer housing near the speedometer cable attachment point, Figure 3-12. On early models, the LED is powered through the ECM; on later versions, power comes directly from the ignition switch. With the ignition on, the LED directs its invisible infrared light beam to the back of the speedometer cup, which is painted black. The drive magnet, spinning with the speedometer cable, has a reflective surface at one point. As the drive magnet moves into the light from the LED, the light reflects back to the receptive phototransistor. Each time the light strikes the phototransistor, it produces a voltage pulse. These pulses are fed to a buffer switch.

Diagnostic & Service Tip

The LED-generated VSS signal can be blocked by anything that can block the infrared light. Either a dustball or, not infrequently, a ball of grease pushed through the speedometer cable by an overzealous lubricant technician, can shut off the signal. Many experienced technicians check this first when there is indication of a VSS signal failure.

The VSS signal is an analog signal varying in amplitude as well as frequency as the vehicle speed changes. The buffer switch modifies the raw VSS signal to 2,002 digital pulses per mile (different pulse numbers for certain later models). The ECM uses this information to determine when to apply the torque converter clutch.

Some later model vehicles use an electronically operated speedometer with no speedometer cable. These vehicles use a pulse generator device as a VSS. This device bolts to the transmission where the speedometer cable drive would be if it were used. This transmission-driven signal generator rotates a magnet near a coil. As each pole of the magnet swings by the coil, a weak voltage signal, similar to what a magnetic pickup in an ignition distributor generates, is produced in the coil. This signal produces an analog voltage signal proportional to the speed of the car. These signals are fed to a buffer, which converts them to 4,004 digital pulses per mile. These signals then go to the ECM, which calculates vehicle speed from the signal frequency and its own internal clock. Four-wheel-drive vehicles often have two similar sensors, one for transmission output and one for vehicle speed to enable the computer to distinguish when the vehicle is in the low transfer case range.

Ignition Switch

The ignition switch is one of the two power sources for the ECM. When the ignition turns on, the ECM initializes (starts its program) and gets ready to function. The ignition switch also powers most of the actuators the ECM will control in operation (by grounding and completing their circuits). The other ECM power supply comes from the battery through a fuse. This circuit powers the diagnostic memory. The ECM also monitors system voltage through these two inputs.

Park/Neutral (P/N) Switch

The P/N switch consists of another pair of contacts added to the neutral safety switch that were

long a part of vehicles with automatic transmissions (used primarily to prevent the starter from engaging the flywheel in any gears but park or neutral). The computer's P/N switch signals the ECM whether the transmission is in gear or not, and this information is used to help control engine idle speed. On some vehicles the ECM will also not activate any spark advance or EGR commands unless the transmission is in some gear.

Air-Conditioning (A/C) Switch

On CCC cars with air conditioning, the A/C switch connects through a wire to the ECM. When the A/C is turned on or off, the ECM receives that information, which it uses in controlling idle speed.

Idle Speed Control (ISC) Switch

At closed throttle, the throttle lever, pulled by the return spring, presses against the ISC plunger. This pressure closes a set of contact points at the base of the plunger, Figure 3-13. This circuit signals the ECM that it is now in charge of idle speed. Opening the throttle releases the pressure on the plunger and opens the contact points, signaling the ECM not to control idle speed, Figure 3-14.

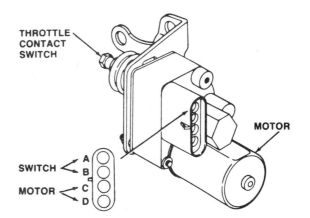

Figure 3–13 Idle speed control motor. (*Courtesy of General Motors Corporation, Service Technology Group*)

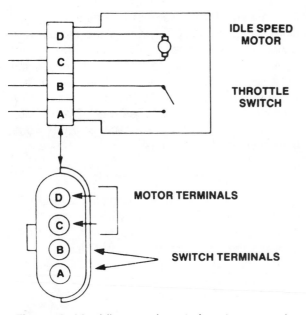

Figure 3–14   Idle speed control motor connector. *(Courtesy of General Motors Corporation, Service Technology Group)*

Transmission Switches

Most transmissions in CCC-equipped vehicles include one or more hydraulically operated electric switches in the valve body. These switches provide the ECM with signals indicating what gear the transmission is in. This information enables it to control the lockup torque converter clutch operation more effectively.

EGR Vacuum Diagnostic Control Switch

Beginning in 1984, most CCC systems included a vacuum-operated switch tied into the vacuum hose between the EGR valve and the EGR control solenoid, Figure 3-15. When vacuum is applied to the EGR valve, the switch closes. If the ECM detects a closed EGR diagnostic switch during starting, idle, or at any other time it has not commanded EGR flow, then it will turn on the check engine light and set a fault code in diagnostic memory. If it sees an open switch during any time it has commanded the EGR to recir-

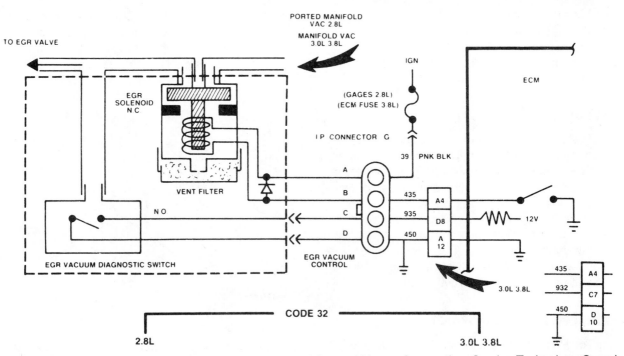

Figure 3–15 EGR diagnostic switch circuit. *(Courtesy of General Motors Corporation, Service Technology Group)*

culate exhaust gas, it will also turn on the check engine light and set the same code.

OUTPUTS

Mixture Control (M/C) Solenoid

In a CCC system, the mixture control solenoid effectively replaces the full power enrichment system that would otherwise be used in the carburetor of a non-CCC vehicle. This solenoid, however, has much greater detailed control of the air/fuel ratio over a much wider range of engine operating conditions than the conventional full power enrichment system does, Figures 3-16, 3-17 and 3-18. The mixture control solenoid gets power directly through the ignition switch. The ECM duty cycles it on and off through a solid-state grounding switch (a power transistor), Figure 3-19. When the ECM grounds the M/C solenoid circuit, the solenoid turns on and drives the solenoid plunger down. When the plunger is driven down, it drives a metering rod (or rods) down into the main metering jet (or jets) to reduce fuel flow. On some

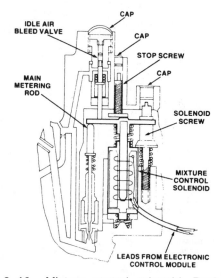

Figure 3–16 Mixture control solenoid, Rochester Dualjet or Quadrajet. *(Courtesy of General Motors Corporation, Service Technology Group)*

carburetors the tip of the solenoid plunger itself serves as the metering rod. When driven down, the solenoid provides a lean mixture; when turned off and lifted by the spring, it provides a full

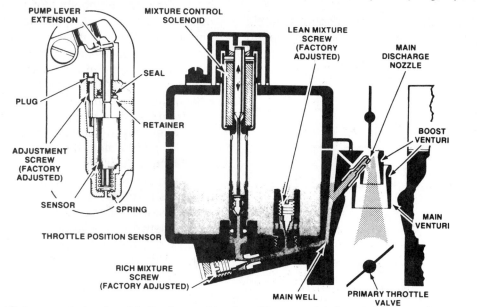

Figure 3–17 Mixture control solenoid, Rochester Varajet. *(Courtesy of General Motors Corporation, Service Technology Group)*

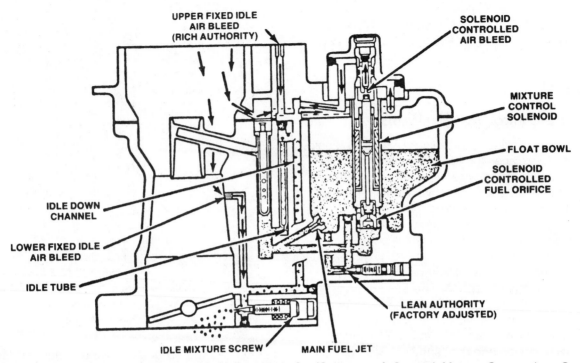

Figure 3–18 Mixture control solenoid, Holley 6510 C. *(Courtesy of General Motors Corporation, Service Technology Group)*

rich mixture.

The ECM cycles the mixture control solenoid ten times per second regardless of engine speed

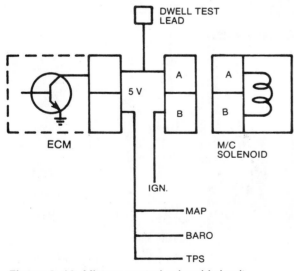

Figure 3–19 Mixture control solenoid circuit

or load (except during shutdown, when the solenoid turns off). The ECM can vary the duty cycle anywhere from 90% to 10% (that is, it can hold the solenoid down leaning the mixture for 90% of the time; or it can let the solenoid up, richening the mixture for 90% of the time, depending on what mixture it calculates is required from the signals from the other sensors), Figure 3-20.

Measuring the Duty Cycle. You can measure the relative time the mixture solenoid is on or off with a dwell meter, Figure 3-21. The wire between the mixture control solenoid and the ECM has a special connector, the dwell test connector, for this purpose. A dwell meter, once connected between the dwell test connector and to ground, and set on the six cylinder (60%) scale, indicates in degrees the relative amount of on-time the M/C solenoid is energized, Figure 3-22. Many newer digital VOMs include a duty-cycle measure that yields a reading directly in percent of on-time.

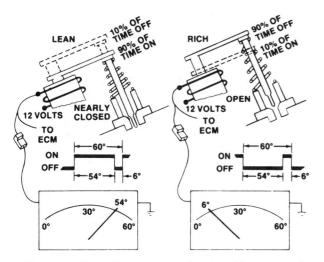

Figure 3–20 Mixture control solenoid duty cycle. *(Courtesy of General Motors Corporation, Service Technology Group)*

Dwell Readings. During engine cranking enrichment or wide-open throttle operation, a dwell reading of 6 degrees is expected (10% duty-cycle or less), for full rich. During engine

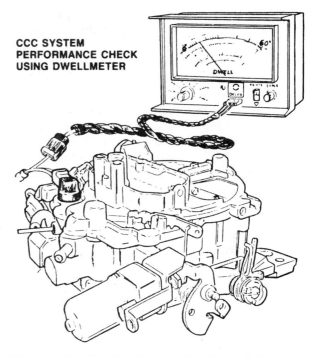

Figure 3–21 Dwell test lead. *(Courtesy of General Motors Corporation, Service Technology Group)*

RELATIONSHIP OF DWELLMETER READINGS TO MIXTURE CONTROL SOLENOID CYCLING

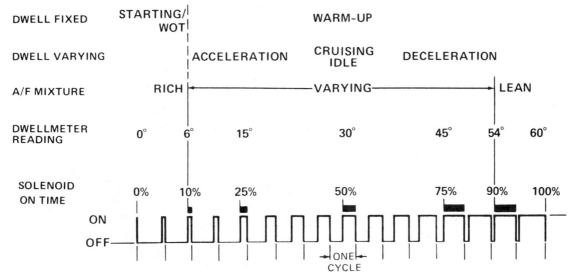

Figure 3–22 Dwell readings. *(Courtesy of General Motors Corporation, Service Technology Group)*

warm-up (open loop) an initial reading of about 10 degrees (about 15%) is considered normal, depending on how cold the engine is to begin with. As the engine warms up, the dwell reading should steadily increase. In open loop, the ECM selects an air/fuel ratio based on engine temperature, load, barometric pressure, throttle position, and engine speed. This produces the best results in terms of exhaust emissions, fuel mileage, and driveability. The fuel mixture, as reflected in the dwell reading, remains constant until some input information changes, or until the system goes into closed loop. When coolant reaches a temperature of about 150 degrees Fahrenheit, the oxygen sensor reaches about 600 degrees Fahrenheit and the required amount of time has passed for the ECM to put the system into closed loop. At that moment, the dwell meter needle begins wagging. It usually wags, or varies, over a range of 5 to 10 degrees. The dwell reading is now taken by selecting the center or average of the range over which it sweeps.

The dwell varies because of the way the ECM responds to the oxygen sensor's feedback signals. When the oxygen sensor reports a lean condition (anything leaner than 14.7 to 1), it reports "lean" to the ECM. The ECM reacts by sending a rich command to the mixture control solenoid. If everything is working properly, this will soon appear as "rich" at the oxygen sensor, and its signal will reflect that condition. The oxygen sensor's rich signal then produces a lean command from the ECM, and so on. In this way, the air/fuel ratio constantly fluctuates back and forth across the 14.7 to 1 stoichiometric ratio—the mixture at which the catalytic converter can most effectively counteract any residual HC, CO, and NO_x in the exhaust. The actual delivered mixture usually remains within a range of about 14.5 to 14.9 to 1.

This ratio, however, should not be taken as invariable. The original definition of *stoichiometric* (the Greek word for *ideal*) meant the combustion state of an engine whose exhaust contained no unburned fuel and no oxygen–a combustion, in short, that perfectly and completely burned all the

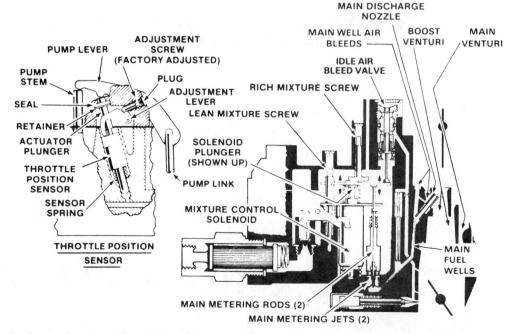

Figure 3–23 Rochester Dualjet or Quadrajet. *(Courtesy of General Motors Corporation, Service Technology Group)*

fuel using all the air the throttle admitted, and no more. This earlier notion focused on power output and fuel economy, whereas the current concept of stoichiometry is: that air/fuel ratio which leaves the exhaust exiting the engine at a chemical and thermal state optimal for reduction and oxidation of the emissions gases by the catalytic converter. The traditional way of describing this is 14.7 to 1, but with oxygenated fuels and various other modifications for seasons and altitudes, this may not be the actual stoichiometric ratio for a given engine under given driving conditions.

At the same time the mixture control solenoid cycles the metering system on the carburetor's main fuel delivery system, it also cycles the idle air bleed valve in the idle speed/low speed system (**bleed** is a controlled leak). The idle air bleed valve, Figure 3-23, is designed as a variable restriction—a sort of unfixed orifice—to determine the amount of air entering the idle/low speed circuit. When the mixture control solenoid metering valves are down, the idle air bleed valve plunger drops and allows maximum air into the idle speed/low speed circuits, driving the mixture lean. When the mixture control solenoid is up, the idle air bleed is also up and allows minimum air into the circuit, driving the mixture rich. With this arrangement, the same mixture control solenoid controls the air/fuel mixture whichever fuel metering circuit is in use.

Electronic Spark Timing (EST)

With the CCC system, the ECM eliminates the vacuum and centrifugal advance mechanisms of the earlier HEI distributors, Figure 3-24 (and along with it most of the failed pickup coils, whose connections often broke from the frequent twisting back and forth). To achieve this, the HEI module was modified to work directly with the ECM. The ECM in the updated system controls timing by signaling the HEI module when to open the primary ignition circuit, therefore turning it off and firing the spark. The HEI module responds only to these spark timing commands from the ECM once the engine has started.

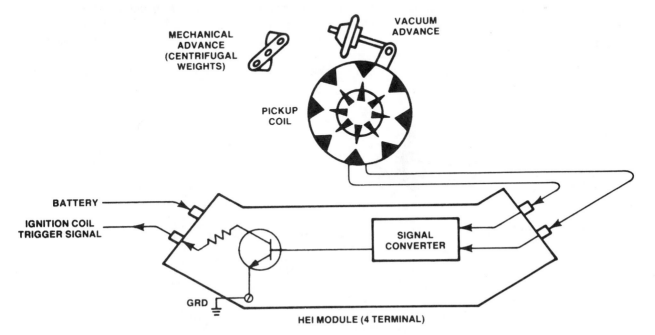

Figure 3–24 HEI circuit with mechanical and vacuum advance. *(Courtesy of General Motors Corporation, Service Technology Group)*

NOTE: In the following discussion and in the accompanying illustrations, the HEI module will be treated as though it has a mechanical bypass relay with a double set of contact points. This is done to make its operation easier to understand. Actually, solid-state components are used to achieve the functions discussed.

During cranking, the bypass relay is in the de-energized or module mode. In this mode, the main switching transistor of the HEI module is connected to the pickup coil, Figure 3-25. The pickup coil signal, after conversion from an analog to a digital signal, turns the switching transistor on and off. This same signal is sent to the ECM as the HEI reference pulse.

When the ECM sees a REF pulse signal of about 200 rpm (calculated from the signal and from its own internal clock), it decides the engine is running. Then it sends a 5-volt signal through the bypass wire to the bypass relay. This causes the double contacts to move. The upper contact disconnects the base of the main switching transistor from the pickup coil and connects it to the EST wire. The lower contact simultaneously disconnects the EST wire from ground. The system is now in EST mode, and timing is controlled by

the ECM. The ECM considers barometric pressure, manifold pressure, coolant temperature, engine speed, and crankshaft position and then sends the HEI module the calculated optimum spark timing command.

NOTE: Some Chevrolet Chevette and Pontiac T-1000 CCC systems do not use an EST subsystem.

Electronic Spark Control (ESC)

Engines with a strong potential for spark knock (particularly higher performance engines) are equipped with ESC. The ESC system becomes a subsystem of the CCC system on the vehicle. It has its own electronic module working in conjunction with the ECM. Its function is to retard ignition timing when detonation (uncontrolled, rapid burning of the fuel charge) occurs. Two slightly different versions of ESC systems have been used.

Of the two ESC systems, the 1981 system, Figure 3-26, passes the EST and the bypass commands from the ECM through the ESC module on their way to the HEI module. If there is spark knock (as reported by the knock sensor), the detonation sensor informs the ESC module that it is occurring. The ESC module then modi-

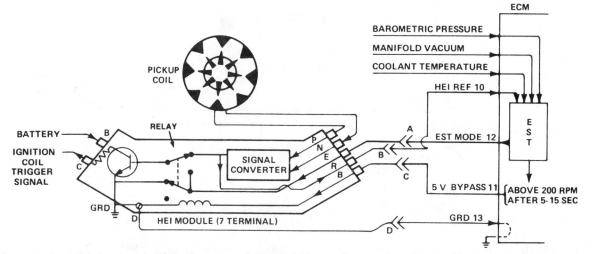

Figure 3–25 HEI circuit with EST. *(Courtesy of General Motors Corporation, Service Technology Group)*

Knock, Detonation and Preignition

Detonation, knock and preignition are sometimes confused. Each of them constitutes effectively too-advanced timing and detracts from engine performance, sometimes also causing damage if the condition persists. Ordinarily the air/fuel mixture burns, rapidly but gradually (over about three thousandths of a second) proceeding from the spark plug and consuming the mixture in the combustion chamber. Detonation occurs if conditions in the combustion chamber reach such a combination of temperature and pressure that the entire charge explodes almost simultaneously. In this case, there is an extremely rapid buildup and loss of combustion pressure doing little torque-generating work. Detonation can easily occur if the spark advance is too great, and the explosion tries to force the piston back down the cylinder at the end of the compression stroke. A high compression engine is more inclined to detonation than a low compression engine, because it has much more heat and compression at the top of its compression stroke. An engine with combustion chamber deposits—which displace empty space and have the effect of raising the compression ratio—will have the same problem. These carbon deposits can also glow red hot and become a source of mistimed ignition themselves. In fact, since detonation also builds up deposits (because the fuel is incompletely burned), it can accelerate this problem.

Preignition occurs when a hot spot in the combustion chamber, either a hot exhaust valve or a hot carbon deposit, ignite the air/fuel mixture before the proper time. A car that diesels after shutdown is suffering from preignition. The effect is the same, but retarding spark on an engine with a combustion chamber hot spot obviously won't change the ignition point. The only repair for that condition is removal of the deposit or repair or replacement of the defective valve.

fies the EST command. Spark timing is retarded about 4 degrees per second until the detonation clears up. When the detonation sensor no longer "hears" knock, the ESC module begins restoring the spark advance it took away. The spark advance is restored at a slower rate than it was removed, approximately 2 degrees per second.

In the second version, used on model year 1982 and later vehicles, Figure 3-27, the EST and bypass commands do not pass through the ESC module. Instead, when the knock sensor produces a voltage indicating knock, the ESC module sends a request to the ECM for it to retard timing. Here's how it works: Ignition voltage is applied to terminal F of the ESC module. The module transfers this voltage, reduced to about 10 volts, to terminal J. It is then carried to terminal L of the ECM. When a detonation signal is sensed on terminal B of the ESC module, the module reduces the voltage at terminal J to below 1 volt. The ECM then retards timing until the signal voltage reappears at terminal L. Either of the ESC systems can be easily tested by tapping on the intake manifold adjacent to the knock sensor with a metal tool, while watching the timing with a timing light or meter. Be sure to consult the vehicle's appropriate service manual for specific directions.

Air Management Valve

As mentioned often before, one of the major goals of the CCC system is making the three-way catalytic converter work at maximum effectiveness. The system achieves this primarily by controlling the air/fuel ratio. On vehicles with a dual-bed, three-way catalytic converter, the system also works to optimize the effectiveness of the converter by introducing air from the air pump into the oxidizing side of the converter. The ECM-controlled air management system is responsible for controlling air injection into either the exhaust manifold or the catalytic converter.

CCC vehicles with dual-bed, three-way catalytic converters also have the appropriate dual air management valve. Those with a single-bed,

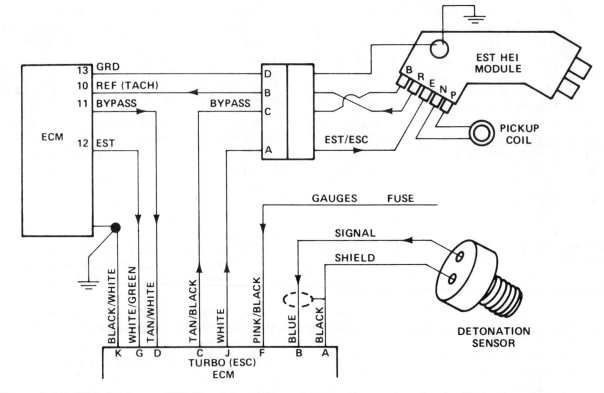

Figure 3–26 ESC circuit, pre-1982. *(Courtesy of General Motors Corporation, Service Technology Group)*

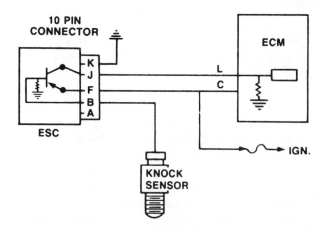

Figure 3–27 ESC circuit. *(Courtesy of General Motors Corporation, Service Technology Group)*

three-way converter use a single air management valve, Figures 3-28 and 3-29.

Single Valve. The single valve system directs air to one of two places. During engine warm-up, air is directed into the exhaust manifold. This will:

- reduce HC and CO by burning them in the exhaust system.
- raise oxygen sensor temperature more quickly from the heat produced by the above process.
- raise the catalytic converter temperature more quickly; the converter must be above 205 to 260 degrees Centigrade (410 to 500 degrees Fahrenheit) to operate effectively.

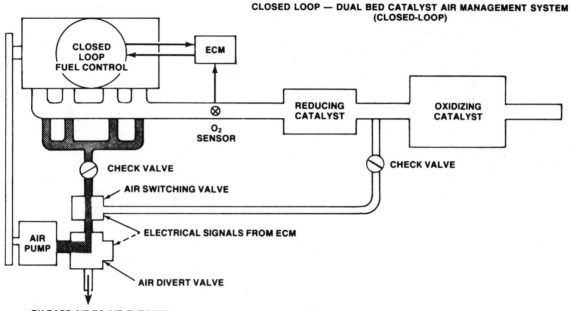

Figure 3–28 Air management with dual-valve and dual-bed catalyst. *(Courtesy of General Motors Corporation, Service Technology Group)*

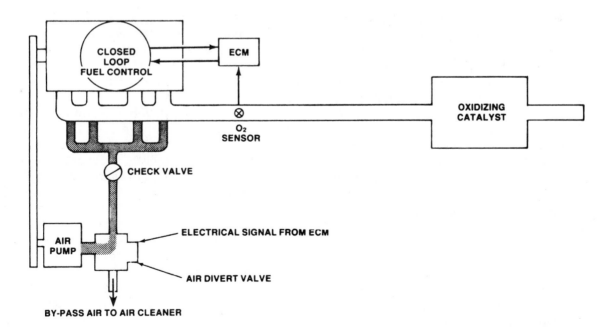

Figure 3–29 Air management with single valve and catalyst. *(Courtesy of General Motors Corporation, Service Technology Group)*

As the engine coolant temperature climbs to 65 degrees Centigrade (150 degrees Fahrenieit), the temperature at which the CCC system goes into closed loop, the ECM commands the divert valve into **divert mode**. Depending on the particular application, the air is diverted either to the air cleaner or to the atmosphere. Had the air continued into the exhaust manifold, the signal from the oxygen sensor would have been inaccurate, not reflecting the results of the combustion mixture's burning. CCC systems with a single-bed converter are always in divert mode except during warm-up.

Dual Valve.   On vehicles with a dual-bed converter, air can be directed to one of three places. During engine warm-up, the divert valve directs air to the switching valve, Figure 3-28. The switching valve directs the air to the exhaust manifold. As the engine approaches closed-loop temperature, the ECM commands the switching valve to direct air to the oxidizing bed of the catalytic converter. Air pumped into the oxidizing bed of the converter has the following effects:

- It prevents the additional oxygen from flowing across the oxygen sensor, which would cause it to send an incorrect signal to the ECM.
- It lowers the temperature in the exhaust manifold. Continued pumping of oxygen into the exhaust manifold after the engine has reached normal operating temperature could produce additional NO_x.
- It makes the oxidizing bed of the catalytic converter operate at maximum efficiency without interfering with the efficiency of the reducing bed (which is upstream of the introduced oxygen and requires some of the residual hydrocarbons to work).

During WOT operation and during deceleration, the ECM commands the divert valve to dump the air, using the divert valve, because there is no other safe place to send it. Either of these driving conditions results in increased amounts of HC and CO in the exhaust system. If during these conditions, air were to continue to

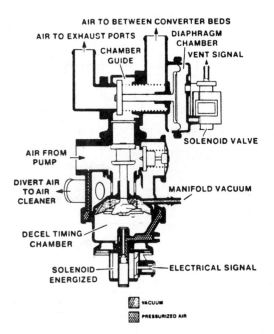

Figure 3–30 Divert mode (deceleration). *(Courtesy of General Motors Corporation, Service Technology Group)*

be pumped into any part of the exhaust system, the converter would quickly overheat. Certain failures within the CCC system, such as in the mixture control solenoid circuit or in the EST system, also cause the ECM to put the air management system into the divert mode.

Air Management Valve Designs.   Many different valve designs have been used on various CCC applications. So a discussion of each different valve is impractical. Most of them, however, are more alike than they are different. They all direct airflow by moving an internal valve to open or block a passage. The valves can be moved by a vacuum diaphragm, an electric solenoid or a combination of the two. Figures 3-30, 3-31 and 3-32 show a typical dual valve in different modes of operation.

In Figure 3-30 the vehicle is decelerating. The high vacuum signal has pulled the diaphragm of the divert valve (lower valve) up. The passage to the switching valve (upper valve) is blocked, and air is diverted to the air cleaner.

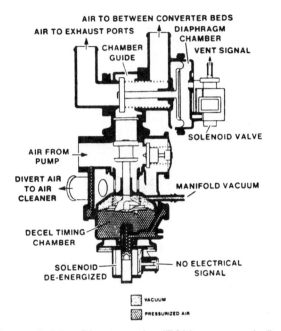

Figure 3–31 Divert mode (ECM commanded). *(Courtesy of General Motors Corporation, Service Technology Group)*

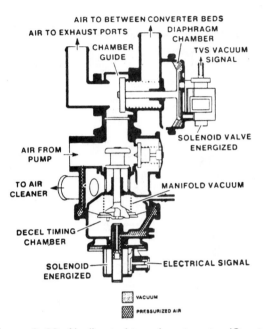

Figure 3–32 Air directed to exhaust ports. *(Courtesy of General Motors Corporation, Service Technology Group)*

The air cleaner serves as a pump silencer. Diverting the injection pump air to it has no effect on the air/fuel ratio since the injection air enters the intake airstream before any metering of air by the throttle.

In Figure 3-31 the ECM has put the air management system into divert mode by de-energizing the solenoid. This allows pump air pressure to be routed to the decel timing chamber. With the help of the manifold vacuum on the other side of the diaphragm, the pressure forces the diaphragm and divert valve up.

Figure 3-32 shows the air management valve directing air to the exhaust manifold during engine warm-up. Manifold vacuum is normally not strong enough to pull the divert valve diaphragm up, and the ECM keeps the divert solenoid energized, blocking pump air pressure from entering the decel timing chamber. The spring holds the divert valve down and allows air to flow up to the switching valve. The switching valve solenoid, energized by the ECM, allows

vacuum to apply to the upper diaphragm. This pulls it and the attached switching valve to the right to block the converter air passage, thus opening the exhaust port passage.

In Figure 3-33, the engine has warmed up, and the CCC system is ready to go into closed loop. The ECM has de-energized the switching valve solenoid, which blocks vacuum to the diaphragm. The spring moves the valve to the left, opening the converter passage and blocking the exhaust port passage.

All air management valves used with CCC systems use a pressure relief valve. At high engine speeds, the pressure relief valve exhausts excess air to reduce system pressure. This can easily be mistaken for a divert mode function.

Pulsair System. Some vehicles use a **Pulsair system** instead of an air injection system, especially on four-cylinder engines. Between each exhaust pulse in the exhaust manifold, a low pressure pulse develops. This low

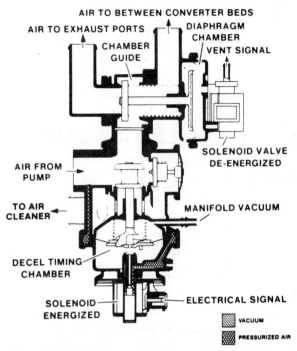

Figure 3–33   Warm engine mode (closed loop). *(Courtesy of General Motors Corporation, Service Technology Group)*

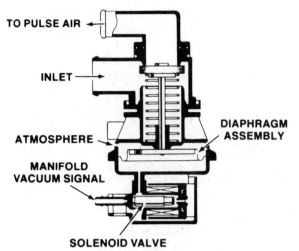

Figure 3–34   Pulsair shutoff valve. *(Courtesy of General Motors Corporation, Service Technology Group)*

pressure can be used to syphon air into the exhaust ports. This air is drawn through a tube with a check valve (usually a reed-type flap). The check valve blocks reverse airflow during the exhaust pulse.

Pulsair Shutoff Valve.   When a Pulsair system is combined with the CCC system, the ECM uses a Pulsair shutoff valve to control air flowing into the exhaust system, Figure 3-34.

During engine warm-up, the solenoid is energized and vacuum applied to the diaphragm. The valve is pulled down, and air is allowed to flow into the exhaust system. With the exception of how air is made to flow, this system is very similar to the single-valve air management system.

Idle Speed Control (ISC)

Idle Speed Control Motor.   To achieve the established goals for the exhaust emission quality, fuel economy, and driveability, an articulated

idle speed control is needed. This is achieved by the use of a permanent magnet field reversible electric motor attached to the side of the carburetor, Figure 3-35.

The ECM normally has a fixed idle speed that it wants to maintain during idle. It moves the

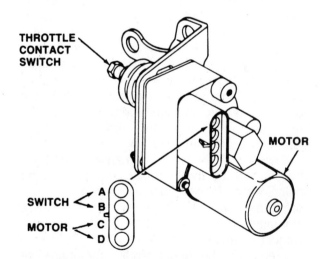

Figure 3–35   ISC motor assembly. *(Courtesy of General Motors Corporation, Service Technology Group)*

throttle as needed with the ISC motor. However, it commands a higher idle speed in response to any one of the following three conditions:

- During closed choke idle, the fast idle cam holds the throttle blade open enough to lift the throttle linkage off the ISC plunger. This allows the ISC switch to open so the ECM does not monitor idle speed. As the choke spring allows the fast idle cam to fall away and the throttle returns to warm idle position, the ECM notes the still low coolant temperature and commands a slightly higher idle speed.
- If the engine starts to overheat, the ECM commands a higher idle speed to increase coolant flow by turning the water pump faster.
- If system voltage falls below a predetermined value, the ECM commands a higher idle speed to increase generator speed and output.

During warm idle, if the automatic transmission is put in gear (forward or reverse), the park/neutral switch signals the ECM of the impending load on the engine. The ECM then commands the ISC motor to extend the plunger. The throttle blade opens at about the same time the transmission engages, and no appreciable change in engine speed occurs. If the air conditioning is turned on, the ECM receives a signal from the air conditioning switch. The ECM commands a wider throttle blade opening to compensate for the additional load of the air conditioning compressor.

Starting in 1982, models with smaller engines came with an ISC relay. The ISC relay uses a solid-state fixed time delay device to keep the ISC motor activated in a retract mode after the key is turned off. This allows the throttle to completely close each time the engine is turned off and thus helps prevent dieseling.

There is no way to adjust idle speed on CCC systems using an ISC motor. Attempting to adjust the idle speed by adjusting the plunger screw merely results in the ECM moving the plunger away to neutralize the adjustment. Idle speed does not change until the ISC plunger motor uses up all its travel trying to compensate for the adjustment, at which point the system is completely out of calibration. Proper calibration can be restored, however by following the **minimum** and **maximum authority** adjustment procedures outlined in the service manual for the specific vehicle. When idle speed driveability problems occur, the ISC system is usually responding to or being affected by the problem, not causing it.

NOTE: Be sure the ignition is turned off before connecting or disconnecting the ISC motor, or damage can result to the ECM.

Idle Stop Solenoid. Some engines do not use an ISC motor. Instead they use an idle stop solenoid. The idle stop solenoid is essentially the same as an antidiesel solenoid or dashpot, used for years and dating back to the first emission control systems. When the ignition is turned on, the solenoid is energized and extends its plunger to hold the throttle lever off the idle stop screw. The extended plunger provides the curb idle position for the throttle blade. When the ignition is turned off, the plunger retracts and allows the throttle blade to move fully closed to prevent dieseling. The same device can be connected to the A/C switch. In this case, the idle stop screw provides curb idle, and the solenoid increases idle speed slightly when the A/C is turned on.

Idle Load Compensator (ILC). Beginning in 1982, the Oldsmobile limited control CCC system used an ILC instead of an ISC motor, Figure 3-36. The ILC is not controlled by the ECM; it is controlled by engine vacuum. The ILC is very similar in construction to a vacuum advance unit on an older distributor. It contains a diaphragm with a spring pushing against one side. The other side attaches to a plunger that extends from the unit in the form of an adjustable screw. The head of the plunger screw acts as the throttle stop to control idle speed. Manifold vacuum is applied to the spring side of the diaphragm. When the automatic transmission is put into gear, the load on

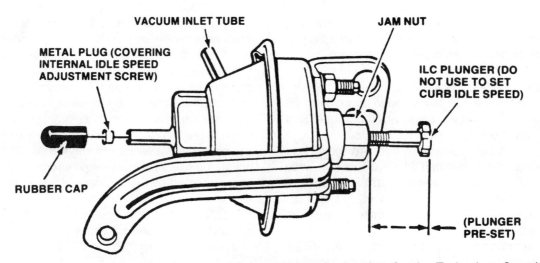

Figure 3–36 Idle load compensator. *(Courtesy of General Motors Corporation, Service Technology Group)*

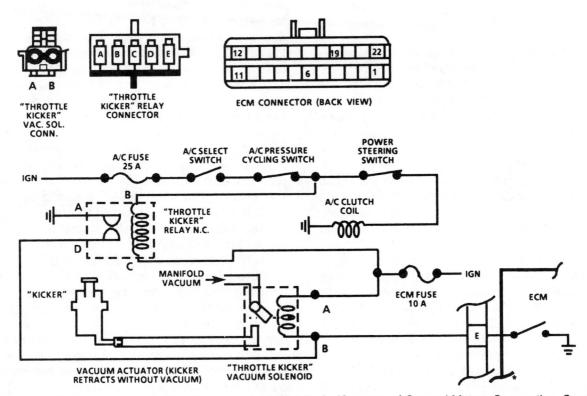

Figure 3–37 Vacuum-operated throttle kicker circuit (2.8-liter). *(Courtesy of General Motors Corporation, Service Technology Group)*

the engine causes vacuum to go down slightly. This allows the spring to push the diaphragm and plunger a little farther, thus opening the throttle a little. The same thing occurs for any load or condition that reduces engine vacuum.

Idle speed adjustments can be made to the ILC; however, the adjustment is made at the vacuum side of the diaphragm, not at the plunger screw. The idle adjusting screw in the stem on the back of the unit changes the preload on the diaphragm spring. Refer to an appropriate service manual for adjustment procedures.

Throttle Kicker. Some later CCC vehicles use either a vacuum or electronic throttle kicker. The functions of a throttle kicker are explained in Figures 3-37 and 3-38.

Torque Converter Clutch (TCC)

The TCC improves fuel economy by eliminating the hydraulic slippage and heat production in the torque converter once a cruise speed is achieved.

Clutch. The major component of the TCC system is the lockup clutch itself. It becomes a fourth element added to a conventional torque converter, Figure 3-39. The clutch plate, splined to the turbine, has friction material bonded to its engine side and near its outer circumference. The converter cover has a machined surface just inside, where the converter drive lugs attach. This machined surface mates with the friction material on the disc when the clutch is applied. When hydraulic pressure is applied to the turbine

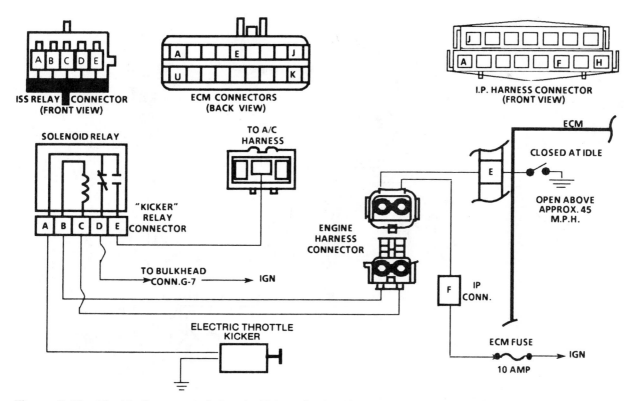

Figure 3–38 Electrically operated throttle kicker circuit. *(Courtesy of General Motors Corporation, Service Technology Group)*

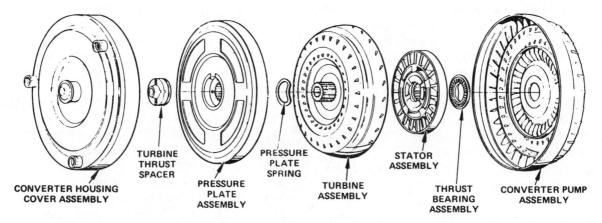

CONVERTER HOUSING COVER ASSEMBLY

TURBINE THRUST SPACER

PRESSURE PLATE ASSEMBLY

PRESSURE PLATE SPRING

TURBINE ASSEMBLY

STATOR ASSEMBLY

THRUST BEARING ASSEMBLY

CONVERTER PUMP ASSEMBLY

Figure 3–39 Torque converter with TCC. *(Courtesy of General Motors Corporation, Service Technology Group)*

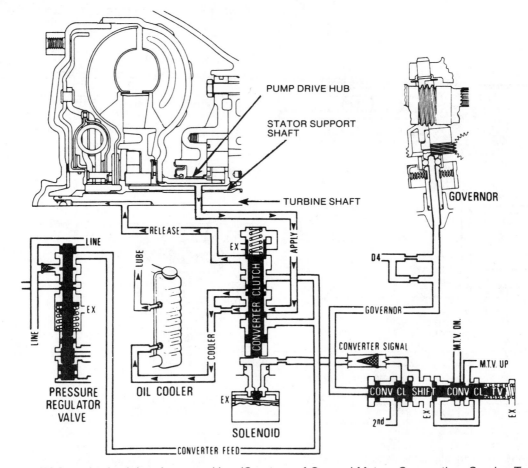

Figure 3–40 TCC apply circuit in release position. *(Courtesy of General Motors Corporation, Service Technology Group)*

side of the disc, the disc is forced against the converter cover. The friction between the clutch's friction material and the cover locks up the unit and causes the torque converter to rotate as a solid unit. The pressure source is converter feed oil coming from the pressure regulator valve in the transmission valve body. In non-TCC applications, converter feed oil is used to charge and cool the converter by circulating through the converter and then the transmission cooler in the radiator. On TCC applications, however, converter feed oil must pass through a converter clutch apply valve on its way to and as it returns from the converter. The apply valve, a small valve in the transmission, controls the direction of oil flow through the torque converter, Figure 3-40.

With the apply valve in its at-rest position and the transmission in neutral, or with the car moving at low speed, converter feed oil is directed into the converter through the release passage by way of the hollow turbine shaft. This oil is fed into the converter between the converter cover and the clutch disc. The oil forces the disc away from the cover and thus releases the clutch. Converter feed oil flows over the circumference of the disc and circulates through the converter. It exits through the apply passage, a passage between the pump drive hub and the stator support shaft.

Two criteria must be met before the clutch applies. The transmission must be ready hydraulically, and the ECM must be satisfied with engine temperature, throttle position, engine load, and vehicle speed. When the transmission is in the right gear (this varies from transmission to transmission), hydraulic pressure is supplied at one end of the clutch apply valve, Figure 3-41. This hydraulic pressure (converter apply signal) has the potential to move the apply valve into the apply position. The apply valve is moved, however, only if the ECM is ready for the lockup clutch to engage.

The ECM controls the position of the clutch apply valve with a solenoid that opens or closes an exhaust port at the converter apply signal end of the apply valve. If the solenoid is not energized, the exhaust port is open and the converter apply signal oil exhausts from the signal end of the apply valve as fast as it arrives. Sufficient pressure does not develop to move the apply valve. Converter feed oil continues to flow into the converter through the release passage, Figure 3-40. When the solenoid is energized, the exhaust port is blocked. The converter apply signal oil develops pressure, and the apply valve moves and is held in the apply position. Converter feed oil now flows into the converter through the apply passage, Figure 3-41. The clutch is applied and prevents oil from exhausting through the release pressure, holding static pressure.

Figure 3-42 shows a typical TCC solenoid control circuit. The power source is the ignition switch, through the gauge fuse. The brake switch is in series on the high-voltage side of the circuit. Anytime the brakes are applied, the solenoid is de-energized. After the current passes through the apply solenoid, it arrives at the 4-3 pulse switch, used only on automatic transmissions with a fourth-speed overdrive. The 4-3 pulse switch momentarily disengages the TCC to allow a smooth 4-3 downshift. It is closed at all other times. From there, the current goes to the ECM, where the circuit is grounded when the ECM is ready to apply the TCC. When the transmission goes into fourth gear, the fourth-gear switch opens to alert the ECM that the transmission is in overdrive. In overdrive the ECM holds the TCC on through a much wider range of throttle position than it does in second or third gear. To put it in engineering language, the ECM holds the TCC on throughout a wider throttle position window. The term *window* refers to a range within a range. For instance, if the TPS signal to the ECM were 0.5 volt at idle and 4.5 volts at WOT, the ECM might select 2.0 to 3.5 as the window in which it would keep the TCC applied.

The TCC test lead, which comes from the low-voltage side of the circuit, can be used to tell when the ECM has electrically applied the TCC circuit or to ground the circuit and thus override the ECM. Its use is covered in more detail under **System Diagnosis and Service** later in this chapter.

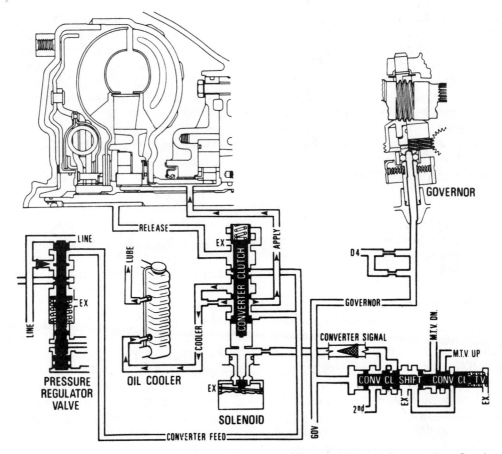

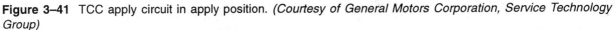

Figure 3–41 TCC apply circuit in apply position. *(Courtesy of General Motors Corporation, Service Technology Group)*

TCCs and Four-Speed Automatic Transmissions

On vehicles using a four-speed automatic transmission, the TCC must be applied while the transmission is in fourth gear. Any failure resulting in the clutch not applying during prolonged fourth gear operation can result in transmission damage as a result of oil overheating. This can occur in a relatively short period of steady-state fourth-gear operation.

Exhaust Gas Recirculation (EGR) Valve

The primary purpose of the EGR valve is to control the production of oxides of nitrogen (NO_x) by lowering combustion temperatures. It does this by metering a certain amount of inert exhaust gas into the incoming air/fuel mixture, slowing its burn rate. The EGR also serves somewhat to control detonation. The pressing need for better fuel economy and driveability, combined with the development of computer-controlled ignition timing has pushed ignition timing to the ragged edge of detonation. Without the cooling effect of EGR

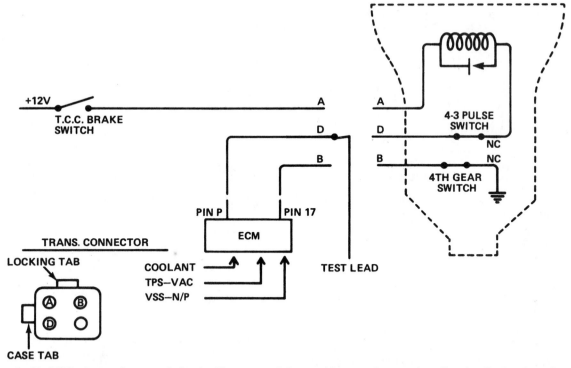

Figure 3–42 TCC electronic control circuit. *(Courtesy of General Motors Corporation, Service Technology Group)*

on combustion temperature, detonation would occur in almost all engines.

Even so, excessive EGR gas causes serious driveability problems, especially when engines are cold. It makes sense to have the ECM in charge of EGR operation to closely monitor and control its application. On CCC engines the EGR is turned on or off or on some later models is modulated (the amount of opening is controlled) by one or more ECM-controlled solenoids. The exact method and amount of EGR control varies considerably with engine and model year application. Generally, however, EGR systems controlled by CCC can be summarized by one of the following statements:

- The ECM operates a solenoid that blocks vacuum when energized. When not energized, it passes ported vacuum to the EGR valve, Figure 3-43.
- The ECM operates a solenoid that bleeds

atmospheric pressure into the ported vacuum signal to the EGR valve when activated. A bleed solenoid can be used to turn the EGR valve off or to partially close (modulate) it, Figure 3-44.

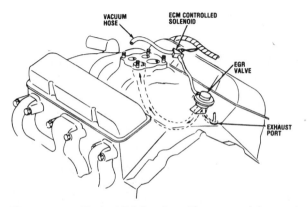

Figure 3–43 Typical EGR valve. *(Courtesy of General Motors Corporation, Service Technology Group)*

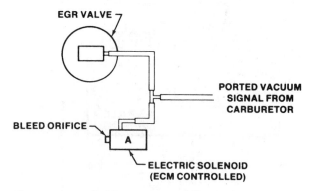

Figure 3–44   EGR bleed solenoid. *(Courtesy of General Motors Corporation, Service Technology Group)*

- The ECM operates two solenoids, one to block or pass ported vacuum to the EGR valve and the other as bleed solenoid to modulate it, Figure 3-45.
- The ECM operates two blocking solenoids in series. Each one is operated according to different sets of engine calibration criteria, Figure 3-46.
- A pulse-width modulating solenoid (rapidly turned on and off with the on-time variable) precisely controls the amount of vacuum allowed to the EGR valve, Figure 3-47.

One other EGR control system that can be

encountered on a CCC vehicle should be discussed at least briefly, the **aspirator**-controlled EGR, Figure 3- 48. The aspirator, mounted in the air cleaner, contains a small venturi. Air from the air pump is directed to the aspirator. During periods of low engine vacuum, the aspirator control valve allows air to pass through the venturi portion of the aspirator to produce a vacuum that keeps the EGR valve open. When engine vacuum is high, the aspirator control valve closes to prevent air from passing through the aspirator venturi. The EGR valve is now operated by ported vacuum.

NOTE: Generally electrical failures in any of the EGR control systems result in the EGR valve being open.

Early Fuel Evaporation (EFE)

The EFE system applies heat to the intake manifold area beneath the carburetor to help evaporate the fuel and to keep it in a vapor state. CCC systems use one of two different types of EFE systems.

Exhaust Heat Type.   An EFE valve (known earlier as a heat riser valve, located where the exhaust pipe connects to the exhaust manifold on one side of the engine) is closed by a vacuum motor, Figure 3-49. This forces exhaust gases

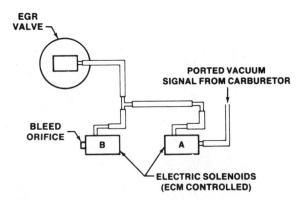

Figure 3–45   EGR bleed and control solenoids. *(Courtesy of General Motors Corporation, Service Technology Group)*

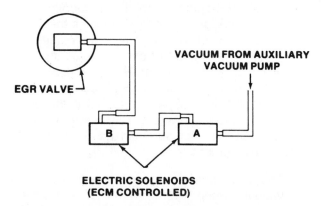

Figure 3–46 Two EGR control solenoids. *(Courtesy of General Motors Corporation, Service Technology Group)*

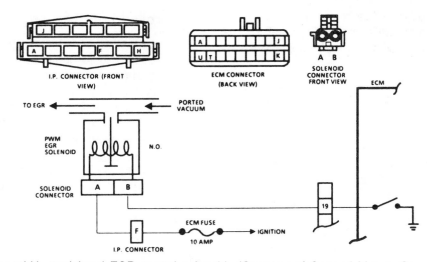

Figure 3–47 Pulse-width modulated EGR control solenoid. *(Courtesy of General Motors Corporation, Service Technology Group)*

through a passage in the intake manifold beneath the space over which the carburetor mounts (the intake manifold plenum). Vacuum to the vacuum motor is controlled by a solenoid under the control of the ECM, Figure 3-50. This system varies from earlier, non-CCC, EFE systems in that they used a **vacuum control valve (VCV)** to control vacuum to the EFE actuator.

Electric Grid Type.   The electric EFE uses a ceramic-encased heating element under the carburetor, Figure 3-51. The heater element is powered by a relay, Figure 3-52. When the ignition is on, voltage is available at one of the normally open relay contacts. Ignition voltage is also applied to the relay coil waiting for ground by the ECM. When coolant temperature is low, the ECM

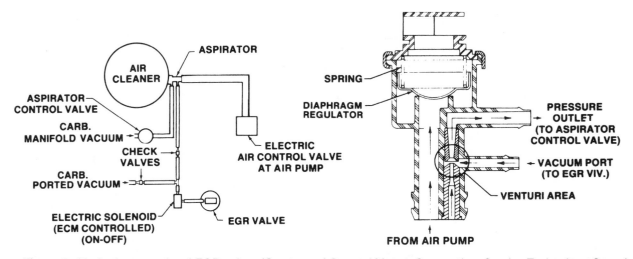

Figure 3–48 Aspirator-assisted EGR valve. *(Courtesy of General Motors Corporation, Service Technology Group)*

grounds the coil and the relay contacts close. This powers the heater element. Electric EFE circuits vary slightly with engine application and model year.

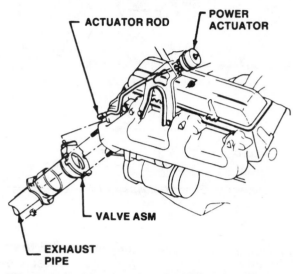

Figure 3–49 Vacuum-actuated EFE valve.*(Courtesy of General Motors Corporation, Service Technology Group)*

Controlled Canister Purge (CCP)

On many CCC vehicles, the ECM also operates a solenoid to control purge vacuum to the **purge valve** on the charcoal canister. When the CCC system is in open loop, or before a predetermined time period has elapsed since engine start-up, or below a specific rpm, the purge solenoid is energized and blocks purge vacuum. In closed loop, after a specified elapsed time and above a specific rpm, the purge solenoid is de-energized and canister purging occurs.

Check Engine Light and Lamp Driver

All CCC systems use a check engine light. If the ECM sees a fault in one of the circuits it monitors for malfunctions, it turns on the check engine light located on the instrument panel. This warns the driver that a malfunction exists. By grounding the test terminal (one of the terminals in the assembly line communication link, ALCL, located under the dash), the ECM causes the check engine light to flash. These flashes indi-

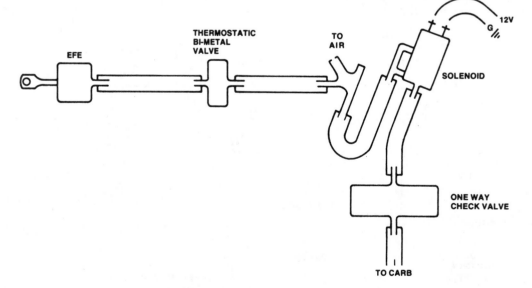

Figure 3–50 EFE vacuum control circuit.*(Courtesy of General Motors Corporation, Service Technology Group)*

cate one or more codes identifying the circuit or circuits in which the fault exists. This is explained more fully under **System Diagnosis and Service** in this chapter.

In 1982 a remote lamp driver was added to the system. It is a separate module, located under the dash in most cases, powering the check engine light with greater reliability.

E-cell

Some Oldsmobile and Chevrolet eight cylinder engine CCC applications incorporate an E-cell, often referred to as green engine calibration unit. The E-cell slightly modifies some engine calibrations during engine break in. It is not necessary to replace the E-cell even if it fails prematurely.

✔ SYSTEM DIAGNOSIS & SERVICE

Self-Diagnosis

The ECM monitors the major input sensors, the mixture control solenoid, the EST and their respective circuits for proper operation. If the ECM sees a fault, such as an open, a short or a voltage value that stays too high or too low for too long, in any of the circuits it monitors, it will turn on the check engine light on the instrument panel and record a code number in its diagnostic memory. Some later models have a service engine soon light and a service engine now light, either of which can be selected depending on the severity of the problem. The code number identifies the circuit in which the fault exists. The check engine light warns the driver that a fault exists. The check engine light should come on, however, anytime the ignition is on without the engine running, as a bulb check.

Diagnostic Memory. A portion of the ECM's random access memory is devoted to diagnostic

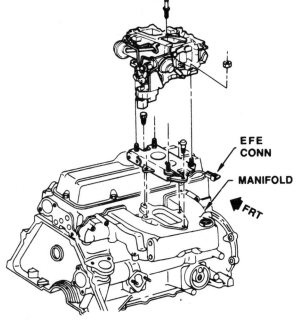

Figure 3–51 Electrically heated EFE valve. *(Courtesy of General Motors Corporation, Service Technology Group)*

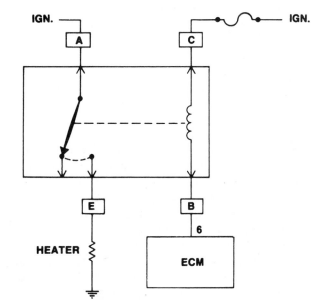

Figure 3–52 Electric EFE control circuit. *(Courtesy of General Motors Corporation, Service Technology Group)*

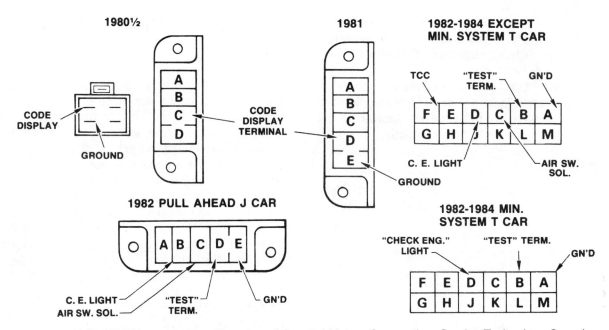

Figure 3–53 ALCL (ALDL) connectors. *(Courtesy of General Motors Corporation, Service Technology Group)*

memory, sometimes referred to as long-term memory. The diagnostic memory enables the ECM to store code numbers, referred to as fault codes, that identify the type of fault and the circuit in which the fault exists. A technician can obtain the stored codes by either of two methods. Grounding the test terminal in the assembly line communications link (ALCL), often referred to as the assembly line diagnostic link (ALDL) puts the system into diagnostics, Figure 3-53. The ECM reads out its stored codes by flashing the check engine light. All codes are stored as two-digit numbers. Two quick flashes followed by a pause of about two seconds represents 2; three more quick flashes represents the second digit—3. Figure 3-54 demonstrates a code 12 being displayed. Once the two-digit number completes, a slightly longer pause separates it from the next two-digit number or code. When a code displays, it repeats twice before moving on to the next code. The second method of obtaining stored codes is to plug a special piece of test equipment, manufactured by any one of several differ-

ent companies and ranging in size from large engine analyzers to small hand-held units, into the ALCL. With the ignition on, the test unit receives any stored codes from the ECM and displays them on its digital screen.

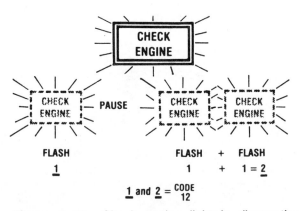

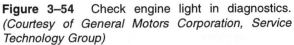

Figure 3–54 Check engine light in diagnostics. *(Courtesy of General Motors Corporation, Service Technology Group)*

NOTE: The ALCL is most often located just under the instrument panel, somewhere between the steering column and the radio.

NOTE: Diagnostic procedures for 1987 and later vehicles are written to include the use of an ALCL tool, often referred to as a "scanner," where it is applicable. Previously, diagnostic procedures were written around the use of a voltmeter to obtain such measurements as sensor readings.

Intermittent Faults. If the ECM sees that a fault has turned on the check engine light and set a code in memory, it keeps the check engine light on until the ignition is turned off, or until the ECM no longer senses the fault. If the perceived fault clears up, the check engine light goes off, but the code will remain in memory. If the fault does not repeat again within the next fifty ignition cycles (starting the engine and turning it off constitutes one ignition cycle), the code will be automatically erased.

Clearing the Memory. The code can also be erased by disconnecting the power supply from terminal R of the ECM for ten seconds. This can be done most easily either by pulling the fuse marked ECM from the fuse panel, disconnecting the negative battery lead, or on some later models, disconnecting a fusible link near the positive battery terminal.

Behavior in Diagnostics

When the test terminal in the ALCL is grounded and the engine is not running, the check engine light should begin flashing code 12. Code 12 indicates no reference pulse is coming from the distributor, which is what you would expect since the engine is not running. Any additional stored codes display in numerical sequence after code 12.

NOTE: The ignition should always be on before grounding the test terminal; otherwise the check engine light does not function properly.

There are some other noteworthy points about code 12. It is not a storable code (it is not stored in memory). It is most often used as an indication that the self-diagnostic function of the ECM is working properly. If you put the system into diagnostics with the engine off and do not see a code 12, the self-diagnostic function is not working and must be repaired before you can continue. The only time the presence of code 12 indicates a fault is if it displays when the engine is running. Then it means the REF pulse is not coming in from the distributor to the ECM.

While the codes are displayed, the ECM also:

- sends a steady 30-degree dwell command to the mixture control solenoid.
- energizes all ECM-controlled solenoids.
- pulses the ISC motor in and out.

If the test terminal is grounded while the engine is running, the ECM will:

- disable the open-loop timer, which maintains the time lapse before the system can go into closed loop.
- take out some of the adaptive enrichment modes.
- send a 30-degree dwell command to the mixture control solenoid if the system is in open loop and is not in an enrichment mode.
- set the EST at a fixed spark advance position.

While in diagnostics (test terminal grounded), the ECM will not store new codes that might be encountered.

The 1982 and later T-car (Chevrolet Chevette and Pontiac T-1000) has no diagnostic memory. It can only display codes concerning faults it currently perceives. To receive fault codes from the minimum function system used on these cars, the engine must be running when the test terminal is grounded, and of course, the fault must be present.

Diagnostic Procedures

The diagnostic procedures are contained in several different charts or groups of charts. They are as follows:

- diagnostic circuit check.
- customer complaint (or in newer manuals, driveability symptoms).
- system performance check.
- diagnostic charts without trouble codes.
- diagnostic charts with trouble codes.
- diagnostic charts on related components.

It is important to follow the directions very carefully when any of the diagnostic procedures or flow charts are used. Failing to do so or taking other shortcuts usually results in inaccurate diagnostic conclusions.

NOTE: When using diagnostic procedures, be sure to use procedures or charts that apply to the specific engine or application and model year. Procedures vary for different engines and are often changed by model year.

Diagnostic Circuit Check.   (Figure 3-55) This chart of procedures should always be used before attempting to diagnose any ECM-controlled system. It is primarily responsible for sorting out why a code is stored. It also aids in discovering why the check engine light is not working properly if it is not. There can be several reasons for a particular code's being stored:

- *Existing problem.* A fault can exist, in which case its corresponding code is called a hard code.
- *Intermittent problem.* A fault can develop and clear up by itself, in which case the resulting code is intermittent.
- *No real problem.* Someone previously working on the car can open a CCC system circuit that the ECM monitors while the ignition is on. This can set a code even though there is nothing wrong with the circuit. Sometimes a

strong radio signal, such as can be encountered near an airport, can set a code. This type of code is often referred to as a phantom code.

It is important that this type of code identification be made. The trouble code charts are written assuming a real fault exists. If one of the trouble code charts is used to pursue a phantom or intermittent code, the chart more often leads to an invalid conclusion because it is beginning with an invalid assumption—that a fault currently exists when in fact none does.

Driver Comment Chart/Driveability Symptom Section.   (Figure 3-56) This chart or section, depending on the model year, acts as a guide to aid the technician in determining the cause of problems that either do not have trouble codes or that show intermittent trouble codes.

System Performance Check.   (Figure 3-57) This procedure should always be used after repairing any part of the CCC system or any component the ECM controls. It verifies that the heart of the system, the fuel mixture control, is working properly. If the fuel mixture is not controlled properly, it refers to another chart for further diagnosis.

A trouble code found in memory and shown by the Diagnostic Circuit Check to be intermittent should be dismissed as a phantom code when no driver complaint or performance problem was identified and the System Performance Check shows satisfactory results.

Diagnostic Charts without Trouble Codes. This is a series of charts contained in the service manual. Any of the charts within the series can assist in finding the cause of any one of several specific fault conditions that do not have a corresponding trouble code. Figure 3-58 is shown as an example. These charts should only be used when referred to by one of the preceding charts (diagnostic circuit check, driver comment or system performance check) or when a specific condition addressed by one of them has been identified. As mentioned previously, improper use of the charts most often leads to mistakes, sometimes costly ones.

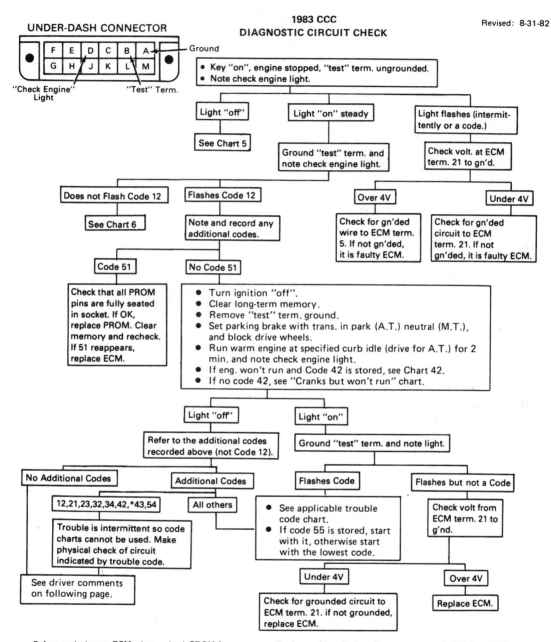

UNDER-DASH CONNECTOR

F	E	D	C	B	A
G	H	J	K	L	M

"Check Engine" Light

"Test" Term.

Ground

1983 CCC
DIAGNOSTIC CIRCUIT CHECK

Revised: 8-31-82

- Key "on", engine stopped, "test" term. ungrounded.
- Note check engine light.

Light "off"

See Chart 5

Light "on" steady

Ground "test" term. and note check engine light.

Light flashes (intermittently or a code.)

Check volt. at ECM term. 21 to gn'd.

Does not Flash Code 12

See Chart 6

Flashes Code 12

Note and record any additional codes.

Over 4V

Check for gn'ded wire to ECM term. 5. If not gn'ded, it is faulty ECM.

Under 4V

Check for gn'ded circuit to ECM term. 21. If not gn'ded, it is faulty ECM.

Code 51

Check that all PROM pins are fully seated in socket. If OK, replace PROM. Clear memory and recheck. If 51 reappears, replace ECM.

No Code 51

- Turn ignition "off".
- Clear long-term memory.
- Remove "test" term. ground.
- Set parking brake with trans. in park (A.T.) neutral (M.T.), and block drive wheels.
- Run warm engine at specified curb idle (drive for A.T.) for 2 min. and note check engine light.
- If eng. won't run and Code 42 is stored, see Chart 42.
- If no code 42, see "Cranks but won't run" chart.

Light "off"

Refer to the additional codes recorded above (not Code 12).

Light "on"

Ground "test" term. and note light.

No Additional Codes

12,21,23,32,34,42,*43,54

Trouble is intermittent so code charts cannot be used. Make physical check of circuit indicated by trouble code.

See driver comments on following page.

Additional Codes

All others

Flashes Code

- See applicable trouble code chart.
- If code 55 is stored, start with it, otherwise start with the lowest code.

Under 4V

Check for grounded circuit to ECM term. 21. if not grounded, replace ECM.

Flashes but not a Code

Check volt from ECM term. 21 to g'nd.

Over 4V

Replace ECM.

Before replacing an ECM, always check PROM for correct application and installation. Also, remove terminal(s) from ECM connector for circuit involved, clean terminal contact and expand it slightly to increase contact pressure and recheck to see if problem is corrected. In case of repeat ECM failure, check for a shorted solenoid or relay controlled by the ECM.

The system performance check should be performed after any repairs to the system have been made.

*It is possible to set a false Code 42 on starting, but the "Check Engine" light will not be "on". No corrective action is necessary.

Figure 3–55 Diagnostic circuit check. Charts 5 and 6, referred to in the diagram, are found in the manufacturer's service manual. *(Courtesy of General Motors Corporation, Service Technology Group)*

DRIVER COMMENT

Revised: 8-31-82

Engine performance troubles (stalling, detonation, surge, fuel economy, etc.)

IF THE "CHECK ENGINE" LIGHT IS NOT ON, NORMAL CHECKS THAT WOULD BE PERFORMED ON CARS WITHOUT THE SYSTEM SHOULD BE DONE FIRST.
IF GENERATOR OR COOLANT LIGHT IS ON WITH THE CHECK ENGINE LIGHT, THEY SHOULD BE DIAGNOSED FIRST.
INSPECT FOR POOR CONNECTIONS AT COOLANT SENSOR, M/C SOLENOID, ETC., AND POOR OR LOOSE VACUUM HOSES AND CONNECTIONS. REPAIR AS NECESSARY.

- Intermittent check engine light but no trouble code stored.
  - Check for a loose connection in the circuit from:
    - Ignition coil to ground and arcing at spark plug wires or plugs.
    - Bat. to ECM term's. C and R
    - ECM terms. A and U to engine ground
    - EST wires should be kept away from spark plug wires, distributor housing, coil and generator. Wires from ECM term. 13 to dist. and the shield around EST wires should be a good ground.
  - Open Diode across A/C Compressor Clutch.

- Loss of long term. memory
  - Grounding dwell lead for 10 seconds with "test" term. ungrounded with engine running should give code 23. This code should be retained in long term. memory after the engine is stopped and restarted. If it is not, ECM is at fault.

- Stalling, Rough Idle, or Improper Idle Speed
  - See idle speed control . . . Page 4A-60

- Denotation (Spark Knock)
  - Check — ESC performance, if applicable . . . Page 4A-55
    MAP or Vacuum Sensor output . . . Pages 2A-20 and 2A-21
    EGR Check . . . Pages 4A-56 and 4A-57
    TPS enrichment operation . . . Chart No. 4
    HEI operation . . . Page 4A-54

- Poor Performance and/or Fuel Economy and Surging
  - See — Carb. on car service — Divisional Service Manual
    EFE check . . . Pages 4A-58 thru 4A-60
    TCC operational check . . . Page 4A-63
    EST diagnosis . . . Page 4A-54
    ESC diagnosis if applicable . . . Page 4A-55

- Poor Full-Throttle Performance
  - See Chart 4 if equipped with TPS

- Intermittent No-Start
  - Incorrect pick-up coil or ignition coil. See "Cranks, But Won't Run" chart.
  - Intermittent ground connections on ECM.

- All Other Comments.
  - Make system performance check on warm engine (upper radiator hose hot)

The system performance check should be performed after any repairs to the CCC system have been made

Figure 3–56 Driver complaint chart. Pages and charts referred to in this illustration are found in the manufacturer's service manual. *(Courtesy of General Motors Corporation, Service Technology Group)*

SYSTEM PERFORMANCE CHECK   Revised: 8-31-82

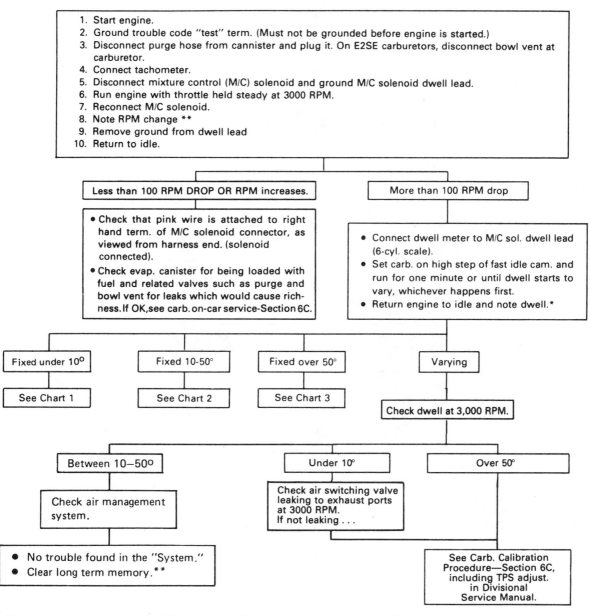

1. Start engine.
2. Ground trouble code "test" term. (Must not be grounded before engine is started.)
3. Disconnect purge hose from cannister and plug it. On E2SE carburetors, disconnect bowl vent at carburetor.
4. Connect tachometer.
5. Disconnect mixture control (M/C) solenoid and ground M/C solenoid dwell lead.
6. Run engine with throttle held steady at 3000 RPM.
7. Reconnect M/C solenoid.
8. Note RPM change **
9. Remove ground from dwell lead
10. Return to idle.

Less than 100 RPM DROP OR RPM increases.

- Check that pink wire is attached to right hand term. of M/C solenoid connector, as viewed from harness end. (solenoid connected).
- Check evap. canister for being loaded with fuel and related valves such as purge and bowl vent for leaks which would cause richness. If OK, see carb. on-car service-Section 6C.

More than 100 RPM drop

- Connect dwell meter to M/C sol. dwell lead (6-cyl. scale).
- Set carb. on high step of fast idle cam. and run for one minute or until dwell starts to vary, whichever happens first.
- Return engine to idle and note dwell.*

Fixed under 10°	Fixed 10-50°	Fixed over 50°	Varying
See Chart 1	See Chart 2	See Chart 3	

Check dwell at 3,000 RPM.

Between 10–50°

Check air management system.

- No trouble found in the "System."
- Clear long term memory.**

Under 10°

Check air switching valve leaking to exhaust ports at 3000 RPM. If not leaking . . .

Over 50°

See Carb. Calibration Procedure—Section 6C, including TPS adjust. in Divisional Service Manual.

---

*Oxygen sensors may cool off at idle and the dwell change from varying to fixed. If this happens running the engine at fast idle will warm it up again.

**If car is equipped with an electric cooling fan, it may lower the RPM when it engages.

Figure 3–57 System performance check. Charts referred to in this illustration are found in the manufacturer's service manual. *(Courtesy of General Motors Corporation, Service Technology Group)*

CHART #3   Revised: 4-21-82

(Rich Exhaust Indication)

DWELL FIXED OVER 50°

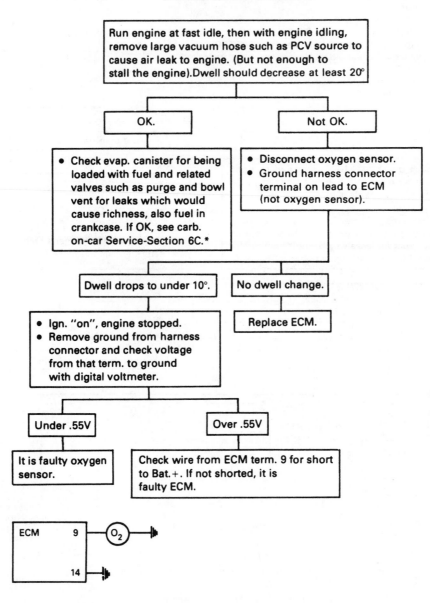

*Use divisional service manual.

Figure 3–58 Typical diagnostic chart. *(Courtesy of General Motors Corporation, Service Technology Group)*

Diagnostic Charts with Trouble Codes. This is a series of charts contained in the service manual. These charts can be used to find the cause of a fault once the fault has been identified by a code stored in the ECM's memory, and after the diagnostic circuit check has been performed to verify that the fault currently exists. Figure 3-59 is an example of this type of chart

Diagnostic Charts on Related Components. This series of charts, contained in the service manual, helps with the diagnosis of components and related circuits controlled by the ECM but not monitored by it for proper operation. These components and circuits include the air management valve and the torque converter clutch. Figure 3-60 shows an example of this sort of chart.

Dwell Diagnosis

If the system is in closed loop and the dwell varies between 10 and 50 degrees on the six cylinder scale, we know the system is able to maintain the desired air/fuel ratio, Figure 3-61. If the dwell is not varying and is between 10 and 50 degrees, the system is not in closed loop. If the dwell stays below 10 or above 50 degrees, some condition is causing the air/fuel mixture to run either too lean or too rich, and is beyond the capacity of the system to compensate. It is possible, more remotely, that the system electronics have malfunctioned and are unable to recognize the delivered air/fuel ratio.

Let's assume, for example, that we are getting a steady dwell of 54 degrees, which indicates a rich condition. To determine whether an electronic malfunction has occurred or whether the engine is actually running rich, run the engine for two minutes at a fast idle to make sure it is in closed loop. Next pull a large vacuum hose (just short of what would kill the engine), and observe the dwell. It should begin to go down. If it does not respond, there is an electronic problem; the system did not recognize the change in the air/fuel mixture.

NOTE: Some dwell meters, particularly swinging needle analog types originally designed to test dwell on contact point ignition, sometimes do not work well on CCC systems because they draw too much current and affect the signal to the mixture control solenoid. If a particular dwell meter causes any change in engine performance when it is connected, do not use it to test CCC mixture delivery.

You do not necessarily need a so-called dwell meter to test the mixture. Duty-cycle is the same measurement as dwell, but in terms of percent of on-time instead of dwell. On a duty-cycle reading, 100% corresponds to 60 degrees on a dwell meter set to six cylinders; 50% corresponds to 30 degrees, and so on. The additional advantage to a duty-cycle measurement is that the tool ordinarily has enough electrical impedance (low draw) that it cannot affect the delivery of the signal in any significant way.

Carburetor and M/C Solenoid Adjustment

Varying degrees of adjustment capacity exist on the four different carburetors used on CCC vehicles. The 6510-C Holley, used on the T-car, has only its idle mixture screw for an external mixture adjustment. Mixture adjustments are made at idle only using a dwell meter. The E2SE Varajet has an idle mixture screw, and inside the bowl is a lean mixture screw, Figure 3-62. The lean mixture screw is a main metering adjustment and is made at 3,000 rpm using a dwell meter. The idle mixture screw is adjusted at idle using a dwell meter. The E4ME Quadrajet and the E2ME **Dualjet** carburetors have much more adjustment capacity. Their adjustments are identical and include idle mixture screws, an idle air bleed valve, and adjustments on the mixture control solenoid itself.

TROUBLE CODE 21 Revised: 2-15-81

THROTTLE POSITION SENSOR CIRCUIT

Check for stuck* or misadjusted** TPS Plunger Repair as necessary.
If OK, proceed:

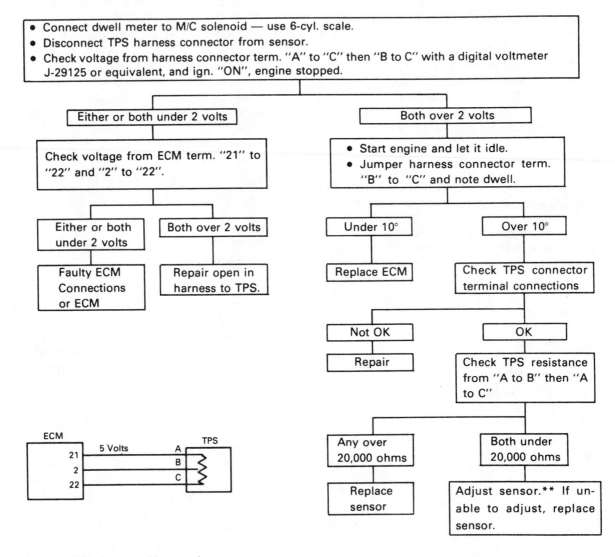

* Use a small blade screw driver on plunger.

** See divisional service manual.

Figure 3–59 Typical trouble code chart.*(Courtesy of General Motors Corporation, Service Technology Group)*

ELECTRIC DIVERTER VALVE CHECK (EDV)   Revised: 5-20-82

3.8L — V6 VIN CODE A NON-TURBO, CALIF.
2.8L V-6 "S" TRUCK, A.T., CALIFORNIA
5.0L VIN CODE H

Check for at least 10" of vacuum at valve with engine idling.

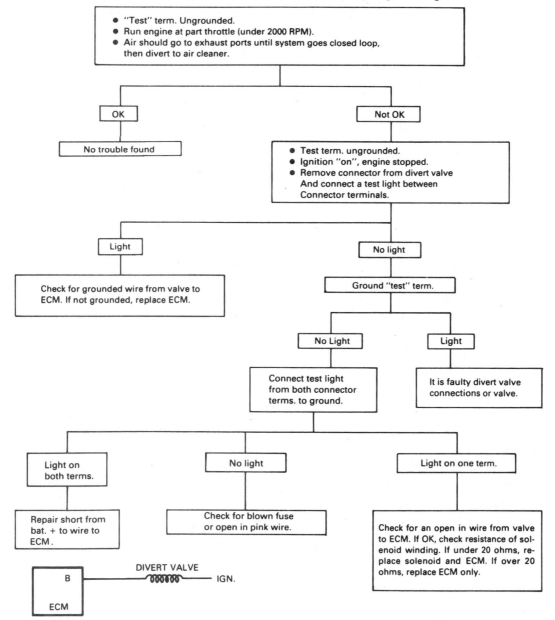

Figure 3–60 Typical diagnostic chart for nonmonitored functions. *(Courtesy of General Motors Corporation, Service Technology Group)*

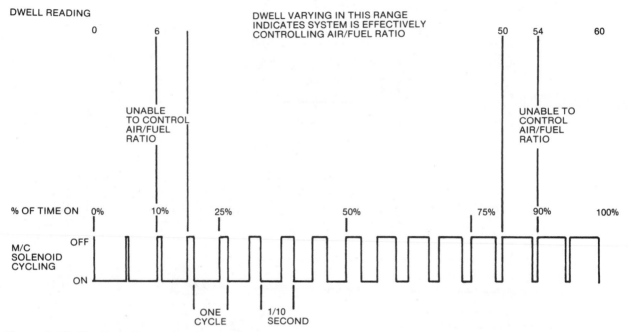

Figure 3–61 Dwell reading and system performance

WARNING: When performing mixture adjustments when the engine is running, either at idle or at higher speeds, be certain that all instrument leads are clear of the coolant fan and the accessory drive belts. Also be sure that you have no loose clothing such as shirt or coat sleeves that can be caught in the machinery, and that the wheels are securely chocked so vehicle movement is impossible.

Since the CCC system's introduction, several revisions have been made in adjustment specifications and procedures. One significant procedure revision was made in 1983 for the E4ME and E2ME carburetors, used from 1981 to 1984. The E4ME and E2ME carburetors included on 1985 and later models have internal changes that prevent the use of earlier adjustment procedures. Although such carburetor adjustments should be performed only when clearly indicated, many driveability problems can be eliminated by merely restoring proper float level and carefully making updated adjustments.

WARNING: Make no attempt to adjust the rich mixture screw on the E2SE carburetor shown in Figure 3-62. Removing and replacing the covering plug produces a potential fuel leak and fire hazard.

ECM and PROM Service

Take considerable care when replacing a PROM. A finger touching one of the PROM's pins can discharge static electricity into it and damage it, Figure 3-63. To insert the PROM into the ECM, position the PROM in the carrier. The carrier only fits in the PROM cavity of the ECM in one position. The PROM, however, fits in the carrier in either of two ways. If the PROM is installed in the

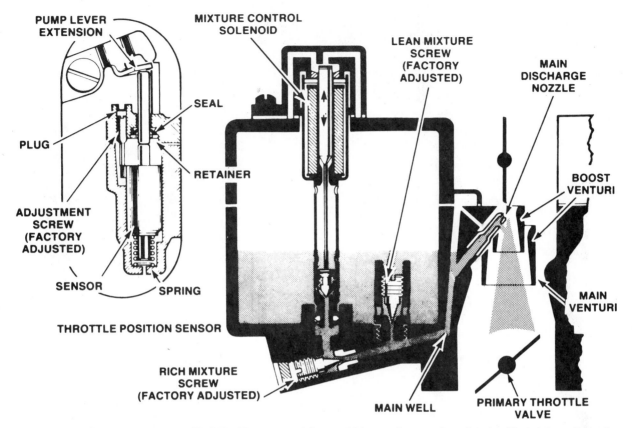

Figure 3–62 Rochester Varajet (E2SE). *(Courtesy of General Motors Corporation, Service Technology Group)*

Figure 3–63 PROM

ECM incorrectly, it will be damaged electrically. To avoid this, identify the small half-circle notch on one end of the PROM. Now look in the PROM cavity and identify a similar notch on one end of the PROM seat, Figure 3-64. To avoid bending the pins when installing the PROM, be sure the PROM is positioned in the carrier with the tips of the pins above the shoulder of the carrier, as shown in Figure 3-65, before placing the PROM and carrier in the PROM cavity. This allows the carrier to guide each pin into place as the PROM is pressed down.

The PROM and ECM are serviced separately. Remove the PROM before exchanging the ECM for a new one. Each PROM encodes a large amount of information for each specific

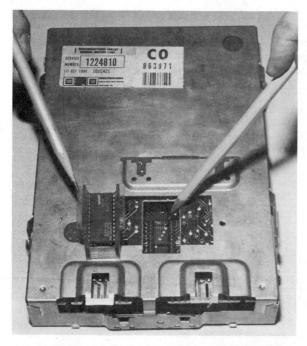

Figure 3–64 PROM, PROM carrier, and PROM socket

Figure 3–65 PROM pins/PROM carrier shoulder

vehicle, including its unique set of accessories. Occasionally, GM issues revised, updated PROMs for earlier vehicles. The PROM number, available through the scan tool, identifies the unit installed. Some driveability problems that do not yield to other diagnostic measures can be corrected with an updated PROM.

Weather-Pack Connectors

All CCC system harness connections under the hood are made with weather-pack connectors, Figure 3-66. These are designed to provide environmental protection for sensitive electrical connections so corrosion and contamination buildup is held to a minimum. Many of the circuits in the CCC system do not carry more than 1 or 2 **milliamps** at about 100 **millivolts** (0.1 volt). Any resistance caused by oxidation or contamination of an electrical connection can have a significant impact on the performance of the system. It is important that the integrity of the weather-pack connectors be maintained. Don't break the seal

1. OPEN SECONDARY LOCK HINGE ON CONNECTOR

2. REMOVE TERMINALS USING SPECIAL TOOL

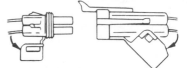

3. CUT WIRE IMMEDIATELY BEHIND CABLE SEAL
4. SLIP NEW CABLE SEAL ONTO WIRE (IN DIRECTION SHOWN) AND STRIP 5.00mm (.2") OF INSULATION FROM WIRE. POSITION CABLE SEAL AS SHOWN.

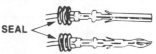

Figure 3–66 Weather-pack connector. *(Courtesy of General Motors Corporation, Service Technology Group)*

by inserting a probe into the connector. Voltage readings can be obtained by opening the connector and temporarily installing short jumper wires made for this purpose (many voltage checks require that the circuit be electrically connected during the test). If the weather-pack connector has been damaged, replace it as shown in Figure 3-66 or solder and tape the connection as shown in Figure 3-67.

Diagnostic & Service Tip

TPS. The TPS is the one sensor that can be adjusted. Sometimes driveability complaints can be cured by adjusting the TPS to exact specifications, which often vary with engine and model year application. Be sure to check the applicable service manual for correct procedures and specifications.

TWISTED/SHIELDED CABLE

1. **Remove outer jacket.**

2. **Unwrap aluminum/mylar tape. Do not remove mylar.**

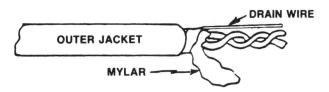

3. **Untwist conductors. Strip insulation as necessary.**

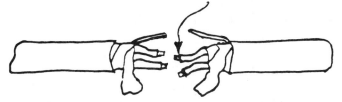

4. **Splice wires using splice clips and rosin core solder. Wrap each splice to insulate.**

5. **Wrap with mylar and drain (uninsulated) wire.**

6. **Tape over whole bundle to secure as before.**

TWISTED LEADS

1. **Locate damaged wire.**

2. **Remove insulation as required.**

3. **Splice two wires together using splice clips and rosin core solder.**

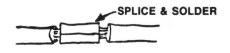

4. **Cover splice with tape to insulate from other wires.**

5. **Retwist as before and tape with electrical tape to hold in place.**

Figure 3–67 Wire repair. *(Courtesy of General Motors Corporation, Service Technology Group)*

Diagnostic & Service Tip

TCC Test Terminal. When using the TCC test terminal, remember the following points:

- Voltage is present when the TCC is not applied and is near zero when the TCC is applied, Figure 3-42, because applying the TCC causes the voltage to drop across the apply solenoid.
- The TCC can be applied by grounding the TCC test terminal.
- Applying voltage to the TCC test terminal damages the ECM if the ECM tries to apply the TCC while the voltage is applied. (On earlier models with five-terminal ALCLs, the TCC test terminal is in the fuse panel, and thus increases vulnerability to this mistake).

Surging Complaints. Surging complaints are occasionally heard concerning CCC vehicles at steady-state speeds around 35 miles per hour or higher. This is a fairly common problem, often caused by the EGR valve, referred to as EGR chuggle. The inert gases introduced in the intake

CCC System Behavior

Some of the behaviors of the CCC system can be amusing or frustrating, depending on whether or not they are understood. For instance, driving up a long hill in the mountains at near full throttle can set a code and turn on the check engine light. Changing altitude rapidly or driving backwards for about a quarter of a mile can have the same kinds of effects. There is nothing necessarily wrong: the system was just not programmed for those kinds of driving conditions.

manifold by the EGR valve cause an increased frequency of cylinder misfires. Although this has been occurring since the EGR valve's introduction in the early 1970s, it was not noticeable to the driver until a clutch appeared in the torque converter, making it a solid coupling. Combined with lower axle numerical ratios, this means the chuggle is now more perceptible. This complaint is sometimes misdiagnosed as a malfunction of the lockup torque converter. To isolate the problem, temporarily disconnect and plug the vacuum line from the EGR valve. Drive to see whether the surge still occurs. If it does, the problem is not EGR chuggle.

> **CAUTION:** Do not leave the EGR valve disconnected. To do so is not only emissions tampering and against the law, it can also result in engine damage from detonation.

Also refer to figure 3-68 for additional terms used on this system.

SUMMARY

We have seen how the torque converter lock-up clutch in an automatic transmission or transaxle works to eliminate the slip that would otherwise occur and improve fuel economy. We reviewed several methods used to control the air/fuel mixture with a mixture solenoid on a feed-back carburetor, fine-tuning the delivered mixture by adjusting the amount of air and fuel in the main metering circuits of the carburetor; and we have seen how a technician can diagnose this component using a dwell meter.

This chapter covered the method used by the CCC system to control the engine's idle speed under various temperature and load conditions. We also learned how the air injection system works, directing injected air to the exhaust manifold, the catalytic converter, or diverting it to the atmosphere when not needed.

Finally, we began considering the computer's

A/C	Air-Conditioning
AIR	Air Injection Reaction
ALCL	Assembly Line Communication Link
BARO	Barometric
ECM	Electronic Control Module
EFE	Early Fuel Evaporation
EGR	Exhaust Gas Recirculation
ESC	Electronic Spark Control
EST	Electronic Spark Timing
HEI	High-energy Ignition
ILC	Idle Load Compensator
ISC	Idle Speed Control
MAP	Manifold Absolute Pressure
M/C	Mixture Control
PCV	Positive Crankcase Ventilation
P/N	Park/Neutral
PROM	Programmable Read-only Memory (engine calibration unit)
PWM	Pulse Width Modulated
TCC	Torque Converter Clutch
TERM.	Terminal
TEST LEAD or TERMINAL	Lead or ALCL connector terminal grounded to obtain a trouble code
TPS	Throttle Position Sensor
VAC.	Vacuum
VIN	Vehicle Identification Number
VSS	Vehicle Speed Sensor
WOT	Wide-open Throttle

Figure 3–68 Explanation of abbreviations

ability to test itself and its circuits—its self-diagnostic capacities. We have seen the gradual development of these capacities from the early flashing LEDs to more informative types of code. This capacity will, as we will see in later chapters, grow in complexity as the computers gain control responsibilities.

▲ DIAGNOSTIC EXERCISE

A vehicle is brought in with surging and other driveability complaints. When put into diagnostic mode, the computer displays trouble codes pointing to the TPS circuit. Describe the steps a technician should follow between finding that trouble code and deciding to replace the TPS.

REVIEW QUESTIONS

1. What input values are used by the ECM to calculate fuel mixture in open loop? (See **Blended Enrichment**).
2. The Hall-effect switch generates what type of signal?
3. What influence does the P/N switch have over engine operation?
4. What two conditions must exist before the TCC is applied?
5. Name at least two ways in which closed-loop operation can be identified.
6. An oxygen sensor voltage of 0.3 volts causes the ECM to issue a _____ command.
7. What inputs does the ECM consider when

calculating the spark timing command?

8. With a dual-bed catalytic converter, where is the injected air directed under each of the following conditions?
   a. during warm-up.
   b. in closed loop.
   c. at wide-open throttle.

9. What results if the air switching valve fails to switch air away from the exhaust manifold when the system goes into closed loop?

10. How does the ECM know when to control idle speed?

11. How does the ECM control TCC application?

12. What can happen if the TCC fails to apply during fourth gear operation with a four-speed automatic?

13. List two likely results if the EGR valve does not open when it should?

14. List two types of EFE system.

15. During what operating mode is the canister purge solenoid de-energized, allowing purge of the stored vapors in the charcoal canister?

16. A dwell reading between 10 and 50 degrees and varying suggests what?

17. How is the CCC system put into diagnostics?

18. Why is a high-impedance digital voltmeter required for many CCC system tests?

19. Why must trouble codes be identified as hard codes before the trouble code charts can be used?

ASE-type Questions (Actual ASE test questions are rarely so product specific).

20. Technician A says that the CCC system responds to the oxygen sensor only when the system is in closed loop. Technician B says the CCC system uses only the oxygen sensor input for fuel control. Who is correct?
    a. A only.
    b. B only.
    c. both A and B.
    d. neither A nor B.

21. Technician A says that when the mixture control solenoid is down, the mixture is rich and the solenoid is turned off (no power is applied to it). Technician B says that when the mixture control solenoid is down, power is applied to it and the ECM is responding to a lean signal from the oxygen sensor. Who is correct?
    a. A only.
    b. B only.
    c. both A and B.
    d. neither A nor B.

22. A car with the CCC system is being tested for dwell. The dwell meter, connected to the mixture control solenoid circuit and set on the six-cylinder scale, indicates a dwell reading of 50 degrees.
    Technician A says the car is running rich. Technician B says the car is running lean. Who is correct?
    a. A only.
    b. B only.
    c. both A and B.
    d. neither A nor B.

23. A car with a TCC is driven with a test light connected to the TCC test terminal. As the car reaches cruising speed, the light goes out. This indicates that:
    a. either the TCC circuit has failed or the fuse has blown.
    b. the TCC has applied.
    c. the TCC has failed to apply.
    d. the TCC has released.

24. Technician A says while in diagnostics the CCC system does not set fault codes. Technician B says that in diagnostics, the CCC system displays fault codes only if the engine is not running. Who is correct?
    a. A only.
    b. B only.
    c. both A and B.
    d. neither A nor B.

Chapter 4

General Motors' Electronic Fuel Injection

OBJECTIVES

In this chapter you can learn:
- ❏ the six most important inputs required for engine calibration control.
- ❏ to understand and define pulse width.
- ❏ to explain how the computer knows engine temperature, engine load, engine speed and throttle position.
- ❏ at least two EFI system failures that will stop the engine.
- ❏ the circuit controlling fuel pump operation.
- ❏ the first two steps of EFI diagnostic procedure.
- ❏ several precautions to prevent damage to the computer system when the vehicle is worked on.

KEY TERMS

Block Learn
CALPAK
Pickup Coil
REF Pulse
TBI (Throttle-body injection)

General Motors' electronic fuel injection (EFI)—not to be confused with the digital fuel injection (DFI) system used by Cadillac beginning in the late 1970s and sometimes also called EFI—is a single-point fuel injection system. It is often called a **TBI (throttle body injection)** system. That name, however, more precisely described a major component of the system, Figure 4-1, the throttle body with its single or double fuel injector. This system was introduced by General Motors in 1982.

There were two versions of the EFI system when it first appeared in 1982: one used a single TBI unit for four-cylinder engines and a second, higher performance version (the Crossfire system) with two separate TBI units for eight-cylinder engines. This latter is not to be confused with TBI

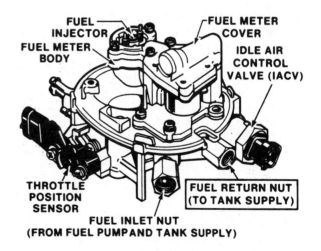

Figure 4–1 TBI unit. *(Courtesy of General Motors Corporation, Service Technology Group)*

systems using two fuel injectors in a single throttle body. The two injector TBI unit was introduced on some V6 engines beginning in 1985.

ELECTRONIC CONTROL MODULE

The ECM is usually above or near the glove compartment. On some models, notably the Pontiac Fiero, it is in the console. On some Corvettes it is in the battery compartment behind the driver's seat. Its inputs and the functions it controls (its outputs) are shown in Figure 4-2.

ECM for 2.5-Liter, EFI Operates Cruise Control

The ECM for selected engines such as the 2.5-liter four-cylinder engine was given more extensive and powerful internal circuitry to enable it to perform additional functions, such as control of the cruise control in 1988.

Throttle Body Backup (TBB)

The throttle body backup (TBB) is a fuel backup circuit within the ECM. It is primarily responsible for providing fuel pulses to the injector solenoid in the event of a general ECM failure, a failure that prevents it from running its program. The ECM used on the 2.0-liter engine uses a removable **CALPAK** (calculation packet) that provides fuel backup. The CALPAK plugs into the ECM in the same way the PROM does.

Block Learn

The EFI ECM has a learning ability called **block learn.** It accepts and corrects for gradually changing input values that result from sensor wear or driving conditions, such as changing altitude. It even learns to slightly modify output commands to complement a particular driver's driving habits. The aspect of this learning ability of most concern to technicians is covered under **System Diagnosis and Service** in this chapter.

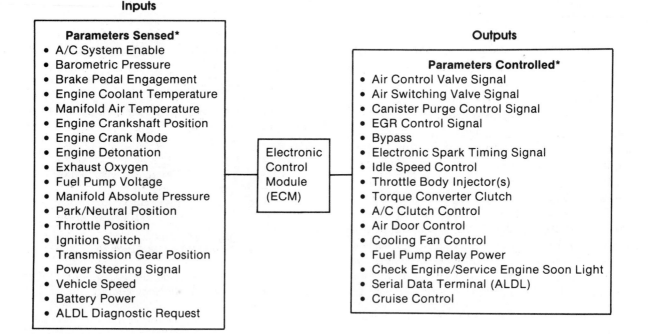

Inputs

Parameters Sensed*
- A/C System Enable
- Barometric Pressure
- Brake Pedal Engagement
- Engine Coolant Temperature
- Manifold Air Temperature
- Engine Crankshaft Position
- Engine Crank Mode
- Engine Detonation
- Exhaust Oxygen
- Fuel Pump Voltage
- Manifold Absolute Pressure
- Park/Neutral Position
- Throttle Position
- Ignition Switch
- Transmission Gear Position
- Power Steering Signal
- Vehicle Speed
- Battery Power
- ALDL Diagnostic Request

Electronic Control Module (ECM)

Outputs

Parameters Controlled*
- Air Control Valve Signal
- Air Switching Valve Signal
- Canister Purge Control Signal
- EGR Control Signal
- Bypass
- Electronic Spark Timing Signal
- Idle Speed Control
- Throttle Body Injector(s)
- Torque Converter Clutch
- A/C Clutch Control
- Air Door Control
- Cooling Fan Control
- Fuel Pump Relay Power
- Check Engine/Service Engine Soon Light
- Serial Data Terminal (ALDL)
- Cruise Control

*Some features not used on all engines.

Figure 4–2 EFI system overview

Keep-Alive Memory (KAM)

The ECM has a KAM like that discussed in earlier chapters. As long as it remains connected to a source of electrical power (the battery), it will retain the information it has accumulated in the previous driving intervals. The amount of power required to maintain this stored information is very small, much less than the battery drain caused by internal forces in the battery itself.

OPERATING MODES

The EFI system features the typical closed-loop and open-loop modes. During open loop, the ECM is programmed to provide an air/fuel ratio most suited to driveability as well as economy and emissions concerns. Of course, like all computerized engine control systems, emissions concerns are primary except in certain safety-related circumstances.

Synchronized Mode

In synchronized mode the injector is pulsed once for each reference pulse from the distributor. To describe it another way, the injector in synchronized mode sprays fuel once for each time a cylinder fires. On dual-TBI units, the injectors are pulsed alternately to avoid fuel pressure pulse problems. All closed-loop operation is in synchronized mode as is most open-loop operation.

Nonsynchronized Mode

In nonsynchronized mode, the injector pulses every 12.5 milliseconds, regardless of distributor reference pulses. On dual-TBI systems, each injector pulses every 12.5 milliseconds, but because the injectors pulse alternately, the engine actually "inhales" a fuel spray pulse every 6.25 milliseconds. Nonsynchronized pulses occur only in response to one or more of the following operating conditions:

- The injector is on or off (open or closed) and time becomes too small for accurate duty-cycle control (about 1.5 milliseconds), as occurs near full throttle or during deceleration. Even though the injector only opens a few thousandths of an inch, it is a mechanical device opened by a magnetic field and closed by a spring. It has a definite, finite maximum speed in contrast to its virtually instantaneous electric current flow. When the injector is required to open or close at a rate that approaches its mechanical response capacity limits, the ECM stops making the injector keep up with the reference pulse and employs the slower mode of operation.

- During *prime* pulses. On the 1982 Crossfire EFI system, once the ignition was turned on and if the coolant temperature was low enough, the ECM commanded the injectors to deliver a spray or two into the manifold to prime the engine for starting. This action was similar to pumping the throttle before starting a carbureted engine and squirting a small amount of fuel into the manifold by way of the accelerator pump. The Crossfire system was, however, discontinued in 1983.

Cranking Mode

The fuel injector(s) delivers an enriched air/fuel ratio particularly suited for starting conditions when the starter cranks the engine. These conditions include high manifold pressure and low airflow velocity, both of which make fuel vaporization more difficult. The actual pulse widths depend on coolant temperature. At -36 degrees Centigrade (-32 degrees Fahrenheit) the pulse width is calibrated for a maximum rich air/fuel, approximately 1.5 to 1. At 94 degrees Centigrade (201.5 degrees Fahrenheit) the pulse width is calibrated for a maximum lean air/fuel ratio, approximately 14.7 to 1. The higher the temperature, the shorter will be the pulse width or injector on-time, unless the engine is overheated. In the case of overheating, pulse width is widened somewhat to provide a slightly richer mixture, required for starting in that condition.

Clear Flood

If an engine floods, that is, if the spark plug electrodes are damp enough with fuel that the electric pulse grounds through the fuel rather than jumping the spark plug gap, thus misfiring, the ECM adopts the clear flood strategy. Depressing the throttle to 80% of wide-open throttle or more during crank indicates to the computer to issue an air/fuel ratio of about 20 to 1. It stays in this mixture mode until the engine starts or until the throttle is closed to below 80% WOT. If the engine is not flooded and the throttle is held at 80% WOT or more, it is unlikely the engine will start with the super-lean mixture.

Real World Problem

While the purpose of the clear flood mode is to allow the engine to start when flooded, a malfunctioning or completely misadjusted throttle position sensor can indicate an 80% or wider open throttle even when the driver has not depressed the pedal at all. This condition will, of course, prevent starting. The most common cause of this kind of problem occurs if the TPS loses ground and cannot reduce the reference signal back to the computer.

Run

As soon as the ECM sees a reference pulse from the distributor indicating 600 rpm or more, it puts the system into open loop. In open loop the ECM does not use information from the oxygen sensor to determine air/fuel mixture. It monitors the oxygen sensor signal to see whether the system is ready to go into closed loop, but its delivered mixture commands are determined from engine coolant, speed, and load information. The following criteria must be satisfied before the engine can go into closed loop:

- The oxygen sensor must produce voltage signals crossing 0.45 volt (450 millivolts). Before it can do this, the oxygen sensor must have reached a temperature of at least 300 degrees Centigrade (570 degrees Fahrenheit).
- The engine coolant temperature must reach a specified temperature, about 65.5 degrees Centigrade (150 degrees Fahrenheit). This temperature varies depending on model and build year.
- A specified amount of time must elapse since the engine was started, regardless of component temperatures. This factor also varies by model.

The values for each of these criteria are encoded in the PROM. When all of the criteria are met, the ECM puts the system into closed loop. At this point, the ECM uses the input from the oxygen sensor to calculate fine adjustments to the air/fuel mixture commands. Of course, if one of the other sensors indicates closed loop criteria have failed, the ECM will return to open loop calculations.

Semi-Closed Loop

To improve fuel economy under certain cruise conditions, the ECM for some throttle body injected engines is programmed to go out of closed loop during some sustained cruise conditions at highway speeds. During these periods, the air/fuel mixture may go as lean as 16.5 to 1. The ECM monitors the following information parameters and will put the system into this fuel control mode only when they are within predetermined values (varying somewhat by model and year):

- engine temperature.
- spark timing.
- canister purge.
- constant (sustained) average vehicle speed.

The ECM will periodically switch back to closed loop to check on engine operating parameters. If the parameters are still within the

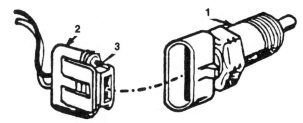

1 **Coolant temperature sensor**
2 **Harness connector to ECM**
3 **Locking tab**

Figure 4–3 Engine coolant sensor. *(Courtesy of General Motors Corporation, Service Technology Group)*

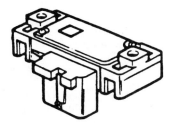

Figure 4–4 EFI MAP sensor. *(Courtesy of General Motors Corporation, Service Technology Group)*

required limits, it will go back to the lean calibration, open-loop mode. Operation in the lean calibration, open-loop mode, and switching back and forth between closed loop and open loop are smooth and should not be detectable to the driver.

INPUTS

Engine Coolant Temperature (ECT) Sensor

This is the same sensor used in the computer command control system and on the other General Motors systems. On later models it has a slightly different appearance, Figure 4-3.

Manifold Absolute Pressure (MAP) Sensor

The MAP sensor on EFI applications, Figure 4-4, is similar to that used on CCC applications, but the ECM is programmed to use the sensor as a barometric pressure sensor also. When the ignition is first turned to run, but before it is turned to start, the ECM takes a reading from the MAP sensor. Since the engine is not running, the manifold pressure is identical with atmospheric pressure. The ECM records this reading as the base barometric pressure and uses it for calculations of fuel mixture while in open loop and for calculating spark timing. This BARO reading is retained and used until the ignition is turned off or until the throttle is opened to WOT. At that time the engine

vacuum goes to nearly atmospheric pressure, and an updated BARO reading is taken.

Oxygen Sensor

This sensor is the same single-wire used on other earlier General Motors systems. Once it reaches operating temperature, the oxygen sensor provides a low-voltage signal inversely varying with the amount of residual oxygen remaining in the exhaust system after combustion and before the catalytic converter.

Throttle Position Sensor (TPS)

The throttle position sensor (TPS) is a variable resistor (potentiometer) mounted on the TBI unit, connected to the end of the throttle shaft, Figure 4-5. On some EFI applications the TPS is

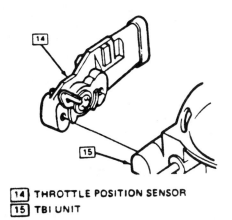

14 **THROTTLE POSITION SENSOR**
15 **TBI UNIT**

Figure 4–5 Throttle position sensor. *(Courtesy of General Motors Corporation, Service Technology Group)*

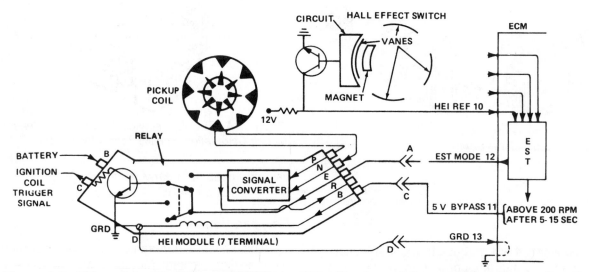

Figure 4–6 HEI module with EST and Hall-effect switch. *(Courtesy of General Motors Corporation, Service Technology Group)*

adjustable. On others there is no way to adjust the TPS.

Distributor Reference Pulse (REF)

The REF, sometimes called the **REF pulse**, tells the ECM what the engine speed is (when factored against the ECM's internal clock) and what the crankshaft position is. On most EFI applications, the distributor **pickup coil** provides the REF signal, as described in earlier chapters. It is important to note that if the REF signal fails to arrive at the ECM while the engine is running, fuel injection shuts off.

The earlier 2.5-liter EFI engines used a Hall-effect switch to supply the REF signal, Figure 4-6. The pickup coil at the distributor was retained for starting, but that was its only function.

Denotation (Knock) Sensor

The knock sensor used on the EFI system is similar to that used on other GM systems. It is a crystal piezoelectric sensor that generates a signal when excited by vibrations characteristic of knock frequencies, Figure 4-7.

Vehicle Speed Sensor (VSS)

Two kinds of vehicle speed sensors have been used on EFI vehicles, similar to those described in the previous chapter, similar in fact to those used by other manufacturers as well. On EFI vehicles the VSS input is also used by the ECM to help identify a deceleration condition. This is discussed further under **Injector Assembly** in the **Outputs** section of this chapter.

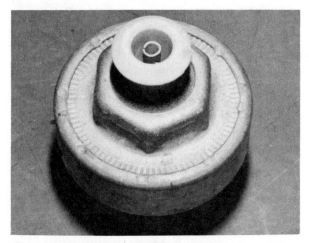

Figure 4–7 Knock sensor

Ignition Switch

The ignition switch is one of the two power sources for the ECM. When the ignition is first turned on, the ECM initializes (starts its program bootup sequence) and is quickly ready to function. The ignition switch also powers most of the actuators the ECM controls.

Park/Neutral (P/N) Switch

The park/neutral switch, usually a part of the shifter assembly, closes when the vehicle is shifted into park or neutral. This provides the ECM with information about whether or not the vehicle is in an active gear. The ECM uses this information to control engine idle speed. If the P/N switch is disconnected or significantly out of adjustment (enough to signal one gear when it is in another), idle quality can suffer while in park or neutral. An out of adjustment P/N switch can also prevent engagement of the starter, as has been the case for many years before computer controls.

Transmission Switches

Most transmissions used in CCC and EFI vehicles employ one or more hydraulically operated electric switches screwed into the valve body. These switches send the ECM signals indicating what gear the transmission is in. This information allows the ECM to properly control engagement and disengagement of the torque converter lockup clutch.

Air Conditioning (A/C) Switch

On EFI cars with air conditioning, the A/C switch includes a wire to the ECM. When the A/C is turned on or off, this wire conveys this information to the ECM, which uses the signal in its determination of the proper idle speed.

Ignition Crank Position

The circuit that energizes the starter solenoid also signals the ECM through a wire. This informs the ECM that the engine is cranking,

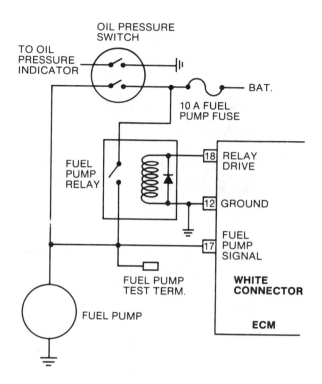

Figure 4–8 Fuel pump control circuit

which the ECM uses in determining fuel injection programs for start-up.

Fuel Pump Voltage Signal

Some EFI systems have a wire from the power side of the fuel pump (positive) to the ECM. This wire indicates to the ECM that the fuel pump is on, Figure 4-8. Not all vehicles include this feature.

OUTPUTS

Throttle Body Injection (TBI) Unit

The throttle body injection (TBI) unit is made of three castings: the throttle body, the fuel meter body, and the fuel meter cover, Figure 4-9. The throttle body contains the throttle bore and valve.

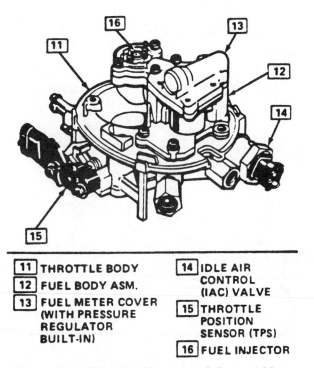

11 THROTTLE BODY	**14** IDLE AIR CONTROL (IAC) VALVE
12 FUEL BODY ASM.	
13 FUEL METER COVER (WITH PRESSURE REGULATOR BUILT-IN)	**15** THROTTLE POSITION SENSOR (TPS)
	16 FUEL INJECTOR

Figure 4–9 TBI unit. *(Courtesy of General Motors Corporation, Service Technology Group)*

It provides the vacuum ports for EGR, canister purge and so on, similar to what is found on the base of a carburetor. Mounted to it are the idle air control motor and the TPS. The fuel meter body contains the fuel injector and the fuel pressure regulator, Figure 4-10.

Injector Assembly

The injector is a solenoid-operated assembly. When energized by the ECM (pulsed by grounding), a spring-loaded metering valve lifts off its seat a few thousandths of an inch. Fuel under pressure passes through a fine screen filter fitting around the tip of the injector and sprays in a conical pattern toward the walls of the throttle bore, just above the throttle blade. This angle and position was chosen to maximize the exposure of fuel droplets to onrushing air, thus optimizing vaporization. The angle of injection sprays the fuel droplets across the incoming air as well as into

The TBI

The fascinating thing about the EFI system is that the TBI unit performs every function a carburetor does, but with greatly increased control. When the TPS moves rapidly, the injector momentarily sprays more fuel (increased pulse width), just as with an accelerator pump. On cold starts the injector sprays additional fuel to duplicate the function of a choke. With the idle speed controller, the throttle body achieves fast idle by simply opening the idle control motor. The TBI requires no mechanical choke, fast idle cam, accelerator pump, and piston and so on. Everything is done electronically. Even better than with a carburetor, on deceleration the TBI unit can lean the fuel mixture maximally, even shutting fuel off altogether in some cases. This can compensate for any fuel that evaporates from manifold walls as a result of the momentarily higher vacuum under deceleration conditions. Hard deceleration—a closed throttle, a sharp drop in manifold pressure and a decrease in vehicle speed—determines the shutoff of fuel.

The major difference is in the control circuits: the carburetor functions as a pneumatic/mechanical/hydraulic computer to meter the air/fuel mixture. The TBI unit, on the other hand, is a dumb actuator, responding entirely to the computer's control commands. It does no controlling itself.

the throttle body, maximizing vaporization by maximizing exposure to passing air.

Fuel Pressure Regulator

Inside the TBI assembly is a fuel pressure regulator, Figure 4-10. The purpose of this regulator is to maintain the fuel pressure within a range that the ECM's pulse-width injection commands will translate into the proper stoichiometric ratio in the combustion chambers. The spring

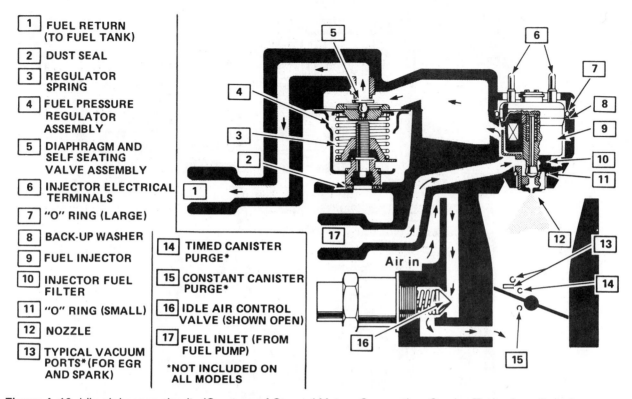

1	FUEL RETURN (TO FUEL TANK)
2	DUST SEAL
3	REGULATOR SPRING
4	FUEL PRESSURE REGULATOR ASSEMBLY
5	DIAPHRAGM AND SELF SEATING VALVE ASSEMBLY
6	INJECTOR ELECTRICAL TERMINALS
7	"O" RING (LARGE)
8	BACK-UP WASHER
9	FUEL INJECTOR
10	INJECTOR FUEL FILTER
11	"O" RING (SMALL)
12	NOZZLE
13	TYPICAL VACUUM PORTS*(FOR EGR AND SPARK)
14	TIMED CANISTER PURGE*
15	CONSTANT CANISTER PURGE*
16	IDLE AIR CONTROL VALVE (SHOWN OPEN)
17	FUEL INLET (FROM FUEL PUMP)

*NOT INCLUDED ON ALL MODELS

Figure 4–10 Idle air bypass circuit. *(Courtesy of General Motors Corporation, Service Technology Group)*

side of the diaphragm is exposed to atmospheric pressure, so fuel pressure varies slightly with changes in atmospheric pressure. At lower atmospheric pressures, the fuel pressure is reduced, causing slightly less fuel to be injected for a given pulse width. Depending on the vehicle, model year and so on, fuel pressure should range from 9 to 13 psi.

On Crossfire systems, the rear TBI unit has the pressure regulator, while the front TBI uses a pressure compensator, Figure 4-11. It works like the regulator except that its function is to make up for the temporary drop in fuel pressure between front and rear units and to keep the pressure identical at both. Such drops are usually caused by the diaphragm and valve in the pressure regulator moving to a more closed position, thus allowing less fuel to the front unit.

Idle Air Control (IAC)

The IAC assembly controls idle speed by controlling the amount of air bypassing the throttle. It consists of a small, reversible electric stepper motor and a pintle valve, Figure 4-12. As the motor's armature turns, the pintle valve extends or retracts depending on the direction the motor is turning. During idle, the throttle blade is in a fixed, nearly closed position, allowing a constant amount of idle air into the intake manifold. The pintle valve allows additional air through the bypass passage. The ECM uses the stepper motor to position the pintle valve for desired idle speed. During warm engine operation, the ECM attempts to maintain a fixed idle speed by adjusting the IAC valve position for load variations (transmission in or out of gear, air conditioner on

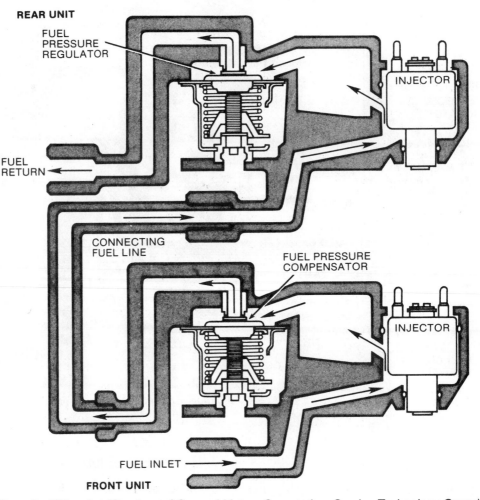

Figure 4–11 Crossfire TBI units. *(Courtesy of General Motors Corporation, Service Technology Group)*

or off, and so on).

If during idle or at low vehicle speed (below 10 mph) engine speed drops below a specified rpm, the ECM puts the IAC into antistall mode. The IAC motor retracts the pintle to allow additional air into the intake manifold to raise engine speed. It momentarily raises engine speed above base idle.

The stepper motor is unique: its armature has two separate windings. The direction in which the armature turns depends on which winding is powered. Power is applied by the ECM

in short pulses. Each pulse rotates the armature about 35 degrees and extends or retracts the pintle valve a corresponding amount. The ECM applies as many pulses as necessary to set the valve where it will achieve the desired idle rpm.

The ECM counts the pulses applied and thus knows where the pintle valve is at all times. There are 12 pulses per revolution and 255 total positions. Fully extended (closed bypass passage) is the reference position: zero. Fully retracted (wide-open bypass passage) is step 255.

To keep an accurate track of the IAC assem-

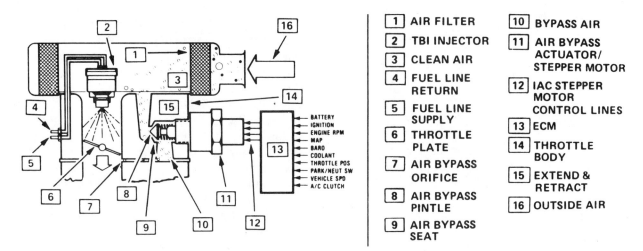

1 AIR FILTER		10 BYPASS AIR	
2 TBI INJECTOR		11 AIR BYPASS ACTUATOR/ STEPPER MOTOR	
3 CLEAN AIR		12 IAC STEPPER MOTOR CONTROL LINES	
4 FUEL LINE RETURN		13 ECM	
5 FUEL LINE SUPPLY		14 THROTTLE BODY	
6 THROTTLE PLATE		15 EXTEND & RETRACT	
7 AIR BYPASS ORIFICE		16 OUTSIDE AIR	
8 AIR BYPASS PINTLE			
9 AIR BYPASS SEAT			

Figure 4–12 IAC motor assembly in TBI unit. *(Courtesy of General Motors Corporation, Service Technology Group)*

bly's position, the ECM references itself frequently. When the engine comes off idle and the car starts to move, the ECM begins looking for a reading from the VSS representing 30 mph. The first time it sees a 30-mph signal, it moves the pintle to the park (fully closed) position to reorient the zero reference. It then moves the pintle to a preprogrammed distance from closed. If for any reason the VSS is disabled, the ECM will not be able to reestablish the correct idle speed.

Diagnostic & Service Tips.

It is not uncommon to find a car in a shop that won't idle properly, but it would not be a natural check to see whether the vehicle speed sensor is working properly if the technician did not know the connection between the two elements of the computer control system. An experienced technician will make sure to check that the speedometer is working on any vehicle that does not idle properly. Note that failure to idle properly can mean idling not at all, too slowly, or too fast.

Fuel Pump Control

Fuel and fuel pressure are supplied by an electric pump inside the fuel tank. The fuel pump assembly contains a check valve preventing the fuel from bleeding back from the fuel line into the tank. If the check valve leaks, the engine will crank longer before starting while the fuel line is purged of vapor.

The fuel pump turns on and off through a fuel pump relay, usually mounted in the engine compartment. The fuel pump relay is controlled by the ECM. Be aware that the fuel pump relay is identical in appearance and usually mounted beside the air conditioning relay, also controlled by the ECM, Figure 4-13. On some models the cooling fan relay is also mounted in the same location and looks just like the others.

When the ignition is turned on, the ECM activates the fuel pump relay, Figure 4-14. If the ECM does not get an ignition REF signal within two seconds, it turns off the fuel pump through the relay and does not turn it back on until it does receive the REF signal. Once the engine is running, the ECM deactivates the relay when the ignition is turned off or anytime the REF signal disappears from the ECM. Note also in

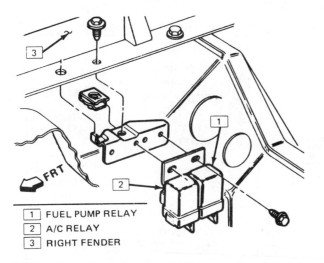

1 FUEL PUMP RELAY
2 A/C RELAY
3 RIGHT FENDER

Figure 4–13 Fuel pump and A/C relay. *(Courtesy of General Motors Corporation, Service Technology Group)*

Figure 4-14 that the oil pressure switch, which operates the oil pressure indicator on the instrument panel, is electrically in parallel with the fuel pump relay. This is done to provide a backup to the fuel pump relay. The engine starts after a fuel pump relay failure if cranked long enough to build about 4 psi oil pressure.

Fuel Pump Test Terminal. A wire connects to the circuit powering the fuel pump. The open end of this wire has a terminal to which a jumper lead can be connected to power the fuel pump. A voltmeter connected to the test terminal quickly shows if either the fuel pump relay or oil pressure switch is supplying power to the fuel pump circuit. This terminal is located in the engine compartment or in the ALDL under the instrument panel, depending on the model and year.

Voltage Correction. During vehicle operation, system voltage can vary considerably as a

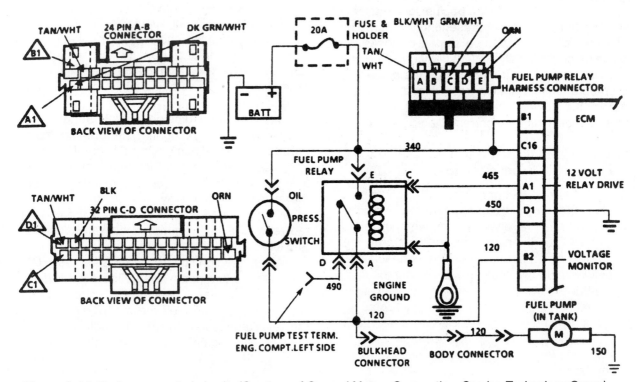

Figure 4–14 Fuel pump control circuit. *(Courtesy of General Motors Corporation, Service Technology Group)*

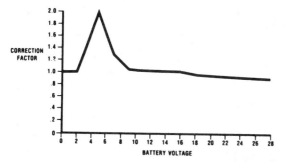

Figure 4–15 System voltage correction graph. *(Courtesy of General Motors Corporation, Service Technology Group)*

result of the various electrical accessories some-times used and sometimes off, and the state of the battery's charge. Variations in the system's voltage can cause differences in the amount of fuel delivered at the injectors because of differences in the amount of time the injectors are open (a higher voltage opens the injector more quickly and vice-versa). This affects the amount of fuel injected during a pulse width. Because of this problem, the ECM monitors system voltage and will multiply the pulse width by a voltage correction factor, Figure 4-15, programmed into its permanent memory. As system voltage goes down, the pulse width increases. As system voltage goes up, the pulse width decreases. If voltage goes down to a criterion value, the ECM increases dwell to maintain good ignition performance by keeping the spark hot and by increasing the idle speed.

Electronic Spark Timing (EST)

EST maintains the optimum spark timing under all conditions of engine load, speed, temperature, and barometric pressure. Two basic functions are incorporated in the system. Dwell control is provided to allow sufficient energy to build the magnetic field in the ignition coil for proper ignition voltage secondary output. Spark is provided at just the right time to start combustion at the moment of peak pressure in the compression stroke. The EST function of the ECM eliminates the vacuum and centrifugal advance mechanisms in the classical distributor. The HEI module accepts spark timing commands from the ECM once the engine is started.

During cranking, Figure 4-16, the solid-state switching circuit connects the main switching transistor of the HEI module to the pickup coil.

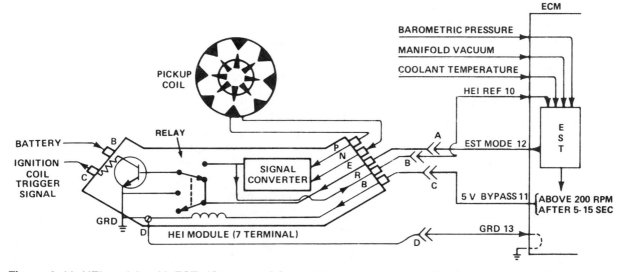

Figure 4–16 HEI module with EST. *(Courtesy of General Motors Corporation, Service Technology Group)*

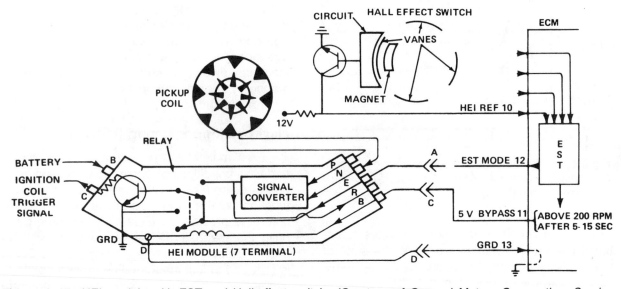

Figure 4–17 HEI module with EST and Hall-effect switch. *(Courtesy of General Motors Corporation, Service Technology Group)*

The pickup coil signal turns on and off the main switching transistor. This state is called module mode and provides no spark advance; the timing is fixed at the reference timing point. On applications that do not use a Hall-effect switch, this same signal, after it is converted from an analog signal to a digital signal by a signal converter circuit in the module, is sent to the ECM as the REF signal. On applications that do use the Hall-effect switch, it supplies the REF signal. Note in Figure 4-17 that terminal R is not used.

When the ECM sees a REF pulse signal corresponding to an engine rpm of about 600 or more, it decides the engine is running. It then sends a 5-volt signal through the bypass wire to the solid-state switching circuit. The switching circuit in turn disconnects the pickup coil from the base of the main switching transistor. The EST wire is simultaneously disconnected from ground and connected to the main switching transistor's base. The system is now in EST mode, and timing is controlled by the ECM. The ECM considers barometric pressure, manifold pressure, coolant temperature, engine speed, and crankshaft posi-

tion. It then sends the HEI module the optimum spark timing command it has calculated.

Electronic Spark Control (ESC)

The Crossfire EFI system is equipped with ESC. Its function is to retard ignition timing when detonation occurs. The system is effectively the same as the later version of ESC described in the **Outputs** section of chapter 3.

Torque Converter Clutch (TCC)

The purpose of the torque converter lockup clutch is to improve fuel mileage. It does this by eliminating the hydraulic slippage and heat production in the torque converter once the vehicle has attained a steady cruise speed. Its operation is essentially the same as the one described in the **Outputs** section of chapter 3.

Manual Transmissions. On manual transmissions, the TCC output terminal on the ECM can be used to operate the upshift light on the instrument panel. The ECM looks at engine speed, engine load, throttle position, and vehicle

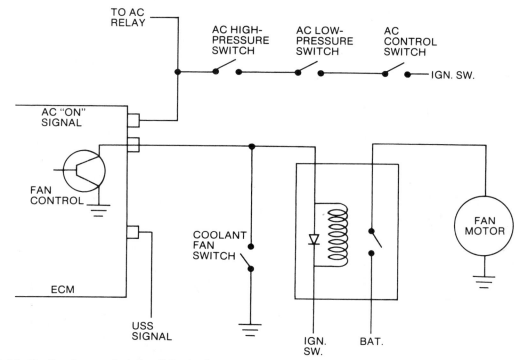

Figure 4–18 Cooling fan control circuit (typical)

speed. Then it alerts the driver when to shift into the next gear for optimum fuel economy. Note that this shift point is strictly for fuel economy, not for maximum engine life. Most experienced mechanics as well as experienced drivers will let the engine turn somewhat faster than the upshift light would indicate.

Cooling Fan Control

Transverse engines almost universally use an electrically powered radiator cooling fan to pull air through the radiator and across the fins when they are moving at speeds too low for ram air to provide sufficient cooling. On most EFI systems, the ECM has some control over the cooling fan operation. The cooling fan can be turned on either by the ECM itself or by a coolant temperature switch. On some applications, the ECM has full control. The fan is turned on and off by a relay operated by the ECM, Figure 4-18. On most vehicles the fan turns on when the coolant tempera-

ture is high, when the vehicle is traveling below 30 mph with the air conditioner on and in some cases when the air conditioning high side pressure reaches a certain criterion point.

Hood Louver Control

The Crossfire EFI system includes an electrically operated air door the ECM can open using a relay and solenoid, Figure 4-19. When the hood louver door opens, it allows fresh air from above the engine hood through a hole in the top of the air cleaner housing to the air cleaner elements. The ECM opens the hood louver door above a specified coolant temperature or under WOT conditions.

Air Conditioning (A/C) Relay

The A/C relay is what actually turns the compressor clutch on and off, Figure 4-20. When the driver turns on the A/C switch, a signal goes to the ECM and power is made available at terminal

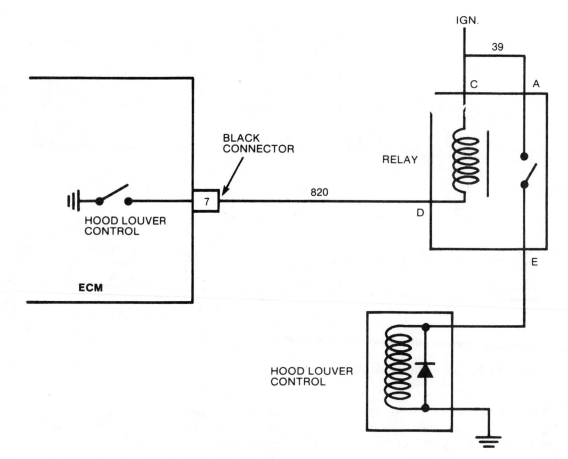

Figure 4–19 Hood louver control circuit

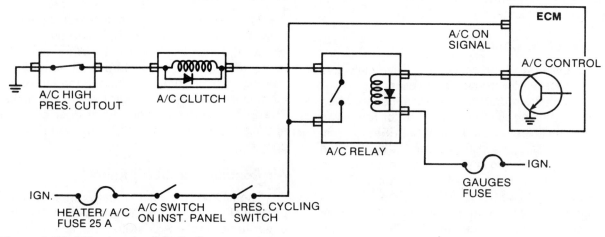

Figure 4–20 A/C relay circuit (typical)

B of the relay. The ECM waits a half-second and then grounds terminal C of the relay. This energizes the relay and engages the compressor clutch (if the engine is at idle, the half-second allows the ECM time to adjust the IAC position and the engine idle speed before the compressor clutch engages). The ECM deactivates the relay at WOT, during heavy engine load, or when the IAC is in power steering antistall mode. As seen in Figure 4-20, the compressor clutch circuit can be opened also by the high pressure switch or the pressure cycling switch, both of which are standard features of the air conditioning system.

Related Emission Controls

On all but the Crossfire EFI system, EGR, canister purge, and air management control are not controlled by the ECM. The EGR, canister purge, and air management controls on the

Crossfire system are essentially the same as those on V8 engines with computer command control (CCC) discussed in chapter 3.

Catalytic Converter. Most EFI engine systems use single-bed, three-way catalytic converters. No supplemental air is introduced into the converter, so no air injection is used. In some cases a Pulsair system is used and works independently of the computer.

EGR. Many four-cylinder EFI systems use a self-modulating EGR valve with no ECM control (most vehicles have a back-pressure EGR, Figure 4-21).

Under the main diaphragm found in a conventional EGR valve is a second diaphragm. A small spring pushes downward on the second diaphragm. A passage in the EGR valve stem allows exhaust manifold pressure to enter into a chamber under the second diaphragm. When exhaust pressure is low, the small spring can

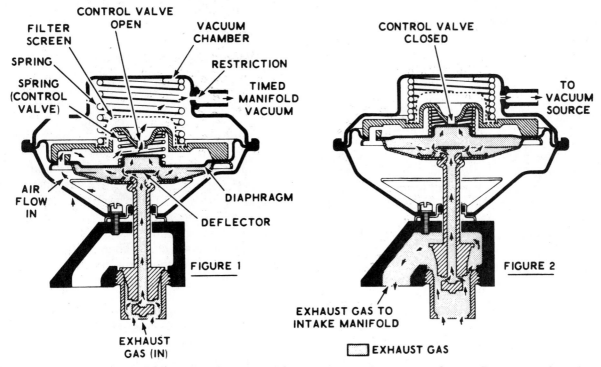

Figure 4–21 Backpressure EGR valve. *(Courtesy of General Motors Corporation, Service Technology Group)*

hold the second diaphragm down in spite of the exhaust pressure pushing it up. Atmospheric pressure flows through a vent in the lower part of the EGR valve housing and through a port in the second diaphragm, into the space between the two diaphragms. If the second diaphragm is down, the atmospheric pressure can flow through another port in the main diaphragm into the vacuum chamber. With atmospheric pressure finding its way into the vacuum chamber, not enough vacuum is developed to lift the main diaphragm and open the EGR valve. As the throttle opening increases, exhaust manifold pressure increases. When exhaust pressure is high enough, the second diaphragm is forced up and closes off the port through the main diaphragm into the vacuum chamber. The vacuum chamber is no longer vented to atmospheric pressure, and sufficient vacuum can now develop to open the EGR valve.

Canister Purge.   On four-cylinder EFI engines, canister purge control is achieved with a thermal vacuum switch blocking purge vacuum until the engine is in closed loop. In closed loop, any additional fuel from the canister can be compensated for by varying the injection quantity, while still maintaining the stoichiometric ratio of 14.7 to 1 air to fuel.

Service Engine Soon Light

All EFI systems use a service engine soon light. On early 1980s vehicles, this was called the check engine light. If the ECM discovers a fault in one of the circuits it monitors for malfunctions, it turns on the service engine soon light on the instrument panel. This warns the driver that a malfunction exists. By grounding the test terminal in the assembly line communication link (ALCL or ALDL), located under the dash, the ECM flashes the service engine soon light. The sequence of these flashes indicates one or more trouble codes identifying the circuit or circuits in which the fault was detected. This is explained more fully under **System Diagnosis and Service** in this chapter.

✔ SYSTEM DIAGNOSIS & SERVICE

Self-diagnosis

The self-diagnostic capacity and procedures for the EFI system are essentially the same as for the CCC system described in the **System Diagnosis and Service** section of chapter 3. An abbreviated discussion pointing out just the variations from the CCC system is what is presented here.

NOTE: Diagnostic procedures for 1987 and later are written to include the use of an ALCL tool, often referred to as a scanner or scan tool. Previously, diagnostic procedures were written around the use of a voltmeter to obtain such measurements as sensor readings. The technician should keep in mind that the diagnostic information collected by either method can show the same things about the engine management system. The advantages of the scanner are basically convenience and speed.

Approach to Diagnosis

When any of the diagnostic procedures are used, it is important to follow the directions carefully. Failing to do so or taking shortcuts usually results in inaccurate conclusions and leads to unneeded and ineffective repairs.

Diagnostic Circuit Check.   When diagnosing a problem on an EFI system, after verifying that all non-EFI engine support systems are working properly, start with a diagnostic circuit check, Figure 4-22. The diagnostic circuit check either verifies that the system is working properly or refers to another chart for further diagnosis. Be sure the diagnostic circuit check and trouble code charts come from the same model year manual as the vehicle you are working on. While some techniques and codes are retained over several years and for various vehicles, others are changed with some frequency. On earlier systems, the diagnostic circuit check is designed to sort out whether or not the fault currently exists or

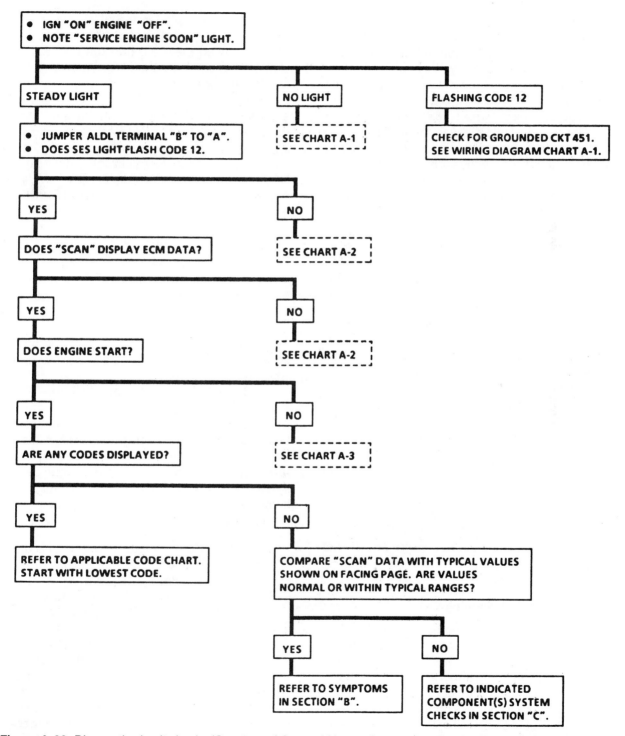

Figure 4–22 Diagnostic circuit check. *(Courtesy of General Motors Corporation, Service Technology Group)*

if it is intermittent, present once but not now. On those systems the trouble code charts are written assuming the fault exists currently. If one of them is used to pursue an intermittent code, the chart most often leads to an invalid diagnosis because it is beginning with an invalid assumption—that a fault currently exists when in fact none does.

For later systems the diagnostic circuit check and many of the trouble code charts are written to allow the trouble code charts to determine whether the code obtained from the diagnostic memory identifies a current fault (a hard code) or an intermittent code.

Field Service Mode. Grounding the test terminal with the engine running puts the system into field service mode. In this mode the ECM uses the check engine light to show whether the system is in open or closed loop and whether the system is running rich or lean, Figure 4-23.

The field service mode was incorporated into the diagnostic circuit check in 1984 and was discontinued when the diagnostic procedures were written around the use of a scanner. The scan tool provides the information the field service mode was designed to provide.

Driver's Complaint Chart. The driver's complaint chart, Figure 4-24, guides the task of finding the cause of problems not identified by trouble codes. It either suggests things to check for certain problems or refers to other, more specific charts. For later models this chart has been expanded into a complete section in the service manual and is referred to as "Symptoms." There are other charts in the later service manuals that assist in diagnosing specific EFI component or system malfunctions.

ECM and PROM Service

Considerable care should be taken when working on a PROM or CALPAK (used only on 2.0-liter engine ECMs). A finger touching one of the pins can discharge enough static electricity into it to damage its memory data, Figure 4-25. The PROM carrier fits into the PROM cavity of the ECM in only one position. The PROM, however, fits into the carrier in either of two ways. If the PROM is installed in the ECM reversed end to end from its correct position, it will be damaged electrically and may also damage the computer. Consult the service manual for detailed installation instructions.

The PROM, CALPAK, and ECM are serviced separately. Therefore, the technician should remove the PROM and CALPAK before exchanging the ECM for a new one. The same ECMs are used on a variety of vehicles; PROMs and CALPAKs are specific to that vehicle model, year, and accessory package. If the CALPAK or PROM is missing or damaged, the engine may not start.

Weather-Pack Connectors

All EFI system harness connections under the hood in the engine compartment are made through weather-pack connectors, as discussed in the **System Diagnosis and Service** section of chapter 3.

Vacuum Leaks. A vacuum leak on an EFI system usually results in increased idle speed while in closed loop.

TBI. The TBI unit has two O-rings, Figure 4-26, between the injector solenoid and the fuel meter body. The O-rings prevent pressurized fuel from leaking past the injector solenoid. A leak at either of these O-rings can result in dieseling and/or an overly rich mixture. A suspected leak can be easily checked. With the engine off and a jumper lead supplying battery voltage to the fuel pump test terminal, inspect the injector for signs of leaks.

WARNING: Fuel Pressure Test. Some driveability complaints require a fuel pump pressure test. Before the fuel system is opened, pressure in the line should be relieved. The system is designed to retain operating pressure in the line after the engine is shut off. Opening the line under pressure causes a considerable spray of fuel, resulting in a fire and personal safety hazard. To relieve the pressure, remove the fuel pump fuse and crank the engine for several seconds. See the service manual for further test procedures.

1983
FUEL INJECTION
FIELD SERVICE MODE CHECK

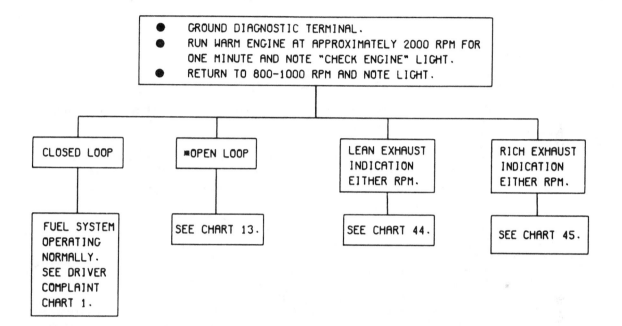

FIELD SERVICE MODE: ENGINE RUNNING, DIAGNOSTIC TERMINAL GROUNDED. "CHECK ENGINE" LIGHT IS "OFF" WHEN EXHAUST IS LEAN AND "ON" WHEN IT IS RICH.

OPEN LOOP: "CHECK ENGINE" LIGHT FLASHES AT A RATE OF 2 TIMES PER SECOND.

CLOSED LOOP: "CHECK ENGINE" LIGHT FLASHES AT A RATE OF 1 TIME PER SECOND.

LEAN EXHAUST INDICATION: "CHECK ENGINE" LIGHT IS OUT ALL THE TIME OR MOST OF THE TIME.

RICH EXHAUST INDICTION: "CHECK ENGINE" LIGHT IS ON ALL THE TIME OR MOST OF THE TIME.

* OPEN LOOP AT IDLE ONLY MAY BE CAUSED BY A COLD OXYGEN SENSOR AFTER A PERIOD OF IDLING. IF THIS IS THE CAUSE, RUNNING ENGINE AT 2000-3000 RPM FOR ONE MINUTE SHOULD WARM IT UP AND GIVE CLOSED LOOP OPERATION AT IDLE.

AFTER ANY REPAIR, CLEAR CODES AND CONFIRM "CLOSED LOOP" OPERATION.

Figure 4–23 Field service mode. *(Courtesy of General Motors Corporation, Service Technology Group)*

1982-1983
CHART 1
FUEL INJECTION
DRIVER'S COMPLAINT

● INTERMITTENT "CHECK ENGINE" LIGHT OR STORED CODES.

NOTICE: DO NOT USE DIAGNOSTIC CHARTS FOR INTERMITTENT PROBLEMS. THE FAULT MUST BE PRESENT TO LOCATE THE PROBLEM. IF THE FAULT IS INTERMITTENT, USE OF THE CHARTS MAY RESULT IN THE REPLACEMENT OF NON-DEFECTIVE PARTS.

MOST INTERMITTENT PROBLEMS ARE CAUSED BY FAULTY ELECTRICAL CONNECTORS OR WIRING. DIAGNOSIS MUST INCLUDE A CAREFUL VISUAL AND PHYSICAL INSPECTION OF THE INDICATED CIRCUIT WIRING AND CONNECTORS.

- POOR MATING OF THE CONNECTOR HALVES OR TERMINALS NOT FULLY SEATED IN CONNECTOR BODY (BACKED OUT "TERMINALS").
- IMPROPERLY FORMED OR DAMAGED TERMINALS. ALL CONNECTOR TERMINALS IN PROBLEM CIRCUIT SHOULD BE CAREFULLY REFORMED TO INCREASE CONTACT TENSION.
- HEI DISTRBUTOR EST WIRES SHOULD BE ROUTED AWAY FROM DISTRIBUTOR, IGNITION COIL, SECONDARY WIRING AND GENERATOR.
- CKT 419- "CHECK ENGINE LAMP" TO ECM, SHORT TO GROUND.
- CKT 451- DIAGNOSTIC CONNECTOR TO ECM, SHORT TO GROUND.
- CKT 450 AND 450R- CHECK ECM GROUND AT ENGINE BLOCK ATTACHMENT.
- ELECTRICAL SYSTEM INTERFERENCE CAUSED BY A DEFECTIVE RELAY, ECM DRIVEN SOLENOID, OR A SWITCH CAUSING A SHARP ELECTRICAL SURGE. NORMALLY, THE PROBLEM WILL OCCUR WHEN THE DEFECTIVE COMPONENT IS OPERATED.
- IMPROPER INSTALLATION OF ELECTRICAL OPTIONS, I.E. LIGHTS, 2 WAY RADIO, ETC.
- OPEN AIR CONDITIONING CLUTCH DIODE.

● STALLING, ROUGH OR IMPROPER IDLE SPEED - SEE CHART 11.

● ENGINE CRANKS BUT WILL NOT RUN- SEE CHART 4 OR 4A.

● HARD STARTING, POOR PERFORMANCE, DRIVABILITY, OR FUEL ECONOMY- SEE CHART 7.

● DETONATION (SPARK KNOCK)
   - ESC PERFORMANCE CHART 10, IF APPLICABLE.
   - EGR CHART 8.

● POOR ENGINE PERFORMANCE WITH AIR CONDITIONING ON.
   REFER TO SECTION 1B.

FOLLOWING ANY REPAIRS OR ADJUSTMENTS, ALWAYS CLEAR CODES AND CONFIRM "CLOSED LOOP" OPERATIONS AND NO "CHECK ENGINE" LIGHT.

Figure 4-24 Driver's complaint chart. *(Courtesy of General Motors Corporation, Service Technology Group)*

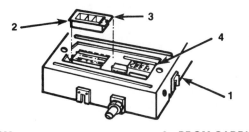

1 ECM
2 PROM (ENGINE CALIBRATOR)
3 PROM CARRIER
4 CALPAC

Figure 4–25 PROM and CALPAK in ECM. *(Courtesy of General Motors Corporation, Service Technology Group)*

Erasing Learned Ability. Anytime battery power is removed from the ECM, its learned memory is lost. This can produce a noticeable change in performance. Learning can be restored

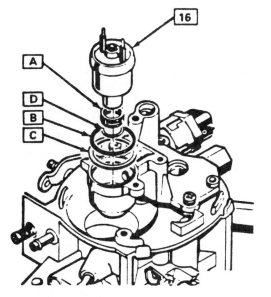

16 FUEL INJECTOR
A FILTER
B LARGE "O" RING
C STEEL BACK-UP WASHER
D SMALL "O" RING

Figure 4–26 EFI injector assembly. *(Courtesy of General Motors Corporation, Service Technology Group)*

by driving the car for usually not more than half an hour at normal operating temperature, part throttle and idle, and at moderate acceleration.

In addition to the key terms, refer to Figure 4-27 for a list of the abbreviations used in this chapter.

SUMMARY

We have covered the most important information inputs the computer uses to calibrate the engine's delivered fuel mixture and spark advance, factors such as engine and air temperatures, engine speed and load, throttle position and vehicle speed. The various ways these sensors work and what use the computer makes of each kind of signal were reviewed. We have learned how fuel pump and ignition circuits work in the system as well as diagnostic precautions and procedures.

The chapter has explained the way the computer makes use of a hard-wired memory, but also builds a body of learned information to correct the original computer instructions and enable the various mixture and advance maps to most effectively optimize exhaust emissions, performance, and fuel economy.

▲ DIAGNOSTIC EXERCISE

1. A car is brought into the shop and won't start. Spark and fuel pressure are good, while there are no codes, hard or intermittent. Spark plug tips are black, and replacing them allows the engine to start with difficulty. Soon, however, it dies again. Fuel pressure, it turns out, drops rapidly after shutdown even with feed and return lines clamped off. What component should be checked first?

2. A car with the EFI system will not hold a consistent idle. It idles very poorly when cold, idles slowly when warm, and stalls when the air conditioner is turned on. What is the most

A/C	Air-conditioning
AIR	Air Injection Reaction
ALCL	Assembly Line Communication Link
BARO	Barometric Pressure
CCP	(Charcoal) Canister Purge
DIAGNOSTIC TEST TERMINAL	Lead or ALCL connector terminal grounded to obtain a trouble code
ECM	Electronic Control Module
EFI	Electronic Fuel Injection
EGR	Exhaust Gas Recirculation
ESC	Electronic Spark Control
EST	Electronic Spark Timing
HEI	High-energy Ignition
IAC	Idle Air Control
MAP	Manifold Absolute Pressure
PCV	Positive Crankcase Ventilation
P/N	Park/Neutral Switch
PROM	Programmable Read-only Memory (engine calibration unit)
TBI	Throttle Body Injection
TCC	Torque Converter Clutch
TPS	Throttle Position Sensor
VIN	Vehicle Identification Number
VSS	Vehicle Speed Sensor
WOT	Wide-open Throttle

Figure 4–27 Explanation of abbreviations

fruitful sequence of tests to determine the cause of the problem? Are there any precautions a technician should observe to prevent damage to components?

REVIEW QUESTIONS

1. What are the six inputs most involved in engine calibration control?
2. What is the secondary function of the MAP sensor? Under what two conditions does it perform this function?
3. Name three functions the VSS influences.
4. What influence does the A/C switch have on engine operation?
5. What introduces fuel to the engine on EFI applications?
6. How is fuel quantity controlled?

7. Describe the ECM's fuel system control response to each of the following conditions on a four-cylinder EFI application:
   a. the ignition is turned on.
   b. no REF signal is sent to the ECM within two seconds.
8. What variable control mechanism determines the amount of air going into the intake manifold during idle?
9. During what two operating conditions can the antistall mode be activated?
10. What signal alerts the ECM to move the IAC valve back to its parked position?
11. What is the power source for most output devices?
12. What type of device is used by the ECM to activate the fuel pump, cooling fan, and A/C clutch?
13. What are the first two steps when pursuing

a driveability complaint on an EFI-equipped vehicle?

14. What is the function of a weather-pack connector?

15. What two functions does the TCC test lead serve?

16. Name six things that should be observed to avoid damage to the ECM.

ASE-type Questions (Actual ASE test questions are rarely so product specific):

17. Technician A says EFI systems using a Hall-effect switch do not use a pickup coil in the distributor. Technician B says the Hall-effect switch, on those EFI systems using it, provides the signal to the HEI module to control the primary ignition circuit switching. Who is correct?
    a. A only.
    b. B only.
    c. both A and B.
    d. neither A nor B.

18. The engine of a car equipped with an EFI system dies while being driven. The car is brought into the garage, and it is determined that the ignition system is operating properly, the fuel tank is at least half full, and there appear to be no problems in the fuel lines. The ECM is working properly, and no fault codes are stored in it. Technician A says the problem could be a fault in the reference pulse wire from the distributor. Technician B says the fault could be a blown fuel pump fuse. Who is correct?
    a. A only.
    b. B only.
    c. both A and B.
    d. neither A nor B.

19. A car equipped with EFI is brought in with a complaint of poor mileage and black smoke during some driving conditions. A fuel pressure check shows the fuel pressure is too high. The most likely cause is:
    a. a faulty fuel pump.
    b. a faulty fuel pressure regulator.
    c. a faulty coolant temperature sensor.
    d. a faulty ECM.

20. Technician A says that an EFI system is put into field service mode by grounding the test terminal with the ignition on and the engine off. Technician B says that the same system is put into field service mode by grounding the test terminal with the engine running. Who is correct?
    a. A only.
    b. B only.
    c. both A and B.
    d. neither A nor B.

Chapter 5

General Motors' Port Fuel Injection

OBJECTIVES

In this chapter you can learn:

❏ several advantages of port fuel injection systems over other types of fuel induction.

❏ the two methods used by General Motors port fuel injection systems to measure intake air mass.

❏ the operation of the cold-start valve.

❏ the operation of the turbocharger wastegate.

❏ the types of EGR diagnostic switch.

❏ how the idle air control valve controls idle speed.

❏ how the computer learns that spark knock has occurred and what it does to correct for it.

❏ how the EGR valve is modulated on PFI systems.

KEY TERMS

AC
C3I
High-Pressure Cutout Switch
Longitudinal
Micron
Normally Aspirated
Pressure Cycling Switch
Thermac
Thermo-time Switch
Wastegate

Excluding the Cadillac digital fuel injection system discussed in Chapter 7, port fuel injection (PFI) represents the fourth generation of General Motors' comprehensive computerized engine control systems. PFI was introduced with limited applications on 1984 models. It followed the electronic fuel control system, General Motors' first closed-loop fuel system of 1978, the CCC system introduced in 1980, and the EFI system introduced in 1982. In 1985 the PFI system was expanded to several engine applications, Figure 5-1. Some General Motors car divisions have chosen special names for the system because of some special feature designated for a given engine or body application. These special features include:

• sequential fuel injection (SFI) on the tur-

bocharged 3.8-liter engine.

• multiport fuel injection (MFI) on most other engines.

• tuned port injection (TPI) on the Corvette 5.7-liter engine. The TPI version includes tuned intake manifold runners, matched in shape, length, and cross-sectional area to help maximize volumetric efficiency.

Because only air moves through the intake manifold, there is no concern about holding fuel in a vapor state as it passes through the runners. Consequently the heated air (**Thermac**) and the early fuel evaporation (EFE) systems are eliminated. Most of the PFI applications do run engine coolant through a passage in the throttle body to prevent the formation of ice around the throttle blade.

ELECTRONIC CONTROL MODULE (ECM)

The ECM is usually located under the instrument panel or behind the passenger kick panel. On the P car (Pontiac Fiero) it is in the console between the seats. It contains a removable PROM and CALPAK. The CALPAK provides calibration for cold-start cranking and for fuel back-up in the event of an ECM failure. The fuel back-up circuit (FBC) provides an operating mode when any of the following conditions occur:

- ECM voltage falls below 9 volts (most likely during cold cranking).
- The PROM is missing or not functioning.
- The ECM is unable to provide its normal computer-operated pulses (COP).

The FBC makes use of the throttle position sensor, the coolant temperature sensor signal, and the engine rpm signal. It is powered through the ignition switch. During FBC mode the engine runs erratically and sets a code 62.

Keep-Alive Memory (KAM)

The ECM has basically the same self-diagnostic capacity and KAM as discussed in earlier chapters. The inputs fed into the ECM and the functions controlled for most systems are shown in Figure 5-1.

OPERATING MODES

The different operating modes of the PFI system control how much fuel to introduce into the intake manifold for the various possible operating conditions.

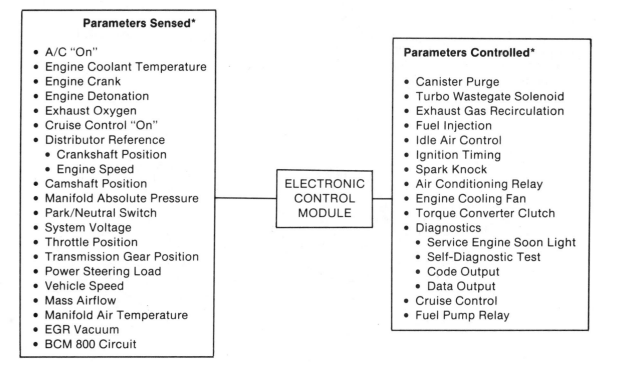

Parameters Sensed*

- A/C "On"
- Engine Coolant Temperature
- Engine Crank
- Engine Detonation
- Exhaust Oxygen
- Cruise Control "On"
- Distributor Reference
  - Crankshaft Position
  - Engine Speed
- Camshaft Position
- Manifold Absolute Pressure
- Park/Neutral Switch
- System Voltage
- Throttle Position
- Transmission Gear Position
- Power Steering Load
- Vehicle Speed
- Mass Airflow
- Manifold Air Temperature
- EGR Vacuum
- BCM 800 Circuit

ELECTRONIC CONTROL MODULE

Parameters Controlled*

- Canister Purge
- Turbo Wastegate Solenoid
- Exhaust Gas Recirculation
- Fuel Injection
- Idle Air Control
- Ignition Timing
- Spark Knock
- Air Conditioning Relay
- Engine Cooling Fan
- Torque Converter Clutch
- Diagnostics
  - Service Engine Soon Light
  - Self-Diagnostic Test
  - Code Output
  - Data Output
- Cruise Control
- Fuel Pump Relay

*Not all systems used on all engines.

Figure 5–1 PFI system overview

Starting Mode

When the ignition is turned on, the ECM turns the fuel pump relay on. If it does not see an ignition signal within two seconds telling it that the engine is cranking, it turns the fuel pump off again. The fuel pump provides fuel pressure to the injectors. As the system goes into cranking mode, the ECM checks coolant temperature and throttle position to calculate what the air/fuel ratio should be for this start-up. The air/fuel ratio ranges from maximally rich, 1.5 to 1, at -36 degrees Centigrade (-32 degrees Fahrenheit), to maximally lean, 14.7 to 1 at 94 degrees Centigrade (201 degrees Fahrenheit). The ECM varies the air/fuel ratio by controlling the time the injectors are turned on, what is referred to as the pulse width.

The 2.8-, 5.0- and 5.7-liter engine PFI applications also use a cold-start injector. This is an additional fuel injector in the intake manifold that improves cold-starting by spraying additional fuel into the intake manifold during cranking, Figure 5-2. It is not ECM-controlled. Spraying the additional fuel for cold-start enrichment into the manifold gives the fuel more opportunity to evaporate in the cold, dense air than if it were sprayed directly into the intake ports. The injector is activated by the cranking system and is controlled by a timing mechanism. The subsystem is described more fully in the **Outputs** section of this chapter.

Clear Flood Mode

The clear flood mode works essentially the same as it does on the EFI system. As long as the engine is turning below 600 rpm and the throttle is held to 80% WOT or more, the ECM will keep the air/fuel ratio at 20 to 1, except on the 2.8-liter engine, which cuts off fuel completely in clear flood. The same troubleshooting considerations apply to this system's clear flood mode as to the system in Chapter 4.

Run Mode

The run mode consists of the open- and closed-loop operating conditions. When the engine starts and the rpm rises above 400, the system goes into open loop. In open loop, the ECM ignores information from the oxygen sensor and determines air/fuel ratio commands based on input from other sensors (coolant tempera-

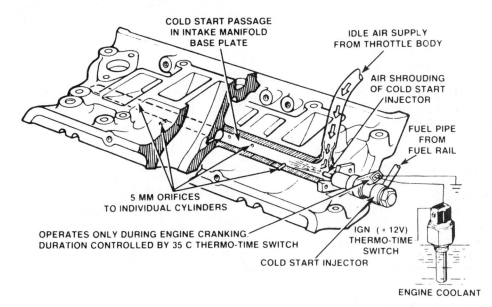

COLD START PASSAGE
IN INTAKE MANIFOLD
BASE PLATE

IDLE AIR SUPPLY
FROM THROTTLE BODY

AIR SHROUDING
OF COLD START
INJECTOR

FUEL PIPE
FROM
FUEL RAIL

5 MM ORIFICES
TO INDIVIDUAL CYLINDERS

OPERATES ONLY DURING ENGINE CRANKING.
DURATION CONTROLLED BY 35 C THERMO-TIME SWITCH

IGN (+12V)
THERMO-TIME
SWITCH

COLD START INJECTOR

ENGINE COOLANT

Figure 5–2 Cold-start injector (typical). *(Courtesy of General Motors Corporation, Service Technology Group.)*

ture, vacuum or mass airflow and throttle position, as well as engine speed if the vacuum sensor is used). The system stays in open loop until:

- the oxygen sensor produces a varying voltage showing that it is hot enough to work properly.
- the coolant is above a specified temperature.
- a specified amount of time has elapsed since the engine last started.

These values vary with engine application and are hard-wired into the PROM. When all three conditions are met, the ECM puts the system into closed loop. In closed loop, the ECM uses oxygen sensor input to calculate air/fuel ratio commands and keeps the air/fuel ratio at a near perfect 14.7 to 1. During heavy acceleration, wide-open throttle or hard deceleration, the system temporarily drops out of closed loop.

Acceleration Mode

Rapid increases in throttle opening and manifold pressure or airflow signal the ECM to enrich the air/fuel mixture. This compensates for the reduced evaporation rate of the gasoline resulting from the higher manifold pressure.

Deceleration Mode

Rapid decreases in throttle opening and manifold pressure or airflow cause the ECM to lean the air/fuel mixture. If the changes are severe enough, fuel is momentarily cut off completely.

Battery (Charging System) Voltage Correction Mode

The battery voltage correction feature on the PFI system is different from that on the EFI system. On the PFI system, when battery voltage drops below a specified value, the ECM will:

- enrich the air/fuel mixture according to a preset formula.
- increase the throttle opening if the engine is idling.

- increase ignition dwell to compensate for a weakened ignition spark should the coil's magnetic field not build sufficiently to generate a hot secondary spark.

Fuel Cutoff Mode

When the ignition is turned off, the ECM immediately stops pulsing the injectors. This is to prevent dieseling as well as to consume the fuel mixture remaining in the ports. Injection is also stopped anytime the distributor reference pulse stops coming to the ECM.

FUEL SUPPLY SYSTEM

Fuel pressure is supplied by an electric pump in the fuel tank, Figure 5-3. The pump is turned on and off by a fuel pump relay controlled by the ECM. A pressure relief valve in the pump limits maximum pump pressure to between 80 and 90 psi. This pressure is realized only if flow stops and the pump is working against static pressure. The filter mounted to the bottom of the pump assembly is a 50-micron filter; downstream is another filter, a 10- to 20-micron in-line filter (one **micron** is one millionth of a meter). Because of the high operating pressure, an O-ring is used at all threaded connections, and all flex hoses have internal steel reinforcement.

Fuel pressure is controlled by the fuel pressure regulator, Figure 5-4, which is usually mounted on the fuel rail in the engine compartment, Figures 5-5 and 5-6. Manifold vacuum is supplied to the spring side of the diaphragm. At light throttle, the high vacuum helps pull the diaphragm up and against the spring. This allows more fuel to return to the tank, thus reducing output pressure. Pressure at the injectors ranges from 34 psi at idle to 44 psi at full throttle. A constant stream of fuel flowing from the tank to the pressure regulator and back to the tank through the return line ensures cool fuel available to the fuel rail and injectors. This plus the operating pressure work to reduce or eliminate the chance of vapor lock.

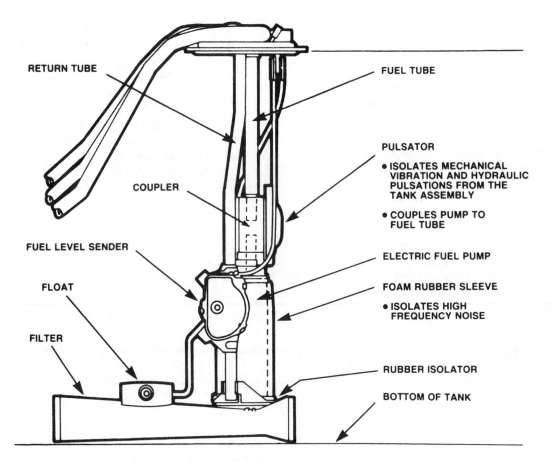

Figure 5–3 Fuel pump and sending unit assembly. *(Courtesy of General Motors Corporation, Service Technology Group.)*

INJECTORS

An injector is installed in the intake port of each cylinder with 70 to 100 millimeters between the tip of the injector and the center of the valve on V-type engines, Figure 5-7. The nozzles spray fuel in a 25-degree conical pattern. O-rings are used to seal between the nozzle and the fuel rail and between the nozzle and the intake manifold. The O-rings also serve to retard the heat transfer from the engine to the nozzle of the injector and to reduce nozzle vibration. The O-rings should be lubricated or replaced as necessary whenever the injectors are removed. A damaged or hardened O-ring allows a fuel or vacuum leak. PFI systems are very sensitive to vacuum leaks.

With the exception of the SFI applications, all injectors are pulsed simultaneously every crankshaft revolution. Fuel sprayed while the intake valve is closed is simply stored in the intake port until the valve opens. The SFI system pulses each injector one at a time in the engine's firing order during the intake stroke.

THROTTLE BODY

The throttle body on a PFI system controls the amount of air that is allowed into the engine's induction system. This is done with a throttle blade and shaft controlled by the accelerator pedal, just

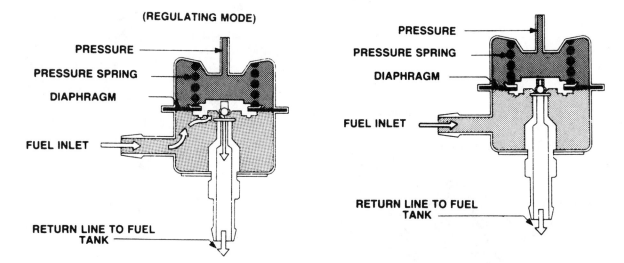

Figure 5–4 Fuel pressure regulator. *(Courtesy of General Motors Corporation, Service Technology Group.)*

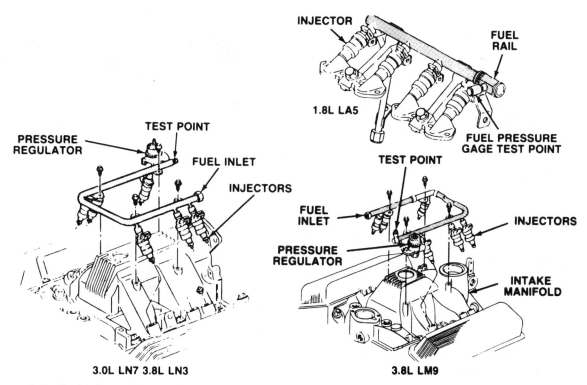

Figure 5–5 Fuel rail: 1.8-liter, 3.0-liter and 3.8-liter engines. *(Courtesy of General Motors Corporation, Service Technology Group.)*

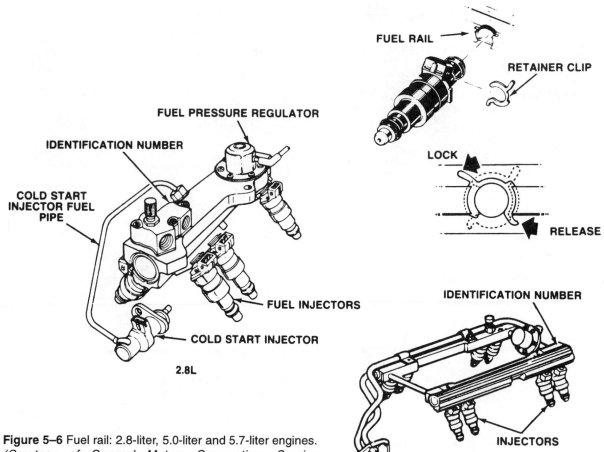

Figure 5–6 Fuel rail: 2.8-liter, 5.0-liter and 5.7-liter engines. *(Courtesy of General Motors Corporation, Service Technology Group.)*

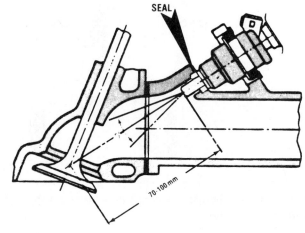

Figure 5–7   Port fuel injection (typical). *(Courtesy of General Motors Corporation, Service Technology Group.)*

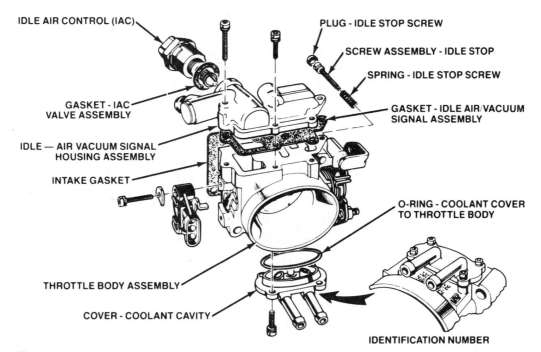

Figure 5–8 Throttle body assembly, 2.8-liter engine. *(Courtesy of General Motors Corporation, Service Technology Group.)*

as in a carburetor or TBI unit, Figures 5-8 and 5-9. In this case, however, only air passes the throttle blade. An idle stop screw determines how nearly closed the throttle blade is at idle. A throttle position sensor keeps the ECM informed as to the throttle valve's position. Idle speed is regulated by an idle air control motor and valve assembly. It varies the amount of air allowed to bypass the throttle blade through a bypass passage and is controlled by the ECM. Most throttle body units have a cavity for engine coolant to flow through to prevent throttle blade icing.

Real World Problem

One of the unexpected problems with port fuel injection throttle bodies has proven to be fuel deposits on the throttle blades. The source of these fuel vapors is what drifts back from the intake ports when the engine is shut down hot. The heat vaporizes most or all of the residual fuel, and the coolest surface it encounters in the

intake path is the throttle blade itself. The problem is that these deposits change the shape and size of the opening in the smallest throttle opening angles. This can adversely affect driveability in unexpected ways, such as hesitation, erratic idle, hard starting, and surging. The throttle blades can be visually inspected for such deposits and cleaned with solvent. This inspection should be a routine maintenance task on cars with this fuel injection system.

NON-ECM EMISSION CONTROLS

A trend is developing whereby in the automotive industry—as computerized engine control systems become more efficient, fewer traditional mechanical emissions control devices are needed. This is evidenced by the following list of non-ECM-controlled emission control devices during the first or second year of PFI production:

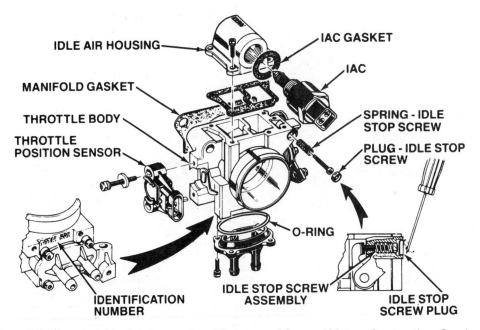

Figure 5–9 Throttle body assembly, 3.0-liter engine. *(Courtesy of General Motors Corporation, Service Technology Group.)*

- 1.8-liter—canister purge, PCV and internal EGR (no EGR valve).
- 2.8-liter—PCV and deceleration valve (admits additional air during deceleration on manual transmission vehicles).
- 3.0- and 3.8-liter—PCV.
- 5.0- and 5.7-liter—PCV.

INPUTS

Coolant Temperature Sensor (CTS)

This is the same sensor used in the computer command control system, the electronic fuel injection system, and other GM electronic engine management systems. The CTS input affects control of:

- air/fuel mixture (in open loop).
- spark timing.
- spark knock control (on some engines).

- engine idle.
- torque converter lockup clutch actuation.
- canister purge (except on the 1.8-liter engine up to 1985).
- EGR (except for the 1.8-liter engine up to 1985).
- cooling fan (on some vehicles, especially with transverse-mount engines).

Mass Airflow (MAF) Sensor

Many General Motors PFI systems either use or have used a MAF sensor. In the mid-1980s, two MAF sensors were used: one on Buick engines and one on Chevrolet engines. Buick engines used a MAF manufactured by AC, a division of General Motors. Chevrolet engines used a Bosch unit. In the late 1980s, the AC unit was discontinued and a unit built by Hitachi of Japan was used on Buick engines. Chevrolet engines were by then no longer equipped with a MAF sensor.

AC. The **AC** unit contains a screen to break up the airflow, Figure 5-10. After the air passes

C3I System

For many years automotive engineers generally felt that it did not make much difference *when* the fuel was sprayed into the intake port as long as precisely the *right amount* of fuel was sprayed. When you consider how fast all of this happens (the intake valve opens and closes twenty-five times per second at 3,000 rpm), it would not seem to make much difference.

With the introduction of the three-coil ignition system (**C3I**) by Buick in 1984—and since extended to most other divisions—it became necessary to employ a sequential fuel injection system. This is because the C3I fires two plugs at one time; one at the regular spark advance position at the end of the compression stroke, the other at the cylinder 360 degrees out of sequence with the first, the cylinder just completing its exhaust stroke. This is the "waste spark" system explained in the **Outputs** section of this chapter. The second "waste" spark serves no purpose in combustion; it is just an unavoidable by-product of the ignition system's electrical design (it is sometimes claimed that the waste spark helps to burn off spark plug deposits). If, however, a pressurized combustible mixture (the first applications of C3I were on turbocharged engines) was just on the other side of the open intake valve, a backfire could occur in the intake manifold. To avoid this, each injector pulse is timed so the fuel is immediately purged from the manifold by its respective cylinder's intake stroke.

However, the additional circuitry and related expense seems to have been proven worthwhile, as GM has rapidly expanded the application of sequential fuel injection to other engines. Ford introduced sequential fuel injection in 1986 on some 5-liter engines and has expanded its use of sequential injection to many other engines since then.

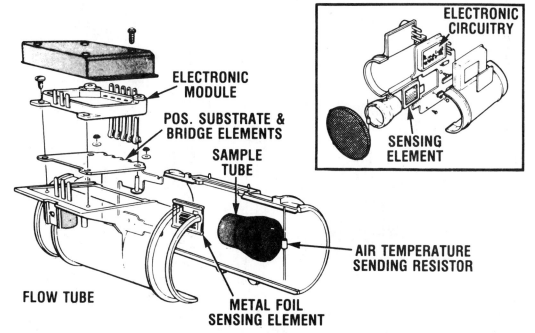

ELECTRONIC MODULE

POS. SUBSTRATE & BRIDGE ELEMENTS

SAMPLE TUBE

ELECTRONIC CIRCUITRY

SENSING ELEMENT

AIR TEMPERATURE SENDING RESISTOR

FLOW TUBE

METAL FOIL SENSING ELEMENT

Figure 5–10 Mass airflow sensor (AC type). *(Courtesy of General Motors Corporation, Service Technology Group.)*

through the screen, it flows over an air temperature sensing resistor. A sample tube then directs some of the air to flow over a heated foil sensing element. The power to heat the sensing element comes from a fuse (a relay as on some applications), Figure 5-11. The circuitry of the MAF sensor controls current flow through the foil sensing element to maintain it at 75 degrees Centigrade above the incoming air temperature, as measured by the temperature-sensing resistor. The power required to keep the sensing foil element 75 degrees Centigrade above incoming air temperature is a measure of mass airflow. This value is sent to the ECM as a digital signal ranging in frequency from 30 cycles per second (30 hertz or 30 hz) to 150 hz (150 hz represents the highest mass airflow rate). The ECM compares the MAF sensor information to a preprogrammed look-up chart and finds airflow in grams per second. Using this value, engine temperature, and engine rpm, the ECM can calculate exactly how much fuel is required to achieve the desired air/fuel ratio. Mass airflow readings are taken and air/fuel

mixture calculations are revised about 160 times per second.

Bosch.   The Bosch MAF sensor works much the same way as the AC unit except that a wire heat element is used instead of a foil sensing element. Also, each time the ignition is turned off after the system has been in closed loop, a separate burn-off module momentarily puts enough current through the wire heat element to make it red hot, Figure 5-12. This burns off any residual accumulation to keep the sensor accurate. It is critical that the sensor wire's surface be clean because any accumulation of deposits will retard its ability to transfer heat.

Manifold Air Temperature (MAT) Sensor

The MAT sensor looks like and essentially is a coolant temperature sensor with its thermistor-sensing element exposed to the intake air instead of being fully enclosed like the CTS. It screws into a hole in the intake manifold or on some vehicles, into the air cleaner housing. The

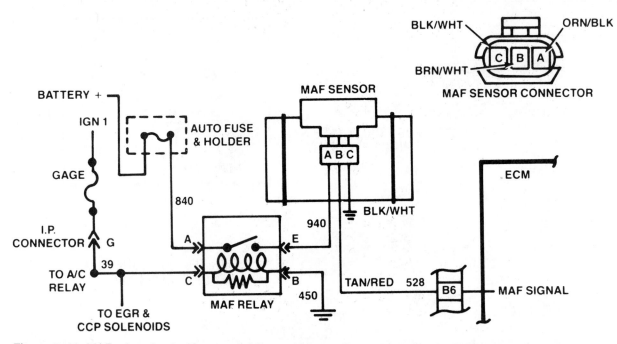

Figure 5–11 MAF relay circuit. *(Courtesy of General Motors Corporation, Service Technology Group.)*

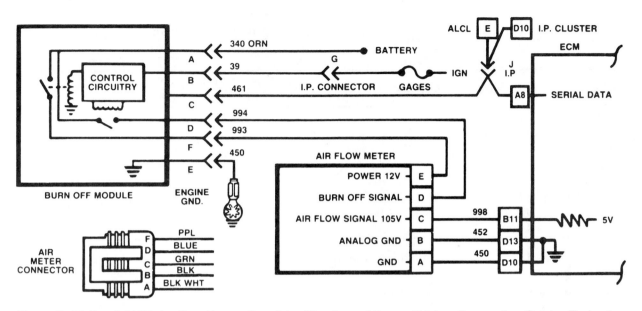

Figure 5–12 Bosch MAF circuit and burn-off module. *(Courtesy of General Motors Corporation, Service Technology Group.)*

perforated nose extends inside. It is sometimes referred to as an air temperature sensor (ATS). While it is not used on all applications, it influences air/fuel mixture, spark timing and idle speed control on those engines that use it.

Throttle Position Sensor (TPS)

The TPS is a potentiometer mounted on the side of the throttle body and attached to the end of the throttle shaft, Figure 5-9. Its input affects most ECM-controlled functions including fuel control, spark timing, idle air control, torque converter lockup clutch engagement, and air conditioning control.

Reference Pulse (REF)

On all but those engines equipped with a distributorless ignition system, the REF signal comes from the distributor pickup coil. Without this signal the engine does not start or run because without it the ECM will not pulse the injectors. A more complete discussion of the REF signal is found in the **Inputs** section of Chapter 3.

The distributorless ignition systems use sensors to provide information to the ignition module relative to crankshaft and camshaft position and speed, Figure 5-13. The C3I (Computer-Controlled Coil Ignition) systems use Hall-effect switches while both the DIS (direct ignition system) and IDI (integrated direct ignition) systems use a permanent magnet pulse generator, Figure 5-14.

C3I. There are three different versions of the C3I system: Type I, Type II, and Type III. Each of the three types uses a Hall-effect switch mounted behind the crankshaft balancer, Figure 5-15. The balancer has three vanes, which are positioned to form a circle, and they extend back toward the engine. As the crankshaft turns, the vanes alternately pass through the Hall-effect switch, producing a signal that is fed to the ECM and the ignition module. When used on the 3.0-liter engine, the Type I system uses a second Hall-effect switch mounted behind the crankshaft balancer with a single vane extending back, nearer the outer circumference of the balancer. This vane acts as a camshaft position sensor.

The Type I or Type II systems, when used

		PM Reluctor	Hall Effect Switches				
		Crank Pos. Sensors		CAM/SYNC Sensors			
System	Engine	On Crank	Harmonic Balancer	Harmonic Balancer	Dist. Hole	Timing Gear	18x
DIS	2.0	X					
	2.5	X					
	2.8	X					
IDI	2.3	X					
C³I-I	3.0		X	X^C			
I or II^A	3.8 SFI		X			X	
I or II	3.8 TURBO		X		X		
III^B	3.800		X			X	X

A — Type I or II may be used on 3.8 liter engines. Type I has molded one piece coil pack. Type II has independent coils.

B — Introduced in 1988, has independent coils.

C — Now called a "sync" sensor rather than a "cam" sensor.

Figure 5–13 General Motors distributorless ignition system applications

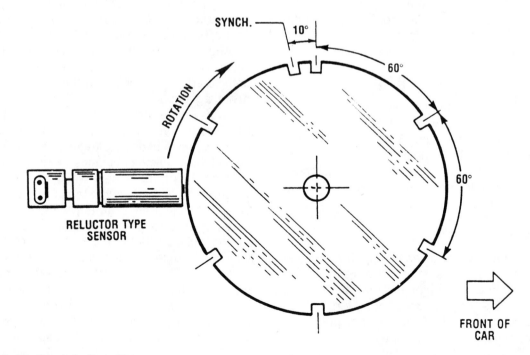

Figure 5–14 Crankshaft position sensor

Figure 5–16 Camshaft position sensor (3.8-liter SFI). *(Courtesy of General Motors Corporation, Service Technology Group.)*

Figure 5–15 Crankshaft position sensor (3.0-liter SFI). *(Courtesy of General Motors Corporation, Service Technology Group.)*

with the 3.8-liter turbo engine, include a Hall-effect switch mounted in what used to be the distributor hole to serve as a camshaft position sensor, Figure 5-16. This sensor is driven by the camshaft at the same speed (unlike the harmonic-balancer mounted sensor, which signals at crankshaft speed). When used on the 3.8-liter non-turbocharged SFI engine, these same two systems use a Hall-effect cam sensor mounted in the timing cover and are triggered by a vane attached to the camshaft timing gear. The cam sensor signal alerts the ECM when the Number One cylinder is at TDC so it can start the firing order over again and time the injector pulses. Loss of this signal will result in a code 41 being set. If the signal is absent or lost during cranking, the engine will not start; if the signal is lost while the engine is running, the engine will continue running, but will not restart. In more recent General Motors literature, the camshaft position sensor is referred to as a "sync sensor."

The Type III system, sometimes referred to as the "Fast Start," also has its sync sensor located in the timing cover. On this system, a third sensor is used. It is a Hall-effect switch and is

located behind the harmonic balancer along with the crankshaft sensor. This sensor, referred to as the "Crank 18x," is triggered by 18 evenly spaced interrupter vanes protruding rearward from near the outer circumference of the balancer. The Crank 18x sensor provides 18 signals per crankshaft revolution to the ignition module which in turn, sends on to the ECM, Figure 5-17.

On this system, the crank sensor is called the "Crank 3x" sensor. Its three interrupter vanes, nearer the center on the balancer, are not symmetrical as they are on the other C3I systems. The windows between the vanes are 10, 20, and 30 degrees wide as shown in Figure 5-17. The signal produced by the leading and trailing edge of each window constitutes a 3x pulse. The number of 18x signals that occurs during each 3x pulse enables the ECM to determine which 3x pulse it is reading. There will be one 18x signal during the 10-degree window pulse, two during the 20-degree window pulse, and three during the 30-degree window pulse. This enables the system to fire a coil for the appropriate cylinders within the first 120 degrees of crankshaft rotation. This system provides the following advantages:

- faster starts, because the ECM can identify cylinder position more quickly.
- more accurate REF signals to the ECM, especially at low speeds.

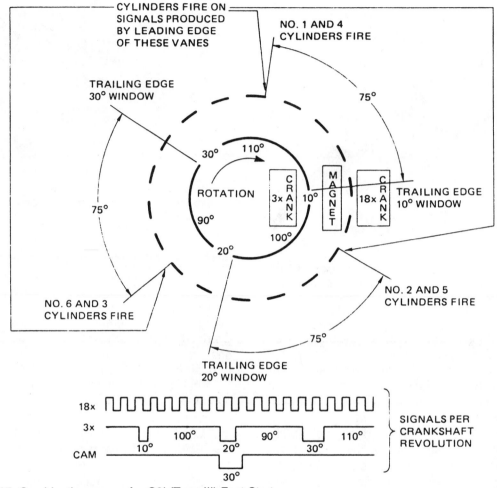

Figure 5–17 Combination sensor for C3I (Type III) Fast Start

- increased run reliability, as the engine can continue to operate without the cam sensor.
- the potential for the ECM to read crankshaft acceleration and deceleration rates.

DIS and IDI. Both the integrated direct ignition (IDI) and the direct ignition system (DIS) use a magnetic pulse generator on the crankshaft instead of the Hall-effect switches for crankshaft reference (position and speed). The crankshaft for those engines which use either the DIS or IDI system has a round steel disc machined into its center. This disc is concentric to the crankshaft's centerline at its center counterweight, Figure 5-18. This disc has seven notches cut into it and is called a reluctor. The term "reluctor" refers to any device that changes the reluctance of a material to conduct magnetic lines of force. The same reluctor is used on both four- and six-cylinder engines. The crankshaft sensor is mounted to the engine block and extends through a hole in the crankcase. Its tip, containing a permanent magnet and a wire coil, is spaced 0.050 inch from the reluctor.

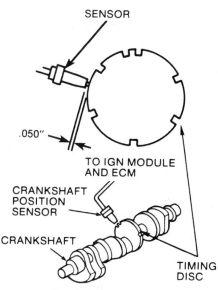

Figure 5–18 Crankshaft timing disc

Six of the reluctor's seven notches are spaced evenly around it at 60-degree intervals, Figure 5-18. The seventh notch is called the "sync" notch and is located between notches 6 and 1, 10 degrees from notch 6 and 50 degrees from notch 1. By being placed at an odd position in relation to the other notches, the sync notch provides an irregular signal that the module can identify. The module uses this signal to synchronize the coil-firing sequence to the crankshaft's position in the engine cycle.

On both the DIS and IDI systems, the notch that produces the signal that results in the appropriate coil being fired during module mode operation is often referred to as the "cylinder event" notch. The crank sensor signal that occurs 60 degrees before the cylinder event notch causes voltage in the REF wire to go low. The cylinder event notch causes voltage in the REF wire to go high. When the engine operates in the electronic spark timing mode, the ECM uses this signal as an input for calculating:

- ignition timing while operating in EST mode.
- fuel injector pulses.

Oxygen Sensor

This oxygen sensor is the same single-wire sensor used on other General Motors vehicles. In some vehicles in later model years, multiwire sensors are used, both to provide independent ground circuits, and to provide electric resistance heat to the sensor to put the system into closed loop more quickly.

Vehicle Speed Sensor (VSS)

The two types of VSS discussed in Chapter 3 are also used on this system.

EGR Diagnostic Switch

On PFI systems that control EGR operation, an EGR diagnostic switch is used to tell the ECM whether the EGR is actually being applied when it is commanded to apply. Two types are used. The four- and six-cylinder engines use the same vacuum-operated switch discussed in Chapter 3.

The 5.0- and 5.7-liter engines use a thermal switch for exactly the same purpose as the vacuum switch used on the smaller engines. It, however, screws into the base of the EGR valve, Figure 5-19. If the EGR valve is open, the heat of the exhaust gases flowing past the heat-sensing tip of the thermal switch, causes the switch to close. The ECM wants to see 12 volts on circuit 935 when the EGR is supposed to be off and less than 1 volt when the EGR is supposed to be on.

Detonation Sensor

Most of General Motors' PFI systems incorporate an electronic spark control (ESC) system using a piezoelectric knock sensor, as described in Chapter 3.

Park/Neutral (P/N) Switch

The park/neutral switch, which can be located near the shifter assembly or in the transmission itself, is a switch that is normally open while driving. When the vehicle is shifted into park or neutral, it closes. The ECM monitors the P/N

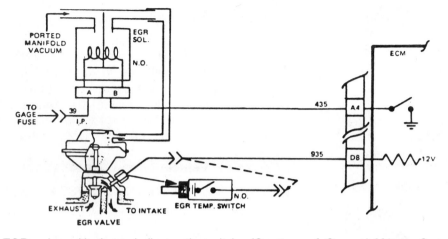

Figure 5–19 EGR valve with thermal diagnostic switch. *(Courtesy of General Motors Corporation, Service Technology Group.)*

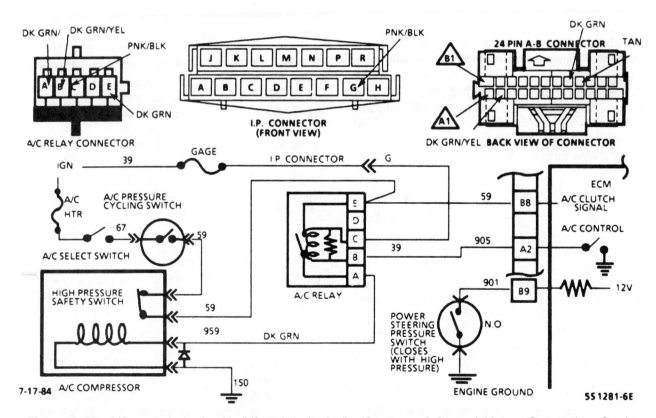

Figure 5–20 A/C control circuit with P/S switch (typical). *(Courtesy of General Motors Corporation, Service Technology Group.)*

switch and thus knows whether the transmission is in gear or not. This information is used to help control engine idle and EGR control.

Air Conditioning (A/C) Switch

On PFI cars equipped with air conditioning, the A/C control switch on the instrument panel connects through a wire to the ECM, Figure 5-20. When the A/C is turned on, the voltage signal travels from the A/C switch to the **pressure cycling switch** to the ECM. Some systems also put the **high-pressure cutout switch** in series in the same circuit. This signal tells the ECM that the A/C switch is on and the pressure cycling switch is closed. The ECM uses this information to turn on the A/C relay and/or adjust the IAC valve.

Transmission Switches

Transmissions used in PFI-equipped vehicles have one or more hydraulically operated electric switches screwed into the valve body. These switches provide the ECM with signals indicating what gear the transmission is in. This information enables it to more effectively control torque converter lockup clutch operation.

Cranking Signal

On some engines, a wire from the starter solenoid circuit feeds a cranking signal to the ECM to alert it that the engine is being cranked. This puts the system in cranking mode, and leads to an enriched air/fuel mixture for easier starting. On other engines, the cranking signal is fed to the cold-start injector in the intake manifold and provides the necessary enrichment through that mechanism. The cold-start injector, on vehicles where it is found, is not controlled by the ECM.

System Voltage/Fuel Pump Voltage

The ECM monitors system voltage through one of its battery voltage inputs. If voltage drops below a preprogrammed value, the system goes into battery voltage correction mode. Some PFI systems have a wire from the positive side of the fuel pump power circuit to the ECM, Figure 5-21. This signal is used to make fuel delivery compensations based on system voltage and causes a code 54 to be stored if pump voltage is lost while the engine is running.

Power Steering (P/S) Switch

A power steering switch is used on PFI systems to alert the ECM when power steering pressure is high enough to significantly affect engine idle performance (because of the momentary high load), Figure 5-22. The P/S switch is normally open and closes in response to high P/S pressure. When the switch is open, feedback voltage to the ECM is about 12 volts; when the switch closes, voltage falls below 1 volt. Also, when the switch closes, the ECM increases idle air bypass and retards ignition timing. On some systems the A/C compressor clutch is also disengaged (also to reduce idle load).

Body Computer Module (BCM) 800 Circuit

Some late-model PFI applications also feature a body computer module, not unlike the Cadillac DFI system. The BCM controls air conditioning and other systems such as a driver information center. The driver information center displays information about fuel economy, time elapsed, distance to destination, time, date and so forth. The BCM shares information with the ECM. For instance, the ECM may send the BCM information about vehicle speed, coolant temperature, or indicate when the vehicle is operating at WOT. The BCM can send the ECM such information as outside air temperature, the temperature in the high-pressure side of the A/C system, or indicate when the rear window defog system is on.

This information is transmitted over a wire between the two computers. The wire is called a data link. If information can be transmitted both ways over the same wire, it is called a bidirectional data link, or it may be called a universal asynchronous receiver/transmitter. General Motors refers to this data link on later model applications as the BCM 800 Circuit. The BCM 800 Circuit also extends to modules that the BCM or ECM communicates with, Figure 5-23. See the

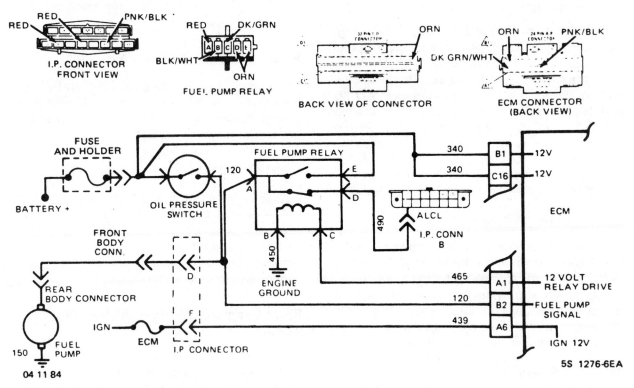

Figure 5–21 Fuel pump control circuit (typical). *(Courtesy of General Motors Corporation, Service Technology Group.)*

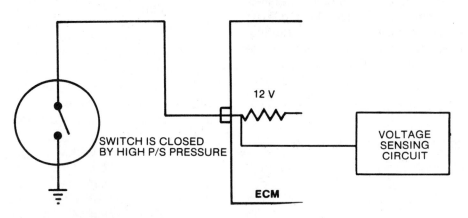

Figure 5–22 Power steering switch (typical)

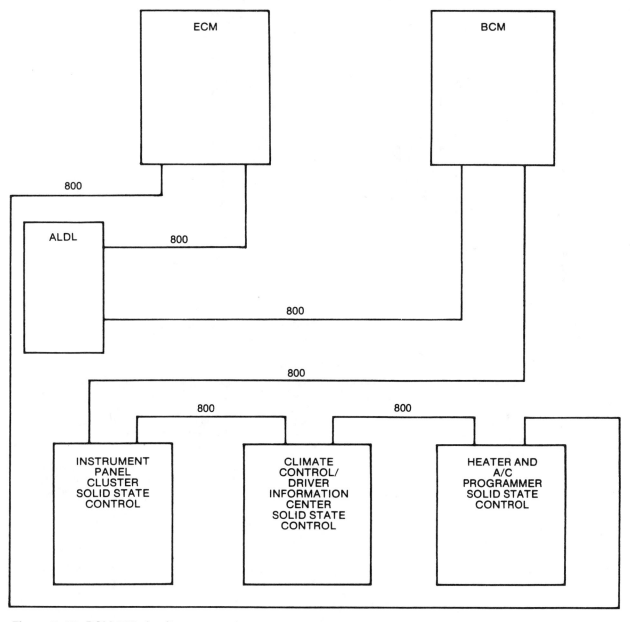

Figure 5–23 BCM 800 circuit

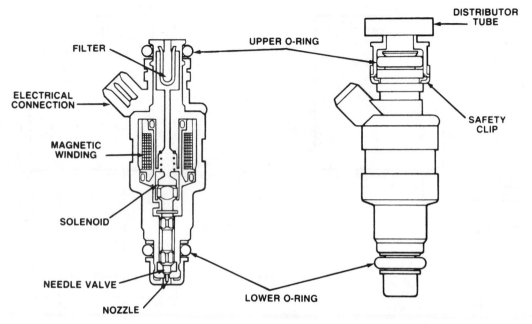

Figure 5–24 Injector (cutaway schematic). *(Courtesy of General Motors Corporation, Service Technology Group.)*

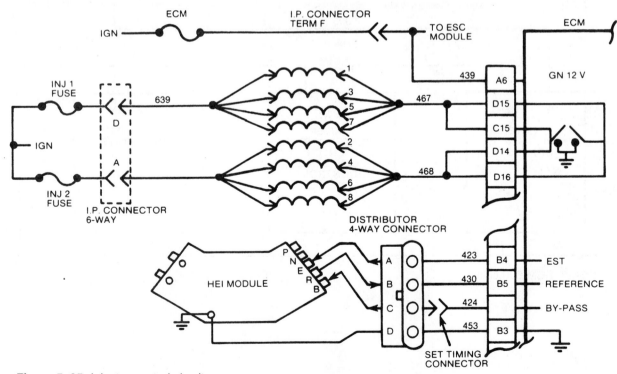

Figure 5–25 Injector control circuit

BCM sections of Chapter 7 for more information about body computers.

Cruise Control

When cruise control is engaged, the ECM is notified by the cruise control module. The ECM uses this information to modify its control of the torque converter lockup clutch.

OUTPUTS

Injectors

The fuel injector is a solenoid-operated nozzle controlled by the ECM, Figure 5-24. The ECM pulses it by grounding the return side (low voltage side) of the injector circuit, Figure 5-25. Fuel under nearly constant controlled pressure is sprayed past the open needle and seat assembly into the intake port of each cylinder. On all except SFI applications, all injectors are simultaneously pulsed once, in each crankshaft revolution. The SFI system pulses each injector individually, once per engine cycle in the firing order, Figure 5-26. During each pulse the injectors are held open for about one or two milliseconds at idle (depending on coolant temperature), to perhaps six or seven milliseconds at WOT (1 millisecond equals 1/1,000 of a second).

Multec Injector. Rochester Products, a division of General Motors, has developed a new style injector for multipoint injection systems, Figure 5-27. It is referred to as a Multec, which is short for "Multiple Technology Injector." This injector features a lower opening voltage requirement (important during cold weather cranking), fast response time, better fuel atomization,

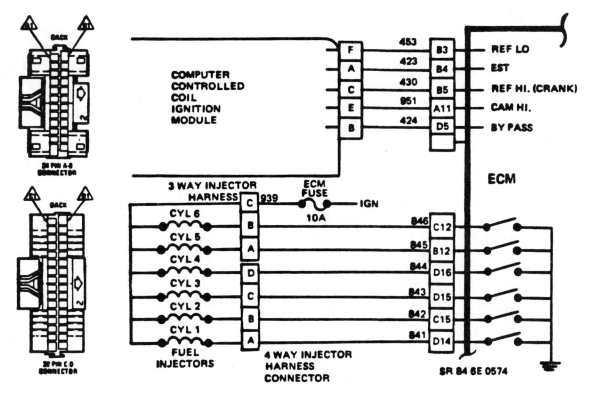

Figure 5–26 Injector control circuit for SFI system. *(Courtesy of General Motors Corporation, Service Technology Group.)*

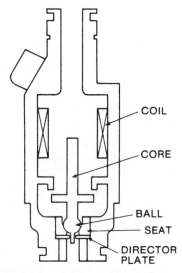

Figure 5–27 Multec injector schematic

improved spray pattern control, and less susceptibility to fouling due to fuel blends or contamination. The traditional pintle, as shown in Figure 5-24, is replaced with a ball shape on the solenoid core. When the ball is lifted off its seat, fuel sprays through the seat opening and past the director plate into the intake port.

Real World Problem

While fuel injectors have no difficulty spraying fuel when they are clean and open, experience has shown that certain brands and blends of fuel are not entirely compatible with the conditions under which the injectors operate. In particular, the lower volatile compounds in the fuel tend to remain on the tip of the injector and in the pintle and seat opening as the engine is shut off. The heat of the engine then hardens the residue into deposits that can either limit the spray quantity, or even degrade the spray pattern so as to prevent proper vaporization. Since this problem is usually not identical at each cylinder of the engine, the quantity sprayed and the quality of the spray patterns gradually begin to vary across the engine's injectors. The system, of course,

does not include measures to identify which cylinder is spraying rich or which is spraying lean. So the computer has to work with the average mixture feedback signal it gets from the oxygen sensor. Once the fuel injectors' delivery volume varies by more than about 3%, however, there is no pulse-width setting that will deliver the correct mixture. In these cases only fuel injector cleaning with special chemicals or fuel injector replacement will clear up the delivery problem.

Cold-Start Injector. The cold-start injector is controlled by a thermal vacuum switch screwed into the engine water jacket, Figure 5-28. When the starter solenoid is engaged, voltage is applied to circuit 806 by way of the crank fuse. Circuit 806 powers a heat element in the **thermo-time switch.** Branching off circuit 806 is circuit 832, which supplies voltage to the cold-start injector. Ground for the cold-start injector is provided by a bimetallic switch in the thermo-time switch. If the bimetallic switch is cool and closed when voltage is applied to the injector, the injector is activated. As soon as voltage is applied to the injector, however, it is also applied to the heat element in the thermo-time switch. If the coolant is at 20 degrees Centigrade (68 degrees Fahrenheit) or below, the bimetallic switch stays closed for a maximum of eight seconds before the heat from the heat element opens the control switch. So the cold-start injector turns on for a maximum of eight seconds or less, depending on coolant temperature, while and only while the engine is cranking. If the engine is started hot, of course, the cold-start injector does not come on at all.

Electric Fuel Pump

Other than providing a higher fuel pressure, the fuel pump for the PFI system operates and is controlled just like the pump for the EFI system. Greater fuel pressure allows finer atomization of the fuel and resists vapor lock problems more effectively.

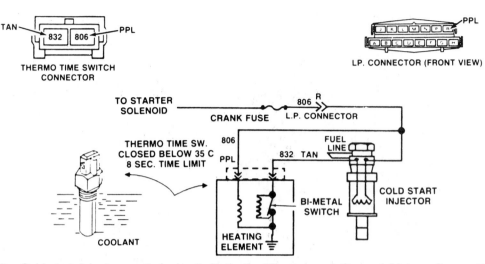

Figure 5–28 Cold-start injector control circuit (typical). *(Courtesy of General Motors Corporation, Service Technology Group.)*

Electronic Spark Timing (EST)/Distributorless Ignition Systems

EST maintains optimum spark timing under all conditions of engine load, speed, temperature, and air density or mass. The distributor type ignition systems used on PFI engines operate the same as those discussed previously, except none of them use Hall-effect switches.

Given the developments we have seen in computerized engine controls, there is no need to continue using a distributor with its tendencies toward wear and mechanical failure because:

- reference signals can be taken directly from the crankshaft and camshaft.
- an ignition module can be mounted almost anywhere in the vehicle.
- spark advance can be more effectively optimized by a fast microprocessor.
- secondary spark distribution can be managed by separate coils and a module.

This should not be very surprising. After all, even the old points and condenser distributors were essentially nothing more than camshaft position sensors, combined with a simple mechanical/vacuum spark advance calculator and an on/off switch.

The term distributorless ignition system has come to be used as a generic label for an ignition system that does not use a distributor and may be abbreviated in some literature as DIS. However, General Motors most often uses the same abbreviation for direct ignition system, one of their specific distributorless ignition systems. General Motors has introduced several different distributorless ignition systems on various engine applications:

- the C3I (computer-controlled coil ignition, of which there are three subtypes).
- the DIS (direct ignition system).
- the IDI (integrated direct ignition).

Each system has its own unique module receiving information about engine speed and crankshaft/camshaft positions from the system sensors. The module monitors this information and passes it on to the ECM. A coil pack, with one ignition coil for every pair of firing-order-opposite cylinders, is mounted on and controlled by the module, Figure 5-29. The module on these systems performs the same function as an HEI

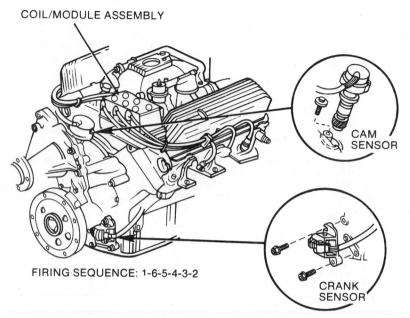

COIL/MODULE ASSEMBLY

CAM SENSOR

CRANK SENSOR

FIRING SEQUENCE: 1-6-5-4-3-2

SENSORS 3.8L TURBO

Figure 5–29 C3I control circuit. *(Courtesy of General Motors Corporation, Service Technology Group.)*

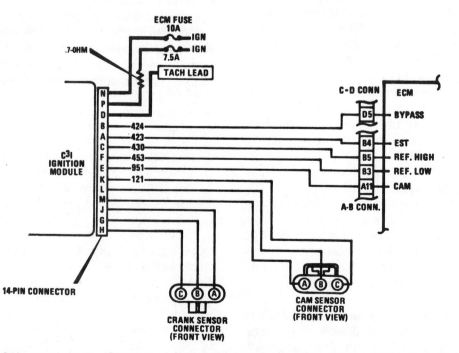

Figure 5–30 C3I control circuit. *(Courtesy of General Motors Corporation, Service Technology Group.)*

module in a distributor type ignition system, with the same REF, bypass, EST, and ground circuits connecting the module to the ECM, Figure 5-30. In this case the module must select and fire each coil in the correct sequence. The module and coil pack may be mounted anywhere on the engine. The end of each coil's secondary winding is connected to a spark plug by way of a spark plug cable, Figure 5-31. Each time a coil's primary circuit opens, that coil fires both of its spark plugs simultaneously.

The spark plugs the coil fires are in companion cylinders; those cylinders arrive at TDC at the same time but opposite each other in firing order. For instance, if number 1 cylinder is being fired at

the end of its compression stroke (assume the firing order is 1-6-5-4-3-2), Number 4 will be at the top of its exhaust stroke and will be fired also. It only takes about 2 to 3 kV (k stands for kilo—"1,000"—and V stands for volts) to fire a spark plug during the exhaust stroke, about the same additional voltage required to jump the rotor gap in a distributor.

The C3I system was introduced in 1984 on the 3.8-liter turbocharged Buick engine. The C3I II system soon followed, and C3I III was introduced in 1988 on the 3800 (simply another name for the same displacement, but a new V6 engine, introduced that year). The ignition module is connected to the ECM by a 14-pin connector, Figure

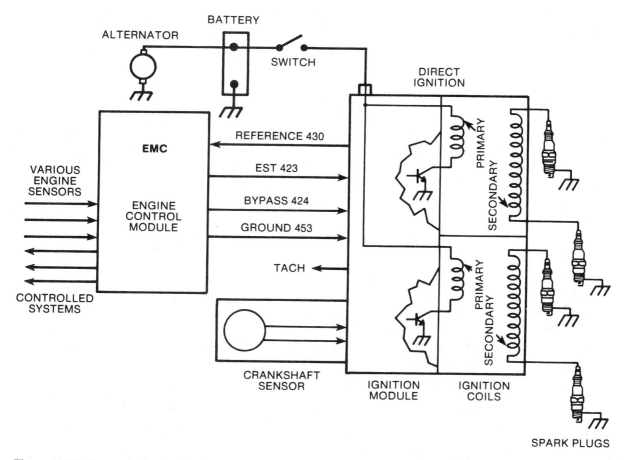

Figure 5–31 Four-cylinder DIS systems schematic. *(Courtesy of General Motors Corporation, Service Technology Group.)*

5-30. Ignition switch voltage at terminal N powers the module; ignition switch voltage at terminal P powers the ignition coils. The 0.7-ohm resistance wire in the circuit to P is to prevent coil overheating during high ambient temperatures. If the ignition is turned on without a cranking signal appearing within one to two seconds, the module will shut off primary circuit current to prevent coil overheating. The ECM has a spark timing range capacity of zero degrees to 70 degrees and will provide dwell times between three milliseconds at high rpm and fifteen milliseconds at low rpm.

On Type I and Type II C3I systems and on DIS and IDI systems, the module has to see a signal from the cam or sync sensor during cranking before it knows what position the crank is in, and which coil to fire. Thus no spark occurs until the module sees the sync signal. Once that signal has occurred, the module will know what position the crankshaft is in and will fire a coil for the next appropriate cylinder event signal, which will be the second cylinder in the firing order. Depending on what position the crankshaft is in when the starting crank begins, the crankshaft may spin more than a full revolution before a spark occurs. On these systems the first cylinder to fire will always be the second cylinder in the firing order. The C3I Type III (Fast Start) system provides slightly faster starting due to its improved sensor input.

IDI. The latest system is called "Integrated Direct Ignition" (IDI) and is used on the 2.3-liter Oldsmobile Quad Four engine. The IDI system is similar to the direct ignition system. It uses the same crankshaft-mounted, seven notch reluctor and crankcase-mounted magnetic sensor, but uses a different module and harness connectors, Figure 5-14. No part of the IDI system is visible from the exterior of the engine. The coil, module, spark plugs, and wires are mounted under the topmost part on the engine, Figure 5-32.

Note that the Quad Four engine and other four-cylinder engines using this system nonetheless use the seven notch plate on the crankshaft. Software hard-wired into the module allows the system to make allowances for the correct number of cylinders.

DIS. During module mode operation (see Chapter 4 "Electronic Spark Timing") on four-cylinder applications, the signal produced by the sync notch tells the ignition module to skip the next signal (produced by the number 1 notch) and to fire the 2-3 coil on the signal produced by the number 2 notch, Figure 5-33. The module will then skip the signals produced by notches 3 and 4 and will fire the 1-4 coil in response to the signal produced by notch number 5. Notches 6 and 7 pass the sensor without the module responding to them and the process begins again. Notice the module is not concerned with cylinder firing order; it is concerned with coil firing order.

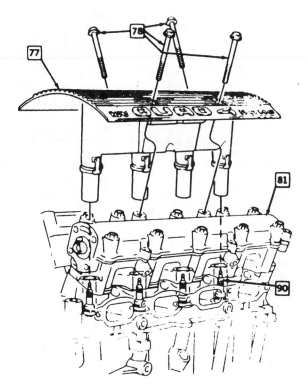

77. **IGNITION COIL AND MODULE ASM.**
78. **BOLTS, IGNITION COIL AND MODULE ASM. TO CAMSHAFT HOUSINGS**
81. **COVER, CAMSHAFT HOUSING (INTAKE SHOWN)**
90. **SPARK PLUG**

Figure 5-32 Quad Four ignition system

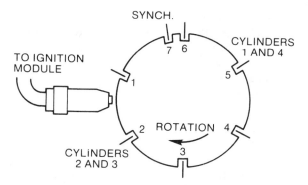

Figure 5–33 Four-cylinder coil firing sequence

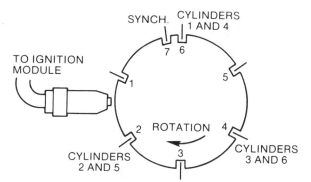

Figure 5–34 Six-cylinder coil firing sequence

On the six-cylinder engine, the signal produced by notch number 7 references the ignition module to skip the signal from notch 1 and respond to the signal from notch number 2 to fire the 2-5 coil, Figure 5-34. The module then skips the signal from notch number 3 and fires the 3-6 coil on the signal from notch number 4. The signal from notch number 5 is skipped and the 1-4 coil is fired on the signal from notch number 6.

Electronic Spark Control (ESC)

All PFI systems except those on the 2.8-liter engines are equipped with ESC. Its function is to retard ignition timing if detonation occurs. Since modern engines are designed to run at the ragged edge of knock, this requires precise controls. The ESC has its own hybrid module that works in conjunction with the ECM, Figure 5-35.

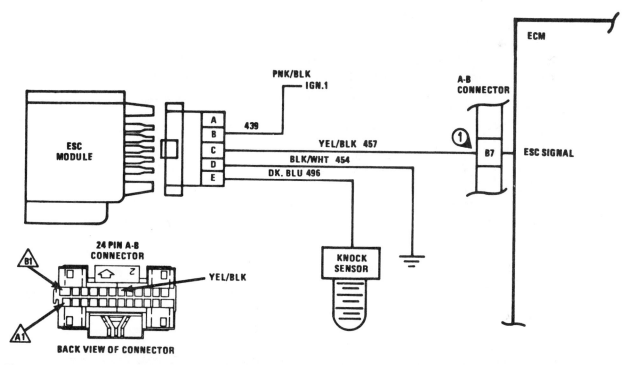

Figure 5–35 ESC circuit. *(Courtesy of General Motors Corporation, Service Technology Group.)*

Its operation is essentially the same as the later style ESC system. The conditions under which ignition timing are retarded in response to spark knock are:

- engine speed is above 1,100 rpm.
- ignition voltage to the ECM is above 9 volts.
- no code 43 is present

Idle Air Control (IAC)

The IAC assembly is a stepper motor and pintle valve just as on the EFI system. It is screwed into the throttle body and controls idle speed, Figure 5-36. With one exception, it works like the IAC used on the EFI system. The exception is that the ECM rereferences its count of the IAC's pintle valve position by moving the valve to its zero position (closed) each time the ignition is turned off, instead of when the ECM gets a 30 mph signal from the VSS as the EFI system does. The ECM then issues a preprogrammed

number of pulses to open the valve to a ready position for the next start-up. The following inputs can affect idle speed performance:

- battery voltage.
- coolant temperature.
- throttle position.
- vehicle speed.
- mass airflow.
- engine speed.
- A/C clutch signal.
- power steering pressure signal.
- P/N switch signal.

Air Injection Reaction (AIR) System

Only some of the PFI systems require an AIR system.

2.8-Liter Engines.   The 1985 2.8-liter engine equipped with manual transmission uses an AIR system with a single-bed, three-way converter. During warm-up the ECM energizes the

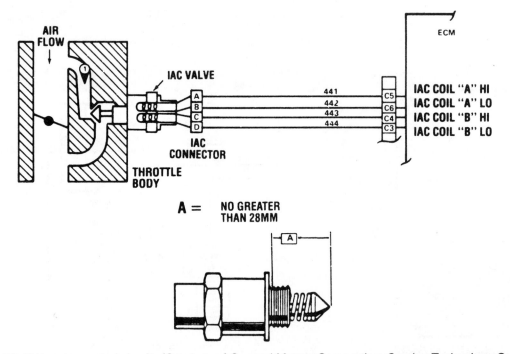

Figure 5–36 IAC motor control circuit. *(Courtesy of General Motors Corporation, Service Technology Group.)*

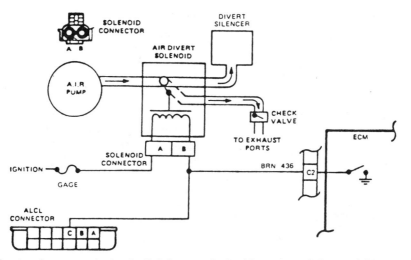

Figure 5–37 Divert valve control circuit (2.8-liter engine). *(Courtesy of General Motors Corporation, Service Technology Group.)*

divert valve solenoid, Figure 5-37, which then allows vacuum to be applied to the divert valve. The pump's air is directed to the exhaust ports. As the engine approaches normal temperature (and just before going into closed loop), the ECM turns off the solenoid, which in turn blocks vacuum from the divert valve. The divert valve then directs the air to the air cleaner (divert mode).

5.0- and 5.7-Liter Engines. These engines use dual-bed, three-way catalytic converters. The air pump pumps air to the system control valve; the control valve, which contains two valves in one housing, directs air to any one of three places, Figure 5-38:

- to the exhaust ports during engine warm-up.
- to the oxidizing chamber of the catalytic converter during closed loop operation.
- to atmosphere (divert mode—during WOT, deceleration, when any failure occurs that causes the ECM to turn on the check engine light or when rpm is high and causes the pressure relief valve inside the control valve to open, though pressure relief is not really a divert mode). Divert mode is intended to prevent converter overheating.

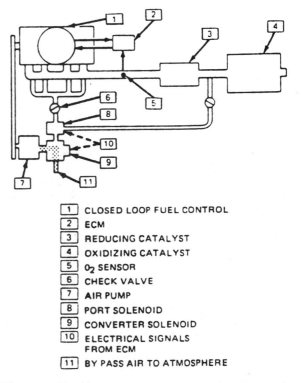

1	CLOSED LOOP FUEL CONTROL
2	ECM
3	REDUCING CATALYST
4	OXIDIZING CATALYST
5	O_2 SENSOR
6	CHECK VALVE
7	AIR PUMP
8	PORT SOLENOID
9	CONVERTER SOLENOID
10	ELECTRICAL SIGNALS FROM ECM
11	BY PASS AIR TO ATMOSPHERE

Figure 5–38 Air management system (5.0-liter and 5.7-liter engines). *(Courtesy of General Motors Corporation, Service Technology Group.)*

Exhaust Gas Recirculation (EGR)

The EGR valve lets small amounts of inert exhaust gas into the intake manifold to lower combustion temperature by slowing combustion, thereby reducing the production of oxides of nitrogen (NO_x). The EGR valve is opened by ported vacuum (manifold vacuum on some applications), which pulls the diaphragm up against the diaphragm spring, Figure 5-39. Three types of vacuum-operated EGR valves are used: the standard ported valve as shown in Figure 5-39, a positive backpressure valve, and a negative backpressure valve. The other two valves are very similar in design to the standard valve, except that they are sensitive to and modulated by exhaust backpressure or a combination of exhaust backpressure and manifold vacuum, respectively. They can be identified by:

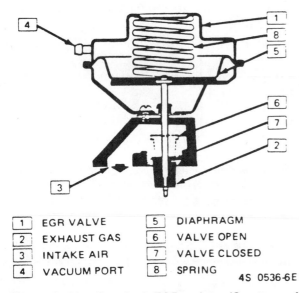

1	EGR VALVE	5	DIAPHRAGM
2	EXHAUST GAS	6	VALVE OPEN
3	INTAKE AIR	7	VALVE CLOSED
4	VACUUM PORT	8	SPRING

4S 0536-6E

Figure 5–39 Standard EGR valve. *(Courtesy of General Motors Corporation, Service Technology Group.)*

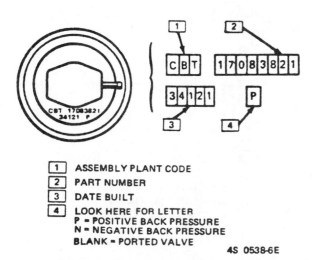

1	ASSEMBLY PLANT CODE
2	PART NUMBER
3	DATE BUILT
4	LOOK HERE FOR LETTER

P = POSITIVE BACK PRESSURE
N = NEGATIVE BACK PRESSURE
BLANK = PORTED VALVE

4S 0538-6E

Figure 5–40 EGR valve identification. *(Courtesy of General Motors Corporation, Service Technology Group.)*

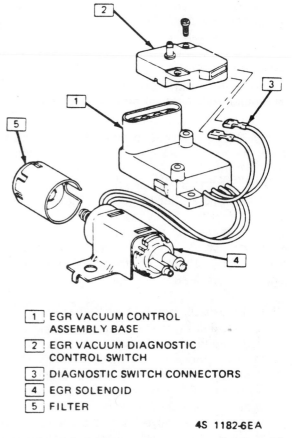

1	EGR VACUUM CONTROL ASSEMBLY BASE
2	EGR VACUUM DIAGNOSTIC CONTROL SWITCH
3	DIAGNOSTIC SWITCH CONNECTORS
4	EGR SOLENOID
5	FILTER

4S 1182-6EA

Figure 5–41 EGR vacuum control assembly. *(Courtesy of General Motors Corporation, Service Technology Group.)*

- a "P" stamped on the top of the positive backpressure valve, Figure 5-40.
- an "N" stamped on the top of the negative backpressure valve.
- a blank space on top of the port valve.

Regardless of the valve used, vacuum to it is controlled by a vacuum control assembly, Figure 5-41. The control assembly solenoid is pulsed many times per second by the ECM. When the solenoid is turned on, vacuum to the EGR port is blocked. The strength of the vacuum signal, which controls how far the EGR valve opens, is controlled by varying the pulse width. The solenoid is fitted with a vent filter which must be replaced periodically. The ECM controls the solenoid using input concerning coolant temperature, throttle position, and mass airflow. The EGR valve is not opened unless the engine is warm and above idle.

Digital EGR Valve. A new concept in EGR valves is employed on the 3800 V6 engine for 1988 and beyond. The digital EGR valve provides the following advantages:

- vacuum is not required for its operation.
- increased control of EGR flow.
- faster response to ECM commands.
- greater ECM diagnostic capacity.

The digital EGR valve is directly operated by solenoids rather than by a vacuum diaphragm.

Instead of trying to regulate EGR flow by controlling how far the valve opens or by duty cycling it, three different valves are used to open or close three different-sized orifices. Each valve can be opened or closed independently of the others, allowing any of seven different increments of exhaust gas flow into the induction system, Figure 5-42. Orifice number one, when open, will flow 14 percent of the maximum EGR. Orifice number two, when open will flow 29 percent of the maximum. Number three will flow 57 percent of maximum when open. By manipulating which valve or combination of valves is open at any given time, EGR flow can be easily and accurately controlled.

The digital EGR assembly consists of the EGR base, the EGR base plate and the solenoid and mounting plate assembly, Figure 5-43. When the base and base plate are fitted together, there is a sealed cavity between them with the base forming the floor of the cavity and the base plate forming the ceiling of the cavity. There are four holes in the base (floor). The center hole is always open and allows exhaust gas to enter the cavity. The other three holes are closed by the three EGR valve pintles, Figure 5-44. The pintles are attached to their pintle shafts by a ball-joint type connection so they can more readily align with the seats they rest on and maintain a more positive seal.

As shown in Figure 5-44, the pintle and its shaft are part of the armature assembly. Each

Increment	Orifice #1 (14%)	Orifice #2 (29%)	Orifice #3 (57%)	EGR Flow
0	closed	closed	closed	0%
1	open	closed	closed	14%
2	closed	open	closed	29%
3	open	open	closed	43%
4	closed	closed	open	57%
5	open	closed	open	71%
6	closed	open	open	86%
7	open	open	open	100%

Figure 5–42 Increments of EGR flow with digital EGR valve system

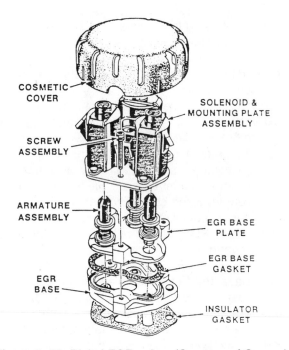

Figure 5–43 Digital EGR valve. *(Courtesy of General Motors Corporation, Service Technology Group.)*

pintle-shaft-armature assembly is held down by its armature return spring, which is attached to the lower portion of the shaft. When one of the solenoid windings in the solenoid assembly is turned on by the ECM, the resulting magnetic field lifts the corresponding armature assembly. This opens the valve and compresses the return spring. When the ECM turns the solenoid off, the spring drives the armature assembly back down and closes the valve.

There are two seals on the pintle shaft. The lower seal is pushed down against the top of the base plate. It prevents exhaust gases in the base cavity from leaking out between the stem and the base plate. On this EGR valve design, ambient air has much less tendency to leak into the induction system by way of the EGR valve stem seal than it does on the traditional EGR valve design. This is because exhaust pressure in the base cavity prevents a low pressure from developing under the seal.

The upper seal is pressed against the bottom of the solenoid assembly. It keeps dirt and dust

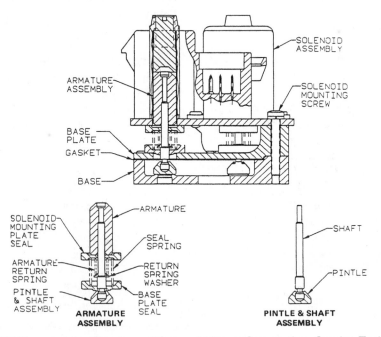

Figure 5–44 Digital EGR assembly. *(Courtesy of General Motors Corporation, Service Technology Group.)*

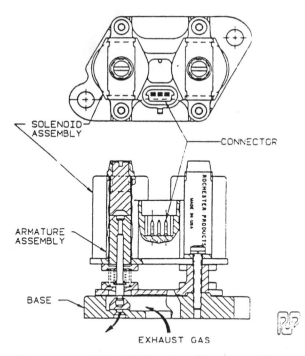

Figure 5–45 Digital EGR valve. *(Courtesy of General Motors Corporation, Service Technology Group.)*

from entering the armature cavity of the solenoid assembly to reduce wear and provide greater reliability. A spring fitted between the seals holds them in place. The return spring also helps hold the upper seal in place.

Some engines use a two-solenoid digital EGR valve, which works the same way except that it only has three different EGR flow rates, Figure 5-45. When the first valve is open, about 33% of the total EGR will flow. When the second valve is open, about 66% of the total will flow. When both valves are open, 100% of EGR flows.

By not relying on vacuum to operate the EGR valve assembly, the ECM can be made to monitor EGR system operation for fault diagnosis without any additional external hardware. If a short or open occurs in any one of the solenoid circuits, a code will be set in the ECM's keep alive memory. Each solenoid has its own assigned code number.

Charcoal Canister Purge

The ECM operates a solenoid that controls vacuum to the purge valve on the charcoal canister. The solenoid is turned on and blocks vacuum to the purge valve when the engine is cold or at idle, Figure 5-46. The ECM turns the solenoid off and thus allows any stored hydrocarbons (fuel) to be purged through the purge valve when:

- the engine is warm.
- the engine has run for a preprogrammed amount of time since it was started.
- a preprogrammed road speed is reached.
- a preprogrammed throttle opening is achieved.

In addition to the purge valve, some engine application canisters also have a non-ECM-controlled control valve, Figure 5-47. It is controlled by ported vacuum. When vacuum is applied to it, it allows canister purging to occur through another port connected to manifold vacuum.

For 1987, the Chevrolet 5.0- and 5.7-liter engines have a revised canister purge strategy. When the engine is operating in the closed-loop mode, the ECM will duty cycle the purge solenoid to control the amount of vapors that are admitted into the engine's induction system, rather than just turning the purge solenoid on to stop purging and off to allow purging. The ECM will use input from the oxygen sensor to determine the volume of vapors to purge from the canister. If a rich condition is indicated by the oxygen sensor, purge volume is reduced. This strategy is intended to provide improved driveability.

Turbocharger

One of the major factors that controls power is how much air and fuel are put into the cylinders. Putting more fuel in is a fairly simple matter. Putting more air in is not so simple (the proper air/fuel ratio must be maintained). The amount of air that can be put into the cylinder is limited by atmospheric pressure unless some means is

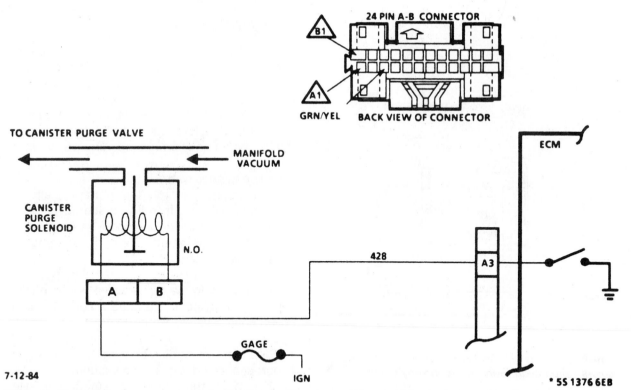

Figure 5–46 Canister purge control solenoid circuit. *(Courtesy of General Motors Corporation, Service Technology Group.)*

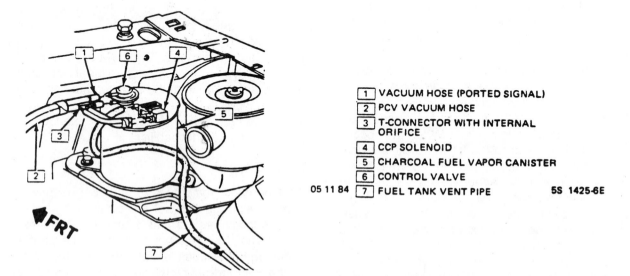

1	VACUUM HOSE (PORTED SIGNAL)
2	PCV VACUUM HOSE
3	T-CONNECTOR WITH INTERNAL ORIFICE
4	CCP SOLENOID
5	CHARCOAL FUEL VAPOR CANISTER
6	CONTROL VALVE
7	FUEL TANK VENT PIPE

05 11 84   5S 1425-6E

Figure 5–47 Canister purge and control valves. *(Courtesy of General Motors Corporation, Service Technology Group.)*

used to force air in at greater than atmospheric pressure. This is accomplished with some type of supercharger of which the turbocharger is the most popular and generally the most efficient.

A turbocharger is a centrifugal, variable-displacement air pump driven by otherwise wasted heat energy in the exhaust stream. It pumps air into the intake manifold at pressures that are limited only by the pressure that can be developed without significantly increasing the air's temperature.

Most **normally aspirated** engines (those that rely on atmospheric pressure to fill the cylinder) only fill the cylinder to about 85% of atmospheric pressure (they actually achieve 85% volumetric efficiency, or VE, at full throttle with the engine at its maximum torque speed). Turbocharging produces VE values in excess of 100%.

Turbocharger Operation. After leaving the manifold, the hot exhaust gases flow through the vanes of the exhaust turbine wheel and thus spins it, Figure 5-48. The more exhaust volume coming out of the engine, the faster the turbine spins. It can achieve speeds in excess of 130,000 rpm. The exhaust turbine drives a short shaft that drives a compressor turbine. The high speed produces centrifugal force to move the air from between the vanes and thus causes it to flow out radially. The air is then channeled into the intake manifold. As air is thrown from the turbine vanes, a low pressure develops in its place. Atmospheric pressure pushes more air through the air cleaner, the mass airflow sensor (if used), and the throttle body.

A criterion turbine speed must be reached before a pressure boost is realized; this speed varies with turbine size and turbocharger design. If turbine speed goes too high, boost pressure (and charge temperature) will also go too high and can cause preignition and engine damage. To limit boost pressure, a **wastegate** is used to divert exhaust gases away from the exhaust turbine and thus limit its speed.

Wastegate. The wastegate opens by a wastegate actuator. A spring in the actuator holds the wastegate closed. When manifold pressure (turbo boost) reaches approximately 8 psi, it overcomes the actuator spring and opens the wastegate. If however, engine operating parameters are favorable (coolant temperature, incoming air temperature, etc.), the ECM pulses a wastegate solenoid, which in turn bleeds off some of the pressure acting on the actuator. When this

Turbochargers

Because a turbocharger puts more air into the cylinder, it raises the engine's compression pressure. Because it does not produce significant boost pressure at low exhaust flow rates, it allows an engine the benefits of low compression during light throttle operation and high compression during heavy throttle operation. Low compression offers lower combustion chamber temperatures, which result in less wear and lower NO_x emissions. High compression provides higher combustion chamber temperatures, which result in more complete combustion and more power for the amount of fuel consumed.

Another way to look at it is in terms of displacement. If an engine is operating at an atmospheric pressure of 14 psi, and the turbocharger is producing a boost of another 14 psi, the cylinders are charged at a pressure of 28 psi. The engine is consuming approximately twice the amount of air it could without the turbocharger. We have effectively doubled the engine's displacement. The engine certainly does consume twice as much fuel, too, but not as much as it would with twice the displacement. This is because the higher compression pressure and temperature produces more complete combustion. The turbocharged engine will also weigh less than an equivalently powerful naturally-aspirated engine, thus saving on the amount of work done.

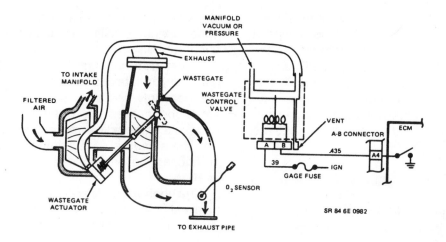

Figure 5–48 Turbocharger and wastegate control (typical). *(Courtesy of General Motors Corporation, Service Technology Group.)*

occurs, boost pressure is allowed to rise to 10 psi before the wastegate opens.

A code 31 is set if:

- an overboost is sensed by the MAP sensor on the 1.8-liter engine.
- the ECM, monitoring the wastegate solenoid circuit operation, sees a malfunction in the circuit while the solenoid is operated between a 5% and a 95% duty cycle on the 3.8-liter engine.

Transmission Converter Clutch (TCC)

Notice that the name has changed slightly from this unit's application on CCC and EFI systems, where TCC meant "torque converter clutch." It is still the same part, however. The purpose of the TCC is still to increase fuel economy. It does so by eliminating hydraulic slippage during cruise conditions and by eliminating heat production in the torque converter, especially during overdrive operation.

Electric Cooling Fan

All General Motors vehicles with transverse mounted engines and a few with **longitudinal** engines (parallel to the center line of the car) are equipped with an electric cooling fan to pull air through the radiator and A/C condenser. Control of the fan varies somewhat with engine application. In all cases, however, the fan is turned on when:

- coolant temperature exceeds a specified value. This can be done by either the ECM or a coolant temperature override switch on some applications and only by the ECM on others.
- when A/C compressor output pressure (head pressure) exceeds a specified value. This is done by a switch in the high pressure side of the A/C system on some engine applications. It is done by the ECM on other applications (on those applications, an A/C head pressure switch feeds head pressure information to the ECM).

On most applications, the ECM turns on the fan anytime the A/C is on and vehicle speed is less than a specified value. Others turn it on under a specified speed whether the A/C is on or off. Once the vehicle reaches a criterion speed, enough air is pushed through the radiator without the aid of the fan, unless overheating or high A/C head pressure conditions exist. This is a fuel economy feature. Study Figure 5-49 as a typical example.

Air Conditioning (A/C) Control

The ECM controls the relay that turns the A/C clutch on and off for two and in some cases three reasons, Figure 5-50:

- When the A/C control switch (on the instrument panel) is turned on, the ECM delays A/C clutch engagement for 0.4 second to allow time to adjust the IAC valve.
- The ECM disengages the A/C clutch during WOT operation.
- On some applications, the ECM turns off the A/C clutch if power steering pressure

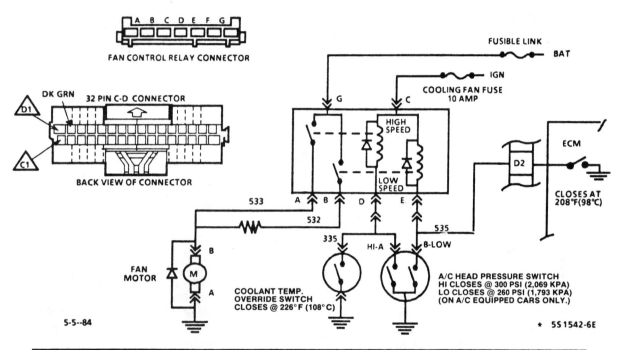

COOLANT FAN CONTROL OPERATING CONDITIONS (WITH A/C)					
A/C SW.	HEAD PRESS.	ROAD SPEED	ENG. TEMP.	FAN SPEED	FAN ON BECAUSE:
OFF/ON	UNDER 260 PSI	UNDER 45 MPH	OVER 98°C	LOW	ECM TURNED ON
OFF/ON	UNDER 260 PSI	UNDER 45 MPH	UNDER 95°C	OFF	
OFF/ON	UNDER 260 PSI	OVER 45 MPH	UNDER 106°C	OFF	
ON	OVER 260 PSI	N/A	N/A	LOW	A/C HEAD PRESS. SW. (LOW) TURNED ON
ON	OVER 300 PSI	N/A	N/A	HIGH	A/C HEAD PRESS. SW. (HIGH) TURNED ON
OFF/ON	N/A	N/A	OVER 106°C	HIGH	COOLANT TEMP. OVERRIDE SW. TURNED ON
WITHOUT A/C					
IGN. SW.		ROAD SPEED	ENG. TEMP.	FAN SPEED	FAN ON BECAUSE:
ON		N/A	OVER 98°C	LOW	ECM TURNED ON
ON		OVER 45 MPH	UNDER 108°C	OFF	
ON		OVER 45 MPH	OVER 108°C	HIGH	COOLANT TEMP. OVERRIDE SW. TURNED ON

Figure 5–49 Coolant fan control circuit (typical). *(Courtesy of General Motors Corporation, Service Technology Group.)*

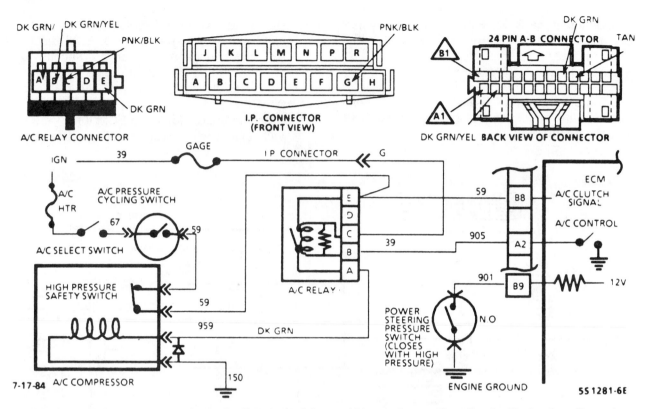

Figure 5–50 A/C control circuit (typical). *(Courtesy of General Motors Corporation, Service Technology Group.)*

exceeds a specified value during idle. On others the power steering switch is in series with either the A/C clutch or the control winding of the A/C relay and disengages the A/C clutch without relying on the ECM.

✔ SYSTEM DIAGNOSIS & SERVICE

Self-Diagnosis

The self-diagnostic capacities and procedures of the PFI systems are essentially the same as those of the CCC system discussed in the **System Diagnosis and Service** section of Chapter 3. In 1988 General Motors made a significant increase in the self-diagnostic capacity of most PFI engine applications. The ECM was designed to monitor more circuits, both input and output. New diagnostic code numbers were also assigned. It is important to note that from year to year, and in some cases even from engine to engine within a single model year, code numbers may have different meanings. Be sure the service literature you use is applicable to the engine being serviced and to the model year of the vehicle.

TCC Test Lead. The TCC test lead is in cavity F of the ALCL. It can be used to monitor the TCC circuit operation with a voltmeter or test light or it can be used to ground the TCC test lead, overriding the ECM.

Fuel Pump Test Lead. On some vehicles the fuel pump test lead is found in cavity G of the ALCL. On others it is found on the left side of the engine compartment. A voltmeter or test light can be connected to this lead to determine whether the fuel pump relay or oil pressure switch has supplied power to the fuel pump power lead. Or, a jumper wire can be connected from the test lead to a 12-volt source. This powers the fuel pump for fuel pressure or injector tests.

WARNING: When powering the fuel pump by the above method, be sure that there are no fuel leaks and avoid causing any sparks. Otherwise there is a serious fire hazard.

NOTE: Diagnostic procedures for 1987 and later are written to include the use of an ALCL tool, often referred to as a scanner or scan tool. Previously, diagnostic procedures were written around the use of a voltmeter to obtain such measurements as sensor readings.

Diagnostic Procedure

Diagnostic Circuit Check. After making a thorough visual inspection, paying particular attention to possible causes of vacuum leaks, and correcting any problems found, perform the diagnostic circuit check. This procedure is outlined in the service manual. It is designed to identify the type of problem causing the complaint and it refers to the next step or chart to use.

Fuel Pressure Test. Some diagnostic charts call for a fuel pressure test. Fuel pressure is critical to the performance of a fuel injection system. The fuel rail has a fuel pressure test fitting that contains a Schraeder valve like that used in an A/C pressure test fitting or in the valve stem of a tire.

WARNING: While the fuel pressure gauge hose is screwed to the test fitting, a shop towel should be wrapped around the fitting to prevent gasoline from being sprayed on the engine. Remember, the fuel is under high pressure and the residual check valve in the fuel pump can hold pressure for some time after the engine is shut off. Spraying fuel can be not only a fire danger, but also a risk to the eyes and nose.

Injector Balance Test

To realize some of the advantages of PFI, the injectors must all deliver the same amount of fuel to each cylinder. The injector balance test tests injector performance and uniform fuel delivery. Check the applicable service manual for specific directions.

Integrator and Block Learn

The integrator serves General Motors fuel injected engines in much the same way that dwell, a measure of duty cycle, serves CCC applications. Let's assume that while in closed loop a particular fuel injected engine is operating with a slight vacuum leak. The oxygen sensor reading indicates a lean mixture until the ECM increases injector pulse width enough to compensate for the vacuum leak and get the air/fuel mixture back to 14.7 to 1. If an ALCL tool were connected to the ALCL with "integrator" selected, a number would be shown on the ALCL tool representing the injector pulse width currently being used. In a normal situation where no compensation is required, that number would be 128. If, as in our example, pulse width has been increased to compensate for a tendency to be lean—the vacuum leak—the number would be above 128. If the pulse width has been decreased to compensate for a tendency to run rich, the number would be below 128.

Block learn functions much like the integrator, except it represents:

- pulse width adjustments that have become a trend over a longer period of time.
- learned behavior on the part of the ECM that has caused it to modify its original pulse width programming.

Once the engine has been running in closed loop long enough for block learn information to be stored in memory (as a number), that information is retained in memory until it is modified by new information or until the ignition has been turned off. Some earlier Buick and Cadillac systems

A

System operating
at design
specifications

heavy
L
O
A
D
light

128	128	128	128
128	128	128	128
128	128	128	128
128	128	128	128

low   hi

RPM

B

Result of slightly
lean tendency

heavy
L
O
A
D
light

128	130	128	129
129	130	130	131
133	136	129	128
135	132	129	130

low   hi

RPM

C

Result of slightly
rich tendency

heavy
L
O
A
D
light

121	124	128	128
119	120	119	121
115	117	116	121
122	119	126	120

low   hi

RPM

Figure 5–51 Block learn information

would hold block learn numbers in memory until battery power is disconnected from the ECM.

Block learn numbers are stored in sixteen different cells within a square block, as shown in Figure 5-51. Each cell represents a different combination of engine speed and load. Remember that block learn only functions during closed-loop operation. If the pulse width that was originally programmed for each of the sixteen combinations of engine speed and load requires no adjustment to maintain a 14.7 to 1 air/fuel ratio, the block learn number in each cell will be 128 as in Figure 5-51A. If for some reason, the engine tends toward a slightly lean condition, the block learn numbers may look similar to those in Figure 5-51B. If due to some fault the engine has a tendency to run somewhat rich, the block learn numbers might look similar to those in Figure 5-51C.

If block learn numbers are above 128 but the integrator is at 128, the system has a lean tendency that is being corrected by block learn with wider pulse widths, and the integrator is not having to make any additional corrections. In other words, block learn has modified "normal" to a wider pulse width and the integrator is not having to correct.

If block learn numbers are 128 but the integrator is above 128, the system is trying to run lean but has not been doing so long enough for block learn to have made any correction.

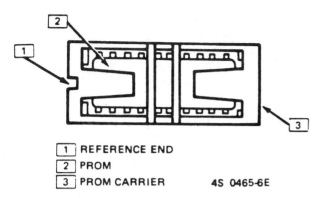

1 REFERENCE END
2 PROM
3 PROM CARRIER   4S 0465-6E

Figure 5–52 PROM and PROM carrier. *(Courtesy of General Motors Corporation, Service Technology Group.)*

If block learn is well above 128 and the integrator is the same or higher, the system is running lean and the ECM is not able to compensate enough to correct it. The same number relationships would occur if the system were trying to run rich, except that the numbers would be below 128.

ECM, PROM and CALPAK Service

If correctly used diagnostic procedures call for an ECM to be replaced, check the ECM and

PROM service and part numbers to be sure they are the right units for that vehicle. If a PROM is replaced, check that it has not been superseded by an updated replacement PROM. This can be done with the aid of the dealer's parts department or with service bulletins. Some aftermarket service manuals contain service bulletins. Carefully remove the PROM and CALPAK from the ECM to be replaced. The service (replacement) ECM does not include a new PROM or CALPAK. Be careful not to touch the pins of either unit with your fingers because static electricity applied to the pins can damage them. Store them in a safe place while they are out of the ECM.

When the PROM is reinstalled, be sure to orient it properly to the PROM carrier, Figure 5-52. While it can be inserted backwards, doing so will destroy it when the ignition is turned on.

Weather-Pack Connectors

All PFI system harness connections under the hood are made with weather-pack connectors.

Diagnostic & Service Tips

IAC Assembly. If the engine is run with the IAC harness disconnected or if the IAC assembly is removed and reinstalled, idle speed will probably be incorrect when it is reconnected. To rereference idle speed, simply start the engine with the IAC harness connected. Allow the engine to warm up and then turn the ignition off. The 3.0- and 3.8-liter engines rereference simply by cycling the ignition on and off.

If the IAC assembly is removed, be sure that the distance from the tip of the pintle valve to the motor housing does not exceed 28 millimeters (1-1/8 inch), Figure 5-53. Otherwise it will be damaged when it is installed. See the service manual for detailed instructions.

When it is installed, use the correct torque specifications. Overtorquing can distort the IAC housing and cause the motor to stick. Complaints of engine stalling as the car comes to a stop can be a result of this condition. Also, if the IAC assembly is replaced, be sure to get the right part. There are three different pintle shapes, and the correct one must be used.

Minimum Idle Speed Adjustment. Minimum idle speed is the speed achieved when the IAC valve is in its closed position. The only air getting into the intake manifold is what goes by the throttle blade (unless there is a vacuum leak). An idle

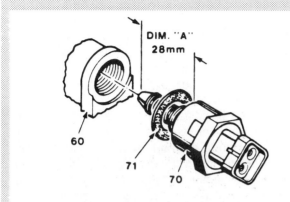

60 IDLE AIR/VACUUM SIGNAL HOUSING
70 IDLE AIR CONTROL VALVE (IAC)
71 IAC GASKET

05 24 84   5S 1455-6E

Figure 5–53 IAC motor. *(Courtesy of General Motors Corporation, Service Technology Group.)*

CYLINDER	WIRE COLOR
1	Black
2	Dk. Green
3	Pink
4	Red
5	White
6	Lt. Green

Figure 5–54 SFI injector wire color codes. *(Courtesy of General Motors Corporation, Service Technology Group.)*

stop screw is used to position the throttle blade. The head of the idle stop screw is recessed and covered with a metal plug. The plug can be removed and the idle stop screw adjusted. However this should only be done if the throttle body is replaced or a service procedure calls for it. This is not a normal service routine adjustment.

To set the minimum idle speed adjustment, the engine should be warm and the IAC valve fully extended (closed). Consult the appropriate service manual for specific directions and specifications.

TPS Adjustment. The TPS on some PFI systems is adjustable. Adjustment is checked by connecting a voltmeter between terminals A and B at the sensor connection with the ignition on. Gain voltmeter access to the terminals by disconnecting the three-wire weather-pack connector and inserting three short jumper leads to temporarily reconnect the circuit. Performing a minimum idle speed adjustment will probably cause the TPS to need adjustment too.

Injector O-Rings. The O-ring that seals between the bottom of the injector and the manifold is a potential source of vacuum leaks. Carefully examine the O-rings and/or replace them when the injectors are serviced.

3.8-Liter Turbo (SFI) Injector Leads. Because the injectors are pulsed individually and in firing order on the 3.8-liter turbocharged engine, the wires that power the injectors must be attached in the firing order. These wires are color-coded as shown in Figure 5-54.

Fuel Line O-Rings. Threaded connections between the fuel pump and the fuel rail use O-rings to reduce the possibility of leaks. When replaced, they should only be replaced with O-rings designed to tolerate exposure to gasoline.

Fuel Flex Hose. The fuel flex hose used on PFI is internally reinforced with steel mesh. No attempt should be made to repair it. It should be replaced if it becomes unusable or questionable.

WARNING: For personal safety, the factory recommended procedures should be carefully adhered to when servicing fuel lines or fuel line connections. The fuel is under high pressure, and failure to comply with factory-recommended procedures can result in a fuel leak and fire.

In addition to the key terms, refer to Figure 5-55 for a list of the abbreviations used in this chapter.

A/C	Air-conditioning
AIR	Air Injection Reaction
ALCL	Assembly Line Communication Link
ATS	Air Temperature Sensor
BARO	Barometric Pressure
C^3I	Computer-Controlled Coil Ignition
CO	Carbon Monoxide
DIAGNOSTIC "TEST" TERM.	Lead or ALCL connector that is grounded to get a trouble code
DVM (10 Meg.)	Digital Voltmeter with 10 million ohms resistance
ECM	Electronic Control Module (controller)
EECS	Evaporative Emissions Control System
EFE	Early Fuel Evaporation
EFI	Electronic Fuel Injection
EGR	Exhaust Gas Recirculation
ESC	Electronic Spark Control
EST	Electronic Spark Timing
HC	Hydrocarbons
HEI	High Energy Ignition
IAC	Idle Air Control
IP	Instrument Panel
IGN	Ignition
MAF	Mass Airflow
MAP	Manifold Absolute Pressure
MAT	Manifold Air Temperature
MFI	Multiport Fuel Injection. Individual injectors for each cylinder are mounted in the intake manifold. The injectors are fired in groups rather than individually.
MPH	Miles Per Hour
N.C.	Normally Closed
N.O.	Normally Open
NO_x	Oxides of Nitrogen
O_2	Oxygen (sensor)
PCV	Positive Crankcase Ventilation
PFI	Port Fuel Injection
P/N	Park/Neutral
PROM	Programmable Read-only Memory (calibrator)
RPM	Revolutions per Minute
SFI	Sequential Fuel Injection. Used on 3.8 L Turbo. Each injector is individually fired.
TACH	Tachometer
TBI	Throttle Body Injector (unit)
TCC	Transmission Converter Clutch
THERMAC	Thermostatic Air Cleaner
TPI	Tuned Port Injection
TPS	Throttle Position Sensor
TVS	Thermal Vacuum Switch
V	Volt
VIN	Vehicle Identification Number
VSS	Vehicle Speed Sensor
WOT	Wide-open Throttle

Figure 5–55 Explanation of abbreviations

SUMMARY

This chapter has focused on port fuel injection systems that use one injector per cylinder and are triggered by the computer either in the firing order of the engine, or in alternating banks of cylinders. We've seen the advantages port injection provides in a more accurately metered fuel delivery, and the elimination of problems associated with distribution of an air/fuel mixture through the intake manifold. We have seen the method General Motors uses employing a cold-start injector to provide an initially richer mixture to aid in starting under low temperature conditions.

In this chapter, we considered the controls on turbocharger boost for vehicles equipped with this accessory. These controls include a mechanism to modify boost using the wastegate as well as a more sensitive spark advance system using a knock sensor.

We have seen the introduction of a stepper motor controlled engine idle speed, using an air bypass to meter the amount of intake air around the closed throttle. We have also covered the EGR system on this system and how the computer monitors and controls thisis exhaust recirculation.

▲ DIAGNOSTIC EXERCISE

1. A port fuel injected car with a turbocharged engine has suddenly developed the following symptoms: much reduced engine power and increased fuel consumption, a tendency to overheat on only moderately warm days and blue smoke in the exhaust under high throttle settings. The computer indicates a code showing a problem in the oxygen sensor circuit. What sorts of problems might be causing these symptoms?

2. A front-wheel-drive car with the port fuel injection system has a radiator cooling fan that runs whenever the ignition key is turned to on, even when the engine is dead cold. What components of the system play a role in operating this fan, and how should they be checked?

REVIEW QUESTIONS

1. What is one of the unexpected problems with PFI systems?
2. What driveability symptoms can be experienced becaused of this condition?
3. How are the fuel lines used on PFI systems different from those used on carburetion systems?
4. What causes the fuel pressure regulator to vary the pressure from 34 psi to 44 psi?
5. Name at least two ECM responses to the power steering switch opening during idle or speeds below 10 mph.
6. What circuit powers the cold-start injector?
7. List two unique features of the C3I system.
8. If the wire from terminal C of the ESC module to terminal B7 of the ECM were to open, what would happen?
9. How is vacuum to the EGR valve controlled?
10. What opens the wastegate on PFI applications?
11. Name at least two conditions that cause the cooling fan to come on.
12. What does the check engine light do if the ECM sees a fault in a circuit it monitors and then the fault clears up?
13. What is the primary function or functions of the diagnostic circuit check?
14. Name at least one precaution that should be taken before connecting or disconnecting the fuel pressure test gauge.
15. What removable part is found in the ECM besides the PROM?
16. What is a clamping diode or resistor?
17. What can be the result of overtorquing the IAC motor assembly when installing it in the throttle body?
18. Name at least one major concern relevant to the O-ring that seals between the injector

tor and the manifold.

19. What is the combustion air source during minimum air adjustment?

ASE-type Questions (Actual ASE test questions are rarely so product specific).

20. Technician A says that on most General Motors PFI systems, the injectors are all pulsed at the same time. Technician B says they are divided into two groups with each group pulsed alternately. Who is correct?
   a. A only.
   b. B only.
   c. both A and B.
   d. neither A nor B.

21. A 5.0-liter tuned port injection engine is hard to start cold. Inspection shows that the cold-start injector is not working. This could be a fault of _____:
   a. the ECM.
   b. the thermal time switch.
   c. either A or B.
   d. neither A nor B.

22. A car with a PFI system is hard to start, especially after sitting for an hour or more. A fuel pressure test shows that fuel pressure is normal but drops to zero quickly after the engine is turned off. The hard-starting problem is most likely the result of _____:
   a. a defective fuel pump.
   b. a plugged injector.
   c. a defective check valve in the fuel pump.
   d. none of the above.

23. Technician A says that the PFI system rereferences the IAC motor each time the VSS produces a signal that equals 30 mph. Technician B says the PFI system rereferences the IAC motor when the ignition switch is cycled. Who is correct?
   a. A only.
   b. B only.
   c. both A and B.
   d. neither A nor B.

24. Technician A says that if the wastegate fails to open, detonation will probably occur. Technician B says that if the wastegate fails to open, piston damage can occur. Who is correct?
   a. A only.
   b. B only.
   c. both A and B.
   d. neither A nor B.

Chapter 6

Recent General Motors' Engine Controls

OBJECTIVES

In this chapter you can learn:

❑ the expanded range of sensors monitored and actuators controlled in the Aurora/Northstar system's powertrain control module (PCM).

❑ the role of quad drivers and output drivers, special clusters of power transistors doing the high current work of the PCM.

❑ the torque management system and one of its applications, traction control.

❑ the four-coil waste spark system used on the vehicle plus the PCM's ignition and fuel start-up strategy.

❑ the role of the downstream, secondary oxygen sensor, the third oxygen sensor on this engine, and what it can show about the catalytic converter.

❑ how the system conducts misfire detection, and what it does about it when found.

❑ the way the two crankshaft position sensors and the camshaft position sensor work together to provide the PCM with a finely discriminated map of the crankshaft and cylinder positions.

❑ about the EEPROM, the updated and improved PROM, storing semipermanent information that the computer can now update.

KEY TERMS

Aurora/Northstar
Background Noise
Diagnostic Trouble Code (DTC)
EEPROM
Engine Metal Overtemp Mode
Hot Wire Air Flow Sensor
Misfire Detection
Powertrain Control Module (PCM)
Primary/Secondary Oxygen Sensor
Quad Driver/Output Driver
Torque Management
Voltage Drop Test
Vortec Injection
Waste Spark

While General Motors builds a large number of different engines of various sizes and configurations, in this chapter we will focus on the engine used in the Oldsmobile **Aurora,** Figure 6-1. The same engine, in a slightly larger displacement, is also used under the name **Northstar** in certain Cadillac models, because it employs a computer control system that includes most of the systems used on other engines. The several components, like superchargers, not used with the Aurora engine are covered separately.

GENERAL SYSTEM FEATURES

The heart of the latest General Motors engine control computer systems lies in the **powertrain control module (PCM),** Figure 6-2. It is a much faster and more powerful computer than those used previously to monitor and control engines for emissions quality, driveability, and fuel economy. In addition, the PCM is responsible for all the diagnostic functions required for compliance with OBD II regulations. The term **power-**

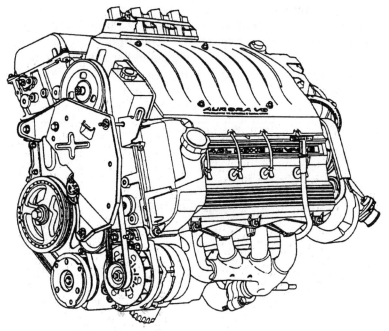

Figure 6–1   Oldsmobile Aurora/Cadillac Northstar engine. *(Courtesy of General Motors Corporation, Service Technology Group.)*

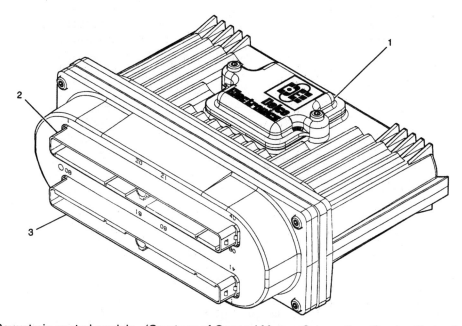

Figure 6–2   Powertrain control module. *(Courtesy of General Motors Corporation, Service Technology Group.)*

train control module applies to the same computer whose predecessors previously were called the ECM (electronic control module). While it has increased functions, like managing certain transmission shift operations, the real reason for the change of name is to conform with OBD II standardized nomenclature regulations. These are detailed more in Chapter 18, so they will not be repeated at length here.

One precaution worth remembering: with the considerably increased number of **diagnostic trouble codes (DTC)** in the OBD II system, technicians are often startled to find a huge number of codes stored. Check to see what circuits are

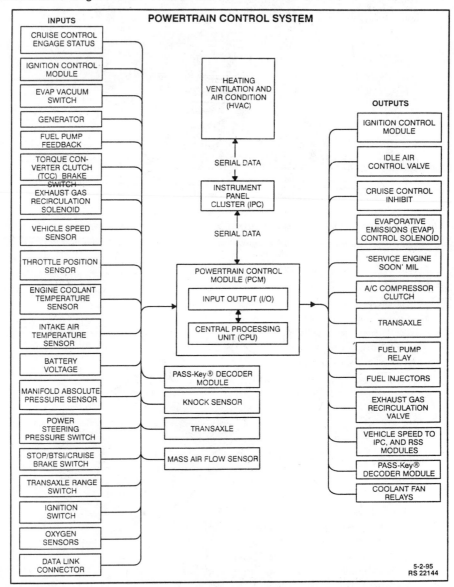

Figure 6–3 Powertrain control system inputs/outputs. *(Courtesy of General Motors Corporation, Service Technology Group.)*

involved before you decide to jack up the windshield wipers and replace the car: often all the codes recorded are for a single circuit or a set of closely related circuits.

Figure 6-3 shows typical inputs and outputs for the system. Among the engine and vehicle functions the PCM controls, are:

- fuel mixture control.
- ignition control.
- knock sensor system (detonation control).
- automatic transaxle shift functions.
- enabling the cruise control.
- regulating the output of the generator (alternator).
- evaporative emission (EVAP) purge (canister purge).
- exhaust gas recirculation (EGR).
- air injection reaction (AIR).
- A/C clutch engagement control.
- radiator cooling fan control.

Sensor Reference Voltage. The PCM also provides a reference voltage to the various sensors and switches it uses, to collect information on how the engine is running. To protect both the PCM and the sensor, this voltage is "buffered" in the PCM before it is sent to the switches and sensors. Many voltmeters, particularly analog voltmeters, may not give an accurate reading of this reference voltage because their impedance is too high. To be sure of accurate measurements of reference voltage (and for that matter, of voltage throughout this system), use only digital voltmeters with a minimum of 10 megohm input impedance, specification J 39200.

CAUTION: To continue a precaution hopefully familiar from previous computer systems, when you work on these vehicles, you must be very careful to prevent damage to components by static electricity through the fingers. While it takes about 4,000 volts of static discharge for a person to feel even a slight zap, less than 100 volts static dis-

charge—1/40th of that—can be enough to damage a computer chip or memory unit.

Quad Drivers/Output Drivers. Internally the PCM has many devices like analog-to-digital converters, signal buffers, counters, timers and special drivers. It controls most components, as car computers have for many years, by completing the circuit to ground (the components' power source usually comes from the ignition switch or some other source). These power transistors are effectively solid-state on/off switches, often called **quad drivers** or **output drivers.** If the switches are surface-mounted in a group of 4, they are quad driver modules; if the switches are surface-mounted in a group of up to 7, they are output driver modules. Certain switches may go unused on a particular vehicle, depending on its model year, accessory list and so on.

PCM Learning

The PCM has a much more advanced learning capacity than previous engine control computers, but resetting the system is relatively similar. Yet, if erroneous data has somehow become recorded in memory, the technician can no longer merely disconnect the battery for a few minutes while the computer's volatile memory ("keep-alive memory") deletes. Changes to the memory data may require the use of a dedicated scan tool to make the deletions. Finer discrimination learning comes as the system performs various tests and experiments, while the vehicle is driven at part throttle, with moderate acceleration. Some vehicles systems also allow re-programming of the control module without removal from the vehicle, a process called "flashing."

Torque Management

Besides the functions normally provided by an engine control computer, this PCM also works to control, that is, reduce, engine torque under certain circumstances. Torque reduction is performed for three reasons:

- to prevent overstress of powertrain components.
- to limit engine power when the brakes are applied.
- to prevent damage to the vehicle or powertrain from certain abusive driving maneuvers.

To calculate whether or not to employ its torque-limiting functions, the PCM analyzes manifold pressure, intake air temperature, spark advance, engine speed, engine coolant temperature, the engagement or disengagement of the A/C clutch through its control circuit, Figure 6-4, and the EGR status. It also considers whether the torque converter is locked up or not, what gear the transmission is in, and whether the brakes are applied. When the numbers add up to torque reduction, the PCM's first strategy is to reduce spark advance to lower engine torque output. In some more extreme cases, the PCM

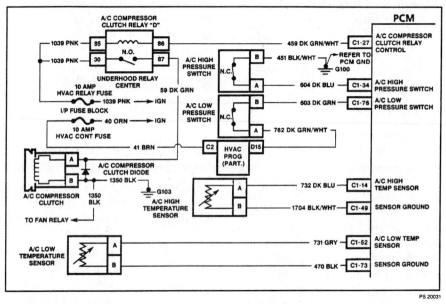

Circuit Description

This system consists of a HVAC control assembly, separate A/C refrigerant high and low pressure switches, separate A/C high and low temperature sensors, a control relay, the compressor clutch and the PCM.

When the HVAC control assembly is placed in the A/C mode, a request signal is sent to the PCM. The PCM will then energize the A/C clutch relay, unless abnormally high or low A/C pressure is detected by the A/C refrigerant pressure switches. The PCM also monitors the A/C low temperature sensors to cycle the A/C clutch "ON" and "OFF." If the PCM detects an A/C evaporator temperature less than 2°C (36°F), the A/C clutch will be prohibited from being enabled. The PCM will also energize the cooling fans "ON" when A/C is requested. Refer to "Electric Cooling Fan(s)," Section "6E3-C12". The A/C control relay is controlled by the PCM so that the PCM can increase idle speed before turning "ON" the clutch or disable the clutch when high engine coolant temperature is detected or high engine RPM is detected.

Diagnostic Aids

- If DTC P1653 is stored as history, it is most likely that one of the A/C pressure switches opened.
- If any PCM DTC(s) are set, perform those diagnostics before using this chart. The A/C clutch is disabled for many DTC(s).

Test Description

Number(s) below refer to the step number(s) on the Diagnostic Table.

3. The outside air temperature must be great enough for the A/C compressor clutch to engage.
5. This step uses the Scan Tool to determine if there is an open or short.
6. This step checks the power feed circuit.
7. This step bypasses the relay.

Figure 6–4 A/C compressor clutch control circuit. *(Courtesy of General Motors Corporation, Service Technology Group.)*

can also shut off fuel to one or more cylinders to further reduce engine power.

There are several occasions when **torque management** is likely to come into play:

- during transaxle upshifts and downshifts.
- when there is hard acceleration from a stand-still.
- if the brakes are applied simultaneously with moderate to heavy throttle.
- if the driver initiates abusive or stress-inducing actions, like shifting gears at high open throttle angles.

In the first two cases, the operation of the torque management will probably not be noticeable.

Traction Control. The ECM is also responsible for the vehicle's traction control system. If the computer learns from the wheel speed sensors that the front wheels are slipping during acceleration, it can individually apply the front brakes to prevent spin. In extreme cases, the torque management system can reduce engine torque also, disabling as many as seven of the fuel injectors to do so, Figure 6-5. This strategy will not be adopted, however, if any of the following conditions occur:

- engine coolant temperature is below -40 degrees.
- the engine coolant level is low (there is a sensor for the purpose).
- the engine is at a speed below 600 rpm.

If the traction control system is disabled for one of the above reasons or a system failure, the TRACTION OFF light will come on in the driver's information panel.

The torque management system can sometimes have some surprising consequences. If a car is running under cruise control and encounters slippery pavement, perhaps slick ice, the wheel speed sensors will report extremely rapid acceleration. The traction control, among its other measures, will shut off the cruise control. Not only will it shut the cruise control off, but it will

disable it, set a code, and leave the subsystem disabled until the code is deleted by a technician. Various other problems, such as a slipping transaxle, can have the same effect.

Information Functions. The PCM also provides the driver with a variety of messages, about engine and transmission oil life or remaining coolant level, for example, some of which are engine-related. As these are not directly related to controlling the combustion in the engine, we will omit them for space reasons.

Ignition and Fuel Start-up. The engine uses a four-coil **waste spark** ignition system, as described in Chapter 5. When the engine is cranked, the ignition control module monitors crankshaft position sensor signals to determine which coil to fire first. Once the module determines the coil timing, it sends a fuel reference

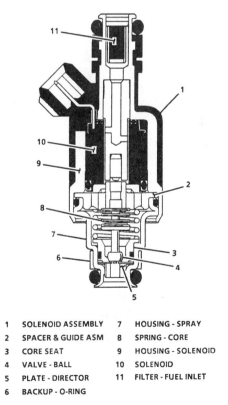

1	SOLENOID ASSEMBLY	7	HOUSING - SPRAY
2	SPACER & GUIDE ASM	8	SPRING - CORE
3	CORE SEAT	9	HOUSING - SOLENOID
4	VALVE - BALL	10	SOLENOID
5	PLATE - DIRECTOR	11	FILTER - FUEL INLET
6	BACKUP - O-RING		

Figure 6-5 Fuel injector. *(Courtesy of General Motors Corporation, Service Technology Group.)*

signal to the PCM. At reception of that pulse, the ECM sprays all eight injectors once simultaneously. It then leaves all the injectors off for two full engine crankshaft revolutions. This provides each cylinder with enough fuel for one power stroke. During those two revolutions, a camshaft position signal should have reached the PCM, and it can begin to trigger the injectors in the proper sequence, Figure 6-6. Should there be no camshaft position signal, the PCM will start fuel delivery in a random pattern, though the chances are 8 to 1 against its being correct. While it will not necessarily be correct, most of the 7 possible wrong sequences can support an engine running at somewhat reduced performance levels.

INPUTS/OUTPUTS

Mass Air Flow (MAF) Sensor

The mass air flow sensor uses a **hot wire airflow sensor** type device to measure the air flow rate, Figure 6-7. It works by maintaining the hot wire at a specified temperature above ambient temperature. A second wire senses the ambient temperature. The hot wire is placed in the

airstream and cooled by the intake air. The current required to keep the wire at the specified temperature corresponds closely to the amount of air entering the intake manifold by mass. Internal circuits in the MAF convert the current required to keep the hot wire at the specified temperature into a frequency signal sent to the PCM. This air flow information is used in the calculation of fuel injector pulse width. This type of air flow sensor has been used for some time, so it will not be described at greater length here.

Manifold Absolute Pressure (MAP) Sensor

The PCM gets engine load and barometric pressure information from the MAP sensor, Figure 6-8. This is the type commonly used. It receives a 5-volt reference signal from the ECM, reduces it in accordance with the sensed pressure, and returns the reduced voltage as an indication of manifold pressure. The MAP signal works inversely to what you might read on a vacuum gauge, that is, the higher the vacuum goes, the lower the output signal voltage will be. The higher-capacity PCM monitors MAP sensor signals against throttle position reports to determine if the MAP sensor is sticking. It will notice if throttle position changes as little as 3.2 degrees in 0.5 second. If the MAP signal does not change by a voltage corresponding to at least 4 kPa during

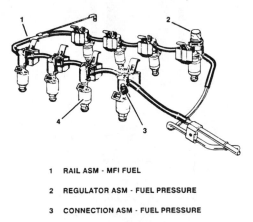

1   RAIL ASM - MFI FUEL

2   REGULATOR ASM - FUEL PRESSURE

3   CONNECTION ASM - FUEL PRESSURE

4   INJECTOR ASM - MFI FUEL

7-1-94
PA 1411AS

Figure 6–6 Fuel rail. *(Courtesy of General Motors Corporation, Service Technology Group.)*

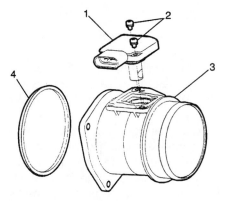

Figure 6–7 Mass air flow (MAF) sensor. *(Courtesy of General Motors Corporation, Service Technology Group.)*

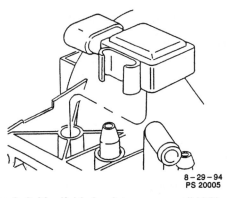

Figure 6–8 Manifold absolute pressure (MAP) sensor. *(Courtesy of General Motors Corporation, Service Technology Group.)*

that period (assuming a base MAP reading of at least 17.3 kPa), it will set a trouble code for the MAP sensor.

The computer uses MAP sensor information to calculate spark advance, fuel mixture (including acceleration enrichment and modulated by

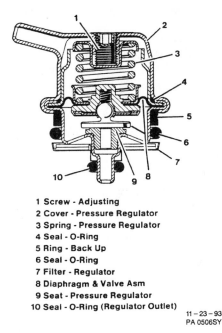

1 Screw - Adjusting
2 Cover - Pressure Regulator
3 Spring - Pressure Regulator
4 Seal - O-Ring
5 Ring - Back Up
6 Seal - O-Ring
7 Filter - Regulator
8 Diaphragm & Valve Asm
9 Seat - Pressure Regulator
10 Seal - O-Ring (Regulator Outlet)

11 – 23 – 93
PA 0506SY

Figure 6–9 Fuel pressure regulator. *(Courtesy of General Motors Corporation, Service Technology Group.)*

the fuel pressure regulator, Figure 6-9), and sometimes to determine whether to employ torque limitation measures.

Temperature Sensors

Intake Air Temperature (IAT) Sensor. The intake air temperature sensor is a thermistor, a variable resistor, that varies its resistance depending on its temperature. Receiving a 5-volt signal from the PCM, it reduces that depending on its temperature, modifying the signal corresponding to the intake air temperature. As sensor temperature goes up, resistance goes down, so when the temperature is high, the signal will go to a low voltage.

Engine Coolant Temperature (ECT) Sensor. The system uses the two-resistance type coolant sensor, Figure 6-10. A 3.65 kilohm resistor and a 348 ohm resistor are in series. As the engine coolant temperature rises to 50 degrees Centigrade (122 degrees Fahrenheit), the total resistance falls to 973 ohms, corresponding to a signal voltage of 0.97 volts. At this temperature, the PCM shifts to the single 348 ohm resistor for its measurements for temperatures above 50 degrees Centigrade. The effect of this telescopes the information available through the single coolant temperature sensor circuit. These resistors are called "pull-up" resistors. The PCM watches for rapid changes in sensor signal

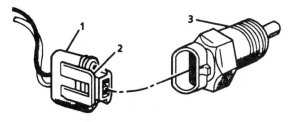

1   JUMPER HARNESS CONNECTOR TO PCM HARNESS
2   LOCKING TAB
3   ENGINE COOLANT TEMPERATURE SENSOR   MS 11345

Figure 6–10 Engine coolant temperature (ECT) sensor. *(Courtesy of General Motors Corporation, Service Technology Group.)*

(except at the 50-degree point), which would indicate a circuit failure of some sort since engine temperature normally changes quite gradually. The information from this sensor, of course, also determines the operation of the radiator cooling fan, Figure 6-11. This operation can be checked through a cooling fan functional check, Figure 6-12.

One of the internal audit tests the computer automatically performs is to check to see whether the engine is warming quickly enough. After the engine has run 4.25 minutes, it checks to see that the coolant temperature is 5 degrees Centigrade (41 degrees Fahrenheit) or greater. If it is not, or if the coolant temperature falls below that temperature for more than 3 seconds, a code is set. The timer ratchets backwards during torque management or traction control activity. Ordinarily, the only reason that an engine does not reach this temperature within the specified time is because a thermostat is blocked open. This assumes, of course, that the engine has been run under moderate load during the elapsed time. If someone merely started an engine at -40 below and let it idle for 4.25 minutes, it might not gain the 80+ degrees, and a code would be set.

Throttle Position (TP) Sensor

The computer uses TPS information for calculating idle speed, fuel mixture, spark advance, deceleration enleanment, and acceleration or WOT enrichment. This sensor returns a proportionate signal: the lower the return voltage, the lower the throttle angle. The computer also compares 1) changes in throttle position with its internal clock to determine how quickly the throttle has been changed, and 2) TPS with MAP information: any significant discrepancy indicates one or both circuits are faulty.

The throttle position sensor on this system is said to be self adjusting, Figure 6-13. What that actually means is that it is not adjustable, but the computer learns what to expect from it and adjusts its calculation tables accordingly. For the 'self-adjustment' the computer assumes that the throttle is closed when the key is turned off with the throttle position signal at the same value twice in succession.

Heated Oxygen Sensor

The heart of any closed loop fuel metering system is the oxygen sensor, and this system

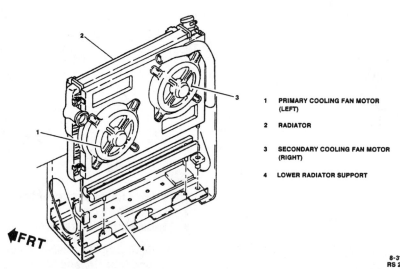

1   PRIMARY COOLING FAN MOTOR
    (LEFT)

2   RADIATOR

3   SECONDARY COOLING FAN MOTOR
    (RIGHT)

4   LOWER RADIATOR SUPPORT

8-31-94
RS 22123

Figure 6–11 Cooling fan configuration. *(Courtesy of General Motors Corporation, Service Technology Group.)*

uses a special, heated sensor (to bring it up to operating temperature as quickly as possible), Figure 6-14. In addition, the computer sends it a reference voltage of 0.45 volts. As long as this is the return voltage, the PCM stays in open loop. Once the oxygen sensor gets warm enough (and newer sensors start to work at about 200 degrees Centigrade or 392 degrees Fahrenheit), the oxygen sensor signal will swing from rich to lean about once every two seconds once the PCM is controlling closed loop. A two-second oxygen sensor cycle is actually fairly long for a functioning sensor: this is the point at which the new Aurora/Northstar system will set a code for a slow

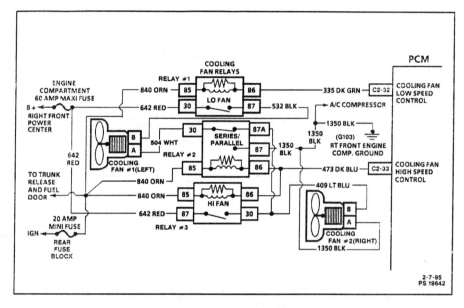

Circuit Description

To determine if a fault is present perform the "Cooling Fan Functional Check". If DTC P1660 is set, or sets during the functional check, it must be diagnosed before proceeding with any of the symptom charts. When the PCM commands low speed fan operation it grounds Cooling Fan Relay 1 which allows current to flow through both cooling fans in a series circuit to ground. If the PCM commands high speed fan operation it grounds all the cooling fan relays (including Fan Relay 1) which changes the circuit to a parallel circuit to ground. If a fault occurs certain symptoms will occur due to the series/parallel circuit design.

PCM will command fan operation when:

LOW SPEED FAN OPERATION

- Engine coolant temperature exceeds approximately 106°C (229°F).
- Transmission fluid temperature exceeds 150°C (302°F).
- A/C operation is requested.
- After the vehicle is shut "OFF" if the coolant temperature at key-off is more than 151°C (304°F) and system voltage was more than 12 volts. The fans will stay "ON" for approximately 3 minutes.

The fans will switch from low to "OFF" when the coolant drops below 102°C (216°F).

HIGH SPEED FAN OPERATION

- Engine coolant temperature reaches 112°C (234°F).
- Transmission temperature is more than 151°C (304°F).
- When certain DTC's set.

The fans will switch from high to low (except DTC's set) when the coolant drops below 106°C (229°F).

Diagnostic Aids

- If the cooling fans operate when commanded "OFF" and DTC P1660 is not set either a cooling fan relay is stuck "ON" or a cooling fan circuit is shorted to power.
- Whenever a repair is completed repeat the "Cooling Fan Functional Check". This will help diagnose possible multiple failures (for example two water contaminated relays).

Figure 6–12 Cooling fan functional check. *(Courtesy of General Motors Corporation, Service Technology Group.)*

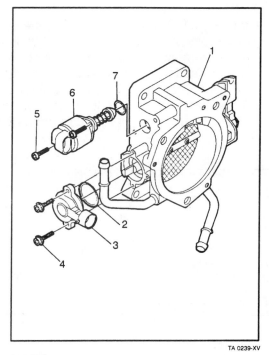

TA 0239-XV

Legend
1. Body Assembly - Throttle
2. O-Ring - Throttle Position (TP)
3. Sensor - Throttle Position (TP)
4. Screw - TP Sensor Attaching
5. Screw - IAC Valve Attaching
6. Valve Assembly - Idle Air Control (IAC)
7. O-Ring - IAC Valve Assembly

Figure 6–13 Throttle position sensor. *(Courtesy of General Motors Corporation, Service Technology Group.)*

oxygen sensor. New sensors typically cycle faster than once per second.

With OBD II comes a second oxygen sensor, functionally equivalent to the first, but downstream of the catalytic converter. Its function is to check on the effectiveness of the converter. If all is well with the catalytic reduction and oxidation processes, things will be fairly uneventful at the downstream oxygen sensor. There should be little or no change of output voltage, provided the converter has removed the hydrocarbons and neutralized (reduced) the NO_x. When the second oxygen sensor starts reflecting the first, the cat-

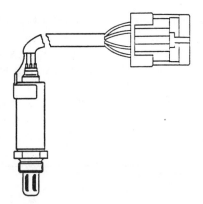

Figure 6–14 Heated oxygen sensor (HO_2S). *(Courtesy of General Motors Corporation, Service Technology Group.)*

alytic converter is not doing anything; and the computer, recognizing this, will set a fault.

NOTE: On the Aurora engine as well as the Cadillac Northstar (and a number of other modern engines) there are two **primary oxygen sensors,** one on each cylinder bank. The individual oxygen sensors, obviously, report on the air/fuel mixture and combustion in the cylinders on that side, but there is nothing in principle different from a single sensor system. Keep in mind, however, that the system uses *three* oxygen sensors. The computer's trouble code recording capacity includes monitoring of each sensor and each sensor's heater circuit separately.

The PCM also monitors the oxygen sensor's heater circuit and reports opens or shorts in it as different trouble codes. Likewise, it compares closed-loop voltage cycling rates as a test of oxygen sensor function. If the cycle falls too long, this is stored as a fault.

Fuel Pump Feedback Circuit. The PCM gets a signal through the fuel pump relay confirming operation of the fuel pump. As on previous systems, the fuel pump is actuated for 2 seconds after the key is turned to ON to prime the injectors for starting. After that 2 seconds, the pump shuts off until there is a crankshaft or camshaft position signal. If the PCM detects a voltage below 2 volts on the fuel pump power circuit, it records that as a fault. The engine, of course, will probably not start because the pump is not turning, or not turning fast enough to pressurize the fuel.

Misfire Detection

Among the most important capacities mandated by OBD II are measures for the **misfire detection** and correction of misfire. A misfire means a lack of combustion in at least one cylinder for at least one "combustion event." A misfire pumps unburned fuel through the exhaust system, and while the catalytic converter can deal with an occasional puff of fuel vapor, a steady stream of it from a dead cylinder would soon overheat it and spill into the atmosphere. The PCM detects misfire by very closely monitoring minute changes in crankshaft speed. At higher speeds, these changes are measured over two engine revolutions for a comparison. If the system detects a dead cylinder, it will shut off the fuel injector to that cylinder, preventing catalytic converter overheating.

Keep in mind that shutting off an injector does have an effect on closed-loop feedback systems. The camshaft and valves still work to move air from the intake manifold to the exhaust manifold. The air pumped through that cylinder contains a higher percentage of oxygen than the exhaust of the air/fuel mixture burned in the other cylinders. Since the oxygen sensor doesn't know where oxygen is coming from, it will report an overlean condition to the computer, which will try to richen the mixture. On most engines, that much enrichment will be beyond the capacity of the system, because no amount of fuel added will burn the pass-through oxygen through the dead cylinder. The computer will set a trouble code and turn on the malfunction indicator light. If the misfire is steady, the computer will consider that an event with a high probability of damage to the catalytic converter. It will flash the service engine soon light until the condition is no longer present or the engine is shut off. Such serious events will be stored in the freeze frame and failure records for later retrieval and diagnosis.

When cylinder misfire is present, check to see whether two cylinders are involved. If there are, look for something shared by both cylinders, like the common coil or exposure to the same vacuum leak.

Knock Sensor (KS) Circuit

An engine can withstand a small amount of spark knock for a short time, and this computer factors that into its spark advance calculations. The knock sensor must report a detonation event lasting longer than 99.99 milliseconds for 3 continuous cycles. The response then is to retard spark until the knock disappears. The basic ignition (spark advance) map for the engine assumes a fuel with an octane equivalence of 87, but the system can manipulate spark (more or less advanced) to accommodate a higher or lower detonation resistance with different fuels. In general, the system will drive the spark advance right to the edge of knock to maximize the fuel economy and reduce emissions problems.

The computer's knock sensor circuit is also used to detect other knock sensor faults. Because the sensor is fundamentally a kind of microphone, and the computer sorts out what kinds of noise are knock and what is just ordinary engine noise, the **background noise** can be used to detect some circuit faults. If the background noise does not rise in an expected way with engine speed and load (as reflected by TPS and rpm through crank and cam sensors for a given period of time), it will record a knock sensor fault, even if there is no knock.

Crankshaft Position Sensors

The system uses two crankshaft sensors, Figure 6-15, on the front of the engine block between cylinders 4 and 6. Crankshaft position sensor A is in the upper crankcase, and crankshaft position sensor B is in the lower crankcase. Both sensors extend into the block and are sealed with O-rings. There are no adjustments.

The pickups work like distributor pickups. As

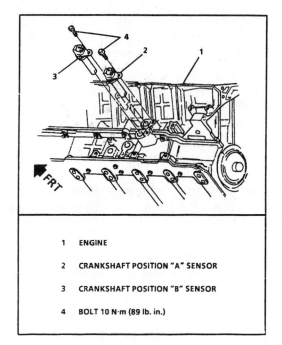

1	ENGINE
2	CRANKSHAFT POSITION "A" SENSOR
3	CRANKSHAFT POSITION "B" SENSOR
4	BOLT 10 N·m (89 lb. in.)

Figure 6–15 Crankshaft position sensors A and B. *(Courtesy of General Motors Corporation, Service Technology Group.)*

the reluctor on the crankshaft moves over the sensor, the coil generates an alternating, polarity-reversing signal that appears (after it goes through the buffer circuit) to the computer as an on/off/on/off signal (almost like that of old-fashioned contact points!). The reluctor ring under the sensors has 24 evenly spaced notches and an additional 8 unevenly spaced notches for a total of 32, Figure 6-16. As the crankshaft rotates once, each sensor reports 32 on/off signals. But since the sensors are positioned 27 degrees of crankshaft rotation apart, this creates a pattern of on/off signals that enables the ignition control module to identify the crankshaft position.

Camshaft Position Sensor (CAM) Signal

The camshaft position sensor is on the rear cylinder bank at the front of the exhaust camshaft, Figure 6-17. It extends into the cylinder head and is sealed with an O-ring. There are no adjustments.

As the camshaft turns, a steel pin on its drive sprocket passes the magnetic pickup of the sensor, Figure 6-18. This generates an alternating, polarity-reversing signal similar to the crankshaft position sensor signal. This on/off signal, in time with the camshaft, occurs once every other crankshaft revolution.

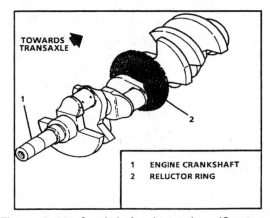

| 1 | ENGINE CRANKSHAFT |
| 2 | RELUCTOR RING |

Figure 6–16 Crankshaft reluctor ring. *(Courtesy of General Motors Corporation, Service Technology Group.)*

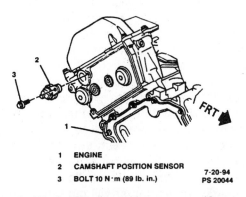

1	ENGINE
2	CAMSHAFT POSITION SENSOR
3	BOLT 10 N·m (89 lb. in.)

7-20-94
PS 20044

Figure 6–17 Camshaft position sensor. *(Courtesy of General Motors Corporation, Service Technology Group.)*

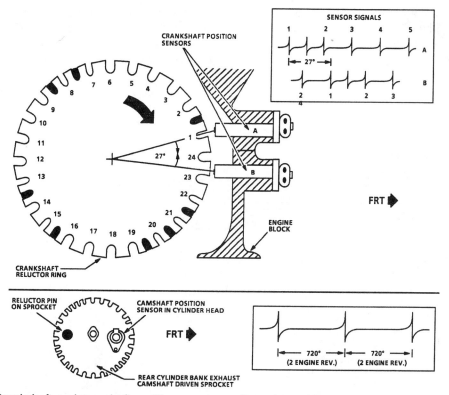

Figure 6–18 Crankshaft and camshaft position sensor configurations. *(Courtesy of General Motors Corporation, Service Technology Group.)*

Ignition Control Module

The ignition control module (IC) is on top of the rear camshaft cover, Figure 6-19. It performs these functions:

- monitoring the on/off pulses from the two crankshaft position sensors and the camshaft position sensor.
- generating a 4x and 24x reference signal sent to the PCM for ignition control.
- generating a camshaft reference signal sent to the PCM for fuel injection control.
- providing an ignition ground reference for the PCM.
- providing for the PCM's control of spark advance.
- providing for "module mode," a limited means

of controlling spark advance without PCM input, fundamentally a limp-home running mode.

The ignition module is not repairable; replacement is the only option if it develops a problem.

Ignition Coils

The system employs four coils to fire eight spark plugs in a waste spark ignition system, as explained in Chapter 5, Figure 6-20. On this system the coils are replaceable individually rather than as a complete coil pack.

Exhaust Gas Recirculation (EGR)

The linear EGR valve used in this system can be tested by the PCM during specific driving operations. When the proper test conditions are met, the PCM cycles the EGR from off to on and back, meanwhile monitoring the MAP signal. Obviously, opening another source of intake into the manifold should raise the intake manifold pressure somewhat (or similarly lower vacuum). The PCM repeats the test a number of times, comparing results to what is hard-wired in its program. If the results do not agree, it sets a code.

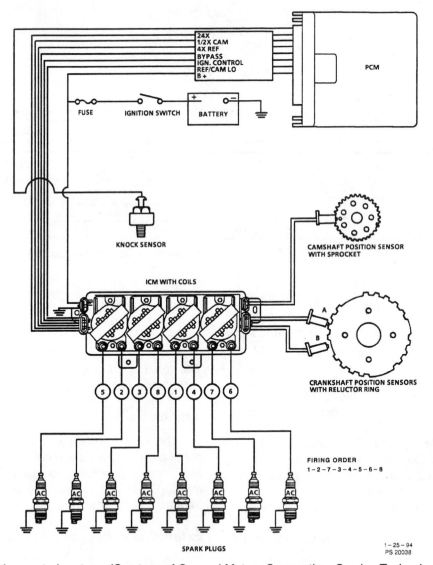

Figure 6–19 Ignition control system. *(Courtesy of General Motors Corporation, Service Technology Group.)*

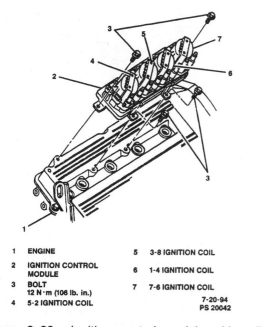

1	ENGINE	5	3-8 IGNITION COIL
2	IGNITION CONTROL MODULE	6	1-4 IGNITION COIL
3	BOLT 12 N·m (106 lb. in.)	7	7-6 IGNITION COIL
4	5-2 IGNITION COIL		7-20-94 PS 20042

Figure 6–20 Ignition control module with coils. *(Courtesy of General Motors Corporation, Service Technology Group.)*

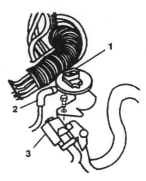

Figure 6–21 EVAP purge vacuum switch. *(Courtesy of General Motors Corporation, Service Technology Group.)*

Catalytic Converter Efficiency Test

As explained in the section on oxygen sensors in this chapter, the PCM compares the signal from the two primary oxygen sensors with the single oxygen sensor downstream of the catalytic converter. If the converter is working properly, there should be very little variation in the signal produced by the **secondary,** downstream **oxygen sensor.** Should the secondary sensor produce a signal similar to those generated by the primary oxygen sensors, the PCM sets a code indicating a failed catalytic converter test.

Evaporative Emission Test

The PCM monitors the function of the evaporative emissions system by monitoring the vacuum switch in the line between the canister and the solenoid, Figure 6-21. The charcoal canister is almost identical to those used in many vehicles for many years. When the engine draws 12 inch-

es of vacuum, the switch is supposed to open and allow venting of the stored fuel vapors. The PCM monitors this by following the amount of time the vacuum switch is continuously open or closed during purging. When purging is supposed to occur, the PCM starts a timer. If the switch stays closed for 9 seconds, the test is failed; if the switch opens for 2 seconds continuously, the test is passed.

Idle Air Control System

The idle air control system uses a stepper motor that works the same as those on earlier systems, Figure 6-22. In this system, the PCM matches the position it sets the IAC motor to, with the achieved idle speed. If the idle speed is higher or lower than the desired speed by 80 rpm, the test is failed and a code is set.

Power Steering Pressure Switch

Normally the power steering switch is closed, but when hydraulic boost pressure in the P/S system rises from 450 to 600 psi, it opens. The purpose of the switch is to inform the PCM when to increase idle speed slightly to compensate for the additional load during, for example, parking maneuvers. The PCM monitors the switch information against its other inputs and sets a code if the switch is ever open at a vehicle speed above 45 mph as sensed by the vehicle speed sensor,

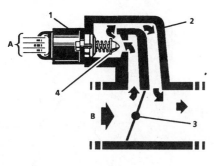

1	VALVE ASSEMBLY - IDLE AIR CONTROL (IAC)
2	BODY ASSEMBLY - THROTTLE
3	VALVE - THROTTLE
4	PINTLE - IAC VALVE
A	ELECTRICAL INPUT SIGNAL
B	AIR INLET

Figure 6–22 Idle air control (IAC) valve. *(Courtesy of General Motors Corporation, Service Technology Group.)*

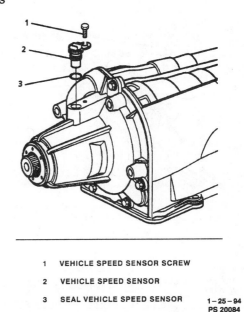

1	VEHICLE SPEED SENSOR SCREW	
2	VEHICLE SPEED SENSOR	
3	SEAL VEHICLE SPEED SENSOR	1 – 25 – 94 PS 20084

Figure 6–23 Vehicle speed sensor. *(Courtesy of General Motors Corporation, Service Technology Group.)*

Figure 6-23. This is a speed at which little or no power steering boost should be required.

PCM Self-Test and Memory Test

Like computers in vehicles previously, this computer checks its internal circuits continuously for integrity. Likewise, it checks its nonremovable **EEPROM** for the accuracy of its data. In computer terms, it tests the checksum of its files against what they are supposed to be and sets a code if they are different. Besides the hard-wired memory, the PCM also monitors its volatile keep-alive memory. If that has been improperly changed or deleted, this sets a code. Note: Such a code will be set when the battery is disconnected for more than a brief instant.

PCM Memory and the EEPROM

Instead of the fixed, unalterable PROM of previous years, the PCM has an electrically erasable programmable read-only memory (EEPROM) it uses to store a large body of information. While the EEPROM is soldered into the PCM and is not service-removable, it can store data, like the PROM, without a continuing source of electrical power. The PCM uses it as a storage bank for its throttle position/learned idle control tables, its transaxle adapt values, its transmission oil life information, its cruise control learning, and more. The EEPROM includes several areas available to store this data, and the PCM will move the data to a good location if there is damage to an earlier place. It will store a code if internal damage prevents the EEPROM from storing the information. The same data is stored in the keep-alive memory, so even if there is an EEPROM failure, the driver will probably not notice any difference. If the EEPROM defect code is set, the PCM must be replaced since the EEPROM is not removable.

Pass-Key Decoder Module

As part of an antitheft system, the PCM gets information from the pass key decoder module to identify the key used for the ignition switch and steering lock. If the wrong key is used or if the system fails, the PCM disables the fuel injection system to prevent theft of the vehicle.

Real World Problem

While the PCM on this system has become much more complex than earlier, simpler engine control computers, it has also become much more reliable. Despite the widespread belief by many technicians that almost any driveability problem is caused by "the computer," in fact the PCM itself is very rarely the cause of driveability or emissions problems. Even when the trouble codes indicate something in the computer, it is important to check the ground circuits for the computer and the engine itself. Because of the very low current/low voltage signals used for these components, it is a much more revealing test to check **voltage drop** rather than resistance when trying to eliminate problems. All ohmmeters are of questionable accuracy as they approach zero resistance, but voltmeters remain quite accurate even down to millivolts. Learn how to substitute voltage drop tests for resistance tests if you plan to do successful work on engine control computers.

✔ SYSTEM DIAGNOSIS & SERVICE

WARNING: These vehicles are equipped with safety airbags. Any technician working on or around the circuits involved in the airbag system should review the sections in the service manual before performing any service on the vehicle. Failure to do so can result in personal injury from unexpected airbag deployment, as well as expensive damage to the vehicle.

The first step in any problem diagnosis is a careful visual inspection. As mentioned before, one of the most important points of this inspection is to check for grounds for the engine and the computer. Familiarize yourself with those locations (they vary with the vehicle) and learn to use the voltage drop test to check for resistance. While checking for grounds is often described, as here, as a visual inspection, it is, of course, nothing of the sort. Use voltage drop measurements to make these checks.

Very many intermittent problems are the result of poor ground connections. Check for leaking vacuum lines, loose connectors and each of the ordinary problems that might disable any car, computer-controlled or not. Check for damaged or collapsed air ducts, air leaks around the throttle body or sensor mating surfaces. Check the spark plug cables for cracks, hardness, proper routing, and continuity. Most problems should be discovered in the course of a visual inspection.

Check that the PCM and MIL (service engine soon) lights are working properly. They should come on with the ignition switch as a bulb check. The next step is to check for stored codes. Connect a scan tool to the OBD II connector (which is supposed to be visible from a squatting position at the driver's door). Following the instructions on the tool, read out any stored codes and record them.

Make sure the engine is at a comparatively low temperature. The Aurora/Northstar engines include a feature whereby at temperatures above 130 degrees Centigrade (265 degrees Fahrenheit), what is called **engine metal overtemp mode,** the PCM will selectively disable four of the cylinders at a time to keep the engine temperature from reaching destructive levels. This injector disabling will be felt by the driver as a loss of power and roughness, of course. Verify the customer complaint and compare the stored codes to those found in the appropriate section of the manual.

The PCM includes a "snapshot" feature that records all the relevant sensor and actuator states at the time of any recorded fault. Often this stored information will point to the cause of the problem. The scan tool can also call up the "freeze frame" or "failure records" data for intermittents. You can check to see that the PCM is properly storing malfunctions by unplugging the MAP sensor at engine idle until the service engine soon light comes on. It should store a code that remains in memory after the ignition switch is cycled off; otherwise the PCM itself is faulty.

OTHER DEVELOPMENTS

One of the more interesting late developments in the General Motors fuel injection systems is the Vortec injection system. Combining the manufacturing economy of throttle body injection and the individual injectors of port fuel, the **Vortec injection system** uses a central injector body with individual hoses and injectors for each cylinder. The injectors are merely spring-loaded nozzles; all the injection pressure is metered through the injector body itself.

SUMMARY

In this chapter we covered the newly developed extended sensors and actuator ranges for the new PCM used on the Aurora/Northstar engine. We learned about the electronic mechanism of the output drivers, successors to the earlier quad drivers, that work similarly to energize the actuators by providing ground circuits.

We looked at the torque management system, which serves not only to control traction slip but also to cushion transmission shifts. The four-coil waste spark system was explained in some detail, indicating how the direct ignition system works firing opposite pairs of spark plugs simultaneously, one just before the power stroke and the other at the end of its exhaust stroke. We learned the sensor techniques the computer uses to identify which cylinders are the appropriate ones to fire at a given crankshaft position.

The chapter included the beginnings of the

OBD II standardized self-diagnostic system. This requires the addition of the downstream, secondary oxygen sensor, which monitors the effectiveness of the catalytic converter. We learned how a very precise measurement of crankshaft speed enabled the computer to detect misfire, and what sorts of countermeasures the computer uses to control emissions quality and protect the catalytic converter should there be such a misfire. And we learned about the EEPROM and how it is different from, and a successor of, the older, hard-wired PROM. We began to understand the diagnostic procedures for this system and OBD II systems generally.

▲ DIAGNOSTIC EXERCISE

A vehicle with the Aurora/Northstar engine is brought into the shop with much reduced power shortly after start-up. All three oxygen sensors report a lean mixture, though the spark plugs are all black with carbon. Even before you pull the OBD II codes, what kinds of problems could cause these symptoms?

REVIEW QUESTIONS

1. Explain why the latest engine control computer systems use multiple oxygen sensors, upstream and downstream of the catalytic converter.
2. How do these latest systems detect misfire, and what measures does the system employ in response? Are there any unintended consequences to the countermeasures?
3. Describe the expanded role in the knock sensor in the late model systems.
4. What is the "engine metal overtemp mode," and what purpose does it serve?
5. Explain how quad drivers and output drivers work. The circuits are arranged to protect

the computer; explain how these actuators are wired.

ASE-type questions (Actual ASE test questions rarely are so brand-specific.)

6. Technician A says a PROM is read-only, that no additional information or changed information can be stored in one. Technician B says the latest versions of General Motors PROMs can be reprogrammed, a process sometimes called "flashing." Who is right?
   a. A only.
   b. B only.
   c. both A and B.
   d. neither A nor B.

7. The Aurora/Northstar engine management system includes torque management functions. Technician A says this is to prevent wheel spin under conditions of reduced traction, such as ice or gravel. Technician B says the torque management program can prevent damage to the engine and drivetrain. Who is right?
   a. A only.
   b. B only.
   c. both A and B.
   d. neither A nor B.

8. A car with the Aurora/Northstar engine "hunts" between torque converter lockup and disengagement at a steady cruise speed. Technician A says this may be caused by low transmission fluid level. Technician B says a defective thermostat might cause the problem. Who is right?
   a. A only.
   b. B only.
   c. both A and B.
   d. neither A nor B.

9. Technician A says the Aurora/Northstar with the OBD II diagnostic system allows actuators to be turned on through the scan tool. Technician B says the system runs tests on its own sensors and actuators frequently in the course of normal driving. Who is right?
   a. A only.
   b. B only.
   c. both A and B.
   d. neither A nor B.

10. When the Aurora/Northstar engine first starts, the computer sprays each fuel injector once to prime the intake runners with fuel. On a specific car, the camshaft position sensor signal does not reach the computer. Technician A says the engine cannot start, though it would continue running if the camshaft sensor circuit failed while underway. Technician B says the computer will try random fuel injection sequences until it finds one that will work. Who is right?
    a. A only.
    b. B only.
    c. both A and B.
    d. neither A nor B.

11. Technicians A and B are monitoring the coolant temperature sensor signal on an Aurora/Northstar engine, and comparing it to a reading they get from an independent thermocouple measurement of engine temperature at the thermostat housing. Just at 50 degrees Centigrade, the signal suddenly changes much higher. Technician A says this means the coolant temperature sensor is defective. Technician B says it means the computer has detected a fault and is ignoring the signal, substituting information from the intake air temperature sensor. Who is right?
    a. A only.
    b. B only.
    c. both A and B.
    d. neither A nor B.

12. Checking the oxygen sensor signals, Technicians A and B find that all three are generating a fluctuating signal switching above and below 0.45 volt about once a

second, but not together. Technician A says they will have to check the shop manual to see how to resynchronize the sensors. Technician B says they are all working properly and that something else has failed. Who is right?

a. A only.
b. B only.
c. both A and B.
d. neither A nor B.

Chapter 7

Cadillac's Digital Fuel Injection

OBJECTIVES

In this chapter you can learn:
- ❑ the functions of the ECM controls.
- ❑ the sensors that feed information to the ECM.
- ❑ the functions of the BCM controls.
- ❑ the relationship between the ECM and the BCM.
- ❑ the features of the self-diagnostic system.
- ❑ how to put the system into diagnostics and retrieve any stored codes as displays.
- ❑ how to put the system into any of its self- diagnostic features (switch test, ECM data, BCM data, cooling fans override or ECM outputs cycling).

KEY TERMS

Accumulator
Ampule
Anode
Cathode
Grid
Modulated Displacement
Propagation
Vacuum Fluorescent

Cadillac introduced digital fuel injection (DFI) in the middle of model year 1980. By 1981 the system had replaced all other fuel injection systems on eight-cylinder Cadillac engines except for the diesels. Before DFI, Cadillac used a system called simply electronic fuel injection (a Bendix-patented, single-function, analog computer-controlled fuel injection system). The Cadillac EFI system was used from 1975 through early 1980. In 1979 and 1980, some California 6.0-liter engines came with an early version of what would become the computer command system, then called a computer-controlled catalytic converter. Just to be precise, any further use of the abbreviation EFI in this book refers to the GM TBI system. Since 1981, all four- and six-cylinder Cadillacs have used either CCC or EFI systems.

In 1987, Cadillac introduced a two-seat sports car, the Allante. It featured the 4.1-liter engine introduced earlier on other Cadillacs, but with a sequential port fuel injection system. This system is similar to the nonturbo sequential PFI system on the Buick V6, except for the following:

- It uses an oxygen sensor in each exhaust manifold. Two oxygen sensors allow the system to control the air/fuel ratio more closely. It can adjust the pulse width on one side of the engine independently of the adjustments on the other side.
- It uses a distributor and the HEI ignition system as the DFI system did, as opposed to a computer command control system, Figure 7-1.
- A single, 180-degree vane, Hall-effect switch is used in the distributor as a camshaft position sensor for referencing injector pulses.

225

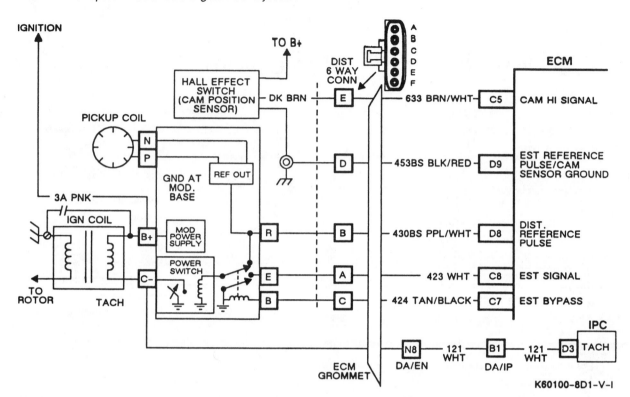

Figure 7–1 4.5-liter PFI/HEI ignition system. *(Courtesy of General Motors Corporation, Service Technology Group.)*

The 4.1-liter engine was later enlarged to 4.5-liters, but it remained basically the same engine. The newer PFI system replaced the DFI system on increasing numbers of Cadillacs until 1990, by which time all 4.5-liter engines used the PFI system. Also beginning in 1990, Cadillac used the 5.7-liter TBI (throttle body injection) engine as used in Chevrolet Suburbans as an option for a vehicle with the towing package in full-sized, rear-wheel-drive cars.

The Cadillac DFI system works very similarly to the EFI system, which probably came directly from Cadillac DFI, minus a few features. DFI uses a dual throttle body injection (TBI) unit with two injectors. Do not, however, confuse a dual throttle body unit with the Crossfire EFI system. The Crossfire system uses two separate, one-barrel TBI units above an open plenum. The DFI's dual throttle body is configured like that of a conventional two-barrel carburetor, Figure 7-2.

An electric fuel pump in the tank provides fuel pressure and delivery. The pressure is then controlled by a pressure regulator in the TBI unit. Each injector pulses alternately with the other, with one pulse corresponding to a cylinder firing during normal driving.

The two unique features of the Cadillac DFI system are:

- the type of functions it controls and monitors such as air conditioning and heating, fuel consumption information, and cruise control, in addition to engine parameters.
- the DFI system has an elaborate self-diagnostic capacity regularly expanded since its introduction.

In model year 1981, the DFI system also con-

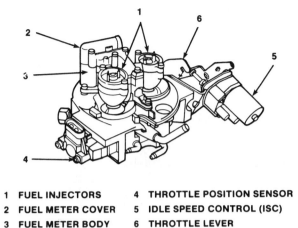

1 **FUEL INJECTORS**   4 **THROTTLE POSITION SENSOR**
2 **FUEL METER COVER**   5 **IDLE SPEED CONTROL (ISC)**
3 **FUEL METER BODY**   6 **THROTTLE LEVER**

Figure 7–2 TBI unit. *(Courtesy of General Motors Corporation, Service Technology Group.)*

trolled the **modulated displacement,** or 8-6-4 system, in which selected cylinders were disabled (by preventing the rocker arms from opening the valves) when power requirements were low. Under acceleration, all eight cylinders are used. During steady-state driving, when minimal power is required, only six cylinders are used; during very low power demand or deceleration, only four cylinders are used. The system was used only on the 6.0-liter engine, discontinued in 1982 for all but limousine and commercial chassis (mostly hearse, some ambulance) vehicles. At that time, the 4.1-liter engine was first introduced. The 8-6-4 system was often regarded as unreliable, troublesome to repair and less than successful at managing emissions and fuel economy objectives.

Cadillac Allante

Cadillac introduced its two-seat sportscar, the Allante, in 1987, with the 4.1-liter engine and a sequential, port-injected fuel system. The system is similar to the nonturbocharged sequential PFI system on Buick V6, except that it uses an oxygen sensor in each exhaust manifold. This "stereo" oxygen sensor system allows the computer to control the air/fuel ratio more closely since it can control the pulse width on one side of

the engine independently of what it does on the other side.

In 1985, Cadillac introduced a body computer (body computer module, BCM) on the new front-wheel-drive C-body car ("C" designates a particular size car body in GM's nomenclature) to work with the ECM. The body computer took over all A/C and related functions, data display, and some diesel engine controls for vehicles with such engines. Although our focus here is on engine controls, we include a brief unit on the BCM later in this chapter because of its interconnections with the engine control computer.

ELECTRONIC CONTROL MODULE (ECM)

The ECM for the DFI system has greater capacity than the computers used on other General Motors' systems of similar model years. It is located on the right side of the instrument panel and contains, like similar models, a removable PROM. All inputs fed into it and all functions it controls are shown in Figure 7-3.

ECM "Learning" Capacity

As driving conditions change and as a vehicle wears with age and use, sensor inputs change. Not all of them change in direct proportion to each other. For instance, if a car that is normally driven at or near sea level starts on a trip that takes it to a significantly higher altitude, the ECM will see:

- engine speed and road speed still match up as they normally do (provided the transmission torque converter lockup clutch still applies), but the throttle position required to maintain a given speed is different.
- engine vacuum is lower than it normally is at the lower altitude throttle position.

If the trip is long enough that the new combination of values is consistent over time, the ECM

1988 DIGITAL FUEL INJECTION

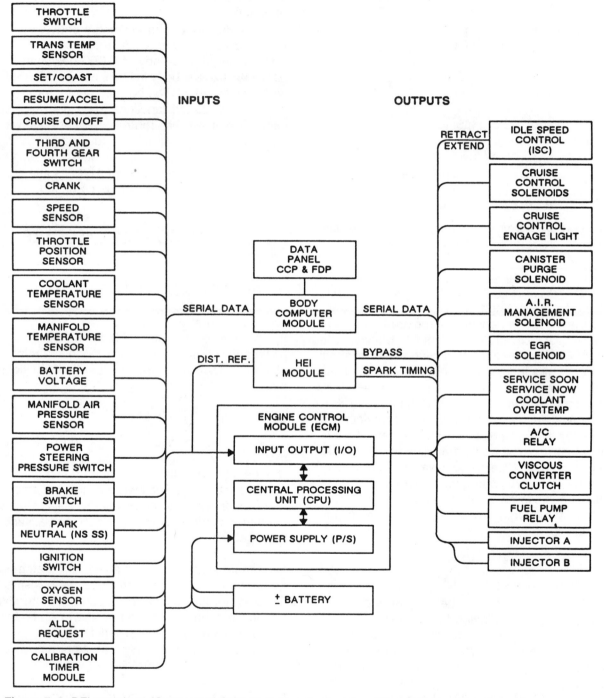

Figure 7–3 DFI overview. *(Courtesy of General Motors Corporation, Service Technology Group.)*

will accept them as normal and will issue spark timing and other commands to provide optimum driveability, emissions performance, and fuel economy under the changed conditions. If battery power is removed from the ECM, its volatile memory of what it had learned is erased, and performance will be noticeably different—and worse—until it can relearn the same data. If the battery has been so disconnected, drive the car at normal operating temperature, under part and moderate throttle conditions until normal performance is relearned.

Keep-Alive Memory (KAM)

The computer has two kinds of memory, ROM (read-only memory) and RAM (random-access memory). The ROM is encoded on the PROM and cannot be changed by the car computer in the course of its activity. The RAM is also called the KAM or keep-alive memory. The basic instructions for running the engine are encoded, hard-wired on the PROM, and this does not change. As the computer operates the engine, however, it builds a copy of this information in its KAM, and it can change this copy as it learns that other, slightly different settings will more successfully optimize the driveability, emissions, and fuel economy performance. It stores these other settings in the keep-alive memory. If the KAM is erased or the data is somehow corrupted, the computer returns to the ROM data until it can relearn the appropriate KAM data.

OPERATING MODES

The ECM operates the DFI system in one of several different operating modes, depending on the prevailing driving conditions. While there are some similarities with other GM engine management systems, DFI is a special system.

Starting Mode

When the ignition is turned on, the ECM also turns on the fuel pump relay and checks engine

coolant temperature and throttle position. Once the engine starts cranking, the computer pulses the injectors to produce an air/fuel ratio ranging from 1.5 to 1 at a temperature of -36 degrees Centigrade (-33 degrees Fahrenheit) to 14.7 to 1 at a temperature of 94 degrees Centigrade (201 degrees Fahrenheit).

Clear Flood Mode

If the engine floods, that is, if the spark plugs become so wet with fuel they short to ground without spark, pressing the throttle pedal to the floor (or to 80% or more of throttle range) and cranking the engine will cause the injectors to pulse so briefly that an air/fuel ratio of 25.5 to 1 is delivered. This ratio is maintained until either the engine starts or the throttle is closed to below 80%.

Run Mode

Most driving is done in this mode, which includes open- and closed-loop conditions. When the engine is first started and exceeds 400 rpm, the system goes into open loop. The ECM calculates air/fuel requirements based on coolant temperature, manifold pressure, throttle position and engine rpm (in that order of importance). Oxygen sensor inputs are not used in open loop.

The ECM puts the system into closed loop once it sees:

- a varying voltage from the oxygen sensor crossing the 0.45 volt measure. This means the oxygen sensor is hot enough to work and is reflecting the mixture delivered to the intake manifold.
- coolant temperature above 70 degrees Centigrade (158 degrees Fahrenheit).
- at least sixty seconds have elapsed since engine start-up. This allows the air and fuel delivery systems as well as all the sensors to stabilize.

The ECM uses oxygen sensor information to determine air/fuel mixture once the system is in closed loop to determine fuel injector pulse width.

The injector opens for only a few milliseconds even during maximum fuel demand.

Acceleration Mode

Whenever the ECM sees the throttle position change rapidly open and this change is confirmed by a like rapid change in the manifold pressure, the ECM takes the system quickly out of closed loop and puts it into acceleration mode. The higher manifold pressure when the throttle opens slows the evaporation of fuel, so more fuel is needed than makes for a 14.7 to 1 air/fuel ratio. As long as the throttle position sensor indicates WOT (or more precisely 80% of WOT), the system will stay in acceleration enrichment mode until the throttle opening and engine load reduce, as reflected by the intake manifold vacuum.

Deceleration Mode

When the ECM sees the throttle close rapidly from the TPS return signal and when manifold pressure confirms the throttle has closed, the ECM reduces the amount of fuel injected. If the deceleration is considerable, the ECM will shut off fuel entirely. During the period of normal running that preceded the deceleration, a certain amount of fuel condensed on the intake manifold walls and floor. When the intake manifold pressure suddenly drops under deceleration, that condensed fuel suddenly evaporates, overly richening the delivered air/fuel mixture. The fuel shutoff is an attempt to correct for that as much as possible.

Battery Voltage Correction

If battery (charging system) voltage drops, the ECM will sense this and will make the following corrections to maintain the best possible operating efficiency:

- to increase injector pulse width to maintain the proper air/fuel ratio.
- to increase idle rpm.
- to increase ignition primary circuit dwell time to maintain satisfactory spark performance.

INPUTS

All the DFI inputs are listed in this section, but some of them are discussed more fully under the main General Motors chapters when they are shared. These include:

- coolant temperature sensor (CTS).
- throttle position sensor (TPS).
- oxygen sensor.
- ignition switch.
- crankshaft position signal.

Others are discussed as required.

Manifold Absolute Pressure (MAP) Sensor

The earliest DFI systems used Bendix-built MAP and BARO sensors, located in the passenger compartment. Later versions adopted the standard General Motors units.

Distributor Reference Signal (REF)

The DFI system uses the HEI (and successor systems) pickup coil in the distributor as the source of the ignition base reference pulse.

Manifold Air Temperature (MAT) Sensor

The MAT sensor is similar to that used on other GM products. It measures the air/fuel mixture's temperature in the intake manifold.

Throttle Switch

This is the same as the ISC switch discussed in Chapter 3 where more can be learned about it. The DFI system uses an idle speed control motor like the one used on CCC systems instead of the idle air control motor used by EFI and PFI systems.

Battery Voltage

The ECM monitors battery voltage to determine when it needs to go into voltage correction mode, as described in this chapter, in the operating modes section.

Park/Neutral (P/N) Switch

The park/neutral switch indicates to the ECM when the transmission is in park or neutral as opposed to being in a forward or reverse gear. The computer uses this information to control idle speed, cruise control, and the torque converter lockup clutch or viscous converter clutch system. The front-wheel-drive C-body cars, first introduced in 1985, use the 440-T4 transaxle, including a viscous lockup torque converter clutch. The transaxle and converter lockup clutch were both introduced in the 1985 model year (see **Viscous Converter Clutch** in the **Outputs** section of this chapter).

Third- and Fourth-Gear Switches

These switches screw into passages in the valve body; hydraulic pressure activates them. When the transmission shifts into third gear, the third-gear switch opens. When the transmission shifts into fourth gear, the fourth-gear switch opens, Figure 7-4. By monitoring these switches,

the ECM knows whether the transmission is in third, fourth or one of the lower gears. The ECM uses this information to control the viscous converter clutch.

Viscous Torque Converter Lockup Clutch (VCC) Temperature Sensor

A thermistor-type temperature sensor is located in the transmission to monitor fluid temperature, Figure 7-4. This information affects the ECM's control of the viscous torque converter lockup clutch. In general, this information is used chiefly when the computer decides to leave the viscous torque converter locked up to prevent overheat damage under sustained high-load cruise conditions (like hill climbing or trailer towing) in which it would ordinarily disengage the clutch.

Generator Monitor

General Motors, for some reason, prefers the word *generator* to the word *alternator,* a preference found throughout the company's technical

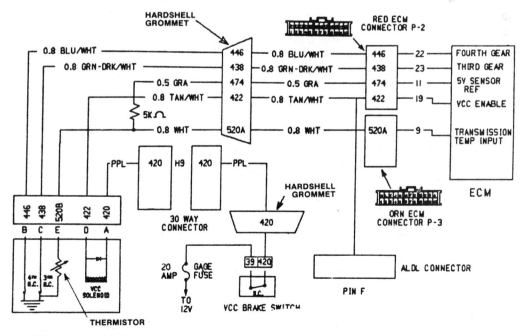

Figure 7–4 VCC control circuit. *(Courtesy of General Motors Corporation, Service Technology Group.)*

literature. The named device, however, is the same device everyone else calls an **alternator**. On most GM computer-controlled engine management systems, the computer monitors "generator" voltage through a wiretap into the fuel pump power circuit (ECM terminal 17, red connector in Figure 7-5). Although this circuit is not directly wired to the generator (alternator), it reflects charging system voltage as most components on the system perceive it, a reliable reflection of the actual alternator output.

If the ECM sees less than 10.5 volts or more than 16, it turns on the service now light and sets a code 16 in its diagnostic memory. On vehicles with a body computer, all engine function codes are preceded by the letter E.

Power Steering (P/S) Anticipate Switch

If power steering pressure reaches a specified pressure, it opens a hydraulically actuated switch. With the ignition on, battery voltage passes through the switch to the ECM. If the switch opens, causing the signal voltage at the ECM to drop to zero during idle, the ECM increases the throttle blade opening to support the additional engine load. If the switch opens above 40 mph, the ECM turns on the service engine soon light and records code 40. This switch does not appear on DFI systems before model year 1985.

Cruise Control Enable (On/Off Switch)

When the cruise control switch is turned on (upper left corner of Figure 7-6), it supplies a 12-volt signal to the ECM. This makes the ECM operate the cruise control system. After 1986 model year, Eldorado and Seville cruise control operation was directed by the body computer.

Cruise Set/Coast Switch

If the driver pushes the set/coast switch on the cruise control, a 12-volt signal is sent to the ECM, Figure 7-6. If the cruise control system is in a coast mode, the ECM will engage and maintain the current vehicle speed. If it is in an engaged

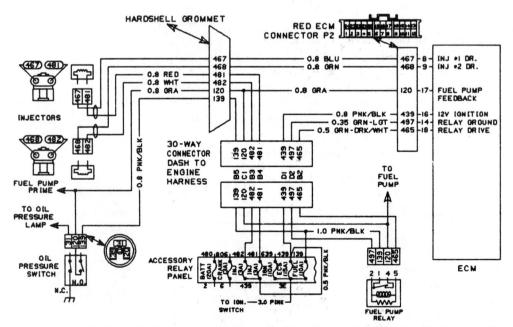

Figure 7–5 Fuel pump control circuit. *(Courtesy of General Motors Corporation, Service Technology Group.)*

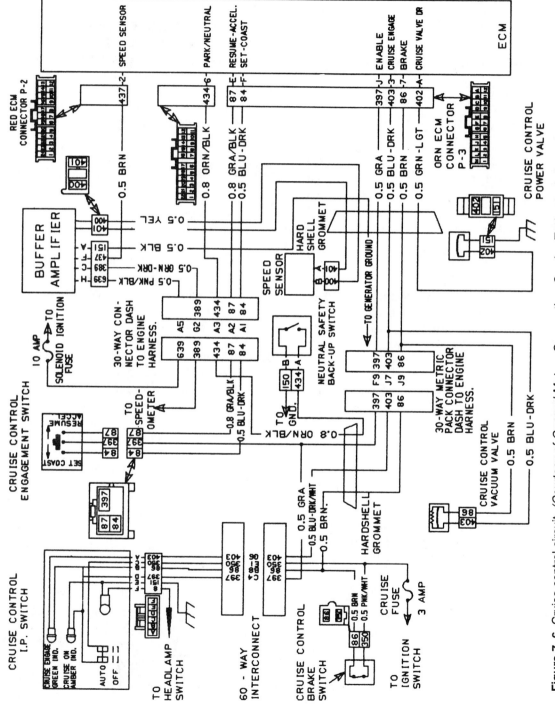

Figure 7–6 Cruise control circuit. *(Courtesy of General Motors Corporation, Service Technology Group.)*

mode when the switch is pressed, the ECM will allow vehicle speed to coast down until the button is released.

Cruise Resume/Accelerate Switch

When the resume/accelerate switch is moved toward resume, a 12-volt signal is sent to the ECM, Figure 7-6. The ECM resumes a controlled preset speed that was disengaged because the brakes were applied. If the switch is moved toward accelerate while the system is engaged at a particular speed, the ECM opens the throttle and accelerates the vehicle as long as the switch is held. Once the switch is released, the ECM maintains the speed at which it was released.

Cruise Control Brake Switch

Whenever the ignition switch is on, it supplies battery voltage to the brake switch, Figure 7-6. When the switch is closed (when the brake is applied), voltage feeds through the switch to the ECM. When the brake is applied, the switch opens. The voltage signal at the ECM disappears, and the ECM knows the brake is applied and modifies its outputs accordingly. Among these output commands is one to disengage the cruise control if it is engaged.

ALDL Request

When someone grounds the test terminal of the ALDL (assembly line data link, most often referred to as assembly line communications link) under the dash, the ECM responds by putting the electronic spark timing at a fixed timing advance. This mode is the one used to check ignition timing.

Serial Data

Information to the ECM reflected on the display panels or from the body computer comes via a data link. It is most properly referred to as serial data. This information can concern a driver-selected heater or air conditioning temperature. It can show fuel mileage, or it can come as information requested from the BCM.

OUTPUTS

Injectors

Each of the two injectors uses an electric solenoid, Figure 7-7. When this solenoid is pulsed by a signal from the ECM, the injector's fuel valve lifts off its seat by a few thousandths of an inch. Fuel sprays at a controlled pressure between 9 and 12 psi. The amount of fuel sprayed is a function of how long the valve stays open (in "pulse width," or the time the valve is open. The "width" in question is the width along the horizontal time line of a graph).

Each of the injectors has its own power supply circuit and fuse. Voltage comes through the ignition switch, and the ECM grounds each injector individually for whatever period of time it calculates is appropriate.

Fuel Pressure Regulator. The pressure regulator provides an operating pressure ranging between 9 to 12 psi at the fuel injectors. It functions identically to the standard GM fuel pressure regulator, discussed Chapter 4.

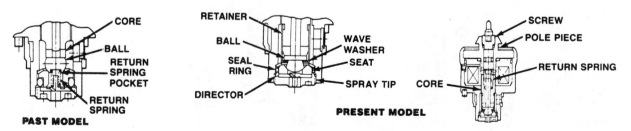

Figure 7–7 Injector comparison. *(Courtesy of General Motors Corporation, Service Technology Group.)*

Fuel Pump Operation

Fuel and fuel pressure come from a twin-turbine electric pump inside the fuel tank. The pump turns on and off following a fuel pump relay under the instrument panel. Its operation is similar to that used in the rest of the General Motors systems and described in Chapter 4. Unlike that system, however, the DFI fuel pump did not use an in-line oil pressure switch as a backup feature.

Electronic Spark Timing (EST)

The EST works the same way as on the EFI system discussed in the **Outputs** section of Chapter 4, except that DFI does not use a Hall-effect switch.

Idle Speed Control (ISC)

The Cadillac DFI system uses the same kind of idle speed control motor that the regular General Motors CCC system uses. While the motor works and you adjust it the same way as those discussed in that chapter, the ECM uses a somewhat different strategy to control it. The ECM uses information from the following sensors and switches to control throttle blade position in any of the three operational modes (some of these are inputs that identify the operational mode):

- VSS.
- TPS.
- throttle switch.
- ignition reference signal.
- A/C clutch signal.
- CTS signal.
- MAP sensor signal.
- P/N switch.
- ignition switch.

During Cranking and Initial Warm-up. During start-up cranking, the throttle is held in a predetermined position. Once the engine starts, the throttle is held in a predetermined fast idle position to allow intake manifold conditions to stabilize. If the engine is started hot, the fast idle period will be very short. If the engine is started cold, the fast idle period will be longer until the engine reaches operating temperature.

Curb Idle. When the engine is idling at normal operating temperature, curb idle is maintained at a predetermined speed. If engine load changes because of the A/C clutch cycling on and off, the transmission going into gear or neutral, or because of momentary high power steering pressure load, the ECM adjusts throttle opening to maintain the predetermined engine speed.

Deceleration. During deceleration, manifold pressure goes very low (or intake manifold vacuum goes very high, an equivalent description), and any liquid fuel in the manifold evaporates very quickly in the reduced pressure. This would result in a very rich air/fuel mixture and high HC and CO exhaust emissions, unless enough air is let into the manifold to mix with the fuel vapors, as well as to reduce the vacuum. The ECM considers vehicle speed and manifold pressure to calculate the degree of deceleration and determine how much throttle opening is required to prevent the rapid fuel evaporation. During very hard deceleration, for example, if descending a hill in manual second, the ECM can also shut off fuel injection entirely.

AIR Management

The air injection reaction (AIR) system on Cadillac's DFI is basically the same as for the rest of the General Motors vehicle systems. The DFI system uses a slightly different strategy to control it, however. The ECM de-energizes the divert solenoid and puts the system into divert under any of the following conditions:

- acceleration or WOT operation. (Some DFI AIR systems go into divert mode any time the vehicle speed exceeds 60 mph.)
- deceleration.
- extremely cold weather start-up.
- if there are certain system failures that also turn on either of the service engine lights.
- high rpm, which can cause air pump pres-

sure to exceed the calibration of the pressure relief valve built in to the control valve. Though this is not a controlled divert function, it has similar results.

Notice that any control valve solenoid or related electrical circuit failure also results in a fail-safe condition in which the system stays in divert mode. Such a failure should leave the valve in the safest operating mode, one that is unlikely to damage the catalytic converter. Of course, a mechanical failure in the valve, should it leave the air injection directed to the converter, will not have this fail-safe feature.

Torque Converter Clutch (TCC) or Viscous Converter Clutch (VCC)

The TCC is a hydraulically-actuated lockup clutch inside the transmission torque converter. When actuated, it eliminates the slippage in the torque converter to improve fuel economy and reduce the heat generated by the turbulence in the converter's automatic transmission fluid. When the lockup clutch is released, the torque converter works normally (though for some vehicles, engineers have used a torque converter with more converter slip than before to maximize torque multiplication and smoothness, knowing

they could eliminate all the slip once the vehicle was underway at cruise speed). Application and release of the clutch are controlled primarily by the ECM. Read the General Motors Chapters 3 and 4 for more information about converter lockup clutches generally. Some parameters by which the Cadillac DFI system controls the clutch are slightly different, and those are discussed here.

The VCC, introduced on the 1985 Cadillac front-wheel-drive C-body cars, works the same way as the TCC does except that it does not provide a 100% lockup between the engine and the transmission input shaft. The VCC clutch is made up of a rotor sandwiched between two parts, called the cover and the body, Figure 7-8. The rotor and body surfaces facing each other have a series of fins that mesh together with a small space between each fin, Figure 7-9. The effective distance between them is very small, and the shared area is very large. This space is filled with a viscous silicone fluid that drives the rotor. When the VCC is applied, it allows about 40 rpm slippage at 60 mph, just enough to eliminate some of the torsional vibration between the engine and the transmission. The viscous clutch lockup is controlled the same way as a TCC clutch except that transmission fluid temperature plays a larger role in its control.

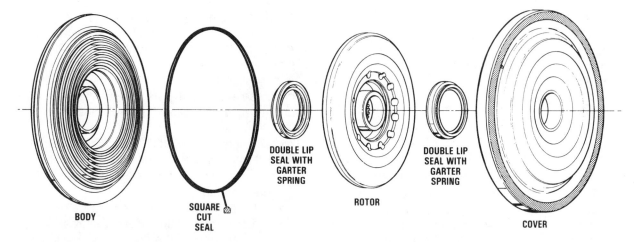

BODY **SQUARE CUT SEAL** **DOUBLE LIP SEAL WITH GARTER SPRING** **ROTOR** **DOUBLE LIP SEAL WITH GARTER SPRING** **COVER**

Figure 7–8 Viscous converter clutch (exploded view). *(Courtesy of General Motors Corporation, Service Technology Group.)*

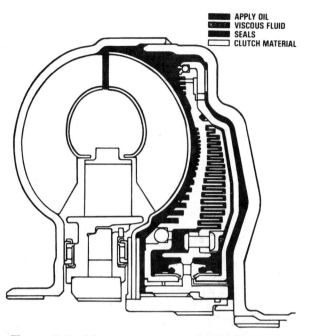

APPLY OIL
VISCOUS FLUID
SEALS
CLUTCH MATERIAL

Figure 7–9 Viscous converter clutch. *(Courtesy of General Motors Corporation, Service Technology Group.)*

TCC/VCC Control Parameters. The ECM grounds the TCC/VCC apply solenoid circuit when:

- the criterion speed is reached according to the vehicle speed sensor (VSS). On some cars this can be as low as 24 to 36 mph depending on throttle position and transmission oil temperature.
- the engine has reached normal operating temperature. If the engine coolant temperature is 18 degrees Centigrade (64 degrees Fahrenheit) or more when the engine starts, the ECM will apply the converter clutch when the coolant reaches 60 degrees Centigrade (140 degrees Fahrenheit) for VCC vehicles. If the coolant is colder than 18 degrees Centigrade at start-up, the ECM uses a fixed-time delay before applying the clutch. The purpose of this delay is to improve low-temperature driveability until conditions allow

lockup.
- the TPS signal is within a predetermined window. This window varies depending on what gear the transmission is in, among other factors. The TPS signal is compared to the vehicle speed signal, and the ECM selects different apply speed thresholds for different combinations of TPS and VSS.

Two other signals can disengage the lockup clutch on VCC vehicles. These are the brake switch and the overtemperature protection switch, Figure 7-10. Most TCC applications do not use an overtemperature switch. These switches are not controlled by the ECM but are in the apply solenoid circuit. Each must be closed before the circuit can close to the ECM. Both are normally closed during driving operations; but if the brake switch is opened when the brake is depressed or the overtemperature protection switch opens (when the transmission oil temperature exceeds 157 degrees Centigrade (315 degrees Fahrenheit), the lockup clutch disengages.

On the 1987 and later Allante, the ECM also controls the transmission 2-3 and 3-4 shifts. The ECM does this through two solenoids it controls in the transmission's valve body. The solenoids block or open passages allowing oil to move under pressure to the 2-3 and 3-4 shift valves.

Exhaust Gas Recirculation (EGR)

When the EGR valve opens, it allows a measured amount of exhaust gas (mostly carbon dioxide and nitrogen, neither of them capable of sustaining flame) into the intake manifold. This inert gas displaces a certain amount of combustible intake mixture and has the effect of slowing the **propagation** of the flame through the combustion chamber. This slower flame burns at a lower temperature and thus reduces the production of oides if nitrogen (NO_x)

The Cadillac DFI system uses a negative backpressure EGR valve, Figure 7-11. This type of valve has some self-modulation capacity. Exhaust pressure from the exhaust crossover

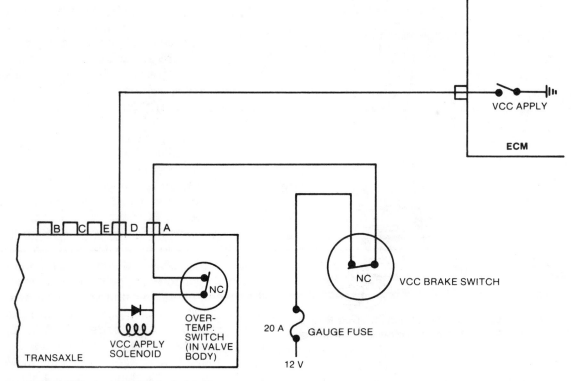

Figure 7–10 VCC electrical control circuit

(see early fuel evaporation in Chapter 3) passage in the intake manifold is sensed through a small hole in the lower portion of the valve itself. The hollow valve stem allows exhaust backpressure to get to a chamber beneath a second diaphragm below the main diaphragm. This pressure helps a small spring push the lower (second) diaphragm up and block a bleed hole in the main diaphragm. When the bleed hole is blocked, ambient air vented into the space between the two diaphragms no longer bleeds into the vacuum chamber above the main diaphragm. This allows the buildup of sufficient vacuum in the chamber (using manifold vacuum as the source) to pull the diaphragm up and open the EGR valve. When the valve opens, the same hole that admitted exhaust pressure also allows manifold vacuum in.

If the net result of the engine vacuum versus exhaust backpressure is a positive or near positive pressure, the lower diaphragm pushes up and opens the valve. If, however, the net result is a negative pressure, the lower diaphragm stays down, and the bleed valve opens. This keeps the EGR valve itself closed.

EGR Modulation. In addition to the EGR control provided mechanically by the negative backpressure EGR valve, the Cadillac DFI system on the 4.1-liter engine also controls vacuum to the EGR vacuum port. It does this with a duty-cycled solenoid that bleeds atmospheric pressure into the vacuum supply hose to the EGR valve when on, causing the valve to close, Figure 7-12. When the solenoid is turned off, it allows manifold vacuum to the EGR valve and causes it to open. By duty-cycling the solenoid, the ECM can control the amount of EGR opening and thus the

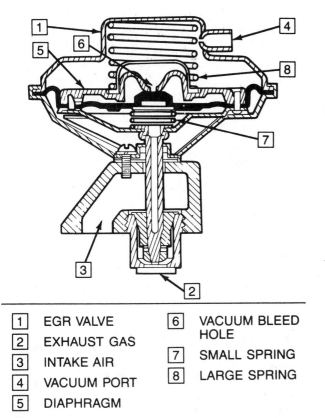

1	EGR VALVE	6	VACUUM BLEED HOLE
2	EXHAUST GAS	7	SMALL SPRING
3	INTAKE AIR	8	LARGE SPRING
4	VACUUM PORT		
5	DIAPHRAGM		

Figure 7–11 Negative backpressure EGR valve. (Courtesy of General Motors Corporation, Service Technology Group.)

EGR flow rate. The ECM varies the flow rate for different driving conditions to provide the best possible driveability characteristics while still maintaining control of the production of NO_x. A 10% duty cycle provides full vacuum to the EGR and a 90% duty cycle vents all vacuum to it, effectively closing the valve. The ECM uses information from the following sensors to determine the prevailing driving conditions and the amount of EGR needed:

- engine coolant temperature (CTS signal—no EGR when coolant temperature is low because there would be no NO_x production at those temperatures anyway).
- throttle position sensor (TPS signal).

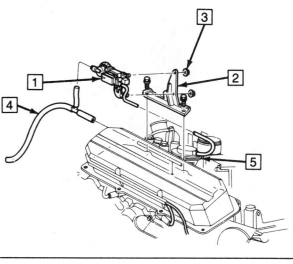

1	EGR SOLENOID
2	BRACKET
3	NUT
4	HOSE ASM.
5	EGR VALVE

Figure 7–12 EGR solenoid. (Courtesy of General Motors Corporation, Service Technology Group.)

- manifold pressure (MAP signal).
- idle control throttle switch (ISC signal—no EGR flow is needed at idle because combustion temperatures are not high enough to generate NO_x).
- ignition reference signal (REF).

Charcoal Canister Purge

Like most engine management systems, the Cadillac DFI system uses an evaporative emissions control system consisting primarily of a charcoal canister pneumatically plumbed to the vapor area of the fuel tank. The tank vents through the canister, and fuel vapors are stored in the activated charcoal in the canister. Once the engine is running in closed loop, the ECM can open the line to the intake manifold and allow vacuum to purge the fuel vapors in the canister, Figure 7-13. Because the ECM has no way to tell how much fuel, if any,

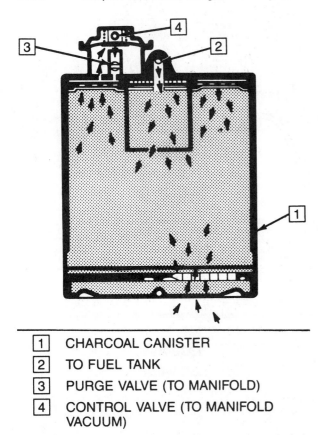

1. CHARCOAL CANISTER
2. TO FUEL TANK
3. PURGE VALVE (TO MANIFOLD)
4. CONTROL VALVE (TO MANIFOLD VACUUM)

Figure 7–13 Charcoal canister. *(Courtesy of General Motors Corporation, Service Technology Group.)*

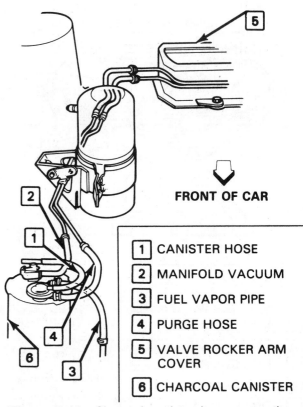

FRONT OF CAR

1. CANISTER HOSE
2. MANIFOLD VACUUM
3. FUEL VAPOR PIPE
4. PURGE HOSE
5. VALVE ROCKER ARM COVER
6. CHARCOAL CANISTER

Figure 7–14 Charcoal canister hose connections. *(Courtesy of General Motors Corporation, Service Technology Group.)*

fuel, if any, is stored in the canister, it waits until the system is in closed loop to allow any canister purge to occur, so it can compensate for any extra fuel or air introduced by the canister.

The canister purge is directly controlled by two valves, the control valve and a solenoid-operated valve in the vacuum line going to the control valve, Figure 7-14. The control valve contains a diaphragm and spring. The spring pushes the diaphragm down and closes the passage that would otherwise let air and fuel vapors to flow through the purge port. When manifold vacuum is applied to the top of the diaphragm, it lifts and opens the passage to the purge port. By turning the solenoid on or off, the ECM controls vacuum to the control valve. When the solenoid is ener-

gized, vacuum to the control valve is blocked. The ECM turns the solenoid off and allows purge when:

- the system is in closed loop and off idle.
- no faults are present triggering codes 13, 16, 44 or 45.

An electrical failure in the solenoid or its circuit allows purging to occur anytime there is enough intake manifold vacuum. This can cause:

- an overrich condition during warm-up or idle.
- an overlean condition during warm-up or idle (if the canister has no fuel vapors stored).
- dieseling after shutdown.

Early Fuel Evaporation (EFE)

The purpose of the EFE system is to apply heat to the area just below the throttle valves to prevent throttle blade icing and to increase fuel evaporation during the crucial engine warmup period. This helps to:

- reduce the richness requirement of the air/fuel ratio.
- improve driveability.
- reduce hydrocarbon and carbon monoxide emissions.

Rear-wheel-drive vehicles with Cadillac DFI have used exhaust heat for this purpose, not under the control of the ECM. There is a valve in the exhaust pipe on one side of the engine where it bolts to the exhaust manifold. A vacuum actuator containing a spring and a diaphragm opens and closes this valve. A thermal vacuum switch controls vacuum to the actuator diaphragm. When coolant temperature is below 49 degrees Centigrade (120 degrees Fahrenheit), the thermal vacuum switch routes vacuum to the EFE actuator. Vacuum applied to the actuator pulls the diaphragm against the spring and closes the EFE valve. This forces exhaust gases from that side of the engine through a passage in the head through the intake manifold EFE passage. The exhaust crosses beneath the plenum to the other head, where it travels through another passage to the exhaust pipe on that side. Once the coolant temperature goes above 49 degrees Centigrade, a temperature-actuated vacuum control blocks vacuum to the actuator. The spring pushes the diaphragm back down and opens the EFE valve. Exhaust gas is no longer forced through the EFE passage.

Real World Problem

The pivots on any valve in an exhaust passage are obviously subject to considerable thermal stress. It is not uncommon on older vehicles to find the EFE valve has seized in either the open or the closed position (ordinarily closed). On a cold engine, the valve should move easily by hand, against the spring actuator. If an EFE valve is seized closed, it can cause excessive exhaust backpressure and hence, poor fuel economy. It can also cause oil coking on the adjacent areas of the intake manifold over the cam valley.

The layout of front-wheel-drive cars makes this arrangement unsatisfactory, so they employ a ceramic-covered electric heating element under the TBI unit as a source of EFE heat, Figure 7-15. The heat element is powered through a relay (EFE relay) in turn controlled by the ECM. The ECM activates the relay when all of the following conditions exist:

- manifold charge temperature (MAT signal) is lower than 75 degrees Centigrade (167 degrees Fahrenheit).
- engine coolant temperature (CTS signal) is below 106 degrees Centigrade (223 degrees Fahrenheit).
- battery voltage is above 10 volts (in conditions of undercharge, the heating element

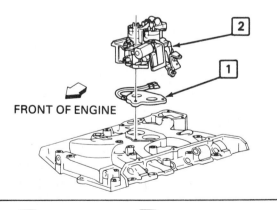

| 1 | EFE HEATER ASSEMBLY | 2 | THROTTLE BODY ASSEMBLY |

Figure 7–15 EFE grid. *(Courtesy of General Motors Corporation, Service Technology Group.)*

could easily rob more important components of electric power).

If any of these conditions is not met, the ECM turns the EFE relay off.

If the EFE is off during driving, the ECM will turn it on if all of these conditions are met:

- MAT is below 38 degrees Centigrade (100 degrees Fahrenheit).
- CTS indicates below 106 degrees Centigrade.
- battery voltage is above 12 volts.

Once any of the above conditions is no longer met, the ECM turns the EFE relay off.

One other set of conditions causes the ECM to turn on EFE heat. If throttle position is open more than 30 degrees (zero is closed and 90 degrees is WOT) and MAT is less than 60 degrees Centigrade (140 degrees Fahrenheit), the ECM will turn on the EFE relay for at least fifteen seconds to prevent throttle blade icing.

Cruise Control Vacuum Solenoid and Cruise Control Power Valve

When the ignition is turned on, the switch provides electric power to (among other things) the cruise control on/off switch and to the bleed switch, Figure 7-16. The cruise control on/off feeds power to:

- the resume/acceleration and set/coast switches.
- the cruise-on indicator light.

The brake switch feeds power to:

- the ECM (as a signal the brake is not applied).
- the cruise control vacuum solenoid.
- the cruise-engaged indicator light.

When the driver depresses the cruise control set/coast switch (at speeds above 25 mph), the selected speed is recorded in the ECM's memory, and the ECM grounds the cruise vacuum solenoid. The solenoid routes vacuum to the cruise control vacuum servo, which actually sets the throttle position through a vacuum servo diaphragm and spring, thus controlling engine speed. As vacuum is applied, the throttle opens. As vacuum is vented, the spring closes the throttle. The amount of vacuum applied to the servo is determined by the cruise control power valve, controlled in turn by the ECM. The ECM supplies a pulsed voltage signal to the power valve.

NOTE: This is one of the very few instances in which the ECM supplies the power for a component instead of providing the ground. In this case, the reason for the difference is to insure safety in case of a short circuit: if something goes wrong with the system, the cruise control will turn off rather than open the throttle any farther.

As the power valve turns on and off, the servo is alternately exposed to vacuum or atmospheric pressure. The ECM pulses the power valve at whatever rate is necessary to make the indicated speed from the VSS match the speed recorded in its memory. As the ECM begins to control vehicle speed, it also turns on the cruise control engaged light.

When the cruise control system engages, holding a set speed, it can disengage in any of three ways:

- The cruise control on/off switch can be turned off, in which case the selected speed is erased from the ECM's memory.
- The brake pedal can be depressed. This causes the ECM to disengage the cruise control vacuum solenoid and vent the servo to atmospheric pressure. Moving the resume/accelerate switch to the resume position and releasing it after the brake is released causes the system to reengage.

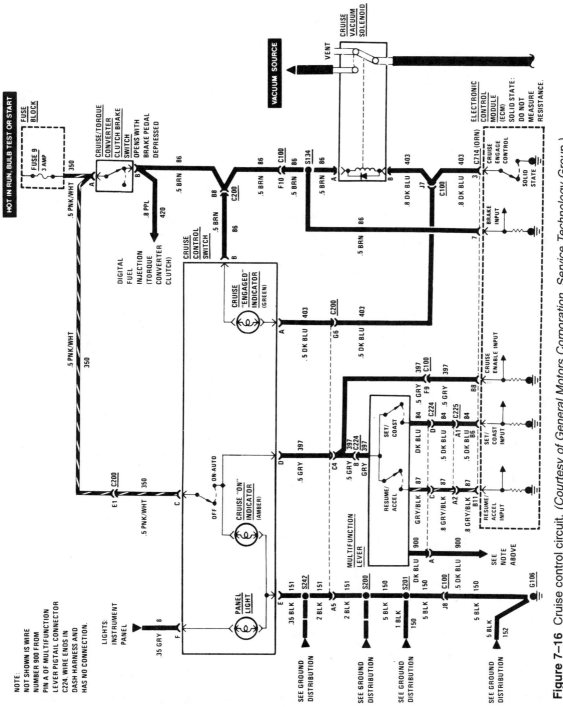

Figure 7–16 Cruise control circuit. *(Courtesy of General Motors Corporation, Service Technology Group.)*

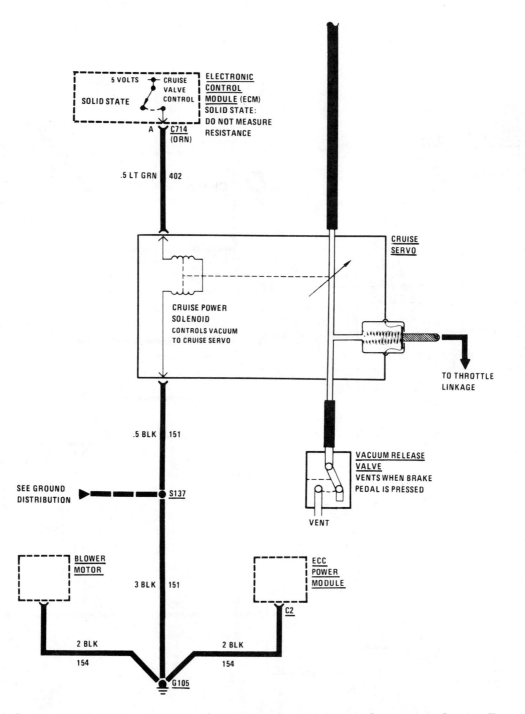

Figure 7–16 Cruise control circuit, continued. *(Courtesy of General Motors Corporation, Service Technology Group.)*

- Many faults that cause the ECM to turn on a service engine light also cause it to disable the cruise control system.

If the resume/accelerate switch is moved and held while the system is turned on, the cruise control system will accelerate the vehicle until the switch is released. The system will then engage whatever speed the vehicle was traveling when the switch was released. If the set/coast switch is depressed and held while the system is on, it will disengage and the throttle will close. The system will then reengage on whatever speed the vehicle is traveling when the set/coast switch is released.

Air-Conditioning (A/C) Relay and Cut-Out

Later model DFI systems use an A/C compressor control relay to turn the compressor clutch on and off. The relay control coil is itself grounded by the ECM, Figure 7-17. The ECM, however, gets instructions from the body control module (BCM) when to ground or unground the relay. The climate control panel receives input from the driver. This information is used to determine the operating mode for the climate control system and is sent to the BCM by way of the serial data line. If the BCM determines that cooling is required, it will use its serial data line to the ECM to command that the compressor be turned on.

The BCM receives input from the A/C high side and low side temperature sensors plus the A/C low pressure switch. It determines whether to turn the compressor on or off according to temperature indications. It will also command the compressor to turn off if the low pressure switch opens, indicating the refrigerant quantity is low enough that continued operation might damage the compressor.

The ECM also has logic circuits that can control compressor engagement based on input from the TPS, the CTS, the VSS, and the power steering pressure switch.

Earlier DFI systems did not use an A/C relay, and the only A/C-related activity of those EFI sys-
tems was to turn the compressor off during WOT operation. On those systems, the A/C compressor was energized and controlled directly by the electronic climate control (ECC) module or the BCM, depending on the model and build year. When the ECM sees a WOT signal from the TPS, it commands the ECC module or the BCM to disengage the compressor. This provides more engine power for acceleration and reduces the heat load on the engine. Refer to **Compressor Clutch Control** and **Programmer and Power Module** in the **Body Computer Module** section of this chapter for more information on compressor clutch control.

Service and Coolant Overtemperature Lights

If either of the two modules, the ECM or the BCM, sees a fault in any of the circuits they monitor, they will turn on one of the diagnostic indicator lights, Figure 7-18. The light the computer illuminates depends on the nature of the problem detected.

BODY COMPUTER MODULE (BCM)

The BCM is a microcomputer just as is the ECM. It has the basic components familiar from the engine computer (ROM, RAM, PROM, input/output interface, etc.), and about the same calculating power as the ECM; however, it controls different functions in the vehicle. On the 1985 C-body Cadillac, where it was introduced, it performs the following functions:

- controls the electronic climate control (air conditioning system).
- controls the electric cooling fans drawing air through the A/C condenser and radiator.
- controls the power windows, trunk release, and electric roof.
- provides and displays information for the driver.
- controls the information display panel dim-

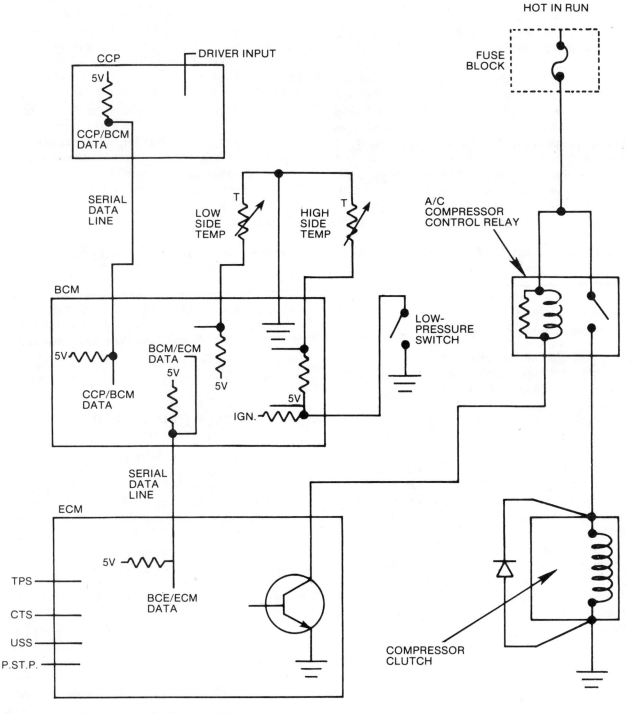

Figure 7–17 Compressor clutch control circuit

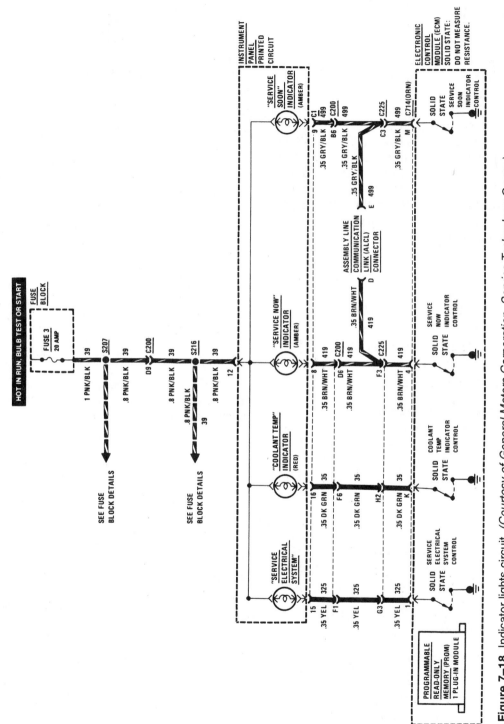

Figure 7–18 Indicator lights circuit. *(Courtesy of General Motors Corporation, Service Technology Group.)*

ming for visual clarity in different driving conditions.

• monitors the BCM system for faults, storing codes to identify the faults and in some cases provides fail-safe measures to compensate for a system failure.

Besides its sensors and switches for input information and its actuators to activate, the BCM and the ECM interact with each other. In some cases information in the ECM is sent by a data link to the BCM as input information, and it in turn can send outputs to the ECM as an ECM input. Figure 7-19 shows the kind of information the two computers exchange.

Electronic Climate Control (ECC)

To maintain the driver-selected temperature in the passenger compartment, the BCM selects the most appropriate air inlet and outlet modes and blower speed. It also monitors the status of the various ECC components.

ECC Inputs

Input information the BCM needs to control the ECC system comes through several sources, Figure 7-20.

Climate Control Panel (CCP). The driver selects the temperature desired and the operating mode (automatic, high fan, low fan, economy, front window defog, rear window defog, etc.) on the CCP. The CCP communicates this information in digital form through a data link to the BCM.

ECM. The ECM informs the BCM when the throttle is fully opened or when the coolant temperature goes above a specific value. In either case, the BCM commands the compressor clutch to disengage.

Temperature Sensors. Thermistors located in specific locations provide information about outside air temperature, inside air temperature and the temperature of the A/C refrigerant on

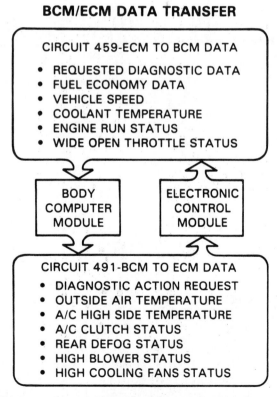

BCM/ECM DATA TRANSFER

CIRCUIT 459-ECM TO BCM DATA

• REQUESTED DIAGNOSTIC DATA
• FUEL ECONOMY DATA
• VEHICLE SPEED
• COOLANT TEMPERATURE
• ENGINE RUN STATUS
• WIDE OPEN THROTTLE STATUS

BODY COMPUTER MODULE

ELECTRONIC CONTROL MODULE

CIRCUIT 491-BCM TO ECM DATA

• DIAGNOSTIC ACTION REQUEST
• OUTSIDE AIR TEMPERATURE
• A/C HIGH SIDE TEMPERATURE
• A/C CLUTCH STATUS
• REAR DEFOG STATUS
• HIGH BLOWER STATUS
• HIGH COOLING FANS STATUS

Figure 7–19 BCM/ECM interaction. *(Courtesy of General Motors Corporation, Service Technology Group.)*

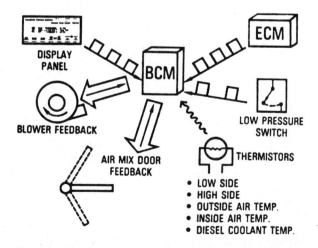

Figure 7–20 BCM information sources. *(Courtesy of General Motors Corporation, Service Technology Group.)*

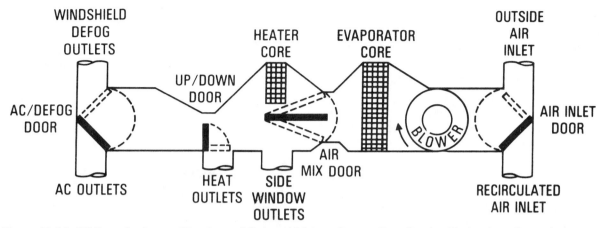

Figure 7–21 ECC mode doors. *(Courtesy of General Motors Corporation, Service Technology Group.)*

both the low and high pressure sides of the A/C system. A switch located in the low pressure side of the system notifies the BCM if the A/C pressure goes too low.

Air Mix Door Position.   A potentiometer attached to the air mix door in the air distribution system housing provides feedback information to the BCM, and identifies where the door is. The air door mix blends heated and refrigerated air to produce the desired air temperature, Figure 7-21. By monitoring voltage at the blower motor the BCM can determine the blower speed and increase or decrease it as needed, Figure 7-22.

ECC Outputs

Program Number.   When the engine first starts, the BCM compares inside air temperature to the driver-selected temperature. The computer then calculates what operational mode the ECC should use to achieve the air temperature selected. In its calculations, it factors in the effect of the outside air temperature. This calculation produces a program number between zero and 100. A low number means the inside air temperature is higher than the selected temperature and calls for cooling. A high number means just the opposite. The program number dictates blower speed and air distribution system door positions, Figure 7-21.

As inside air temperature comes closer to the

selected temperature, the BCM updates the program number, the blower speed decreases, and the computer adjusts the position of the air distribution system doors. When the temperatures are the same, the program number reaches an equi-

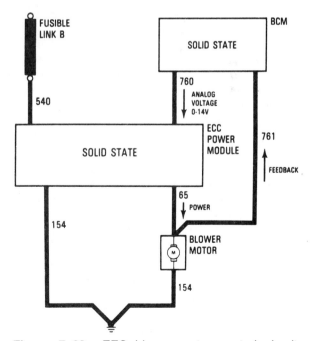

Figure 7–22   EEC blower motor control circuit. *(Courtesy of General Motors Corporation, Service Technology Group.)*

librium, and the BCM maintains the program number at this equilibrium point. The program number is displayed on the CCP only when the system is set to its diagnostic mode by a technician.

Air Mix Door. An electric motor moves the air mix door through the connecting linkage. All the other doors move by vacuum motors controlled by vacuum solenoids. To determine what the air mix door position should be, the BCM checks:

* the engine coolant temperature (information provided by the ECM) to determine what the heated air temperature will be.
* the air conditioning refrigerant temperature on the low pressure side to determine what the cold air temperature will be. If the A/C is not operating, the BCM uses the outside air temperature instead.
* the program number already calculated.

The BCM uses the air mix door feedback signal to see whether the door moved to the commanded position. Because the linkage between the air mix door motor and the door itself is adjustable, it must be carefully adjusted to make sure its position coincides with the BCM-commanded position. The appropriate shop manual provides information for this adjustment.

Compressor Clutch Control. The BCM controls the compressor clutch. Before applying the clutch, it checks that:

* the outside air temperature is above zero degrees Centigrade (32 degrees Fahrenheit).
* engine coolant and/or A/C system high pressure side refrigerant temperatures are not too high.
* refrigerant pressure (indicating the amount of refrigerant charge) is not too low.
* the throttle is not at the WOT position as signalled by the TPS.

When the compressor is engaged, the BCM constantly rechecks these factors and disen-

gages the clutch if any of them fall above or below the threshold values.

To cycle the clutch, the BCM looks at outside air temperature and low pressure refrigerant temperatures. The compressor clutch engages when the low side temperature exceeds 10 degrees Centigrade (50 degrees Fahrenheit). The clutch disengages again when the low side refrigerant temperature drops to -1 degree Centigrade (30 degrees Fahrenheit).

Programmer and Power Module. The BCM uses two interface devices for many ECC functions: the ECC programmer and the ECC power module, Figure 7-23. The BCM sends instructions for air door positions, heater water control, rear window defog, and compressor clutch engagement information through a data link to the programmer. The rear window defogger is a heat element in the rear window glass powered through a relay. The programmer commands the actuators for each of these functions, Figure 7-24, except the compressor clutch. The compressor clutch on/off signal is sent from the programmer to the power module, and the power module actually engages or disengages the clutch. The instructions to turn the compressor on or off, which the BCM sends to the programmer, are also sent to the ECM so it can anticipate the change in engine load and adjust engine speed,

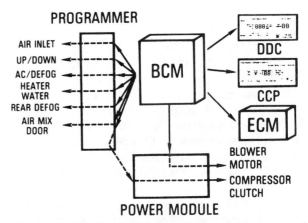

Figure 7–23 BCM interface devices. *(Courtesy of General Motors Corporation, Service Technology Group.)*

an adjustment that comes into play only at idle speed of course.

The BCM sends the blower motor instructions directly to the power module in the form of a variable voltage. The power module amplifies this signal and keeps it proportional to the voltage received from the BCM. It then issues that voltage as a command to the blower motor, Figure 7-22.

Electric Cooling Fans

Inputs. The BCM receives inputs for cooling fan operation from two sources: the A/C high pressure side thermistor, located in the condenser outlet, and from the engine coolant temperature information provided by the ECM.

Outputs. The speed of the two coolant fans is controlled by a pulse width modulated (PWM)

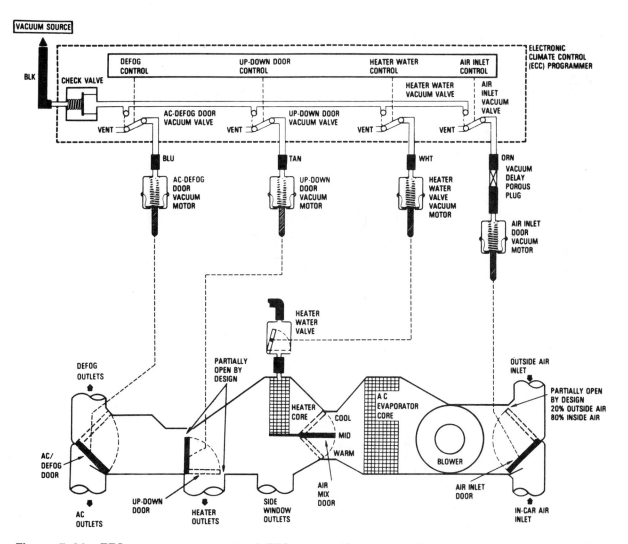

Figure 7–24 EEC programmer control of EEC mode. *(Courtesy of General Motors Corporation, Service Technology Group.)*

signal produced by the BCM and sent to the fan control module, Figure 7-25. The module completes or opens the ground side of the fans' power circuit in response to the BCM signal. The PWM signal is cycled once every 32/1000 of a second. If the signal voltage is high during only a short portion of the on-period, the fan motors run at low speed, Figure 7-26. This is because their power is supplied in short pulses and they spend most of each revolution coasting, though the momentum of the blades keeps their speed near constant. The greater the proportion of each period during which the pulsed on-voltage is high, the faster the fan motors turn.

The fans run at 40% full speed when either the coolant temperature reaches 106 degrees Centigrade (223 degrees Fahrenheit) or when the A/C high pressure refrigerant temperature reaches 61 degrees Centigrade (142 degrees Fahrenheit). The fans run at full speed when either the engine coolant temperature reaches 116 degrees Centigrade (241 degrees Fahrenheit) or the A/C high side refrigerant temperature reaches 72 degrees Centigrade (162 degrees Fahrenheit). They turn off completely when either the coolant temperature drops to 102

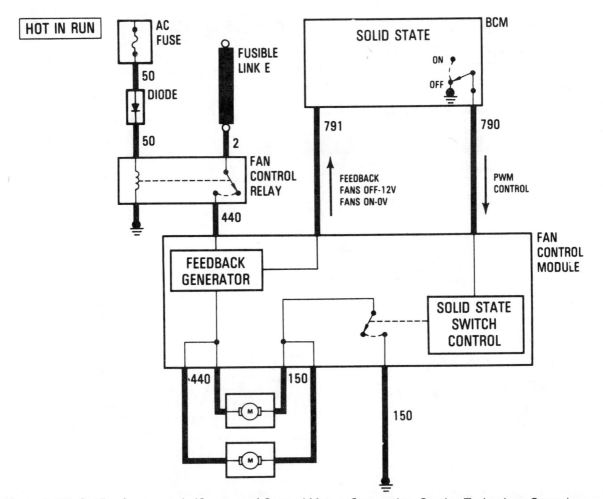

Figure 7–25 Cooling fans control. *(Courtesy of General Motors Corporation, Service Technology Group.)*

degrees Centigrade (216 degrees Fahrenheit) or the A/C high side refrigerant temperature drops to 57 degrees Centigrade (135 degrees Fahrenheit). The lower turn-off temperatures prevent the fans from dithering on and off if one of the temperatures hovers at the on-value.

The feedback signal sent to the BCM from the fan control module enables the BCM to monitor fan motor realization of the commands it was sent. This signal is nearly 12 volts when the fans are off or stalled and near zero volts when they are on or disconnected. The BCM compares feedback voltage to the commands it sent previously; if they do not match, it sets fault code F41.

Retained Accessory Power

If the ignition is turned off (locked) or turned to the accessory position, the BCM provides power to the power windows, electric roof and the trunk release for ten minutes or until either a door opens or the courtesy lights turn off, Figure 7-27. Once the ignition is off, the BCM grounds the retained power accessory relay for ten minutes. This relay then connects circuit 300 to the battery voltage from a constantly powered source. Circuit 300 powers the trunk release, power windows, and electric roof. If a door switch, the courtesy light switch, or the ignition switch turns on, the BCM ungrounds that relay, and the relay reconnects circuit 300 to the ignition run terminal. Circuit 300 will then be powered only when the ignition switch is turned on.

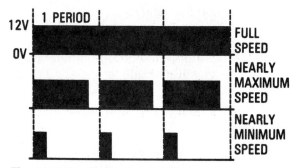

Figure 7-26 PWM command to cooling fan motors. *(Courtesy of General Motors Corporation, Service Technology Group.)*

Driver Information and Display

The BCM displays information about fuel consumption and temperature either on the fuel data center (FDC) or on the CCP. The driver selects the information he or she wants displayed by pressing the appropriate button on one of the display panels. On vehicles with a digital display instrument cluster, the driver can choose to have the information displayed in either English or metric units by moving the English/metric switch on the digital cluster. Vehicles not equipped with the digital cluster display in either English or metric, depending on whether or not a wire (circuit 811) leading from terminal J2-13 of the BCM is connected to ground. If the wire is grounded, the

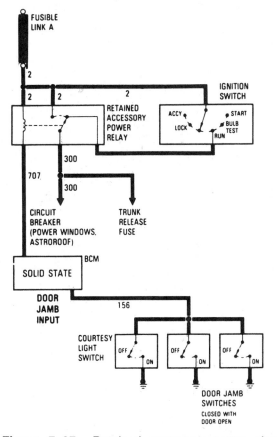

Figure 7-27 Retained accessory power circuit. *(Courtesy of General Motors Corporation, Service Technology Group.)*

information will display in metric units; if it is open, the information will be in English units.

Fuel Data Inputs. The BCM gets information from the fuel gauge sender (a standard, variable potentiometer), the gauge fuse, and the ECM. The potentiometer measures fuel level just as fuel gauge senders have done for many years. An arm with a float on it moves the movable contact of the potentiometer as the fuel level changes. By comparing return voltage from the potentiometer to a voltage reading taken at the gauge fuse, the BCM can calculate the fuel level, Figure 7-28.

Vehicle speed and distance traveled (this information can be easily calculated from VSS information because 4004 VSS pulses equal 1 mile using the pulse generator-type VSS) plus injector pulse width information are sent from the ECM to the BCM through the data link.

Fuel Data Outputs. Normally the BCM displays fuel level through the FDC as explained above. However, on signalled request it can also display:

- *Instantaneous fuel mileage.* If the INST button on the FDC panel is pressed, the BCM looks at fuel consumed (calculated from the injector pulse width data) and vehicle speed. From these two pieces of information it can instantly calculate the current achieved fuel mileage at that moment.

- *Average fuel mileage.* The BCM constantly receives and updates its information regarding fuel used and distance traveled. This information is stored in memory. If the driver presses the reset button on the FDC panel, however, all the previous information is erased and the computer accumulates new data. If the driver pushes the AVG button, the BCM will calculate fuel mileage for the distance traveled since the last reset.

- *Fuel range.* When the RANGE button is pressed, the BCM checks remaining fuel from the fuel level information, calculates the average fuel use for the last 25 miles, and calculates a distance that can be traveled with the remaining fuel. Obviously, if driving speed or other conditions significantly changes so fuel consumption goes up or down, this calculation will be incorrect.

- *Fuel used.* When the driver requests FUEL USED, the BCM calculates the fuel used since the last reset. This does *not* indicate fuel used since the last fill-up unless that coincides with the last reset.

Temperature Display. Upon driver request, the BCM displays outside temperature as reported to it by the outside temperature thermistor. If outside temperature is not requested, the CCP displays inside, passenger compartment temperature.

Instrument Panel Display Dimming

The characters displayed by the FDC, the CCP, the radio, and the optional digital cluster are

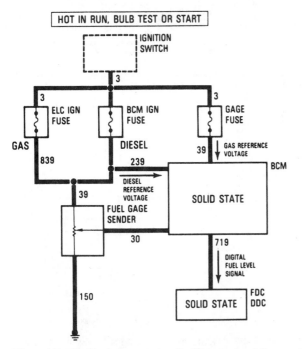

Figure 7–28 Fuel gauge circuit. *(Courtesy of General Motors Corporation, Service Technology Group.)*

displayed by a group of small **vacuum fluorescent (VF)** glass tubes. **Anodes** (conductors on which there is a positive potential voltage) are placed on one side of the tube so they form all of the vertical and horizontal bars of the alphanumeric characters. The anodes are coated with a fluorescent material. A series of thin, tungsten-coated wire strands are placed opposite the anodes on the other side of the tube. This side then serves as the **cathode** (a conductor on which there is a negative potential voltage). Between the anode and the cathode, a fine wire mesh called the **grid** is placed. The tube **(ampule)** is evacuated and then filled with argon or neon gas.

As current passes through the cathode, it becomes hot and causes the tungsten to give off a cloud of electrons. The BCM applies an amplified 17-volt positive potential to selected anodes. The electrons given off by the cathode are attracted to and bombard the energized anodes. As they strike an anode's fluorescent coating, it glows visibly and thus provides the digital display. The more electrons striking the fluorescent material, the more brightly it glows.

During daylight hours the BCM adjusts the VF displays to full brightness for maximum visibility. As the sunlight fades, the VF displays need less brightness to be seen in the darker surroundings. When the parking lights are turned on (either with the headlights or independently), the BCM gets that information as an input signal, Figure 7-29. When the BCM learns the headlights or parking lights have been turned on, it uses the headlight switch rheostat to determine how bright to make the display. The driver can set this rheostat throughout its range to select VF display brightness.

The BCM controls the display brightness by controlling a pulse-width-modulated voltage to the grid in the ampule. When the BCM applies a modulated voltage to the grid, it attracts or filters some of the electrons given off by the cathode and prevents them from striking the energized anodes.

Diagnostic Testing and Fail-Safe Actions

The BCM monitors many of its sensor and output circuits for proper operation, such as the blower motor and the cooling fan motors. It also uses sensors specifically designed to report problems such as the low-pressure refrigerant switch in the A/C **accumulator.** If the BCM sees such a fault (voltage in a given circuit or from a sensor that is out of the expected range) in any of its related circuits, it will record a code in its diagnostic memory. It may turn on the service air conditioning light, or it may initiate a fail-safe action, depending on the circuit with the fault,

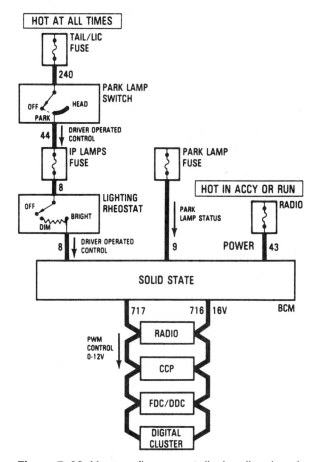

Figure 7–29 Vacuum fluorescent display dimming circuit. *(Courtesy of General Motors Corporation, Service Technology Group.)*

Figure 7-30. For example, every time the BCM turns on the A/C compressor clutch, it checks the A/C low-pressure switch first. Should the switch be open, indicating low pressure, it will not engage the clutch. If the switch stays open for thirty seconds, it will set code F48 and the following fail-safe actions are initiated:

- The service air conditioning light turns on for thirty seconds. The light will come on again every time the driver selects AUTO on the CCP and the next time the ignition switch is turned on.
- The BCM disables the compressor clutch until the ignition is turned off.
- The BCM switches the ECC system into ECON instead of AUTO if the AUTO mode is selected.

Keep in mind the purpose of fail-safe operation is to protect system components and that the codes initiating fail-safe measures can't be set while the system is in diagnostics. So if a code is cleared while the system is in a diagnostic mode, the fail-safe measures will be defeated and it is possible to cause damage to a system component.

✔ SYSTEM DIAGNOSIS & SERVICE

The self-diagnostic function of the DFI system works fundamentally the same way as other General Motors computer control systems, but it is more elaborate, monitoring and controlling more functions. When first introduced in mid-1980 (the same system remained throughout model year 1981), its self-diagnostic capacity was like that of a CCC system. Since then, however, the systems capacities have greatly expanded. As the ECM and/or BCM control the systems for which they are responsible, each continuously monitors operating conditions for faults. They compare the operating conditions against expected, preprogrammed operating condition standards. By doing this, certain circuit

and component faults can be identified. When a fault is identified, a two-digit trouble code number, either with an E or an F prefix, is stored in the

ECM DIAGNOSTIC CODES

CODE	MALFUNCTION
■■ E12	NO DISTRIBUTOR SIGNAL
□ E13	OXYGEN SENSOR NOT READY (CANISTER PURGE)
□ E14	SHORTED COOLANT SENSOR CIRCUIT
□ E15	OPEN COOLANT SENSOR CIRCUIT
■■ E16	GENERATOR VOLTAGE OUT OF RANGE (ALL SOLENOIDS)
□ E18	OPEN CRANK SIGNAL CIRCUIT
□ E19	SHORTED FUEL PUMP CIRCUIT
■■ E20	OPEN FUEL PUMP CIRCUIT
□ E21	SHORTED THROTTLE POSITION SENSOR CIRCUIT
□ E22	OPEN THROTTLE POSITION SENSOR CIRCUIT
□ E23	EST/BYPASS CIRCUIT PROBLEM (AIR)
□ E24	SPEED SENSOR CIRCUIT PROBLEM (VCC)
□ E26	SHORTED THROTTLE SWITCH CIRCUIT
□ E27	OPEN THROTTLE SWITCH CIRCUIT
□ E28	OPEN THIRD OR FOURTH GEAR CIRCUIT
□ E30	ISC CIRCUIT PROBLEM
■■ E31	SHORTED MAP SENSOR CIRCUIT (AIR)
■■ E32	OPEN MAP SENSOR CIRCUIT (AIR)
■■ E34	MAP SENSOR SIGNAL TOO HIGH (AIR)
□ E37	SHORTED MAT SENSOR CIRCUIT
□ E38	OPEN MAT SENSOR CIRCUIT
□ E39	VCC ENGAGEMENT PROBLEM
□ E40	OPEN POWER STEERING PRESSURE CIRCUIT
■■ E44	LEAN EXHAUST SIGNAL (AIR & CL & CANISTER PURGE)
■■ E45	RICH EXHAUST SIGNAL (AIR & CL & CANISTER PURGE)
□ E47	BCM - ECM DATA PROBLEM
■■ E51	ECM PROM ERROR
▼ E52	ECM MEMORY RESET INDICATOR
▼ E53	DISTRIBUTOR SIGNAL INTERRUPT
▼ E59	VCC TEMPERATURE SENSOR CIRCUIT
▼ E60	TRANSMISSION NOT IN DRIVE
▼ E63	CAR SPEED AND SET SPEED DIFFERENCE TOO HIGH
▼ E64	CAR ACCELERATION TOO HIGH
▼ E65	COOLANT TEMPERATURE TOO HIGH
▼ E66	ENGINE RPM TOO HIGH
▼ E67	CRUISE SWITCH SHORTED DURING ENABLE

ECM AND CRUISE CONTROL COMMENTS:

■■	TURNS ON "SERVICE NOW" LIGHT
□	TURNS ON "SERVICE SOON" LIGHT
▼	DOES NOT TURN ON ANY TELLTALE LIGHT
( )	FUNCTIONS WITHIN BRACKETS ARE DISENGAGED WHILE SPECIFIED MALFUNCTION REMAINS CURRENT (HARD)

E16 & E24 DISABLE VCC FOR ENTIRE IGNITION CYCLE

E24 & E67 DISABLE CRUISE FOR ENTIRE IGNITION CYCLE

CRUISE IS DISENGAGED WITH CODE(S) E16, E51 OR E60 - E67

Figure 7–30 BCM diagnostic codes. *(Courtesy of General Motors Corporation, Service Technology Group.)*

computer's memory. Codes with an F prefix are stored by the BCM, Figure 7-30. Codes with an E prefix are stored by the ECM, Figure 7-31. When the system is in diagnostic mode, the stored code displays either on the CCP or the FDC, depending on the year and model of the car.

Types of Trouble Codes

Each computer can store a trouble code in either of two ways: as a current mode (sometimes called a hard code because it refers to a problem that currently exists) or as a history code (representing a fault that occurred at one time during engine operation but is not present at the time of diagnosis). When the system goes into

BCM DIAGNOSTIC CODES

CODE	CIRCUIT AFFECTED
▼ F10	OUTSIDE TEMP SENSOR CKT
▼ F11	A/C HIGH SIDE TEMP SENSOR CKT
▼ F12	A/C LOW SIDE TEMP SENSOR CKT
▼ F13	IN-CAR TEMP SENSOR CKT
▼ F14	DIESEL COOLANT SENSOR CKT
▼ F30	CCP TO BCM DATA CKT
▼ F31	FDC/DDC TO BCM DATA CKT
▼ F32	ECM-BCM DATA CKT'S
▼ F40	AIR MIX DOOR PROBLEM
▼ F41	COOLING FANS PROBLEM
☑ F46	LOW REFRIGERANT WARNING
☑ F47	LOW REFRIGERANT CONDITION
☑ F48	LOW REFRIGERANT PRESSURE
▼ F49	HIGH TEMP CLUTCH DISENGAGE
▼ F51	BCM PROM ERROR

☑ TURNS ON "SERVICE AIR COND" LIGHT
▼ DOES NOT TURN ON ANY LIGHT

COMMENTS:
F11 TURNS ON COOLING FANS WHEN
    A/C CLUTCH IS ENGAGED
F12 DISENGAGES A/C CLUTCH
F14 & F32 TURN ON COOLING FANS
F30 TURNS ON FT. DEFOG AT 75° F
F41 TURNS ON "COOLANT TEMP/FANS"
    LIGHT WHEN FANS SHOULD BE ON
F47 & F48 SWITCHES FROM "AUTO"
        TO "ECON"

Figure 7–31 ECM diagnostic codes. *(Courtesy of General Motors Corporation, Service Technology Group.)*

diagnostic mode, all of the codes display once in numerical order from low to high; then they repeat. This time, however, only the current, hard codes display. From 1982 through 1984, DFI systems show all codes in the first two passes through the display and show only current hard codes in a third pass-through.

Service Lights and Fail-Safe

Faults the ECM recognizes (E-prefixed codes) and that require immediate attention turn on the service engine now light. Faults the ECM recognizes and those that need to be called to the driver's attention but are less urgent, bring on the service engine soon light. Certain F codes pertaining to the ECC bring on the service air conditioning light.

If a fault that might cause damage to a circuit or component that can result in unacceptable performance during continued vehicle operation, the ECM or BCM will activate a fail-safe strategy. The fail-safe action is designed to either deactivate the faulty component or compensate for its malfunction or both.

Diagnostic Procedure

Before beginning a diagnostic procedure on a DFI- equipped vehicle, be sure first that you:

- have at least a general understanding of the operation of the ECM and/or the BCM and any of the subsystems in which the problem or problems seem to occur.
- are at least familiar with the self-diagnostic features of the system.
- have access to a service manual listing the diagnostic charts. Reading the codes without knowing what they mean will not help the technician.
- do a thorough visual inspection of the system in which the fault seems to occur and any related systems. Check for loose or broken vacuum lines, pinched or broken wires, and loose connectors.

SELF-DIAGNOSTIC SYSTEM CHECK

• IF YOU HAVE NOT REVIEWED THE BASIC INFORMATION ON HOW TO USE THE COMPUTOR SELF-DIAGNOSTICS, GO TO THE INTRODUCTION OF THIS SECTION.

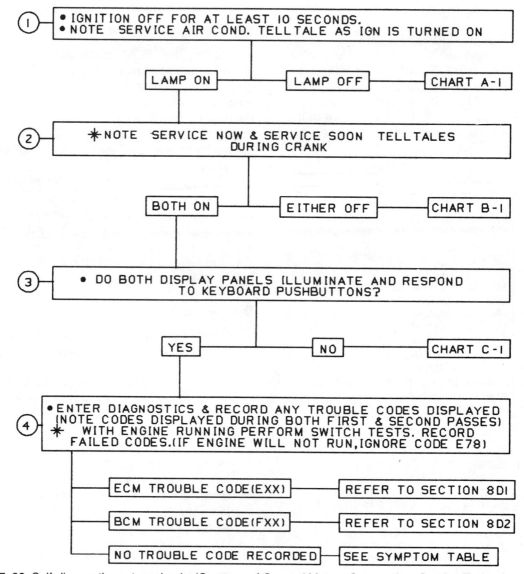

Figure 7–32 Self-diagnostic system check. *(Courtesy of General Motors Corporation, Service Technology Group.)*

Diagnostic Guide. Having complied with the suggestions above, continue your diagnostic procedure by answering the questions in the next four paragraphs:

- *Are the on-car self-diagnostic systems working?* If the service air conditioning, service engine now, or service engine soon lights are not working properly, or if either of the data display panels fails to fully light up or respond properly, the self-diagnostic system check will refer to the appropriate chart for correcting the problem, Figure 7-32.
- *Are trouble codes displayed?* If any trouble codes are displayed, go to the trouble code chart of the same number. If the fault is intermittent (the code was not repeated on the second pass, or third pass for earlier models), observe any notes concerning intermittent codes in the text that accompanies each trouble code chart. The procedure for displaying the trouble codes is explained later in this section. If no trouble codes are stored, proceed to the next paragraph.
- *Do all switch tests pass?* After the trouble codes have been displayed, perform the switch test as explained under **Switch Test** in this section. If any switch tests do not pass, go to the code chart identified by the corresponding number as the diagnostic display number (under **Switch Tests** in Figure 7-33) for the switch that did not pass. If all the switches do pass the test, go to the next paragraph.
- *Is the fuel system controlling air/fuel mixture correctly?* To answer this, use the field service mode, which is a part of the DFI system check, Figure 7-34. If the fuel system is not controlling air/fuel mixture properly, the DFI system check will refer to the appropriate chart. If the fuel system is controlling mixture properly, the DFI system check will refer to the symptom charts.

Using these simple steps at the beginning of the diagnostic procedure saves time and often prevents unnecessary parts replacement resulting from misdiagnosis.

Entering Diagnostic Mode. To enter diagnostics, turn the ignition on, then simultaneously depress the off and warmer buttons on the CCP. Hold the buttons until all display panel segments of both display panels light up, Figure 7-35. If any of the digit segments do not light up (they should all make squared-off computer 8's), the panel will have to be replaced. An 8 displayed by a panel with two segments inoperative can appear as a 3.

Display of Trouble Codes. After the display panel segment check ends, the FDC displays 8.8.8 followed by ..E. The ..E indicates that the first pass of the ECM codes will be displayed next and that all ECM codes will display, Figure 7-35. After the first pass, ..EE displays and indicates the second pass of the ECM codes is coming, in which only the hard codes will display. If all stored ECM codes are intermittent, the ..EE will not display and the second pass will not occur. If code E51 (a PROM fault) displays, it will continue to display until the diagnostic mode is exited or until the PROM fault is repaired and the code clears. While code E51 displays, the system will not advance to another diagnostic feature.

If no ECM codes are stored, the 8.8.8 will be followed by ..F, indicating that BCM codes come next. The same sequence occurs. The first pass displays all stored BCM codes followed by ..FF, and the second pass displays only hard codes. If no BCM hard codes are stored, the ..FF will not display and the second pass will not occur.

After all stored codes display, or if there are no stored codes, .7.0 will display. This indicates the system is ready for selection of the next self-diagnostic feature. There are several different diagnostic tests a technician can select at this point:

- ECM switch tests.
- ECM data display.
- ECM output cycling.
- BCM data display.
- ECC program override.
- cooling fan override.
- exit diagnosis or clear codes and exit diagnosis.

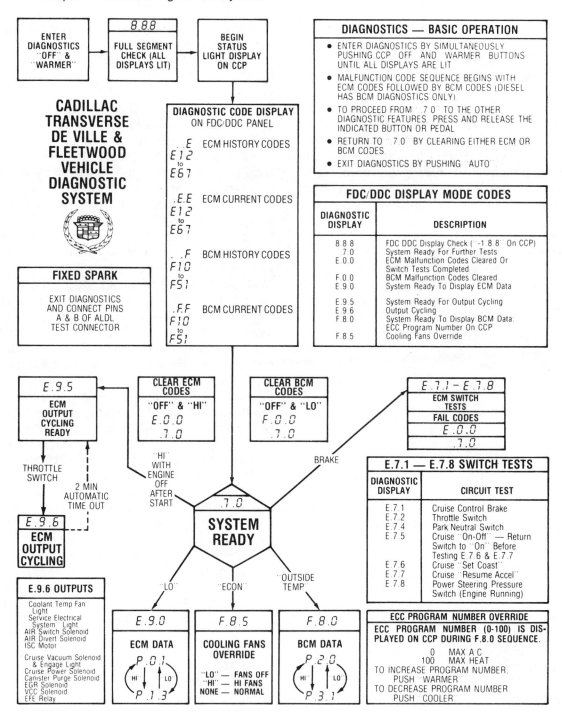

Figure 7–33 Diagnostic chart, part 1. *(Courtesy of General Motors Corporation, Service Technology Group.)*

DFI SYSTEM CHECK

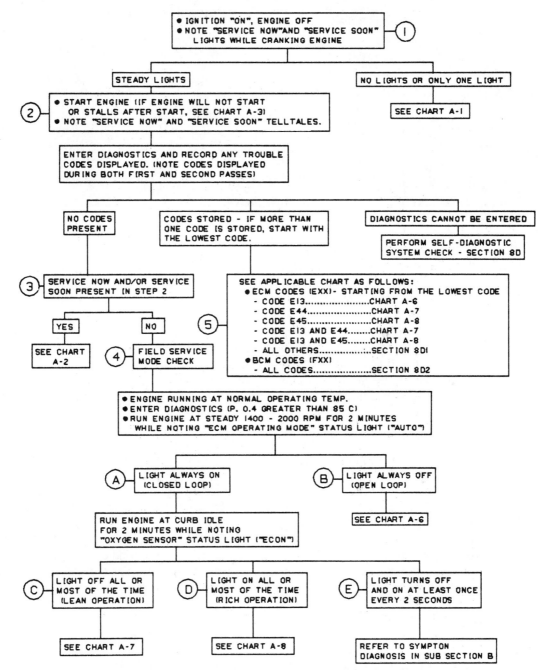

Figure 7–34 DFI system check. *(Courtesy of General Motors Corporation, Service Technology Group.)*

Figure 7–35 Display panel in segment check

ECM Switch Tests. The engine must be running to perform this test sequence. When the FDC displays .7.0, depress and release the brake pedal. The FDC display should advance to E.7.1. When it does, cycle the cruise control brake switch by tapping the brake pedal again within ten seconds. As the switch cycles, the ECM monitors its operation to be sure it is functioning properly. If the switch does not cycle within ten seconds, the ECM will record it as having failed. At the end of ten seconds or after the switch has cycled properly, the display advances to the next switch test code. The same sequence repeats for each code and switch listed under **Switch Test** in Figure 7-33.

After the switch test sequence completes, the ECM displays on the FDC the code of any switch that did not pass. The failed switch code remains on display until the switch circuit is repaired and retested. After the test sequence has completed and all switches have passed, E.O.O will display, followed by .7.0 (meaning the system is ready for another diagnostic feature selection).

ECM Data Display. This mode is often called engine parameters. In this mode the ECM displays the values it receives from selected sensors or switches. It can be entered by pressing the Lo button on the CCP while .7.0 displays on the FDC. When this mode is entered, E.9.0. appears on the FDC. The parameter codes, the parameters they represent, the value range and the units in which the parameters are expressed are shown in Figure 7-36 under **Engine Data Display.** To advance to the next higher parameter code number, press the Hi button on the CCP. Pushing the Lo button returns to the next lower parameter code number.

These parameter values can be used as follows:

- They can be compared to those of a vehicle known to be operating properly to see whether they are what they are supposed to be.
- They can be watched to see that they change as operating conditions change (for instance, the TPS parameter should change as the throttle moves).
- Some of them are called for in the diagnostic charts, where their values are interpreted as acceptable or faulty.

This diagnostic mode can be exited at any time and the display returned to the selection point..7.0. by clearing the ECM or BCM codes

E.9.0 ENGINE DATA DISPLAY

PARAMETER NUMBER	PARAMETER	PARAMETER RANGE	DISPLAY UNITS
P.0.1	Throttle Position	-10 - 90	Degrees
P.0.2	MAP	14 - 109	kPa
P.0.3	Computed BARO	61 - 103	kPa
P.0.4	Coolant Temperature	-40 - 151	C
P.0.5	MAT	-40 - 151	C
P.0.6	Injector Pulse Width	0 - 99.9	ms
P.0.7	Oxygen Sensor Voltage	0 - 1.14	Volts
P.0.8	Spark Advance	0 - 52	Degrees
P.0.9	Ignition Cycle Counter	0 - 50	Key Cycles
P.1.0	Battery Voltage	0 - 25.5	Volts
P.1.1	Engine RPM	0 - 6370	RPM ÷ 10
P.1.2	Car Speed	0 - 255	MPH
P.1.3	ECM PROM I.D.	0 - 255	Code

F.8.0 BCM DATA DISPLAY

PARAMETER NUMBER	PARAMETER	PARAMETER RANGE	DISPLAY UNITS
P.2.0	Commanded Blower Voltage	-3.3 - 18.0	Volts
P.2.1	Coolant Temperature	-40 - 215	C
P.2.2	Commanded Air Mix Door Position	0 - 100	%
P.2.3	Actual Air Mix Door Position	0 - 100	%
P.2.4	Air Delivery Mode 0 = Max A C 4 = Off 1 = A C 5 = Normal Purge 2 = Intermediate 6 = Cold Purge 3 = Heater 7 = Front Defog	0 - 7	Code
P.2.5	In-Car Temperature	-40 - 102	°C
P.2.6	Actual Outside Temperature	-40 - 93	°C
P.2.7	High Side Temperature (Condenser Out)	-40 - 215	°C
P.2.8	Low Side Temperature (Evaporator In)	-40 - 93	°C
P.2.9	Actual Fuel Level	0 - 19.0	Gallons
P.3.0	Ignition Cycle Counter	0 - 99	Key Cycles
P.3.1	BCM PROM I.D.	0 - 255	Code

ECM PROM I.D.

ECM PROM I.D. is Parameter .1.3 of Engine Data and is displayed as a numerical code as follows:

```
                                    X  X  X
FINAL DRIVE RATIO ─────────────────────────┘
  2   3.33:1
     (2.97:1 Effective Ratio)
EMISSIONS SYSTEM ───────────────────────┘
  1 = Federal
  2 = California
  3 = Export
  4 = Altitude
ECM PROM CALIBRATION ───────────────┘
Number varies with individual calibration.
```

BCM PROM I.D.

BCM PROM I.D. is Parameter .3.1 of BCM Data and is displayed as a numerical code as follows:

```
                              X   X  X
ENGINE SYSTEM ──────────────────────┘
  Blank = Gas
     1 = Diesel
BCM PROM CALIBRATION ───────────┘
Numbers vary with individual calibration.
```

ECM STATUS LIGHT DISPLAY					
LIGHT ON	IN 4th GEAR	VCC ENABLED	CLOSED THROTTLE	RICH	CLOSED LOOP
LIGHT OFF	NOT IN 4TH GEAR	VCC DISABLED	OPEN THROTTLE	LEAN	OPEN LOOP
INDICATOR	[symbol]	[symbol]	Off	Econ	Auto
FUNCTION	4TH GEAR INPUT	VCC OUTPUT	THROTTLE SWITCH INPUT	OXYGEN SENSOR INPUT	ECM OPERATING MODE

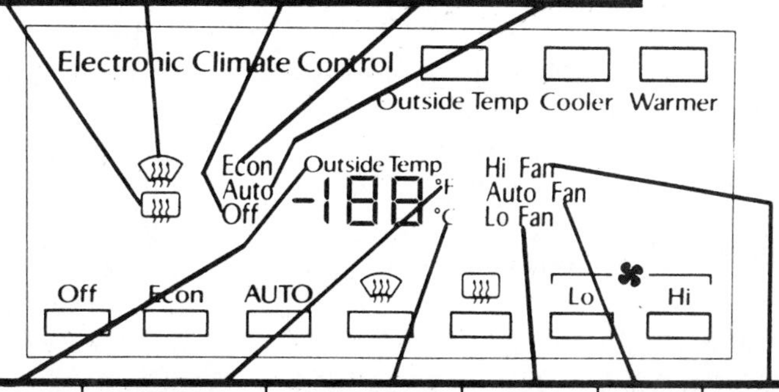

Electronic Climate Control

BCM STATUS LIGHT DISPLAY						
FUNCTION	A/C CLUTCH OUTPUT	COMPRESSOR LOW PRESSURE SWITCH INPUT	HEATER WATER VALVE OUTPUT	A/C-DEF MODE DOOR OUTPUT	COOLING FANS STATUS	UP/DOWN MODE DOOR OUTPUT
INDICATOR	Outside Temp	°F	°C	Lo Fan	Auto Fan	Hi Fan
LIGHT ON	ENERGIZED	OPEN (LOW PRESSURE)	CLOSED (NO WATER FLOW)	A C	FANS RUNNING	UP
LIGHT OFF	DE-ENERGIZED	CLOSED	OPEN	DEF	FANS OFF	DOWN

Figure 7–36 Diagnostic chart, part 2. *(Courtesy of General Motors Corporation, Service Technology Group.)*

(press Off and Hi for ECM codes or Off and Lo for BCM codes).

ECM Output Cycling. In this diagnostic mode, the ECM cycles all of its output actuators every three seconds with the exception of the cruise control power valve, which cycles continuously. Cycling continues for two minutes; then the system automatically reverts to ECM Output Cycling Ready (E.9.5 displays on the FDC). While the actuators cycle, you can observe that each component operates properly. To enter this output cycling mode ..7.0 must be displayed and the engine must be running.

- Turn the cruise on/off switch to the on position. This allows the cruise outputs to cycle.
- Turn the engine off. Within two seconds turn the ignition on.
- Press Hi on the CCP (E.9.5 displays on the FDC).
- Open and close the ISC throttle switch by depressing and releasing the throttle. The ECM output cycling mode is activated (E.9.6 displays).

This diagnostic mode can be exited at any time and the display returned to the selection point ..7.0 by clearing the ECM or BCM codes (press Off and Hi for ECM codes, or Off and Lo for BCM codes).

BCM Data Display. As shown in Figure 7-36 under **BCM Data Display,** this mode displays the values sent to the BCM from most of its information sources as well as the value of some of its outputs. Figure 7-36 also shows the maximum range of each parameter and the units of measurement.

To go into the BCM data display mode, .7.0 must be displayed on the FDC. Depress and release the outside temp button on the CCP. The FDC displays F.8.0 briefly and then goes to parameter code P.2.0 or P.3.1 (the first and last parameters in the sequence). The parameter code displays for one second. Then the parameter value displays for nine seconds. This sequence repeats with the same parameter until you select another.

To advance to another parameter, push the Hi button on the CCP; to move back to a lower parameter code number, push Lo on the CCP.

These parameter values can be used as follows:

- They can be compared to those of a vehicle known to be operating properly to see whether they are what they are supposed to be.
- They can be watched to see that they change as operating conditions change (for instance, the commanded air mix door position should change as the selected temperature changes).
- Some of them are called for in the diagnostic charts, where their values are interpreted as good or at fault.

This diagnostic mode can be exited at any time and the display returns to the selection point, .7.0., by clearing the ECM or BCM codes (press Off and Hi for ECM codes, or Off and Lo for BCM codes).

ECC Program Override. During the BCM data display mode, the CCP displays a two-digit number representing the ECC program number. This number indicates the level of heating or cooling "effort" by the ECC system as explained in the BCM section of this chapter. The program number automatically changes according to changes ordered by the BCM following changes of inside temperature and/or the selected temperature changes.

The automatic calculation of the program number can, however, be overridden by pushing the warmer or cooler buttons (lower right of Figure 7-36). While in the BCM data display mode, holding the warmer button pushed in manually forces the program number up to a maximum of 100; 100 commands maximum heating. Pressing and holding the cooler button drives the program number down to a minimum of 0, which commands maximum cooling.

This override of the automatically calculated program number allows you to set it anywhere between zero and 100 to observe the reaction of the ECC system and the BCM data parameters. Once the program number has been overridden, it continues overridden until the BCM data display mode is cancelled.

Cooling Fan Override. This feature allows you to override the automatic control of the cooling fans' speed and manually command either high fans or fans off. To activate this feature, the FDC must be displaying .7.0. Depress and release the econ button on the CCP. The FDC then displays F.8.5. To command high fans, press and hold the Hi button on the CCP until the fans achieve full speed. Releasing the Hi button returns the fans to BCM automatic control. To command fans off, press and hold the Lo button on the CCP until the fans stop. Releasing the Lo button returns the fans to BCM automatic control. This feature can be exited at any time and the display returned to the selection point, .7.0, by clearing the ECM or BCM codes (press Off and Hi for ECM codes or Off and Lo for BCM codes).

Exit Diagnosis. Exit diagnosis by pushing AUTO on the CCP.

ECM/BCM Status Light Display. While in diagnostics, the ECC mode indicators on the

A/C	Air-conditioning
AIR	Air Injection Reaction
ALDL	Assembly Line Data Link
BARO	Barometric Pressure
CCP	Climate Control Panel
DFI	Digital Fuel Injection
ECC	Electronic Climate Control
ECM	Electronic Control Module (controller)
EECS	Evaporative Emissions Control System
EFE	Early Fuel Evaporation
EGR	Exhaust Gas Recirculation
EST	Electronic Spark Timing
FDP	Fuel Data Panel
HEI	High Energy Ignition
ISC	Idle Speed Control
MAP	Manifold Absolute Pressure
N.C.	Normally Closed
N.O.	Normally Open
O_2	Oxygen (sensor)
PCV	Positive Crankcase Ventilation
PFI	Port Fuel Injection
P/N	Park/Neutral
PROM	Programmable Read-only Memory (engine calibrator)
TBI	Throttle Body Injection (unit)
THERMAC	Thermostatic Air Cleaner
TPS	Throttle Position Sensor
TVS	Thermal Vacuum Switch
VIN	Vehicle Identification Number
VSS	Vehicle Speed Sensor
WOT	Wide-open Throttle

Figure 7–37 Explanation of abbreviations

CCP indicate that the mode in which the selected systems operated, either by the ECM or BCM, are operating (bottom portion of Figure 7-36). The operational mode is indicated by whether or not the designated light is on.

In addition to the key terms in this chapter, refer also to Figure 7-37 for a list of the abbreviations used.

SUMMARY

This chapter has covered the development of the Cadillac DFI system, which parallels in many ways the other General Motors systems. Sensors and actuators follow the patterns in the regular CCC and successor systems, but the Cadillac system includes a variety of driveability enhancement measures specific to that brand. Perhaps most notable is the interconnection with the body control computer, in charge of climate control and other features.

We covered the extended self-diagnostics of both computers, including the capacity to test-activate various outputs. Thus the technician can use the switch test, can access the ECM and BCM data, can override the cooling fans' circuits, and other tests.

▲ DIAGNOSTIC EXERCISE

A car is brought in with a complaint of running hot under virtually all driving conditions and poor power and fuel mileage. What is the most time-economical sequence of diagnostic steps to review how the engine is running and identify the problem? How is this changed if the computer self-diagnostics indicate no stored faults, current or memory?

REVIEW QUESTIONS

1. How is the ECM's learned memory erased?

2. How is its learned memory restored?
3. The ECM monitors generator voltage to be sure that it is between _____ and _____ volts.
4. What is the function of the P/S switch?
5. Name at least one nonengine-related function the ECM controls.
6. What is the acceptable range of fuel pressure for a Cadillac DFI system?
7. What is the function of the black wire in the four-wire distributor connector?
8. List nine sensors or switches that are important for the ISC motor to operate under all driving conditions.
9. How does the ISC motor respond during deceleration?
10. List three criteria that must be met before the ECM will apply the TCC or VCC.
11. In addition to using a negative backpressure EGR valve, how is the EGR flow rate controlled?
12. List three things that turn off the cruise control on a Cadillac DFI system.
13. List six functions of the BCM.
14. What two inputs does the BCM use to control the coolant fans?
15. What does the term *fail-safe* mean?
16. How is the DFI system put into diagnostics?
17. After all of the trouble codes are displayed, .7.0 displays. What does that indicate?
18. What does the switch test determine?
19. What occurs during the ECM data display mode?
20. What is the most practical use of the ECM outcycling mode?
21. While in diagnostic mode, what is the function of the status light display?

ASE-type Questions (Actual ASE test questions are rarely so product-specific.)

22. A Cadillac with DFI is brought into the shop running very rough. Technician A says the cause could be a blown injector fuse. Technician B says this cannot be the problem because if the injector fuse were blown,

no fuel would be injected into the manifold, and the engine could not run at all. Who is correct?

a. A only.
b. B only.
c. both A and B.
d. neither A nor B.

23. Technician A says when a DFI system is running in module mode, ignition timing is controlled by the pickup coil and HEI module. Technician B says when the system is operating in module mode, the engine is at cranking speed or a system failure has occurred. Who is correct?

a. A only.
b. B only.
c. both A and B.
d. neither A nor B.

24. Technician A says if voltage is applied to the divert valve solenoid, AIR system air will be directed to the air cleaner. Technician B says under those conditions the air will be directed to the switching valve. Who is correct?

a. A only.
b. B only.
c. both A and B.
d. neither A nor B.

25. Technician A says the viscous converter clutch works just like the torque converter clutch and is nothing more than a name change. Technician B says the VCC works similarly to the TCC, but that it is designed to allow a small amount of converter slippage even when applied. Who is correct?

a. A only.
b. B only.
c. both A and B.
d. neither A nor B.

26. Technician A says the EFE system on rear-wheel-drive DFI vehicles is not computer controlled and is designed to prevent throttle blade icing. Technician B says the EFE system on front-wheel-drive DFI vehicles is computer controlled and is designed to aid fuel evaporation during engine warm-up. Who is correct?

a. A only.
b. B only.
c. both A and B.
d. neither A nor B.

Chapter 8

Ford's Microprocessor Control Unit

KEY TERMS

Crowd
Idle Tracking Switch
Initialization Mode
Self-Test
Thermactor
Throttle Kicker (TK)

Originally the Ford microprocessor control unit (MCU) system's purpose was just to control the air/fuel mixture and the actuation of the Thermactor air injection. While that purpose did not change, a number of additional functions were gradually added on some engine applications. The MCU system uses a dual-bed catalytic converter first produced in 1980 for California cars with 2.3-liter engines only. In 1981 it was expanded to include:

- passenger cars with a 2.3-liter engine and the Holley 6500 two-barrel carburetor (up to model year 1983).
- California specifications trucks with the 4.9-liter engine (up to model year 1984).
- passenger cars with the 4.2-liter engine (engine was discontinued after the 1982 model year).
- passenger cars with the 5.0-liter V8 engine (up to model year 1984).
- passenger cars and trucks with the 5.8-liter engine (until 1984).

In 1982 and 1983, Ford used the MCU system on California passenger cars with the 3.8-liter engine. In 1983 and 1984, the company used the system on 2.0- and 2.3-liter Ranger trucks. Most recently, Ford has used the MCU system on 1984 through 1990 police cars with the 5.8-liter engine.

Two main characteristics distinguish the MCU system: First, it does not determine ignition timing (though on some engine applications it can retard timing to prevent spark knock). Second, most of its sensors and actuators employ engine vacuum both for signals and for activation. Figure 8-1 provides an overview of the Ford MCU system.

MCU MODULE

The module for the MCU system is under the hood and uses no separate calibration unit or PROM, Figure 8-2. It gets electrical power directly from the ignition switch (as long as the key is in

MCU System									
Inputs					Outputs				
Engine					Engine				
2.3	4.9	3.8	8 Cyl.		2.3	4.9	3.8	8 Cyl.	
X	X	X	X	Exhaust Oxygen Sensor	X	X	X	X	Fuel Control Device
X	X	X	X	Tach Signal Engine Speed	X	X	X	X	Thermactor Bypass Sol.
X	X	X	X	Low Coolant Temp. Sw.	X	X	X	X	Thermactor Divert Sol.
		X	X	Warm Coolant Temp. Sw.		X	X	X	Canister Purge Sol.
		X	X	High Coolant Temp. Sw.		X	X	X	Spark Retard
X	X	X	X	Closed Throttle Sw.			X	X	Throttle Kicker
	X	X	X	Crowd/Mid. Throttle Sw.					
X	X	X	X	WOT Sw.					
		X	X	Knock Sensor					

Figure 8–1 MCU system overview

the RUN position) and uses no power relay. There is no need for a constant power supply because the module has no keep-alive memory (KAM). The module operates in three fuel control modes: initialization mode, open-loop mode, and closed-loop mode.

Initialization Mode

As soon as the ignition is turned to the run position, the module activates and begins its **initialization mode.** The initialization mode provides a rich air/fuel mixture (between 12.8 or 13.2 to 1) for starting the engine. During warm or moderate temperature starts, initialization mode is maintained for only a few seconds after the engine starts to allow the induction system to stabilize. During colder starts (coolant temperature below about 10 degrees Centigrade, 50 degrees Fahrenheit), the initialization mode continues until the engine warms up to approximately this transfer temperature, usually not more than a minute or so.

Open-Loop Mode

If the engine and the EGO sensor are not warm enough to go into closed loop immediately after initialization mode, the module puts the system into open loop (of course, technically the

engine is in open loop anytime it is not in closed loop, and it's not in closed loop during initialization mode). In open loop the air/fuel mixture is richer than the optimal 14.7 to 1, but is leaner than the initialization mode provides. Besides the warm-up period, the system stays in open loop during:

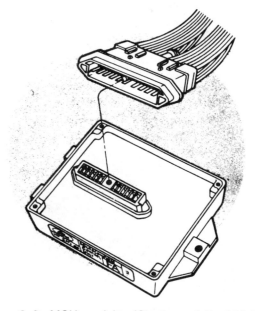

Figure 8–2 MCU module. *(Courtesy of Ford Motor Company.)*

- idle.
- deceleration.
- heavy throttle operation (this can be defined as a throttle opening providing a vacuum less than 10 inches Hg, including high speed cruise, acceleration, hill climbing, and WOT operation).

During open loop, the MCU module sends a fixed command to the mixture control device. The command varies, however from one open-loop operating condition to another (for example, the command for an idle condition is different than that for WOT or deceleration). During engine warm-up, the command also changes as the engine coolant temperature increases, ratcheting the mixture leaner as the engine gets warmer and can run satisfactorily with that mixture. In each of the open-loop operating modes, the module commands an air/fuel mixture just rich enough to provide good driveability.

Closed-Loop Mode

As the engine comes to normal operating temperature conditions (determined by either a timer circuit in the module or a temperature switch, depending on the module's calibration),

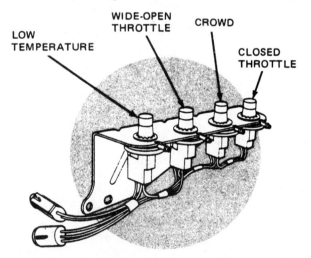

Figure 8–3 Vacuum switch assembly (4.9-liter California engine). *(Courtesy of Ford Motor Company.)*

the module puts the system into closed loop. To go into closed loop, the engine coolant must have reached a temperature of at least 53 degrees Centigrade (125 degrees Fahrenheit), the exhaust gas oxygen sensor must have reached 346 degrees Centigrade (650 degrees Fahrenheit) or more, the engine must be operating at part throttle, light load. If lowered to idle speed while in closed loop, the system automatically drops into open loop. On some systems, however, the control system can stay in closed loop for approximately three minutes before it drops into open loop. This depends principally on the time before the oxygen sensor and catalytic converter take to cool down to a temperature too low to perform their intended functions.

INPUTS

Most of the MCU system's sensors are on/off electric switches triggered by vacuum, temperature, or mechanical linkage positions. The switches then send on or off signals to the MCU module to indicate their state. For example, consider the low-temperature vacuum switch used on the 4.9-liter engine, Figure 8-3.

Vacuum to this switch is controlled through a temperature-controlled, ported vacuum switch. As long as coolant temperature is below 35 degrees Centigrade (95 degrees Fahrenheit), the ported vacuum switch allows vacuum to the low-temperature switch, and its contacts are held open, Figure 8-4. Once coolant temperature exceeds 35 degrees Centigrade, vacuum is blocked to the low-temperature switch and the contacts close, signalling the MCU module.

The module applies 12 volts through an internal resistor to the insulated contact of the low coolant temperature switch. The other contact is permanently grounded. When vacuum is applied to the diaphragm, the circuit to ground is not complete, and there is no voltage drop across the resistor. The voltage signal to the sensor circuit remains at 12 volts. Once the vacuum is cut off,

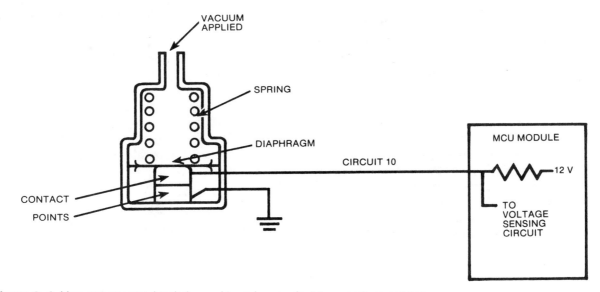

Figure 8–4 Vacuum-operated switch used as a low-coolant temperature sensor

the points close, and the completed circuit to ground drops the voltage across the resistor. This drop allows the voltage signal to the sensor circuit to fall to almost zero volts. This indicates to the module the coolant temperature has reached 35 degrees Centigrade.

Exhaust Gas Oxygen (EGO) Sensor

The EGO sensor on the MCU system works the same way as on other Ford systems. At a certain temperature (approximately 650 degrees Fahrenheit), the oxygen sensor begins to work as a very low current, low voltage battery cell. It generates a signal corresponding to the difference in oxygen between the ambient air and the exhaust gases. Once in closed loop, the system will use this signal to fine tune the air/fuel ratio.

Tach Signal

The tach signal comes through a circuit from the distributor side of the ignition coil to the MCU module. This circuit provides information on engine speed and for spark timing and other system calculations.

Low Coolant Temperature Switch

The low coolant temperature switch for the 4.9-liter engine was described at the beginning of this section. The low coolant temperature switch for another engine group, including the 2.3-liter, is similar except that it is mounted on top of the ported vacuum switch, Figure 8-5.

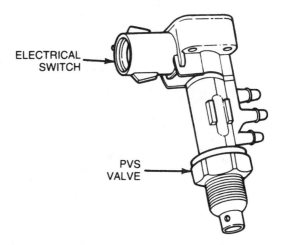

Figure 8–5 Low-coolant temperature switch. *(Courtesy of Ford Motor Company.)*

Mid- and Dual-Temperature Switches

The 3.8-liter V6 and eight-cylinder engine groups use two switches providing information about three different levels of coolant temperature, Figure 8-6. The dual-temperature switch (which is *not* vacuum operated) uses one sliding and one stationary contact. These contacts push together to complete the signal circuit once the coolant reaches 13 degrees Centigrade (55 degrees Fahrenheit). A wax pellet expands as the coolant temperature goes up and contracts as it goes down. This pellet moves the contacts in the nose of the switch. As the coolant temperature continues to increase, the wax pellet continues to expand and the sliding contact continues to move. At 113 degrees Centigrade (235 degrees Fahrenheit), the coolant temperature sensor contacts separate again.

This mid-temperature switch looks and functions like the low-temperature switch in Figure 8-5, except that it closes at 53 degrees Centigrade (128 degrees Fahrenheit). Let's see how the sensors work together with the module:

Below 13 degrees Centigrade:
• The module knows the coolant is below 13

degrees Centigrade as long as both switches are open.

From 13 to 53 degrees Centigrade:
• Once the coolant temperature goes above 13 degrees Centigrade—but before it reaches 53 degrees Centigrade—the dual-temperature switch closes while the mid-temperature switch remains open; so the module "knows" the coolant temperature is within the 13-to-53-degrees-Centigrade range.

From 53 to 113 degrees Centigrade:
• As the coolant warms above 53 degrees, the mid-temperature switch closes, and the dual-temperature switch stays closed; so the module "knows" the coolant temperature is between 53 degrees Centigrade and 113 degrees Centigrade.

Above 113 degrees Centigrade:
• If the coolant reaches a temperature above 113 degrees, the dual-temperature switch is open but the mid-temperature switch is closed; and the module "knows" the coolant temperature is above 113 degrees Centigrade.

Coolant temperature information signals affect all of the modules' output command calculation determinations—those affecting driveability, fuel economy, and emissions.

Diagnostic & Service Tip

Students might notice that using this combination of two on/off switches to convey four states of temperature information is essentially a digital or binary method of collecting information. This is just what later systems do with simpler temperature sensing arrangements than these two sensors with the vacuum lines and multiple circuits. More complicated and capable computers allow a simpler set of monitors.

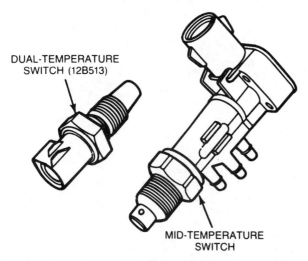

Figure 8–6 Mid- and dual-temperature switches. *(Courtesy of Ford Motor Company.)*

Throttle Position Switches

The 2.3- and 2.0-liter engine group uses a normally closed **idle tracking switch,** mechanically opened by the throttle linkage when the throttle closes, Figure 8-7. When the throttle opens this switch, the voltage in its feed circuit rises and signals the module that the engine is at idle. This engine group also uses a vacuum-operated, normally closed WOT switch. The WOT switch gets manifold vacuum, but the vacuum signal is routed through a thermal vacuum switch in the air cleaner, Figure 8-8. If the engine is warm, vacuum gets to the WOT switch and holds it open. Once the throttle opens far enough for vacuum to drop to the switch's setpoint, it closes, and the module "knows" to adjust the air/fuel mixture for WOT operating conditions—much richer than for ordinary cruise.

The WOT switch also performs the function of a low-temperature indicator. If the engine is started cold and the air temperature in the air cleaner is also low, the thermal vacuum switch in the air cleaner, controlling vacuum to the WOT switch, is closed and blocks vacuum to the WOT switch. Remember that the low coolant temperature switch used on the 2.3-liter engine groups closes at 35 degrees Centigrade or higher. Because of the effectiveness of the heated air inlet system (deriving its hot air from the very rapidly heated exhaust manifold), the air cleaner temperature reaches the thermal vacuum switch's setpoint long before coolant temperature reaches 35 degrees Centigrade. So, if the WOT switch is closed and the low temperature switch is still open, the module will "know" that engine temperature is below a value of about 10 degrees Centigrade (50 degrees Fahrenheit) as measured by coolant temperature. The module can then command an air/fuel ratio appropriate for low temperature starting and for the brief period after a cold start before the heated air inlet system raises the inlet air temperature to the thermal vacuum switch's setpoint.

The throttle position switches for the 4.9-liter engine consist of three switches, Figure 8-3.

These are the closed throttle switch, the **crowd** switch and the WOT switch. *Crowd* refers to a steady-state, heavy-throttle condition, less than WOT, but the kind of load associated with hill climbing, trailer towing or sustained acceleration, such as passing. *Crowd* also refers to a load reflecting an intake manifold vacuum of 10 inches Hg or less. These switches are part of a four-switch assembly; the fourth is the low coolant temperature switch discussed earlier. All these switches are normally closed.

The closed throttle switch connects to ported vacuum. Once the throttle opens, vacuum holds this switch open. As with all ported vacuum, the faster the engine runs and the wider open the throttle (the more air passing through the throttle throat), the higher it goes. If the throttle closes, vacuum shuts off, and the switch closes.

The crowd switch and the WOT switch are each connected directly to manifold vacuum. As the throttle opens, vacuum drops. Once vacuum drops to about 10 inches Hg, the crowd switch closes. Once vacuum drops further to around 2

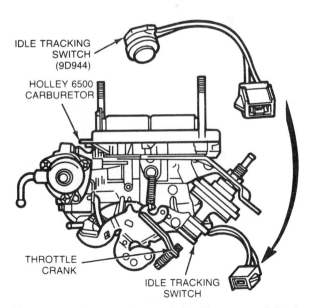

Figure 8–7 Idle tracking switch. *(Courtesy of Ford Motor Company.)*

or 3 inches Hg, the WOT switch closes. In this way, the module keeps track of engine load.

The throttle position switches for the eight-cylinder engine group (also called the zoned vacuum switches) consist of a three-switch assembly, Figure 8-9. They provide essentially the same information in the same way for the eight-cylinder engine module that the other three switches provide for the 4.9-liter engine module.

Knock Sensor

Only engines in the eight-cylinder group are equipped with knock sensors, Figure 8-10. The sensor contains a piezoelectric crystal that produces a voltage signal whenever it is excited by a vibration of the frequency characteristic of spark knock. Whenever that signal is received, the module retards spark advance to prevent detonation.

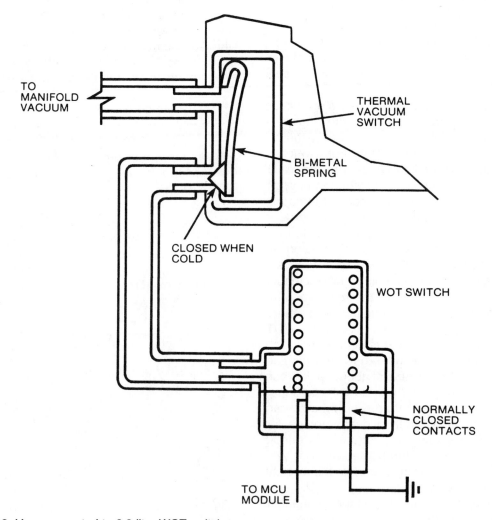

Figure 8–8 Vacuum control to 2.3-liter WOT switch

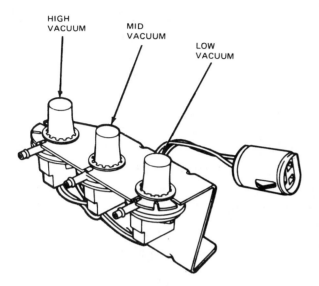

Figure 8–9 Zoned vacuum switch assembly (for eight cylinder applications). *(Courtesy of Ford Motor Company.)*

KNOCK SENSOR (SOME 8-CYLINDER CALIBRATIONS) (12A699)

Figure 8–10 Knock sensor. *(Courtesy of Ford Motor Company.)*

OUTPUTS

Fuel Control Devices

Holley 6500 (Early 2.3-Liter Engine). The Holley 6500 carburetor was modified slightly for feedback operations for use with the MCU system. The main metering jet was reduced in diameter so much that it cannot alone provide a rich enough mixture for normal engine operation. Its delivered mixture is higher than 14.7 to 1 air to fuel. To bring the mixture up to the richness needed, the carburetor includes what resembles a power valve circuit to provide additional fuel to reach a 14.7 to 1 air/fuel mixture for closed loop operation or a richer mixture for open loop, Figure 8-11. This adjustable jet, however, is not a power valve at all; it is a feedback system that is part of the main metering circuit. If the metering rod is pushed down by the foot above it, additional fuel flows through it into the main metering circuit. The farther it is pushed down, the more fuel is admitted. This is just the opposite of an older carburetor's

power valve system, in which a spring lifts a metering rod and richens the mixture under heavy load conditions. This mixture control device does not respond to load at all, but only to the commands of the computer to obtain the proper fuel mixture. Of course, the computer may calculate the need for a richer mixture under specific load conditions, but there is no direct pneumatic/mechanical connection between load and mixture as with the older carburetor's power valve.

The foot that sets the position of the metering rod is controlled by a diaphragm and spring. Vacuum applied above the diaphragm raises the foot. A spring pushes up the metering rod, and fuel flow is restricted. Vacuum above the diaphragm is controlled by a vacuum regulator solenoid, Figure 8-12. This solenoid is mounted away from the carburetor, usually in the right rear corner of the engine compartment, with vacuum hoses going to its vacuum source and to the carburetor. One of the three hoses connected to this mixture control solenoid vents to ambient atmospheric pressure.

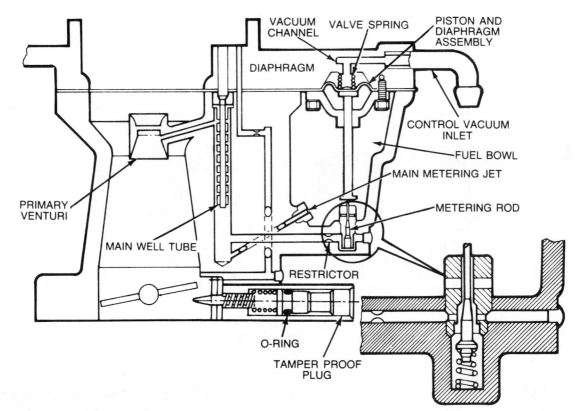

Figure 8–11 Feedback carburetor (Holley 6500). *(Courtesy of Ford Motor Company.)*

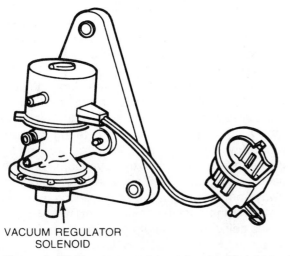

VACUUM REGULATOR
SOLENOID

Figure 8–12 Vacuum regulator solenoid. *(Courtesy of Ford Motor Company.)*

Real World Problem

The vacuum regulator solenoid as well as the piston and diaphragm it controls are subject to all the problems of every other vacuum-actuated component: cracked or porous vacuum hoses, blockage from carbon deposits or, in cold weather, ice, and misrouted hoses. An experienced technician diagnosing a mixture-related problem on one of these systems will make a check of the vacuum system an early part of his troubleshooting routine.

As the vacuum regulator solenoid is cycled by the computer, it alternately connects the vacuum chamber above the diaphragm to vacuum or to atmospheric pressure. The solenoid cycles ten times per second according to commands from the MCU module, but the proportion of time its circuit is grounded or ungrounded with each cycle (its duty-cycle or dwell) can be varied to get different mixtures. In this way the module can control the amount of vacuum above the diaphragm, the position of the metering rod, and thus the air/fuel mixture fed to the engine. A vacuum of 5 inches Hg produces a full lean mixture and a vacuum of 0 Hg produces full rich. Should the engine run at an intake manifold vacuum of close to zero, of course, there will be insufficient vacuum to lean the mixture, but it is at just these conditions that the engine requires the richest mixture. The system does not include a vacuum reservoir to retain vacuum during times of high engine load.

Carter YFA IV (4.9-Liter and Late 2.3-Liter Engines). The fuel control solenoid on this carburetor, Figure 8-13, cycles the same way as the unit just described, except there are no vacuum lines connecting it to the carburetor. Instead, it is directly bolted to the carburetor. As it cycles, however, it controls not the amount of fuel metered into the main fuel circuits of the carburetor, but the amount of air allowed into the same circuits. Obviously, to control the air/fuel mixture one can vary either the amount of fuel or the amount of air; either will change the ratio. This carburetor system varies the amount of air passing through the air bleed or emulsion circuit in the main metering system. This control circuit and carburetor were carried over to some applications of the EEC IV system.

Motorcraft 7200 VV (Variable Venturi) (3.8-Liter and Eight-Cylinder Engine Group). The mixture control device for this carburetor system is called a feedback carburetor actuator. It uses a stepper motor to control the air bleed system. This carburetor and system are also used on EEC III and are discussed in that chapter.

Thermactor Control

Thermactor air control is governed by the TAB and TAD solenoids, and a Thermactor air control valve. It is essentially like the system used on EEC II and EEC III systems discussed in Chapter 9. The module's programming for control of Thermactor air, however, is somewhat different. Thermactor air is delivered as indicated during the following operating conditions:

- prolonged initialization mode—to atmosphere (bypass mode).
- open-loop warm-up—to exhaust manifold.
- WOT or crowd (open-loop)—to exhaust manifold.
- extended idle or deceleration (open-loop)—bypass.
- closed loop—to catalytic converter.

Canister Purge

A normally closed solenoid-controlled valve governs the canister purge function. This valve is located in the purge hose running between the charcoal canister and the manifold vacuum source, just as in the EEC II and carbureted

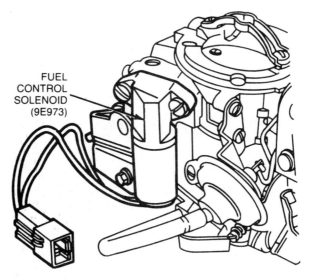

FUEL CONTROL SOLENOID (9E973)

Figure 8–13 Fuel control solenoid. *(Courtesy of Ford Motor Company.)*

EEC III systems. The module activates the solenoid and allows canister purge under these conditions:

- after a predetermined time has elapsed from engine start-up.
- when the engine is within a predetermined rpm window.
- when the engine is within normal operating temperature limits.

Spark Retard

Two different spark retard systems are used on Ford MCU systems.

4.9-Liter Engine. This system does not use a knock sensor. The module is programmed to recognize conditions likely to produce spark knock. When these conditions are met, the module eliminates vacuum advance by activating a solenoid-controlled dump valve that vents the distributor vacuum advance vacuum hose, thereby removing all vacuum advance. Notice this measure occurs under the preprogrammed conditions, whether knock actually occurs or not.

3.8-Liter and Eight-Cylinder Engine Group. Most engines in this group have a knock sensor and the universal ignition module. This module is one of three different modules used with various applications of the Duraspark II igni-

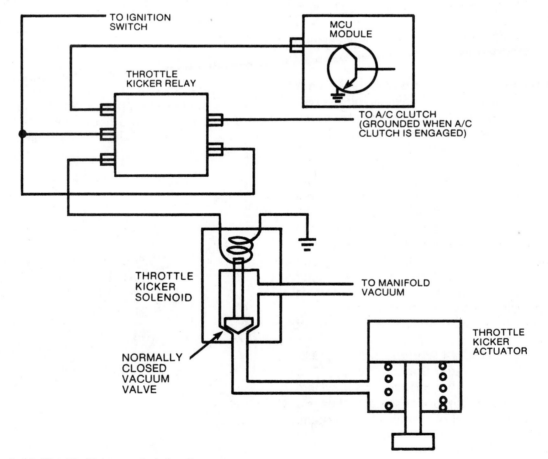

Figure 8–14 Throttle kicker control circuit.

tion system, and it has the capacity to retard ignition timing in response to a command from the MCU module.

Throttle Kicker (TK)

The **throttle kicker (TK)** is similar to those used on other EEC systems, and for that matter for other makes besides Ford. The TK solenoid controls vacuum to the actuator, Figure 8-14. When 12 volts is applied to the solenoid, vacuum is routed to the TK actuator. Vacuum extends the TK actuator's plunger slightly to increase throttle opening when the engine is idling. Grounding the TK relay coil enables the relay to supply voltage to the TK solenoid. The relay is grounded either when the A/C clutch is engaged or by the MCU module. The TK actuator is activated during idle conditions when:

- the engine is cold and warming up.
- the engine is overheated.
- the A/C compressor clutch is engaged.

✔ SYSTEM DIAGNOSIS & SERVICE

The MCU module includes a self-diagnostic capacity limited to faults observed by the module when in the diagnostic mode. It has no fault memory and thus does not monitor and record service codes related to sensor or actuator circuits during normal driving. It does not store any information about malfunctions for later diagnostic recall.

Functional Test

The functional test (sometimes called the **Self-Test**) is divided into three sections: visual inspection, key on/engine off, and engine running.

Visual Inspection. The visual inspection is included as a part of the functional test to encourage the technician to check for obvious conditions such as loose wires, disconnected, broken or missing vacuum hoses, faulty vacuum hose routing, low coolant temperature, and faulty spark plugs or wires. Any such problems can cause the driveability problem for which the vehicle is being serviced. Some can even cause the module to misread the results of the problem as a faulty sensor or actuator circuit. The mistaken service code generated by the module can often result in needless component testing by the technician and needless replacement of MCU system parts.

Key On/Engine Off. In this portion of the functional test, the module tests some of its sensor circuits electrically for proper operating condition, that is to say, for voltages and resistances in the expected ranges. This test is triggered by grounding the trigger terminal in the self-test connector, Figure 8-15 and then turning the ignition on. Service codes can be read either by watching the needle pulses on an analog voltmeter or by readout display on a Star Tester. The other end of either tester is connected to battery ground.

Engine Running. In this portion of the functional test, the module tests its input and output circuits for proper operation under dynamic conditions. The procedure for entering this portion of the tests varies somewhat depending on the engine application and model year of the vehicle. Check with the service manual for the specific vehicle to be sure to follow the specific test procedure steps required.

Service Codes. Service codes are displayed by either sweeps of the analog voltmeter needle or by a digital readout on the Star Tester,

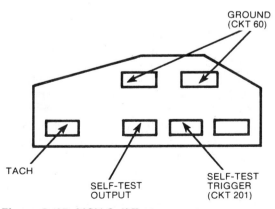

Figure 8–15 MCU Self-Test connector.

just as for the EEC III and EEC IV systems. For the MCU system, however, no special diagnostic tools are required beyond either the Star Tester or the analog voltmeter.

Diagnostic & Service Tip

If you are using an analog voltmeter to read out trouble codes on this system, secure connections to the self-test connector and to ground are required. Otherwise there is no way to distinguish a brief disconnection of the meter from a brief shutoff of the output signal. Alligator clips and spade connectors should be used. Many technicians have been misled when a temporary interruption of voltmeter connection produced what appeared to be a different trouble code. This sort of temporary interruption can be caused by merely moving the probe.

Service Procedures

Service procedures for circuits identified by a specific service code are described in the **Subroutines** section of the vehicle's appropriate service manual. Driveability problems with no associated service code are covered under the **Diagnostic Routine** section of the same service manual.

Diagnostic & Service Tip

Technicians often find the early, simpler forms of computer control systems much more frustrating than later, more complex forms. Somehow they "ought" to be easier to diagnose and repair than later computer controls, but most people's experience is to the contrary. Partly this is because there is such limited self-diagnostic capacity in the early systems, or in many cases none at all, as with the Ford MCU system. In this system, you often find an intermittent problem that has disappeared at the time of diagnosis.

Partly it is because even when new, these systems did not control engine combustion as effectively to maintain driveability as later systems could. This marginality when new may be the reason early computer-controlled cars are often scrapped earlier than one would expect: the costs and difficulty of repair can quickly exceed the value of a fifteen-year-old car. What's more, the combination of aging mechanical systems like vacuum actuators, with more primitive computers that cannot monitor the vehicle's response to commands, can lead to multiple, unrelated driveability problems on the same car. This problem is not unique to Ford, of course; all carmakers had the same difficulties in the early systems. A technician must simply be patient and follow the diagnostic sequences prescribed to avoid missing the problem.

SUMMARY

This chapter has covered the early Ford microprocessor control unit (MCU) system, its operating modes, the different conditions under which it operates in open or closed loop, and its most common sensors and actuators.

Unlike more complicated later systems, the MCU system uses switches for sensors. Thus temperature information is conveyed by a series of on/off switches that place the reading within a certain range rather than at a specific temperature. Instead of a graduated throttle position sensor using a variable potentiometer, the system sees open, closed, and intermediate throttle position.

Similarly, the system employs vacuum for many of its actuators and sensors, instead of electronic circuits. The computer controls only the air/fuel mixture in response to the oxygen sensor signal and the Thermactor air injection. The computer has limited self-diagnostics, no accumulative learning, and very limited diagnostic capacities.

▲ DIAGNOSTIC EXERCISE

This system makes extensive use of vacuum and vacuum-actuated components. Can you list what kinds of misdiagnosis would result from hose porosity or torn vacuum diaphragms in each of the vacuum circuits involved in the system? What kinds of driveability, emissions and fuel economy problems would result? What kinds of false trouble codes would the module present?

REVIEW QUESTIONS

1. What are the three operating modes of the MCU system?
2. What is the most common sensing device used by the MCU system?
3. What is the reference voltage used by the MCU system?
4. Indicate where Thermactor air should flow during the following operating conditions:
   warm-up
   WOT or crowd
   extended idle
   closed loop
5. What is the source of the tach signal?
6. What method is used to anticipate and prevent spark knock on the 4.9-liter engine?
7. What method is used to retard timing on the 4.9-liter engine?
8. Which MCU engines used a knock sensor?
9. What are the two segments of the MCU system's self-diagnostic mode?
10. What type of memory does the MCU module have?

ASE-type Questions (Actual ASE test questions are rarely so product-specific.)

11. Technician A says that service codes obtained from an MCU system always refer to problems that the computer perceives as currently occurring. Technician B says service codes obtained from an MCU system can refer to a problem the computer recognized and stored while the car was being driven. Who is correct?
    a. A only.
    b. B only.
    c. both A and B.
    d. neither A nor B.
12. The functional test of the MCU system consists of three segments. In what order do these segments occur?
    a. Visual inspection, engine running test, key on/engine off.
    b. Engine running test, key on/engine off, visual inspection.
    c. Visual inspection, key on/engine off, engine running test.
    d. Key on/engine off, visual inspection, engine running test.
13. Technician A says that the key on/engine off segment of the MCU system's functional test tests the sensor circuits electrically. Technician B says the engine running test segment of the MCU system's functional test tests the actuator circuits electrically. Who is correct?
    a. A only.
    b. B only.
    c. both A and B.
    d. neither A nor B.

Chapter 9

Ford's EEC I, EEC II, and EEC III

OBJECTIVES

In this chapter you can learn:
- ❏ The engine applications of the EEC I, EEC II, and EEC III systems.
- ❏ The functions the EEC I system controls.
- ❏ The functions the EEC II system controls.
- ❏ The functions the EEC III system controls.
- ❏ The self-diagnostics of the EEC I and EEC II systems.

KEY TERMS

Aneroid
Barometric Pressure Sensor
Default Mode
Duraspark II

Ford introduced the EEC I system in 1978 using it on the 1978 and 1979 5.0-liter Lincoln Versailles. EEC II was introduced in 1979, limited to full-size Fords sold in California and Mercurys sold in all fifty states, all with 5.8-liter engines. Ford expanded the system as EEC III in 1980 with somewhat more extensive control applications, but its employment was still limited to the V8 engines. In 1981, central point fuel injection replaced feedback carburetors as part of the EEC III system on some engines. In 1982 EEC III was introduced for some light trucks.

Many of the component parts described in this chapter are also found on the later EEC IV system. This chapter covers only those controls, sensors, and actuators found on the early systems. Because the earlier systems are simpler, it is easier to learn the basics of the Ford systems from them. Elements that are carried over into the later systems will only be repeated when there is a material difference in their construction or function.

ELECTRONIC ENGINE CONTROL (EEC) I

The most important factor to remember about the earliest Ford system, the EEC I, is that it does not control the air/fuel mixture, but only spark advance. The engines controlled use the variable venturi (VV) carburetor, which has no computer-actuated mixture control on these systems. There is, therefore, no oxygen sensor and no closed loop operational mode. This system was the last, best effort of the mechanical carburetor system to maintain driveability while optimizing emissions.

Electronic Control Assembly (ECA)

The ECA is like later combustion control computers except that it:

- performs fewer functions.
- has no self-diagnostic capacity.

- uses an externally mounted calibration unit to fit the unit to specific vehicles, accounting for curb weight, axle ratio, high altitude use and so forth, Figure 9-1. This is similar to a General Motors PROM, except that it is much simpler.
- includes a fuel octane adjustment switch to retard ignition timing either 3 or 6 degrees if spark knock occurs (later systems use an "octane rod" on the distributor to do the same thing).
- provides a 9-volt reference voltage to its sensors.

Default Mode.   If the ECA fails, it stops sending commands to the three actuators it controls. Ignition stays fixed at base timing; the Thermactor dumps air into the atmosphere as long as the engine is running, and the EGR valve stays closed. Obviously, neither driveability nor emissions quality will be at the desired levels, but the vehicle can be driven at reduced power and greater fuel consumption until it is diagnosed and repaired.

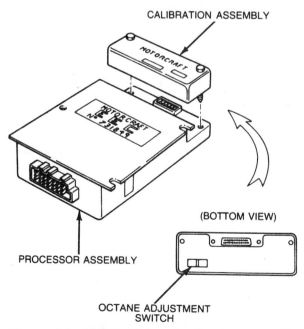

CALIBRATION ASSEMBLY

(BOTTOM VIEW)

PROCESSOR ASSEMBLY

OCTANE ADJUSTMENT
SWITCH

Figure 9–1   EEC I ECA and calibration assembly. *(Courtesy of Ford Motor Company.)*

The ECA includes a power relay to protect it from reversed polarity, a protection retained on later Ford systems. Both the relay and the ECA itself are on the passenger side of the instrument pedal, near the brake pedal.

Inputs

The ECA gets input from these sensors:

- engine coolant temperature (ECT) sensor.
- manifold absolute pressure (MAP) sensor.
- barometric pressure (BP) sensor.
- throttle angle position (TAP) sensor.
- crankshaft position (CP) sensor.
- intake air temperature (IAT) sensor.
- EGR valve position (EVP) sensor.

The ECT, TAP, and EVP sensors are very similar to the corresponding units on the later Ford systems. The IAT sensor is in the air cleaner and works similarly to the corresponding sensor EEC IV uses as either an ACT or VAT sensor.

MAP Sensor.   The MAP sensor uses an **aneroid,** a sealed accordion-like capsule with a measured amount of gas that expands and contracts with changes in the surrounding pressure. As the pressure outside the aneroid increases, it compresses; as the pressure outside decreases, the aneroid expands. The aneroid is situated in a housing directly connected to intake manifold vacuum, Figure 9-2. With changes of intake manifold vacuum, the expansion and contraction of the aneroid moves the wiper of a potentiometer to signal the ECA with a return voltage proportional to manifold pressure.

BP Sensor.   The **barometric pressure sensor** works just like the MAP sensor, except it is vented to the atmosphere rather than plumbed to the intake manifold. Its signal depends exclusively on the barometric pressure and the altitude of the vehicle.

CP Sensor.   The crankshaft position sensor consists of a four-lobe pulse ring on the output end of the crankshaft, and a probe with a permanent magnet and a wire coil at its tip. This sensor is

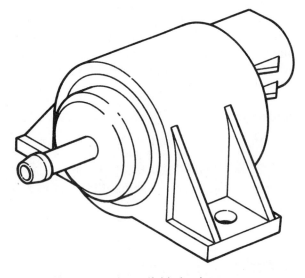

Figure 9–2 Aneroid manifold absolute pressure sensor. *(Courtesy of Ford Motor Company.)*

essentially a pickup coil removed from the distributor and relocated at the rear of the crankshaft, Figure 9-3. Of course, a pickup clocking crankshaft position monitors a shaft turning twice as fast as a conventional distributor, but the computer

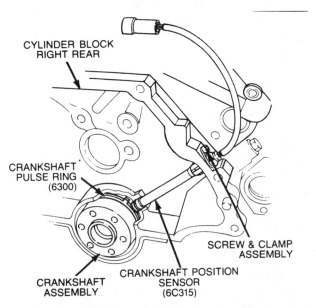

Figure 9–3 Crankshaft position sensor. *(Courtesy of Ford Motor Company.)*

does not need to "know" what part of the engine cycle a particular cylinder is in, compression or exhaust. Since this system uses a (relatively unconventional) distributor to determine which cylinder gets the next spark, the computer need only calculate the appropriate spark advance.

The pickup probe extends into the back of the engine block leaving a small air gap between its tip and the pulse ring on the crankshaft. As the crankshaft turns and each lobe of the ring passes under the magnet in the probe's tip, the air gap is suddenly much smaller. This causes a sudden increase in the strength of the magnetic field at the magnet's pole. This sudden change in the magnetic flux field generates a voltage signal in the coil around the magnet. As the crankshaft continues to turn, this steady stream of voltage polarity-reversal signals is transmitted to the ECA, each pulse indicating a piston is at 10 degrees before top dead center, the ignition reference pulse.

Outputs

EEC I controls three output circuits:

- ignition timing.
- Thermactor air control.
- EGR flow.

Ignition Timing. The ECA monitors coolant temperature, intake air temperature, engine speed, throttle position, and engine load (as reflected from the intake manifold absolute pressure) to calculate the proper ignition timing. Low intake air temperature sensor signals, for example, allow increased spark advance; and higher temperatures call for less spark advance. If intake air temperatures rise above 90 degrees Fahrenheit, the spark advance is also retarded to avoid ignition knock.

If the engine coolant temperature sensor indicates overheating during idle, however, the spark advance is increased to raise the idle speed, turn the water pump faster, and obtain more coolant flow (EEC I does not directly control

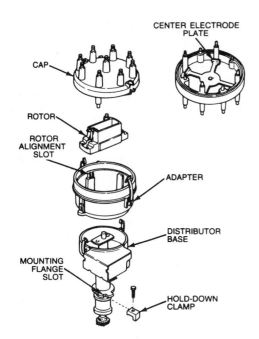

Figure 9–4 EEC I distributor assembly. *(Courtesy of Ford Motor Company.)*

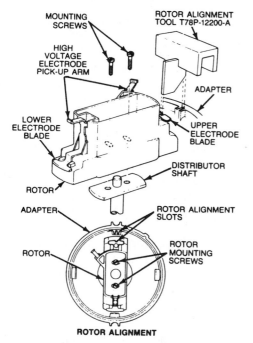

Figure 9–5 EEC I distributor rotor. *(Courtesy of Ford Motor Company.)*

idle speed through airflow, so advancing the timing is the only strategy to increase engine, and thus, water pump speed and coolant circulation). If the throttle opening increases rapidly, spark advance is retarded. Spark advance is also retarded if the manifold pressure increases (intake manifold vacuum declines) even if there is no change in throttle position. This might occur if a vehicle started up a hill and the engine slowed. The ECA's spark timing command is sent to the standard **Duraspark II** ignition module, which interrupts primary coil current to fire the appropriate spark plug. The ECA governs a range of as much as thirty degrees of spark advance compared to the approximately twenty degrees afforded by pre-EEC systems.

Revised Distributor Design. The system also uses a new distributor directing the spark from the coil to the appropriate cylinder, Figures 9-4, 9-5, and 9-6. The new distributor has a notch in its mounting flange that indexes on a lug on the engine block, so once the distributor's base timing is set it cannot be changed, Figure 9-4. The original EEC I system included an enclosure for mounting a future miniaturized ignition module, a feature not carried over to EEC III and not actu-

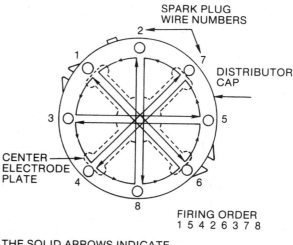

THE SOLID ARROWS INDICATE A WIRE TERMINAL BEING FIRED. THE HOLLOW ARROWS INDICATE THE DIRECTION OF ROTOR ROTATION.

Figure 9–6 EEC I distributor firing order

ally used until the EEC IV systems (when it was redesigned anyway).

Because of the 30-degree wide range of timing advance, a new distributor cap was designed, Figure 9-6. The firing order of the engine does not correspond, as an engine's firing order usually does, to the sequence of distributor electrode terminals counting counterclockwise from cylinder number one. The actual sequence fires to alternating spark plug cables 135 degrees apart. To achieve this, Ford engineers used a two-piece distributor cap, with upper and lower level plug electrodes in the top half with alternate upper and alternate lower electrodes spaced 90 degrees apart.

The center electrode, receiving the high voltage secondary current from the coil, does not end in a carbon button, as is traditional in distributors, but in a large "X" or cross, making the coil's secondary voltage available above each of the lower cylinder electrodes, Figure 9-6. The high voltage current comes from the tips of these electrodes to the distributor rotor top pickup arms.

The rotor, too, is a unique design to facilitate the widely-separated cap electrodes, Figure 9-5. There is a transfer pickup at each end of the rotor. When one of the rotor pickups arms aligns with an end of the distributor cap's coil electrode, the current jumps that gap and flows to the appropriate spark plug. The rotor's pickup arms are not symmetrical; one of them conducts the high voltage vertically from the coil electrode to the lower spark plug electrodes, the other conducts the high voltage down and 45 degrees to the left to the higher spark plug electrodes.

This design means the sequence of high voltage electric discharges through the distributor cap occurs at intervals 135 degrees apart rather than in circumferential sequence.

Let's follow one sequence to make it clearer: suppose cylinder number one has just fired. The current has passed from the coil, through the X-shaped center electrode, down the vertical rotor arm and to the lower plug electrode that goes to cylinder one. The firing order of this engine is 1-5-4-2-6-3-7-8, so the next plug to fire is number 5. The distributor rotates counterclockwise 45 degrees between cylinder firings, of course. Now the rotor's lower electrode arm points to the number three terminal. The pickup arm for the lower electrodes, however, is between the spokes of the center plate, so no secondary current can flow.

However, the pickup arm for the upper electrode blade at the opposite end of the distributor is offset 45 degrees from the pickup point. That pickup arm aligns with one arm of the center electrode plate and conducts the coil output voltage to terminal number five (one of the higher plug electrodes) through a second air gap.

In each case, the second air gap each spark must jump increases the coil secondary output by 2 or 3 kV beyond what the spark plug gap and normal single distributor gap would require. Spark plug firing voltage continues delivered alternately from opposite ends of the rotor every 45 degrees of rotor rotation and from distributor cap terminals separated by 135 degrees between plug firings, Figure 9-6.

Thermactor Air Control. The EEC I system employs only a single-bed catalytic converter, so the Thermactor system does not supply air to the converter during warm engine operation. Air is supplied to the exhaust manifolds only during low-temperature conditions and is dumped into the atmosphere once normal operating temperature is reached. The ECA controls these operational modes with a solenoid-operated vacuum control valve, the Thermactor control solenoid. This normally closed valve controls manifold vacuum to the Thermactor bypass valve, Figure 9-7. The Thermactor pump delivers air to the bypass valve directly.

When the sensor in the air cleaner indicates to the computer that underhood temperature is low, the ECA grounds the Thermactor control solenoid circuit. The solenoid opens the vacuum valve, and vacuum is routed to the bypass valve. The bypass valve then routes Thermactor air into the exhaust manifolds to help burn off residual

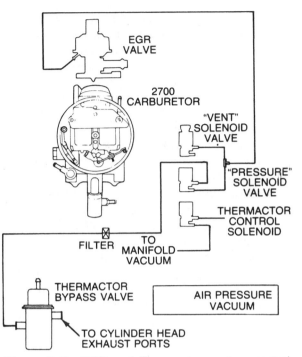

Figure 9–7 EGR and Thermactor system control. *(Courtesy of Ford Motor Company.)*

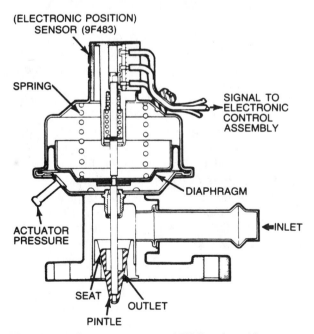

Figure 9–8 Pressure-operated EGR valve. *(Courtesy of Ford Motor Company.)*

hydrocarbons. Once the engine warms to operating temperature, the ECA shuts off the solenoid, and blocks vacuum from the bypass valve. Thermactor air is then dumped to the atmosphere. The ECA also dumps Thermactor air during WOT conditions.

EGR Flow. The EEC I EGR (exhaust gas recirculation) valve opens by air pressure instead of by vacuum, Figure 9-8. An external tube supplies exhaust gas to the EGR valve connected to an exhaust manifold. To improve driveability and to extend EGR valve life, the temperature of the EGR gas is reduced by running the supply tube through a heat exchanger, the EGR cooler, Figure 9-9. Coolant flows from a heater hose through the EGR cooler.

The ECA controls EGR flow through two solenoid valves, the normally closed pressure valve and the normally open vent valve, Figure 9-10. The ECA monitors information from its sen-

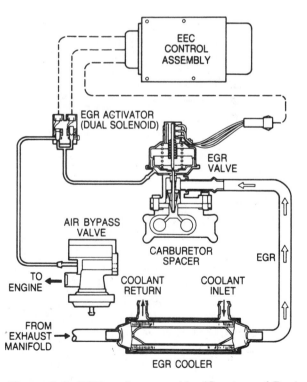

Figure 9–9 EGR cooler assembly. *(Courtesy of Ford Motor Company.)*

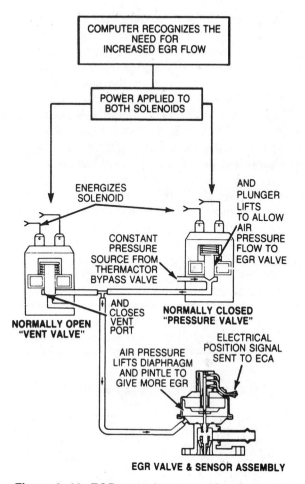

Figure 9–10 EGR control system. *(Courtesy of Ford Motor Company.)*

sors and calculates how much EGR flow is needed. For example, suppose the inputs indicate increased EGR flow is needed: the ECA then energizes both solenoids. The normally closed pressure valve opens to apply pressure from the Thermactor system to the lower side of the EGR valve. The normally open vent valve closes to prevent the pressure escaping to the atmosphere. Now the EGR valve opens wider until the EVP sensor signals the computer that the EGR valve is at the desired position. Then the ECA de-energizes the pressure solenoid. It closes and traps the existing pressure under the EGR

valve diaphragm, keeping the EGR valve in a fixed position. This condition persists until operating conditions change as signaled by the various sensors (throttle position and intake manifold pressure changes) or until some of the trapped pressure leaks out and the EVP sensor signals the ECA the change in the EGR valve's position. If the ECA determines to decrease EGR flow, it de-energizes both solenoids, closing the pressure source and venting the existing pressure. If barometric pressure decreases, either because of weather changes or changes of the altitude of the vehicle (by driving up a hill), EGR flow reduces. The computer cuts off all EGR flow if the coolant is below 21 degrees Centigrade (70 degrees Fahrenheit) or above 110 degrees Centigrade (230 degrees Fahrenheit). Normally—except in the temperature extremes just described—the two solenoids are each cycled rapidly to maintain the calculated EGR valve delivery.

✔ SYSTEM DIAGNOSIS & SERVICE

Early automotive engine control computers like the ECA in the Ford EEC I system had no self-diagnostic capacity. Special test equipment and procedures were developed and are listed in the appropriate Ford service manual for identifying and correcting driveability or other engine-related problems an owner may report. The special tests employ special tester Rotunda No. T78L-50-EEC-I (or its equivalent) in series with the ECA harness. There is a disconnect in the harness for just this purpose. Once the tester is installed, all signals to or from the ECA pass through the tester, not unlike a breakout box. A special DVOM (Rotunda No. T78-50) connects to the first special tool and provides the readings for each test in the sequence.

The first of the two special test sequences is performed after the engine warms up and with the ignition off. Eight test readings check the condition of each of the sensors, the power relay, the

Thermactor control solenoid, and part of the ECA's internal circuitry. In the second test sequence, the technician checks spark advance, EGR valve position and Thermactor air mode with the engine running at 1,600 rpm in park, with vacuum from an external source applied to the MAP sensor. This second test sequence checks the remaining ECA circuitry.

The results of each reading and check are recorded and compared to test specifications provided with the tester. If all the results are within the specification limits, the system is functioning properly; if not, diagnostic procedures in the service manual guide diagnosis and repair of the fault in the identified circuit.

With these detailed tests, the problems of diagnosing EEC I failures are not too difficult, even in the absence of any self-diagnostic aids. While individual tests of sensors and actuators are required, these are fewer in number and simpler in operation—hence easier and quicker to test—than the components employed on later Ford systems.

Real World Problem: Rotor Alignment

The EEC I-governed engine cannot run properly unless the distributor rotor is properly aligned. This alignment is critical not because it might affect spark timing (it cannot), but because improper alignment can shorten the spark duration and combustion burning and thus affect driveability, fuel economy, and emissions quality. This shortening can result from the distributor rotor moving away from the central plate too far for a spark to propagate.

A special rotor alignment tool, Figure 9-5, can be used to position the rotor precisely, though many experienced mechanics can manage this alignment either with homemade tools or by eye. This critical rotor alignment is true for EEC I, EEC II, and EEC III systems, even though EEC III uses a slightly different rotor.

Rotor Alignment

Technicians who regularly service Ford vehicles have discovered correct rotor alignment is very critical to good engine performance. This dimension is one of the first things experienced technicians check responding to complaints of poor engine performance on EEC I, EEC II, or EEC III system vehicles.

EEC II

The EEC II system is similar to EEC I with some minor differences in sensors and with significantly more control responsibility. The ECA and power relays are similar except the ECA includes more internal circuitry to perform the additional functions. It also provides a 9-volt reference signal to each of its sensors. New with the EEC II are an exhaust oxygen sensor, and an electronically controlled air/fuel mixture device. Hence with EEC II we find the first Ford system that can operate in closed-loop mode. Closed-loop means, on Ford systems as on others, that the computer controls the fuel/air mixture in response to feedback signals from the oxygen sensor, which in turn responds to the delivered mixture—hence *closing* the loop.

The "default" mode is now called limited operational strategy (LOS), as we will see in EEC IV. Other systems call this state the limp-home mode or the fail-safe mode or something similar. This mode is triggered by an electrical failure in some critical component or circuit, and in it all actuators are left in their deactivated mode. Driveability, fuel economy, and exhaust emissions quality are all compromised in that state.

Inputs

EEC II sensors include:

- engine coolant temperature (ECT) sensor.

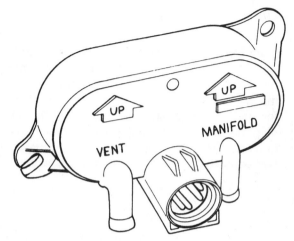

Figure 9–11 BARO and MAP sensor. *(Courtesy of Ford Motor Company.)*

- manifold absolute pressure (MAP) sensor.
- barometric pressure (BP) sensor.
- throttle position (TP) sensor.
- crankshaft position (CP) sensor.
- exhaust gas oxygen (EGO) sensor.
- EGR valve position (EVP) sensor.

ECT. All temperature-related information comes from the ECT, which is identical to the previous system's sensor. EEC II uses no IAT (intake air temperature sensor) as the computer is able to optimize air/fuel mixture adequately without that information.

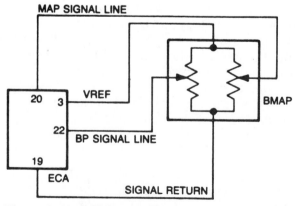

Figure 9–12 BARO and MAP sensor electric circuit. *(Courtesy of Ford Motor Company.)*

MAP and BP Sensors. The MAP and BP sensors work like those on the earlier EEC I system, except they are both in one housing, called a B/MAP sensor, Figures 9-11 and 9-12.

TP Sensor. The TP sensor works the same as on the earlier system, but the word *angle* is dropped from its name. Like its predecessor, the EEC II TP is a variable potentiometer returning a signal proportional to the throttle opening.

CP Sensor. The CP sensor works the same way as its corresponding sensor on EEC I, except it is now at the more accessible front of the engine and is electrically shielded to protect its signal from electrical interference, Figure 9-13.

EGO Sensor. EEC II is the first of the Ford systems to employ an oxygen sensor, which is carried over to the later systems, also. As on vehicles built by other manufacturers, the oxygen sensor on Ford EEC II vehicles generates a low voltage signal proportional to the difference in residual oxygen in the exhaust gases, compared to the ambient air.

EVP Sensor. The EEC II EVP sensor is the same as that for the previous system.

Outputs

The ECA controls the following functions:

- air/fuel ratio.
- ignition timing.
- idle speed control.

Figure 9–13 EEC II crankshaft position sensor. *(Courtesy of Ford Motor Company.)*

- EGR flow.
- Thermactor air control.
- canister purge.

Each of these functions is either new to EEC II or somewhat different from the earlier versions on EEC I.

Air/Fuel Ratio. This function is new to EEC II. The fuel bowl of a carburetor is vented either through a balanced vent or an external vent to the atmosphere. The atmospheric pressure provides the force to drive the fuel through the jets and emulsion passages into the lower pressure region of the carburetor venturi in the throat. EEC II vehicles use the Model 7200 VV (variable venturi) carburetor, and use this pressure at the top of the fuel in the bowl to control the air/fuel mixture: raising the pressure slightly increases the flow of fuel slightly; reducing the pressure slightly correspondingly reduces the fuel flow. The specific means to do this consists of a restricted carburetor vent plus a vacuum passage allowing carburetor vacuum (called *control vacuum* in this application) to draw air from the bowl area and

thus reduce the applied pressure, Figure 9-14.

The small amount of pressure drawn from the fuel bowl is controlled through a pintle valve controlled by a stepper motor. As the stepper motor extends the pintle valve, the vacuum passage is restricted. This permits more pressure to develop in the fuel bowl through the other vent, and the air/fuel mixture gets richer. If the pintle valve retracts, pressure in the fuel bowl is reduced and the air/fuel mixture gets leaner.

The ECA can selectively energize four separate coils in the stepper motor to run it forward or backward. The coils do not get a steady voltage applied, but rather a series of short pulses. Each pulse rotates the armature a specific number of degrees (hence it is a *stepper* motor). As the armature turns, the pintle valve either extends or retracts depending on which way the armature turns. Thus the ECA can transmit pulses to the stepper motor to achieve whatever air/fuel ratio is required, within its range. The pintle has an extension range of 0.4-inch, and 120 pulses are required to move it from one extreme to the other.

Real World Problem

A frequent problem on Ford variable venturi carburetors, including the feedback versions used with EEC II, is mechanical warping of the top of the carburetor, the air horn. If someone tightens the air cleaner hold-down bolt excessively (perhaps in the belief that it holds up the engine), this can bend the metal very slightly. On many carburetors this would make little difference, but on the variable venturi, any bending of the air horn can cause the fuel metering rod controls to stick. Unfortunately, in many cases the air horn does not return to its original position once the strain is relieved and replacement is the only option.

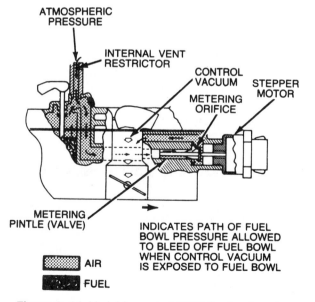

Figure 9–14 Variable venturi 7200 feedback carburetor mixture control. *(Courtesy of Ford Motor Company.)*

Ignition Timing. The EEC II ignition system features the same coil and distributor as EEC I and a new Duraspark III ignition module, designed specifically to work with the EEC II

ECA. The same precautions about rotor alignment apply.

Idle Speed Control. The idle kicker solenoid and idle kicker actuator first appear on EEC II and continue in EEC III and some applications of EEC IV, as we will see in Chapter 10. See this chapter for a full discussion.

EGR Flow. The EGR control solenoids, EGR vent (EGRV), and EGR control (EGRC) function like those in EEC I. However the EGRP (EGR pressure) solenoid used in EEC I was renamed. The word *pressure* was replaced by the word *control.* This name change occurred because the EGRC solenoid on EEC II systems is connected to manifold vacuum and the EGR valve opens by vacuum rather than by Thermactor system pressure. This arrangement continues in EEC III and some EEC IV applications. Experience showed that a vacuum-activated unit was more durable and reliable in service.

Thermactor Air Control. The EEC II system uses a dual-bed catalytic converter and therefore requires an additional Thermactor system mode: the capacity to direct air to the converter in addition to the capacity to direct it to the exhaust manifold or the atmosphere. For this purpose, there are two control solenoids and a combination air management valve: a Thermactor air bypass (TAB) solenoid, a Thermactor air divert (TAD) solenoid, and a combination TAB/TAD air control valve, Figure 9-15. This system continues to EEC III and EEC IV and is discussed in the appropriate chapter.

Canister Purge Solenoid. This function is new for EEC II, in response to new emissions regulations. The charcoal canister stores fuel vapors from the fuel tank to release them during normal operating conditions into the engine for consumption during combustion. The solenoid is a normally closed vacuum valve. The valve is in the purge hose connecting the canister to the intake manifold. When the ECA grounds the solenoid circuit, the solenoid opens and allows manifold vacuum to purge the solenoid's stored fuel vapors. This canister purge system continues with EEC III and some forms of EEC IV. It is discussed in the appropriate chapter.

Real World Problem

If a Ford with EEC II constantly tries to lean the mixture, but always shows a high oxygen sensor output, check whether the charcoal canister purge is either stuck open or whether the canister itself is flooded with liquid fuel. (This can happen when a motorist tries to top off the tank beyond the automatic shutoff.)

✔ SYSTEM DIAGNOSIS & SERVICE

The EEC II system, like its predecessor EEC I, has no self-diagnostic capacity. Diagnosing malfunctions is performed similarly to diagnosing malfunctions on an EEC I system. Special test equipment or its equivalent, listed in the service manual, is required for this diagnosis. Essentially, a technician tries to limit the possible causes to one or a few control circuits and then tests each of them individually.

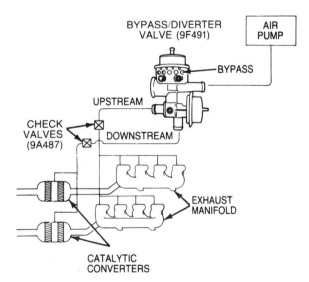

Figure 9-15 EEC II Thermactor control system. *(Courtesy of Ford Motor Company.)*

EEC III

Carbureted EEC III vehicles have the same components as on EEC II. The air/fuel mixture control device on the carburetor and the ECA itself have been modified, however. The ECA became more complicated, and there are some early electronic fuel injection systems with the EEC III system. The fuel injection system is the throttle body type, effectively an electric carburetor, that Ford technical literature most often called central fuel injection. In the Ford EEC III literature, the terms *electronic fuel injection* and *central fuel injection* are used interchangeably to mean the same system. Once the Ford Motor Company went to the EEC IV system, electronic fuel injection usually refers specifically to multipoint fuel injection.

ECA

For EEC III the ECA operates in one of three operating strategies: the base engine strategy, the modulator strategy, and the limited operational strategy.

Base engine strategy covers normal operating conditions, divided into four modes:

- cranking.
- closed throttle operation.
- part throttle operation.
- wide-open throttle operation.

The ECA determines the current operating conditions from the sensor input signals and selects the appropriate actuator commands according to its specific vehicle calibration program.

The modulator strategy compensates for conditions requiring more extreme calibration commands. These conditions are:

- cold engine.
- overheated engine.
- high altitude.

The limited operational strategy is like that of the EEC II system on carbureted vehicles. For CFI-equipped vehicles the LOS keeps the injectors operating, but at a fixed pulse rate to produce a full, rich mixture. LOS is, as we have seen before, similar to a limp-home mode, enabling the vehicle to move under its own power, but with unsatisfactory performance, economy, and emissions quality. Extended driving under these conditions could damage the oxygen sensor and the catalytic converter, as well as produce combustion chamber carbon deposits.

Inputs

Each EEC III sensor receives a 9-volt reference signal from the ECA, which the sensor modifies according to the state of whatever it monitors and returns a corresponding reduced voltage signal to the computer. The only difference between the EEC III sensors and those for EEC II is that the EEC III system with CFI includes an air charge temperature sensor bolted into the number seven intake manifold runner. The ACT sensor carries over to some EEC IV applications and is described in that chapter.

Outputs

With the following exceptions, all the EEC III actuators are exactly like those used for EEC II:

EEC III 7200 VV Air/Fuel Mixture Control. The fuel mixture control works somewhat differently on EEC III than on EEC II. A stepper motor and pintle valve work the same way, though the stepper motor is slightly different. The pintle valve, however, does not control fuel bowl pressure. Instead the pintle valve controls the amount of air bled into the main metering system, a variable main metering air bleed, Figure 9-16. Air enters a passage at the top of the carburetor from inside the air cleaner. This passage leads to an orifice, which the stepper motor and pintle valve control. When the pintle extends, less air is allowed through the orifice; when the pintle retracts, more air passes through. After passing through the orifice, the air channels around the throats of the carburetor to the other side of the

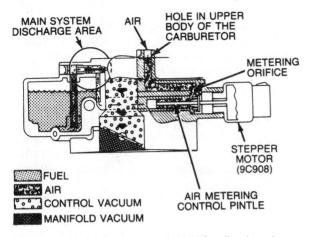

MAIN SYSTEM
DISCHARGE AREA

AIR

HOLE IN UPPER
BODY OF THE
CARBURETOR

METERING
ORIFICE

STEPPER
MOTOR
(9C908)

AIR METERING
CONTROL PINTLE

▨ FUEL
▤ AIR
▦ CONTROL VACUUM
▧ MANIFOLD VACUUM

Figure 9–16 Variable venturi 7200 feedback carburetor, emulsion air control system. *(Courtesy of Ford Motor Company.)*

carburetor, to the main metering system. The more air that bleeds into the main metering system, the less fuel flows into the carburetor throat.

This difference makes use of the fact that to control air/fuel mixture you can vary either the air or the fuel to get the ratio desired. By introducing more air into the emulsion passages, the fuel delivered to the carburetor venturi is itself filled with a greater proportion of small bubbles of air. This effectively leans or richens the mixture as the computer controls the amount of air in those passages.

CFI Injectors. The fuel delivery system and the operation of the CFI injector solenoids for EEC III carry over to some of the EEC IV applications and are described in that chapter.

Second-Generation Distributor. Later EEC III systems have a second-generation distributor, Figure 9-17. It has the upper- and lower-level electrode blades seen in the first EEC distributor rotor, but the rotor shape is changed to conical rather than the original rectangular. The rotor's pickup arms and the cap's center terminal are redesigned, but they still work the same way to pick up and deliver spark alternately every 135 degrees of distributor rotation.

CFI Canister Purge. Four different canister purge systems are used on EEC III CFI systems. Consult the service manual for the specific vehi-

cle for information on how each specific purge system works. The general operation and purpose of each canister purge system is the same.

✔ SYSTEM DIAGNOSIS & SERVICE

The ECA for EEC III *does* have a kind of self-diagnostic capacity, the first in a Ford system. It does not include a malfunction memory as later systems do, but its memory will report faults in the form of two-digit codes for just those problems present when the self-test is run. A quick-test procedure, similar to that for EEC IV, including the self-test is outlined in the service manual. The same manual lists all the necessary test equipment.

SUMMARY

In this chapter, we have covered the early Ford computerized engine control systems. With EEC I, we saw the use of the computer to govern relatively simple elements, like spark advance, EGR flow, and Thermactor air mode. With EEC II we saw the introduction of Ford's first oxygen sensor and the beginnings of closed loop operation. EEC III saw the beginnings of self-diagnostics, at least in the sense of the capacity to report problems currently observed. For each of these systems, we saw Ford's unique new-design distributor, and we looked at the different methods used to govern fuel mixture with both the variable venturi carburetor and the early fuel injection systems.

▲ DIAGNOSTIC EXERCISE

A car with the Ford EEC III system is brought in, running, but poorly. How would you set about determining whether it was in LOS? What elements not controlled by the computer would you inspect? If you discovered that charging system voltage was somewhat low, what would that tell you about the problem?

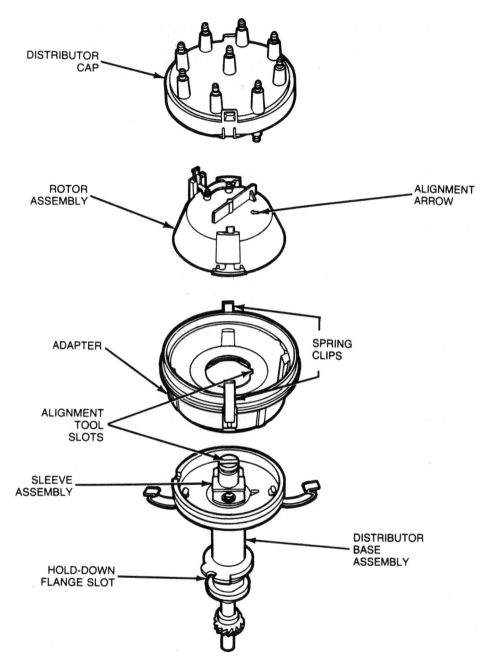

DISTRIBUTOR ASSEMBLY — SECOND GENERATION
(12127)

Figure 9–17 EEC III second-generation distributor design. *(Courtesy of Ford Motor Company.)*

REVIEW QUESTIONS

1. What engines and vehicles use the EEC I system?
2. List the functions that the EEC I system controls.
3. Where is the CP (crankshaft position) sensor on the EEC I system?
4. What kind of device generates the signal for the MAP and BP (manifold absolute pressure and barometric pressure) sensors? What operates the signal-producing device?
5. List the functions that the EEC II system controls.
6. List the functions that the EEC III system controls.
7. What is the sensor reference voltage for Ford EEC I, II and III systems?
8. What is unique about the EEC I EGR valve operation?
9. What fuel-metering device does the EEC I system use?
10. How does the stepper motor and pintle valve in the EEC II system control the air/fuel mixture?
11. How does the stepper motor and pintle valve in the EEC III system control the air/fuel mixture?
12. What kind of service codes do the EEC I, II, and III systems display?

ASE-type Questions (Actual ASE test questions will rarely be so product specific.)

13. Technician A says EEC III systems use the 7200 variable venturi carburetor with a stepper motor for air/fuel mixture control. Technician B says the EEC III systems use CFI for air/fuel mixture control. Who is correct?
    a. A only.
    b. B only.
    c. both A and B.
    d. neither A nor B.
14. An EEC III-equipped car is brought into the shop with complaints of lost power and rough running. Technician A says distributor rotor misalignment could be the problem. Technician B says rotor misalignment cannot be the problem because she has checked the ignition timing, and it is correct. Who is correct?
    a. A only.
    b. B only.
    c. both A and B.
    d. neither A nor B.
15. Technician A says the EEC III system has self-diagnostic capacity. Technician B says the EEC III system does have self-diagnostics, but not the capacity to store codes for intermittent faults. Who is correct?
    a. A only.
    b. B only.
    c. both A and B.
    d. neither A nor B

Chapter 10

Ford's Electronic Engine Control IV (EEC IV)

OBJECTIVES

In this chapter you can learn:
- the three operational strategies of the EEC IV system.
- at least five of the major system sensors.
- at least five of the major actuators.
- the three types of fuel metering devices.
- the controlled components most directly concerned with emissions rather than with engine performance, driveability or fuel economy.
- the major components of the diagnostic procedure.

KEY TERMS

A4LD (Automatic Four-speed Light-duty)
Adaptive Strategy/Calibration Modification
AXOD (Automatic Transaxle Overdrive)
Capacitor
Diagnostic Routine
Electronic Control Assembly (ECA)
Pinpoint Procedure
PIP Signal
Quick-Test
Spout
Star Tester
Thick Film Integrated (TFI)
Torque Converter Lockup Clutch

The electronic engine control IV (EEC IV) system is the fifth generation of Ford's computerized engine control systems. It was first introduced in late 1982. By 1985 the EEC IV system was the only computerized engine control system Ford was using except for a few 5.8-liter police cars and some Canadian models. By 1985 the EEC IV system was also used on most Bronco, Ranger, E-series and F-series light trucks.

The EEC IV system's primary function is to control the air/fuel ratio. This is achieved through either a feedback carburetor, a single-point injection system (which Ford calls central fuel injection) or multipoint injection (which Ford calls electronic fuel injection [EFI]). The throttle body unit for CFI is shown in Figure 10-1 and the EFI unit is shown in Figure 10-2. To control these and in some cases other functions, the computer col-lects information about the engine and vehicle operating parameters from a variety of sensors, Figure 10-3.

ELECTRONIC CONTROL ASSEMBLY (ECA)

The EEC IV system computer (ECA) has much more capacity than previous Ford engine control computers, Figure 10-4. It is the first to include KAM, a keep-alive memory (an expansion of its hard-wired random access memory) enabling it to store codes related to faults that it has previously observed but that are no longer present. This capacity was not true for the first application of EEC IV on the 1.6-liter engine. The number of codes it can recognize is greatly

297

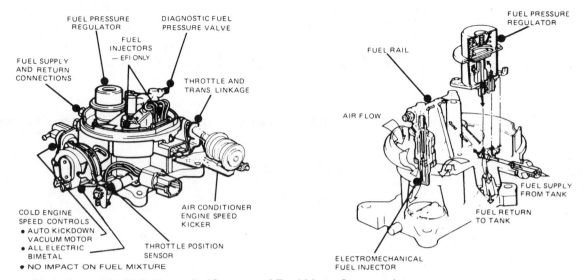

Figure 10–1 Central fuel injection unit. *(Courtesy of Ford Motor Company.)*

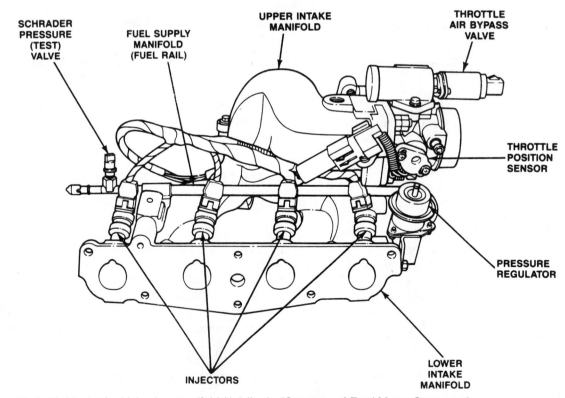

Figure 10–2 Multipoint fuel injection manifold (1.6-liter). *(Courtesy of Ford Motor Company.)*

Inputs		Outputs
Engine coolant temperature		Air/fuel mixture control device,
Manifold absolute pressure and/or		carburetor solenoid or injector
barometric pressure		Ignition timing control
Throttle position		Idle speed control
Engine speed and crankshaft position		Thermactor airflow control
Exhaust gas oxygen	Electronic	Canister purge control
EGR valve position	Control	EGR flow control
	Assembly	Torque converter clutch control
The following sensors are	(ECA)	Turbocharger boost control
unique to specific engines:		A/C and cooling fan controller module
		Wide-open throttle A/C cut-off
Air charge temperature		Inlet air temperature control
Vane air temperature		Variable voltage choke
Vane airflow		Temperature-compensated accelerator
Idle tracking switch		pump
Transmission position		Shift indicator light
Inferred mileage sensor		Fuel pump relay
Knock sensor		
Ignition diagnostic monitor		
Clutch engaged switch		
Brake on/off switch		
Power steering pressure switch		
Air-conditioning clutch	Power	
Ignition switch	Relay	

All components are not used on any one system.

Figure 10–3 EEC IV system overview

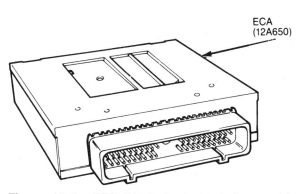

ECA
(12A650)

Figure 10–4 EEC IV electronic control assembly (ECA). *(Courtesy of Ford Motor Company.)*

increased compared to the computers in earlier Ford systems.

Engine Calibration Assembly

The ECA's **engine calibration assembly** contains the necessary programming to fine-tune the ECA's commands to the specific needs of that vehicle and engine's weight, axle ratio, and transmission. The calibration assembly is similar to a GM PROM, except that unlike those systems' unit, the Ford calibration assembly is an integral part soldered into the ECA, and it cannot be removed or serviced separately.

OPERATING MODES

The ECA controls the engine in one of several operational modes or strategies:

Base Engine Strategy. This is the mode the ECA uses to control warm engine calibration through the wide range of operating conditions occurring in normal driving. Prior to 1986, the calibration for warm engine operation was a coolant temperature of 88 degrees Centigrade (190 degrees Fahrenheit) or higher. To reduce the time spent running in open loop conditions, that threshold was lowered in 1986 to about 54 degrees Centigrade (130 degrees Fahrenheit) or higher. This warm-up mode is broken down into four submodes:

- cranking.
- closed-throttle operation.
- part-throttle operation (closed loop).
- wide-open throttle (WOT).

Input information enables the ECA to recognize these operational modes. Once it recognizes the state of the engine, the operational mode, the ECA issues calibration commands to the appropriate actuators, commands designed to produce the best results in terms of emissions, fuel economy, and driveability.

If the engine threatens to stall while in the base engine strategy mode, the ECA will employ an underspeed response feature.

MPG Lean Cruise. When predetermined criteria are met during cruise conditions on some engine applications (beginning in 1988), the ECA will take the system out of closed loop to an even leaner "dead-reckoning" (no feedback) open loop air/fuel ratio. The purpose of this modification is to achieve better fuel economy.

Modulator Strategy. Operating conditions requiring significant compensation to maintain good driveability cause the ECA to modify the base engine strategy. Conditions causing such modifications are:

- cold engine.

- overheated engine.
- high altitude.

Limited Operational Strategy (LOS). When a component failure prevents the system from functioning in a normal strategy, the ECA enters an alternative strategy designed to protect other system components, such as the catalytic converter yet still provide enough driveability to keep the vehicle running until it can be repaired. If the ECA's central processing unit (CPU) fails, the ECA will operate in a fixed mode: there will be no spark advance, no EGR, and air from the Thermactor (Ford's air injection system) will be dumped to the atmosphere.

Adaptive Strategy

This feature, first introduced in 1985, enables the ECA to constantly adjust some of its original calibrations, those programmed into it by engineers based on ideal conditions. Whenever conditions are less than ideal due to variations in manufacturing tolerances, wear or deterioration of sensors or other components that affect the ECA's control of the most critical functions, the **adaptive strategy** puts an adjustment or adaptive factor into the calculation process. Adaptive strategy could also be called a learning capacity or a self-correction ability. The ECA learns from past experience so it can better control the present conditions. Learning starts when the engine is warm enough to go into closed loop and is in a stabilized mode.

Adaptive Fuel Control. This adaptive strategy concerns air/fuel ratio. If the injectors become fouled with deposits or if wear in the pressure regulator causes a slight loss of fuel pressure, the result is less fuel injected into the cylinder than should be during the pulse width the computer has calculated for those conditions. When the system is in closed loop, the EGO sensor reports a lean condition to the ECA. The ECA responds by commanding a wider injector pulse width to maintain the proper 14.7 to 1 air/fuel ratio.

The adaptive fuel control recognizes that a

wider pulse width than what the original calibration designated is required. If this condition continues for several minutes, the adaptive fuel control modifies the original calibration to provide a wider pulse width so the calibration is more on target and less correction is needed to bring the mixture to stoichiometry. Notice that all forms of adaptive strategy, or learning, are in response to signals from the oxygen sensor.

How the Calibration Modification Is Achieved. The ROM (read-only memory, the hard-wired original set of computer instructions) contains an engine load (MAP sensor signal) versus the engine rpm table used as the starting point for calculating pulse width, Figure 10-5. The table is divided into separate cells. Each cell is assigned a base value of "1." Crossreferencing the current MAP value with the current rpm value, the microprocessor identifies the cell that best represents current engine load and speed conditions. The value in that cell is used as a multiplier in the microprocessor's calculations to determine the correct pulse width command for the injectors under the specific conditions then present.

Since this table is stored in the ROM, it cannot be changed or erased. There is, however, another copy of this table in the keep-alive memory (KAM). The microprocessor uses that copy to work from, and the base values in each of the cells of the KAM copy of the table can be modified according to what the computer learns through its adaptive strategy, or learning.

If the pulse width calculation, using the base value in a given cell, does not consistently produce the desired 14.7 to 1 air to fuel ratio as reflected in the oxygen sensor's return signal, the adaptive fuel control will add to or subtract from the base value in the KAM copy of the table, Figure 10-6. If the air/fuel ratio tends to be too lean, the adaptive fuel control adds to the base value. If the air/fuel ratio tends to be too rich, the adaptive fuel control subtracts from the base value. Those familiar with GM engine controls will recognize a cousin of the Block Learn and Integrator values.

At idle, adaptive fuel control works the same way it does at other engine speeds. The microprocessor uses the same table for pulse width calculations at idle, but it uses cells that are used only during idle conditions. Depending on engine application, there are either four or six cells on the MAP versus rpm table for idle pulse width calculations. The table could look something like the

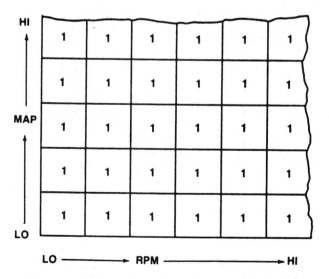

Figure 10–5 Adaptive fuel control table base values

HI	1	1.2	1.1	.95	.9	1
	.975	1	1	1	1.1	1.1
MAP	.95	1	1	1.3	1.2	1.2
	.95	1	1	1.1	1	1
LO	.9	1.1	1.1	1.1	1	.9

LO ⟶ RPM ⟶ HI

Figure 10–6 Adaptive fuel control table with adapted cell values

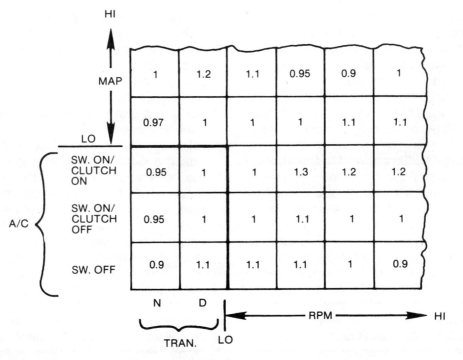

Figure 10–7 Idle fuel control cells on MAP versus RPM table

one shown in Figure 10-7. The idle cells, however, are not selected by MAP and rpm; they are selected by the transmission being in drive or neutral, by the A/C being on or off, and in some cases by the A/C control switch being on or off.

Adaptive Idle Speed Strategy. Adaptive idle speed strategy is used only with CFI and EFI systems, not with carbureted engines. The idle speed strategy calculates the commands sent to the idle speed control actuator. This device controls the amount of air admitted to the induction system bypassing the throttle valve. If the commands sent to the actuator fail to produce the calibrated idle speed, the ECA substitutes a new value in place of the one in the originally calibrated values used for the calculation. The value replaced is obtained from an idle table in the KAM similar to the MAP versus rpm table used for fuel/air mixture and discussed above, Figure 10-8. When idle speed is corrected, the adaptive idle speed strategy monitors the revised value

needed to make the correction and records it in place of the originally recorded value in the KAM idle table.

Adaptive idle speed strategy learning occurs only under the following conditions:

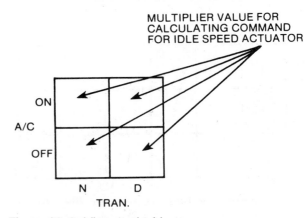

Figure 10–8 Idle speed table

- idle speed is stabilized during base idle conditions (warm engine, automatic transmission in drive, manual transmission in neutral and A/C off).
- system is in closed loop.
- a specific time has elapsed since idle mode began. (This time varies for different engine applications).

Because the information learned by the adaptive strategy is stored in the KAM, it can be used or revised as needed. The learning process works constantly during stabilized, closed-loop conditions.

Other adaptive strategies incorporated into the EEC IV systems since 1985 include adaptive spark control (the computer learns to modify the range and shape of the spark advance map) and adaptive EGO sensor aging (the computer learns how to compensate for the gradually-declining signal frequency of the oxygen sensor). Their operation is similar to those just described.

Adaptive strategy also tests regularly for the possibility that corrupt data (false values) could have gotten into the KAM. If any data corruption is found, those values are erased and replaced with the original corresponding values from the ROM. Learning can then modify the replacement values as needed.

Power Relay

A power relay is used to supply operational power to the ECA, Figure 10-9. The control coil of this relay is powered directly through the ignition switch. Once the ignition switch is turned on, the contact points of the relay, which are normally open, close and supply power to the ECA. The control coil of the relay is in series with a diode. If the battery's power is reversed, the diode will block current through the control coil and will prevent the contact points from closing. This protects the power circuits of the ECA from damage caused by reverse polarity.

Beginning in 1986 on some engines and becoming more widespread in 1987 and later, the power relay was integrated into a unit called the integrated controller. This unit typically contains the fuel pump relay, the power relay, the cooling fan relay (Electro-drive radiator fan), the high speed Electro-drive fan relay and the solid state A/C cutout relay.

The power to the KAM does not come through the power relay, but directly from a separate battery source that is always hot. If this power supply gets disconnected, any service codes stored and all learned adaptive strategy will be erased. Keeping this circuit powered requires very little current (only a few milliamps) and does not discharge the battery even if the vehicle is not operated for several days. In fact, internal current in the battery will discharge it much more quickly than the minuscule KAM current draw.

But because the KAM is not protected by the power relay, its circuitry is designed to be less susceptible to reverse polarity damage. Care should always be taken to prevent such mistakes, however.

INPUTS

Engine Coolant Temperature (ECT) Sensor

The ECT is a thermistor-type sensor, Figure 10-10. Its voltage signal to the ECA normally ranges from 4.5 volts when very cold to 0.3 volt when the engine is hot. Its resistance values are:

- -40 degrees Fahrenheit = 269 kilohms.
- 32 degrees Fahrenheit = 96 kilohms.
- 77 degrees Fahrenheit = 29 kilohms.
- 248 degrees Fahrenheit = 1.2 kilohms.

The information it provides influences the ECA's calibrations commands controlling:

- air/fuel mixture ratio.
- idle speed.
- EGR.
- Thermactor air.

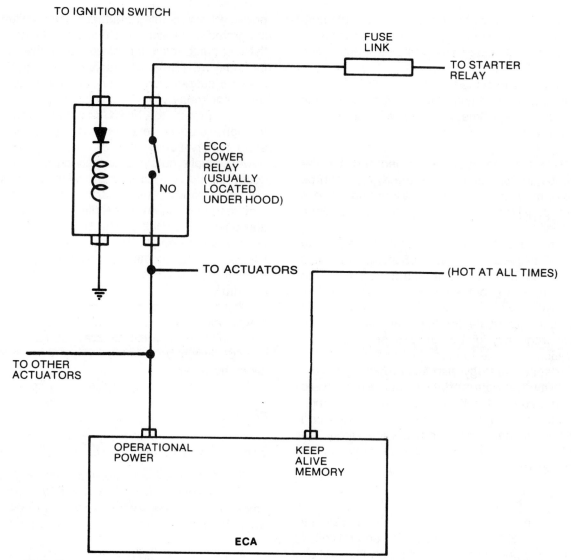

TO IGNITION SWITCH

FUSE
LINK

TO STARTER
RELAY

ECC
POWER
RELAY
(USUALLY
LOCATED
UNDER HOOD)

NO

TO ACTUATORS

(HOT AT ALL TIMES)

TO OTHER
ACTUATORS

OPERATIONAL
POWER

KEEP
ALIVE
MEMORY

ECA

Figure 10–9 Power relay circuit

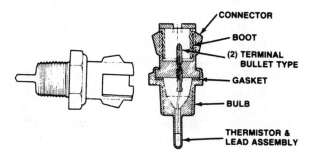

CONNECTOR

BOOT

(2) TERMINAL
BULLET TYPE

GASKET

BULB

THERMISTOR &
LEAD ASSEMBLY

Figure 10–10 Engine coolant temperature sensor.
(Courtesy of Ford Motor Company.)

- canister purge.
- choke voltage.
- temperature-compensated accelerator pump.
- upshift light.

Pressure Sensors

Ford's pressure sensors operate differently than those discussed in earlier chapters. They use a pressure-sensitive variable **capacitor,** Figure 10-11. The capacitor is formed by two conductive plates separated by a thin air space and a thin layer of insulating material. One plate, the positive, is a thin copper film on a round, rigid ceramic bed with a thin film of insulating material

sprayed over the copper. The other plate, the negative, is also a thin copper plate on the underside of a round ceramic disk, but this disk is slightly flexible. The flexible ceramic disk is mounted on top of the rigid ceramic bed with a thin adhesive strip around the entire circumference. The adhesive strip makes an airtight seal and spaces the two copper plates a few thousandths of an inch apart.

This assembly makes up the ceramic body of the sensor and is mounted in a plastic housing. The area below the ceramic body is sealed from the area above it. A small hole through the ceramic bed vents the area between the copper plates to the sealed, reference pressure below

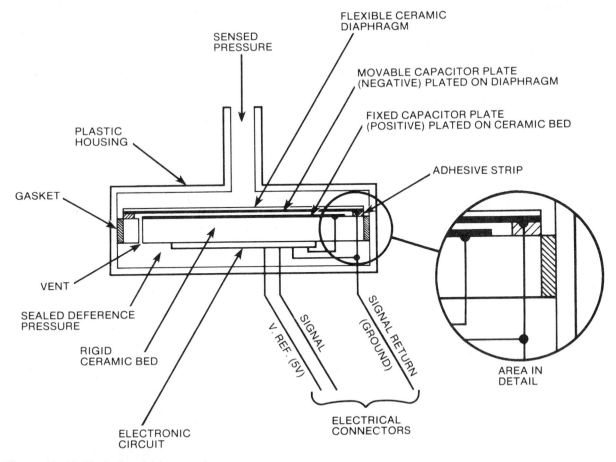

Figure 10–11 Typical variable capacitor pressure sensor

the ceramic bed. The space above the ceramic body is exposed to the monitored pressure.

Electronic circuitry and components are attached to the bottom of the rigid ceramic bed. Each capacitor plate has a wire that connects it to the electronic circuitry. Additional wires connect the circuitry and the negative plate to the ECA.

If the pressure sensed is lower than the reference pressure between the plates, the diaphragm flexes outward, moving the plates farther apart. This *lowers* the capacitance of the capacitor. As the pressure sensed increases, the diaphragm is forced inward, moving the plates closer together. This *increases* the capacitance of the sensor. The electronic circuit, powered by the reference voltage (VREF) from the ECA, measures the capacitance of the variable capacitor and sends a corresponding electrical signal back to the ECA.

Manifold Absolute Pressure (MAP) Sensor. In addition to being a variable capacitor, the MAP sensor used on EEC IV systems is a frequency-generating device, Figure 10-12. An additional chip in the sensor's electronic circuit produces a frequency signal corresponding to the capacitance. The capacitance, as we have seen, is a function of the relative distance of the plates and thus of manifold pressure. This frequency signal is what is sent to the ECA.

If you apply a voltmeter directly to the sensor's output circuit, you will see a steady voltage signal of about 2.5 volts. This will change very little regardless of manifold pressure because it is not the voltage of the output that conveys the information. Actually the signal does change (more or fewer pulses per second) in response to changes in manifold pressure, but the frequency is much more rapid than even the fastest voltmeter can display reading the signal. The meter effectively just gives an average voltage reading, which is not what the computer is looking at.

During the lowest manifold pressure conditions (highest vacuum, lowest load), the sensor's signal frequency is about 92 cycles per second (92 Hz). At WOT, the manifold pressure is nearly equal to atmospheric pressure. At this point the frequency is about 162 Hz. The ECA reads the frequency as a manifold pressure value. The MAP sensor's input affects:

- air/fuel ratio.
- EGR flow.
- ignition timing.

Those EEC IV engine applications using a manifold pressure sensor do not use a separate barometric pressure sensor. Rather the ECA uses a different strategy, using the MAP sensor as a barometric pressure sensor during two specific conditions. During the brief period of time between when the ignition is turned on and the starter starts to crank, the ECA takes a reading from the MAP sensor. The pressure in the intake manifold is atmospheric pressure during this period. The ECA stores this MAP reading in its RAM memory as a barometric pressure reading. In addition, during WOT operation, manifold pressure is almost the same as atmospheric pressure. While there is a slight difference because of intake system aerodynamic friction, this difference is predictable for a specific engine. Thus, during WOT, the ECA will adjust the MAP reading for the difference between manifold pressure and

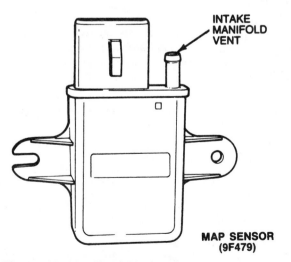

INTAKE MANIFOLD VENT

MAP SENSOR (9F479)

Figure 10–12 Manifold absolute pressure sensor. *(Courtesy of Ford Motor Company.)*

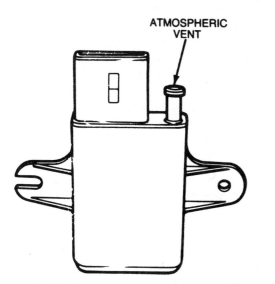

ATMOSPHERIC VENT

Figure 10–13 Barometric pressure sensor. *(Courtesy of Ford Motor Company.)*

atmospheric pressure. This reading will then be used to update the barometric pressure reading stored in RAM. Obviously this will be very close to the start-up reading unless the vehicle has changed altitude without using WOT, or driven so long the weather has changed substantially.

Barometric Pressure (BP) Sensor. The BP sensor is nearly identical to the MAP sensor, both in appearance and in operation, Figure 10-13. Instead of a tube to the intake manifold, there is a vent to atmospheric pressure. The BP sensor's output frequency ranges from 122 Hz at the lowest atmospheric pressure to 162 Hz at the highest.

Real World Problem

Remember that atmospheric pressure and altitude vary inversely, that is, the higher a vehicle goes, the lower the pressure, and the lower in altitude the higher the pressure. The highest pressures will be found at sea level (or at the bottom of Death Valley, below sea level); the lowest pressures will be found at the tops of mountains. These relationships are very easy to reverse in a busy shop environment, with the press of time and need for a quick diagnosis.

The BP sensor is used most often (instead of a MAP sensor) to measure air intake on engines using a vane airflow or a mass airflow sensor. A vane airflow meter responds to a pressure differential across the vane and measures the corresponding airflow into the induction system. With the additional barometric pressure provided by the BP sensor, the ECA can more accurately determine the air charge density. This measurement is even more critical with a turbocharged engine. Finally, even though the supercharged 3.8-liter engine uses a mass airflow meter and can accurately determine air mass without a BP sensor, the system uses a BP sensor anyway to determine the barometric pressure's effect on the air mass. The BP sensor's input affects:

- air/fuel ratio.
- EGR flow.

Throttle Position Sensor (TPS)

The TPS is a variable potentiometer monitoring the position and movement of the throttle butterfly plate. It provides a return reduced voltage signal to the ECA indicating the throttle position and tracing its recent movement.

Linear TPS. A few carbureted EEC IV engine systems use a linear TPS, Figure 10-14. A cam on the throttle shaft contacts a plunger on the TPS and pushes it in as the throttle shaft rotates, Figure 10-15. Most linear TPS sensors can be adjusted to correct to proper idle and WOT voltage outputs with the intervening values correctly adjusted by the same setting.

Rotary TPS. The rotary TPS is functionally similar to the linear type, but the resistor forms an arc around the axis of the throttle shaft instead of a straight line nearby. The wiper pivots from the center of the shaft and more directly parallels the movement of the throttle butterfly plate since there is no tangential "cam" effect. The rotary TPS is mounted on the side of the carburetor or throttle body, so the throttle shaft directly engages and drives the potentiometer wiper, Figure 10-16. Beginning in 1985, all EEC IV systems use the rotary type TPS. Some are

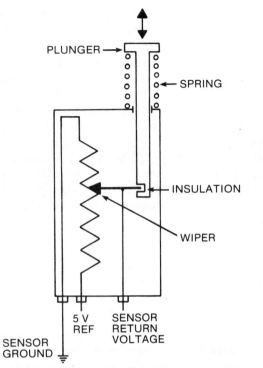

Figure 10–14 Linear throttle position sensor diagram

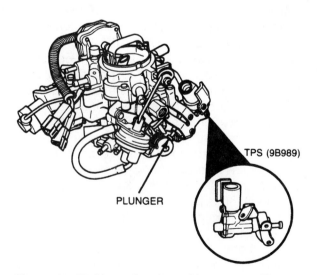

Figure 10–15 Linear throttle position sensor. (Courtesy of Ford Motor Company.)

adjustable; some are not. For either type, the ECA sends a 5-volt reference input to the TPS. Its return signal range is from 0.5 to 4.5 volts. Most rotary TPS sensors in normal operation will transmit a signal between 1.0 volt at closed-throttle idle and 4.0 volts at WOT. The ECA uses the TPS primarily to identify the driving mode (idle, cruise, or WOT), but it also can tell how quickly the throttle has been moved in some applications.

Real World Problem

The most frequent problem by far with throttle position sensors is a bad spot on the variable resistor, causing a high or low voltage spike in the return signal. While this can be observed on a sensitive voltmeter, a lab scope is the most secure way to identify this problem since it will draw a trace of the signal, and any voltage spike will be immediately visible.

The next most frequent problem with throttle position sensors is loss of resistance-free connection through the ground circuit. This will not let the information signal get reduced to the proper voltage, either leaving it at full reference voltage (and causing starting problems because the computer will assume the throttle is wide open and will initiate clear flood mode), or simply raises the output signal corresponding to the voltage drop through the high resistance connection. The latter situation will throw off the signal consistently on the lower range and will make it completely inaccurate in the upper.

Figure 10–16 Rotary throttle position sensor. *(Courtesy of Ford Motor Company.)*

Profile Ignition Pickup (PIP) Sensor

A Hall-effect switch generates the **PIP signal** (see Hall-effect switch in the introductory chapter).

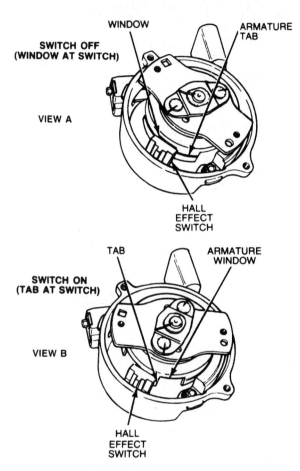

Figure 10–17 Hall-effect switch in distributor. *(Courtesy of Ford Motor Company.)*

This switch is called the PIP sensor and provides the ECA information about engine speed and crankshaft position. On engines with a distributor, the PIP sensor is in the distributor, Figure 10-17.

In 1986, Ford introduced sequential electronic fuel injection (SEFI) on the 5.0-liter engine in passenger cars. In a sequential fuel injection system, the computer pulses each injector individually in the firing order. To do so, the computer must know where the engine is in its cycle, so it can pulse, for example, the number three injector during the intake stroke for cylinder number three. So the ECA can identify the engine's place in the cycle, Ford engineers made one of the PIP sensor trigger vanes narrower than the others. As the narrow vane passes between the magnet and the crystal, a shorter on-time occurs in that part of the signal, Figure 10-18. This shorter pulse is called the signature PIP. The ECA recognizes the signature PIP as the second cylinder in the firing order. Being able to recognize any given cylinder in the firing order, combined with a hard-wired memory of the firing sequence and having a PIP signal counting program, enables the ECA to pulse each fuel injector in the proper sequence.

In 1989, Ford introduced a distributorless ignition system (DIS) on applications such as the 3.8-liter supercharged engine, the 3.0-liter special high-output (SHO) engine, and the 2.3-liter engine used in the Ranger pickup truck, Figure 10-19. On these applications, the PIP sensor is located on the front of the engine, just behind the crankshaft pulley and hub assembly, Figure 10-20. A vaned cup on the back of the pulley triggers the Hall-effect PIP sensor switch, Figure 10-21. Since the

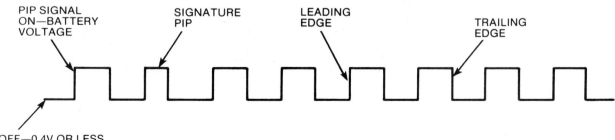

Figure 10–18 PIP signal for 5.0-liter sequential injection engine

switch is mounted on the crankshaft where it monitors two crankshaft revolutions per engine cycle, only one vane is needed for each pair of companion cylinders. As the leading edge of the vane moves between the Hall-effect switch magnet and the crystal, the PIP signal rises quickly to battery voltage, Figure 10-22. The vanes on most engines are configured so a PIP signal goes to battery voltage each time a piston reaches 10 degrees BTDC. As the trailing edge of the vane moves from between the magnet and the crystal, the PIP signal drops to 0.4 volt or less.

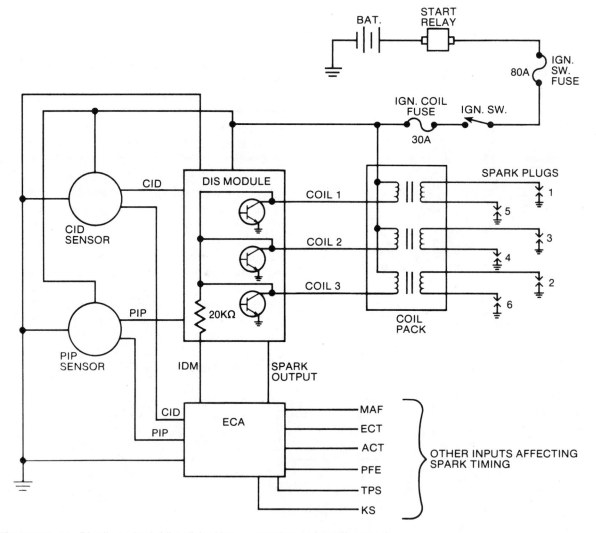

Figure 10–19 Distributorless (direct) ignition, supercharged 3.8-liter engine

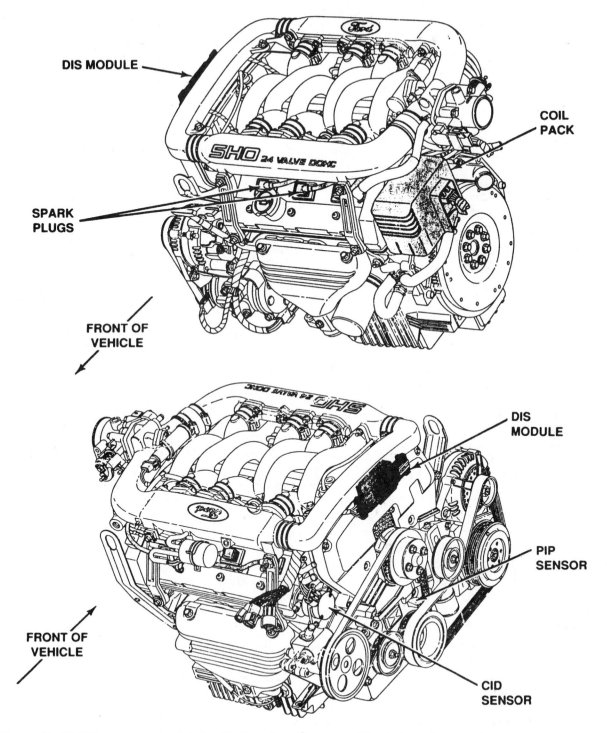

Figure 10–20 DIS components, 3.0-liter SHO engine. (Courtesy of Ford Motor Company.)

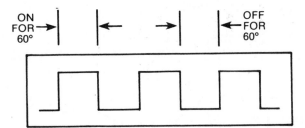

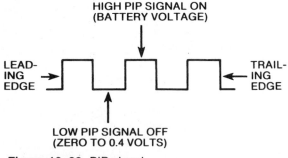

Figure 10–22 PIP signal

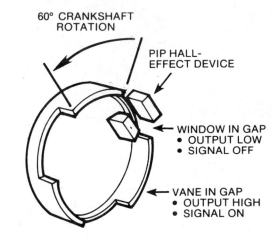

Figure 10–21 PIP sensor, supercharged 3.8-liter and SHO engines

Cylinder Identification (CID) Sensor

Distributorless ignition systems (or an equivalent term, *direct ignition* systems) use a separate coil for each pair of companion cylinders. Companion cylinders are cylinders 360 degrees apart in the 720 degrees of the engine cycle. For each pair of companion cylinders, one approaches the top of its compression stroke while the other approaches the top of its exhaust stroke; the positions reverse next time. This ignition system (one of the waste spark family of ignition systems) fires the spark plugs in each companion cylinder simultaneously, one for its combustion and power stroke, the other (the "waste" spark) at the end of its exhaust stroke. The ECA calculates the proper spark advance and instructs the DIS module when to fire the coil. But the DIS module

must determine which coil to fire. To know that, it must have information about which cylinder is ready for combustion ignition.

The CID sensor is also a Hall-effect switch. On the 3.0-liter SHO engine, the CID sensor is mounted on the right end of the rear head; one of the overhead camshafts drives it, Figure 10-20. On the 3.8-liter engine the CID sensor mounts where the distributor used to be on a more conventional engine, driven directly by the camshaft. On both these engines, the camshaft drives a single vane cup, Figure 10-23. As the leading edge of the vane moves between the Hall-effect switch magnet and crystal, the CID signal rises to battery voltage. This tells the DIS module and the ECA that the number one cylinder is at 26 degrees ATDC on its power stroke. As the trailing edge of the vane moves out from between the Hall-effect switch magnet and crystal, the CID signal drops to 0.4 volt or less. This in turn tells the DIS module and the ECA the number one cylinder is at 26 degrees ATDC on its intake stroke. Either transition, voltage either rising or falling, identifies the position of the number one cylinder. The DIS module uses this information during starting crank to determine which coil to fire first.

Because both of these engine types are sequentially fuel injected (the 3.8-liter only after model year 1988), the ECA uses the CID signal to determine the injector pulse sequence, just as it uses the signature PIP on sequentially fuel injected engines using a distributor.

On the DIS 2.3-liter engine, the CID sensor is combined with the PIP sensor behind the crankshaft pulley and hub assembly, Figure 10-24. The 2.3-liter engine's CID sensor works like those on the 3.0- and 3.8-liter engines with one important exception. Because the single vane triggering the CID sensors is on the crankshaft pulley, it produces two high CID and two low CID signals per engine cycle, Figure 10-25. Therefore, none of the transitions in the CID signal uniquely identifies the number one cylinder. When the signal goes to the high voltage, either the number one or the number four cylinder is at 10 degrees ATDC on its power stroke. The DIS module knows to fire coil 2 in response to the next spark timing command from the ECA. When the signal goes low, either cylinder number three or cylinder number two is at 10 degrees ATDC, and the DIS module knows to fire coil 1 in response to the next spark timing command from the ECA. The CID sensor is sometimes called a cam sensor, though obviously on the four cylinder engine, this is a misnomer.

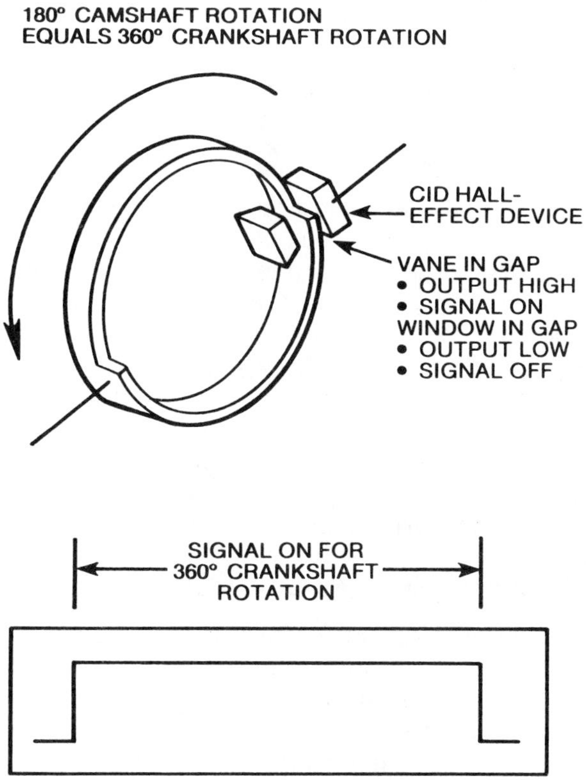

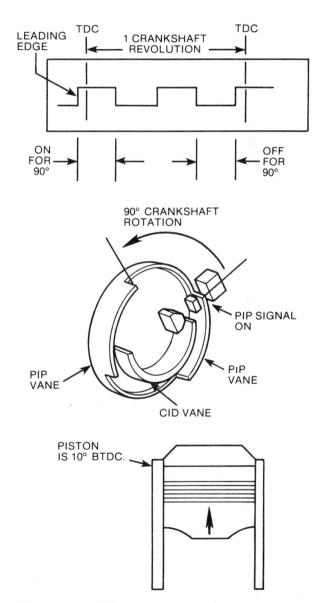

Figure 10–23 CID sensor, supercharged 3.8-liter and 3.0-liter SHO engine

Figure 10–24 PIP sensor, 2.3-liter, dual plug engine

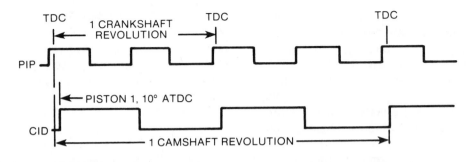

Figure 10–25 CID and PIP sensor signals

Ignition Diagnostic Monitor (IDM)

The IDM, not used on all EEC IV systems, is a wire feeding the tach signal from the negative side of the ignition coil primary windings to the ECA. On a TFI IV ignition system, this wire with a 22 kilohm resistor runs from the coil directly to the ECA, Figure 10-26. On DIS engines, the tach signal is picked up from the DIS module, and fed through a 20 kilohm resistor to the ECA, Figure 10-19. The resistor protects the ECA from the coil's high voltage spike when the primary current is cut off to trigger the spark.

PIP Signal

If the Hall-effect switch did not generate a PIP signal, the engine could not run because there would be no ignition signal: the computer would not know when to fire the plugs. If the switch did generate a PIP signal and sent it to the TFI module but for some reason not to the ECA on a carbureted engine (see Figure 10-30), the engine could run because the TFI module could fire the plugs at base timing. There would be no advance calculation, and driveability, emissions and fuel economy would suffer even though the engine ran. If the same condition occurred on a fuel injected engine, the engine could not run because the ECA would not pulse the injectors without a PIP signal.

The IDM is like an echo of the ECA's spark output **(SPOUT)** command. The ECA compares the IDM signal to the PIP signal to check that:

- the SPOUT command was carried out.
- the calculated timing advance circuit is working properly.

If the ECA sees a fault, it will store a service code in its continuous monitor memory.

Exhaust Gas Oxygen (EGO) Sensor

The EGO looks similar to the oxygen sensing unit used on other computer-controlled engine management systems and works exactly the same way. It generates a low-voltage signal in response to the difference between exhaust residual oxygen and the oxygen in the ambient air. As the fuel mixture controls modify the delivered mixture in response to the oxygen sensor's signal, the oxygen signal in turn completes the closed loop and responds to the new mixture with a corrected feedback signal.

Air Charge Temperature (ACT) Sensor

The ACT sensor measures the temperature of the air in the intake manifold that mixes with the fuel. The sensor uses a thermistor as a temperature-sensing element just as the ECT (engine coolant temperature) sensor does. On the ACT sensor, however, the sensing end of the housing has openings to allow intake air to come into direct contact with the thermistor coils, Figure 10-27. The ACT sensor is screwed into the intake manifold (or into the air cleaner on 2.8-liter engines). The ECA uses ACT sensor information to calculate the air/fuel mixture and spark timing advance. Voltage and resistance values for the ACT sensor are the same as those for the ECT. The ACT sensor is used on most Ford engine applications, but not on all.

EGR Valve Position (EVP) Sensor

The EVP sensor is a linear potentiometer mounted on top of the EGR valve. It is connected to the EGR valve diaphragm and pintle assembly, and it tells the ECA whether the EGR is open, closed, or somewhere in between, Figure 10-28. The ECA sends it a reference voltage of 5.0 volts. With the EGR valve fully closed, the return signal

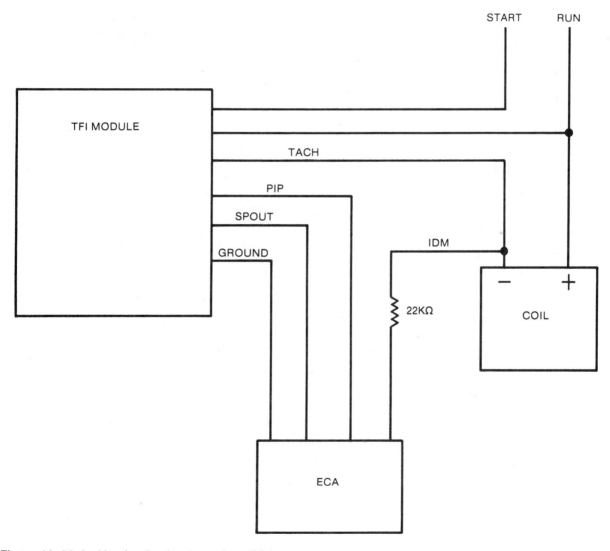

Figure 10–26 Ignition feedback schematic to ECA

voltage is about 0.8 volt; fully open, the return signal voltage is about 4.5 volts. The ECA then uses preprogrammed (hard-wired) values and data from other sensors to convert EGR valve position information to EGR flow rate for any given set of driving conditions. The EVP sensor's input is used to:

- calculate air/fuel ratio.
- calculate spark advance timing.

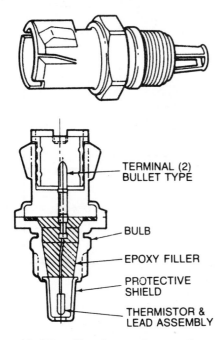

Figure 10–27 Air charge temperature sensor. *(Courtesy of Ford Motor Company.)*

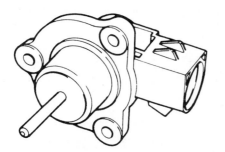

Figure 10–28 EGR valve position sensor. *(Courtesy of Ford Motor Company.)*

- adjust or correct EGR flow.
- set a service code for an EGR valve that fails to open or close when so actuated.

Pressure Feedback EGR (PFE) Sensor. The PFE sensor, Figure 10-29, is a variable capacitor working as described in the preceding section, **Pressure Sensors,** in this chapter. It senses exhaust pressure in a chamber just under the EGR valve pintle, Figure 10-30. When the EGR valve closes, the pressure in the sensing chamber is equal to exhaust backpressure. When the EGR valve opens, pressure in this chamber drops because it is then exposed to intake manifold pressure, which is always lower, while the restriction at the other end of the chamber limits the rate at which pressure from the exhaust system can enter the chamber. The more the EGR valve opens, the more the pressure in the chamber will drop.

The PFE sensor, through its internal electronic circuitry, provides an analog voltage signal to the ECA that the ECA can compare to values on its look-up table to determine EGR flow.

The PFE differs from the EVP in that it monitors actual EGR flow, while the EVP monitors only EGR opening. The EVP leaves the ECA to assume there is no carbon buildup or other restrictions hampering EGR flow. While this may be true for new engines, it is rarely the case for those in service for any length of time.

The ECA uses PFE information to:

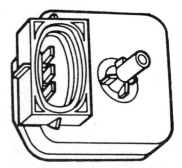

Figure 10–29 Pressure feedback EGR (PFE) sensor

- fine tune its control of EGR valve opening.
- more accurately control air/fuel ratio.
- modify ignition timing.

Vane Meter

The vane meter is used on some multipoint fuel injection applications, such as the 1.6-liter turbocharged and nonturbocharged engines, the 2.3-liter turbocharged engines, and the 2.2-liter engine in the Ford Probe. It fits between the air cleaner and the throttle body, and it measures the velocity of air flowing into the engine's induction system and the temperature of that air. Both of these measurements are made by sensors in the vane meter.

Vane Airflow (VAF) Sensor. The air inlet opening of the vane meter is closed by a spring-loaded door, a vane, Figure 10-31. As the throttle valve opens with the engine running, air moving into the induction system forces the vane open; the more airflow, the wider the vane opens. The vane hinges on a pivot pin. As the vane opens, it moves the wiper of a variable potentiometer, which functions as an airflow sensor for the ECA. The greater the airflow, the higher the VAF sensor's signal voltage (that is, the less voltage drop from the 5 volt reference voltage and the return signal output voltage).

As the vane opens, its compensator flap pushes into a specially designed cavity below the airflow passage. There is a lull or pulse between intake strokes, particularly on four-cylinder engines, and the vane has a tendency to dip closed slightly during those periods. The sealed space behind the compensator flap acts as a damper to reduce vane flutter.

Vane Air Temperature (VAT) Sensor. This sensor is effectively the same as the ACT sensor

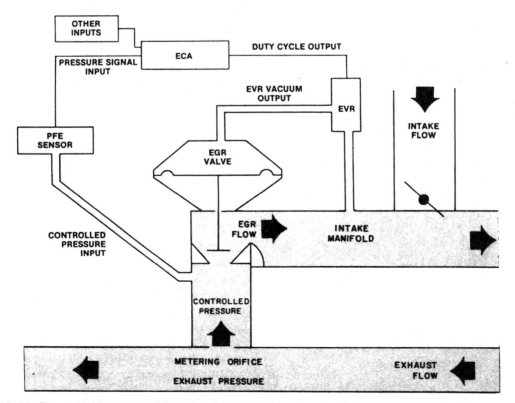

Figure 10–30 Pressure feedback EGR sensor and EGR flow

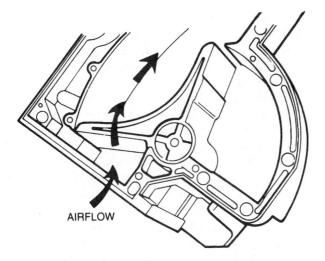

Figure 10–31 Vane airflow sensor. *(Courtesy of Ford Motor Company.)*

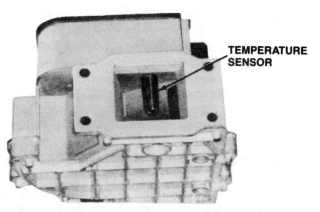

Figure 10–32 Vane air temperature sensor. *(Courtesy of Ford Motor Company.)*

used on some other engine applications, except it is located in the vane meter's air inlet opening, Figure 10-32.

Speed Density Formula.   The ECA compares inputs from the VAF, VAT, and BP sensors (or the MAP sensor, depending on the engine application and the kind of sensors employed), along with those from the TPS and EVP sensors to its hard-wired look-up charts. In this way it can calculate the engine's air intake volume and flow rate. This information, along with estimates of the engine's volumetric efficiency at different engine speeds and loads, is what the ECA uses to calculate exactly how much fuel should be injected. This is the Ford application of the speed density formula (see **Speed Density Formula** in the introductory chapter).

Mass Airflow (MAF) Sensor.   Beginning in 1988, certain engine systems employed a MAF sensor instead of a MAP sensor (see **Mass Airflow Sensor** in Chapter 2). The Ford unit was built by Hitachi of Japan. It is different from the Hitachi unit used by General Motors. The Ford MAP sensor is in the air tube between the air cleaner and the throttle body. A small air sample tube in the top of the MAF sensor air passage

directs air across a hot wire and a cold wire, Figure 10-33. The cold wire includes a thermistor used to measure the temperature of the incoming air. The hot wire wraps around a ceramic element and is coated with a super-thin layer of glass. A module mounted on the MAF sensor body keeps it 200 degrees Centigrade above the ambient air temperature.

Current flow through the sensor varies from 0.5 to 1.5 amps, depending on the air's mass airflow rate. Voltage varies from 0.5 volt at 0.225 kilograms (0.5 lb) of air per minute to 4.75 volts at 14.16 kilograms (31.23 lb) of air per minute. The MAF sensor's input to the ECA affects calculations for:

- air/fuel mixture.
- ignition spark timing.

Idle Tracking Switch (ITS)

In the body of the idle speed control motor assembly (which is used on some carbureted and some CFI engines) is a normally closed switch, Figure 10-34. Whenever the throttle closes, the throttle lever presses against a plunger extending from the nose of the assembly. The pressure forces the plunger to move slightly back into the assembly and to open the switch. When the throttle opens, the switch closes again. The

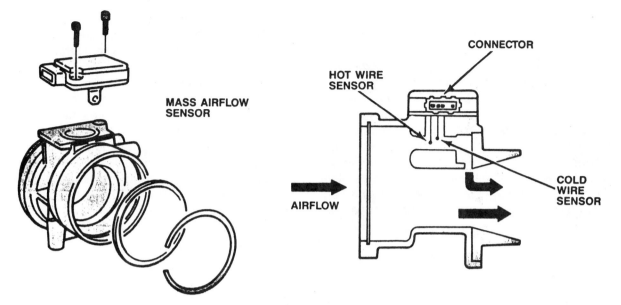

Figure 10–33 MAF sensor. *(Courtesy of Ford Motor Company.)*

ECA monitors the switch and can determine when the throttle is open or closed. It uses this information to control the ISC motor, both to determine when to employ it and what to set it at.

Transmission Switches

The ECA must know whether or not the vehicle is in gear or in park or neutral. This information affects:

- idle fuel and air control strategy for automatic transmission vehicles. (The torque converter loads the engine, even if the vehicle is not moving.)
- response to rapid closing of the throttle. (If the vehicle is in gear, the ECA will issue appropriate commands for deceleration.)

For most automatic transmissions the neutral drive switch, the same switch used to open the starter circuit when the transmission is in gear, serves as an input to the ECA, Figure 10-35. When the transmission is in neutral or park, the switch closes. A 5 volt reference signal is fed through a resistor in the ECA, output pin 30 to ground through the N/D switch. When the switch closes, voltage on the switch side of the resistor is low. By monitoring this voltage, the ECA knows whether the transmission is in gear or not. A diode in the switch prevents battery voltage from being applied to the ECA circuit when the starter solenoid is energized.

The **AXOD (automatic overdrive transaxle)** uses three different switches to identify what gear the transmission is in. They are:

- neutral pressure switch (NPS).

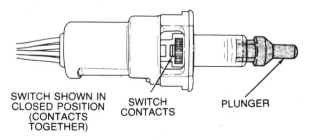
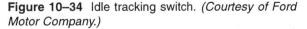

Figure 10–34 Idle tracking switch. *(Courtesy of Ford Motor Company.)*

- transmission hydraulic switch (THS) 3-2.
- transmission hydraulic switch (THS) 4-3.

These are normally open pressure switches screwed into the transmission valve body so they are exposed to the hydraulic pressure applied to specific bands or clutches in the transmission. When hydraulic pressure is applied, the switch closes and grounds the circuit. The ECA supplies voltage to each switch through a resistor, Figure 10-36A. By monitoring the voltage in each switch circuit, the ECA can identify each transmission gear position with the exception that first cannot be distinguished from second, Figure 10-36B.

A transmission temperature switch (TTS) is employed on some AXOD transaxles to alert the ECA if the transmission fluid temperature goes too high, Figure 10-36A. The TTS is normally closed, but it opens at 135 degrees Centigrade (275 degrees Fahrenheit). It also bolts directly into the valve body. If the ECA sees the TTS open during an operating condition in which it normally keeps the **torque converter lockup clutch** disengaged, such as a long uphill climb with a large throttle opening, it will apply the clutch to reduce the heat produced by high-torque turbulence in the torque converter.

Manual transmission applications can be

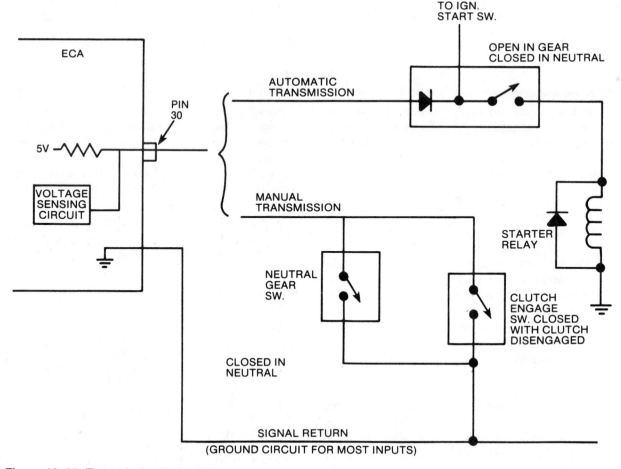

Figure 10–35 Transmission load switches

taken out of gear either by the clutch or by the gearshift. Thus, the system uses two switches, Figure 10-35. A neutral gear switch attaches to the shift linkage. It remains closed in neutral. A clutch engaged switch attaches to the clutch pedal linkage and closes when the clutch pedal is depressed (when the clutch is disengaged).

Because the neutral gear and clutch engaged switches are in parallel, they must both be open for the ECA to determine that the vehicle is in gear. If either switch closes, voltage at pin 30 goes low, and the ECA knows the gear train is disengaged.

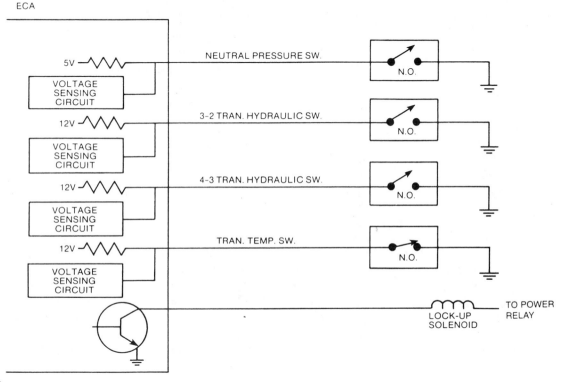

A

Gear Position	Neutral Pres. Sw.	3-2 Tran. Hyd. Sw.	4-3 Tran. Hyd. Sw.
Park/Neutral	Open (5 V)	Open (12 V)	Open (12 V)
First	Closed (0-1 V)	Open (12 V)	Closed (0-1 V)
Second	Closed (0-1 V)	Open (12 V)	Closed (0-1 V)
Third	Closed (0-1 V)	Closed (0-1 V)	Closed (0-1 V)
Fourth	Open (5 V)	Closed (0-1 V)	Open (12 V)
Reverse	Closed (0-1 V)	Open (12 V)	Open (12 V)

B

Figure 10–36 AXOD transmission switches

Inferred Mileage Sensor (IMS)

Beginning in 1985 some light truck EEC IV vehicles came with an electronic module (IMS) containing an E-cell, Figure 10-37. The IMS is energized by the power relay. After a specified amount of ignition-on time, the E-cell opens the IMS's feedback circuit to the ECA. The ECA responds by changing programmed calibrations to direct Thermactor air to the exhaust manifold for longer periods of time. It also changes the EGR flow rate, both measures to compensate for the "inferred" engine wear assumed to have occurred.

The IMS should not be confused with the emission maintenance warranty/extended useful life (EMW/EUL) module. There are two types of these latter units, each used on Ford trucks. The early type used an E-cell depleted after the ignition was on for a total time equivalent to about 60,000 miles. When the E-cell depleted, it triggered an indicator light on the instrument panel. This was supposed to alert the driver to replace the EGR valve and EGO sensor. On later models, the E-cell in the module was replaced by a small microprocessor that could be reset. Replacing the EGR valve or EGO sensor at set intervals is no longer required on other EEC IV systems applications. Replacement is required, of course, when those components fail or degrade beyond useful function.

Knock Sensor (KS)

The EEC IV engine applications using a knock sensor use a standard, piezoelectric type, Figure 10-38. When excited by vibrations in the frequency characteristic of detonation (knock), it sends a signal to the ECA, which can then retard spark timing advance to correct for the condition.

Vehicle Speed Sensor (VSS)

The vehicle speed sensor, a magnetic pulse generator driven by the transmission speedometer output gear, produces 8 cycles for each revolution, or 16 AC signals per revolution, or 128,000 signals per mile to the ECA. The ECA will modify this signal, convert it to a vehicle speed value and store it in memory. In addition to vehicle speed control and depending on vehicle application, this information may be used for the control of the transmission torque converter lockup clutch, coolant fan control, and to identify deceleration conditions.

Brake On/Off (BOO) Switch

Vehicles with the **A4LD (automatic four-speed light-duty) transmission,** such as Rangers and Broncos, use the BOO switch (located in the stop lamp switch assembly) to signal the ECA when the brakes are applied. The A4LD transmission employs an ECA-controlled torque converter lockup clutch. When the switch

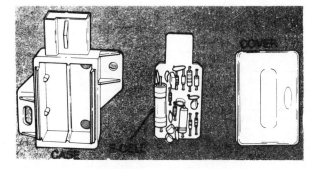

Figure 10–37 Inferred mileage sensor. *(Courtesy of Ford Motor Company.)*

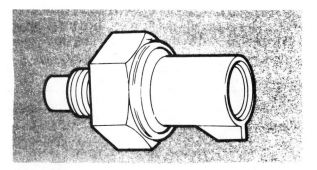

Figure 10–38 Knock sensor. *(Courtesy of Ford Motor Company.)*

closes, the ECA releases the lockup clutch. If the brakes are applied during idle and the transmission is in gear, the ECA may also raise idle speed to compensate for the increased engine load of the torque converter and brake booster. During a prolonged idle with the brakes applied, the ECA may also disengage the A/C compressor clutch.

Power Steering Pressure Switch (PSPS)

The power steering switch is in the high-pressure side of the power steering system. At a pressure of about 400 to 600 psi, the switch closes. When the ECA senses this switch closed during idle conditions, it will raise idle speed to compensate for the additional load caused by the hydraulic steering boost.

Air-Conditioning Clutch (ACC) Signal

On many engine applications, when the A/C turns on, the same voltage signal sent to the A/C compressor clutch is also sent to the ECA. This signal alerts the ECA to raise idle speed slightly to compensate for the added load.

Ignition Switch

When the ignition switch turns on, the power relay activates. The power relay then supplies current to the ECA and most of the circuits the ECA controls. The relay's power to the ECA acts as an input and signals the ECA to begin its programmed functions.

OUTPUTS

Air/Fuel Mixture Control

The EEC IV system employs three different types of fuel metering devices: carburetors (referred to as feedback carburetors), throttle body injectors (referred to as central fuel injection) and port fuel injectors (referred to as multipoint fuel injection).

Feedback Carburetors.   Three different

carburetors have been used with EEC IV engine applications. They each use a duty-cycled solenoid to control the air/fuel mixture within the range available to the control unit. Each feedback carburetor, however, uses a somewhat different feedback control solenoid. Each feedback control solenoid gets power from the power relay, and its function is controlled by the ECA, by grounding the circuit to activate it, or not grounding it to turn it off, Figure 10-39.

The Motorcraft 2150A-2V, Figure 10-40, uses a solenoid that introduces or blocks air from the air cleaner into the fuel vacuum passages, as the solenoid cycles alternately on and off. These passages provide bleed air to the idle and main metering systems and determine the richness of the fuel introduced into the carburetor's venturis. The air introduced by the solenoid is in addition to the air introduced by the fixed air bleeds found in pre-feedback carburetors and retained, in modified form, on these. If the solenoid were turned off completely (a duty-cycle of zero), additional air would be blocked, and the air/fuel ratio would be full rich mixture. If the solenoid were on constantly (a duty-cycle of 100%), the mixture would go full lean. The ECA controls the duty-cycle to optimize the delivered fuel/air ratio. In open loop the duty-cycle is determined by a hard-wired constant value, depending on coolant temperature, engine load, engine speed, and throttle position. In closed loop it varies constantly, depending on the feedback signal from the oxygen sensor.

The Carter YFA-1V carburetor has a feedback control solenoid that looks different from the one used on the Motorcraft 2150A, but its function is identical.

The Holley 6149-1V carburetor is controlled by a remotely mounted feedback control solenoid, Figure 10-41. Its control of the idle circuit air/fuel ratio is much like that of the solenoids for the two carburetors above. Its duty-cycle controls the amount of air introduced into the idle speed circuit, Figure 10-42. The main metering mixture control, however, works differently.

The Holley 6149-1V main metering circuit

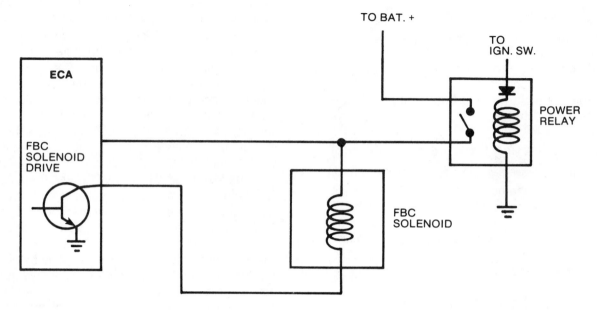

Figure 10–39 FBC solenoid circuit

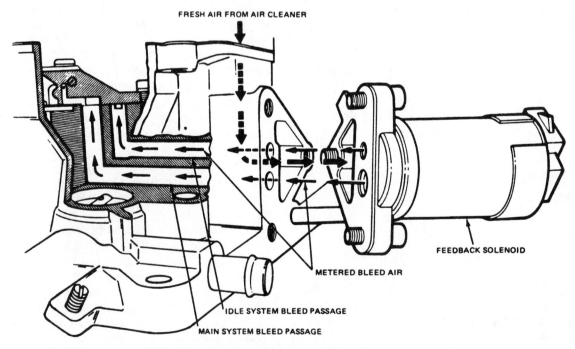

Figure 10–40 FBC solenoid (2150-2V carburetor). *(Courtesy of Ford Motor Company.)*

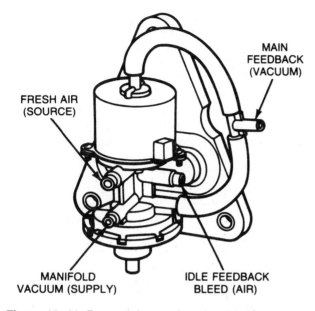

Figure 10–41 Remote duty-cycle solenoid. *(Courtesy of Ford Motor Company.)*

has a solenoid controlling vacuum to a diaphragm in the air horn section of the carburetor, Figure 10-43. The diaphragm is part of a fuel control valve assembly. It can easily be mistaken for a power valve assembly, but its function is quite different. A spring holds the diaphragm down, making the mixture full rich. Extending down from the diaphragm into the fuel bowl is a rod with a foot at the bottom. The foot engages and pushes down on the upward extending tip of a valve (the main metering valve for the feedback system) in the floor of the fuel bowl. With this valve pushed down, it opens fully and additional fuel flows into the main metering system. When the ECA wants to lean the air/fuel mixture, it cycles the solenoid so more vacuum is applied to the fuel control assembly diaphragm. This lifts it and allows the valve in the fuel bowl floor to close partially, thus leaning the delivered mixture.

Central Fuel Injection (CFI). Two kinds of CFI systems are used with EEC IV applications:

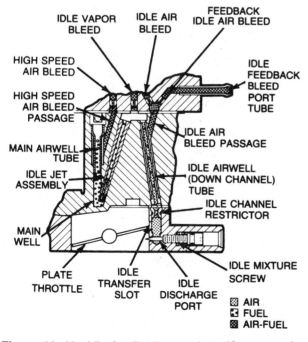

Figure 10–42 Idle feedback metering. *(Courtesy of Ford Motor Company.)*

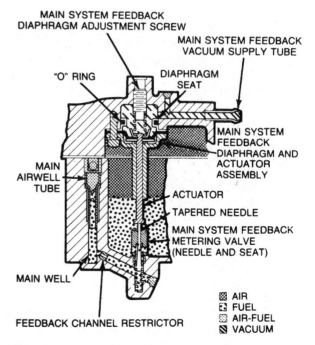

Figure 10–43 Main feedback metering. *(Courtesy of Ford Motor Company.)*

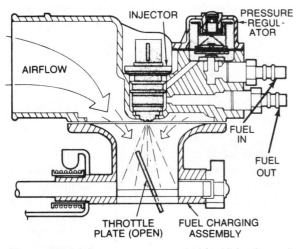

Figure 10–44 Low-pressure central fuel injection unit. *(Courtesy of Ford Motor Company.)*

a high-pressure system and a low-pressure system. Each uses a throttle body unit mounted on the manifold in the place where a carburetor would be. They are, in a real sense, just electronic carburetors, except that all the fuel mixture control functions are provided by the computer.

The solenoid operated injectors are opened by electrical command pulses from the ECA. The injector's control circuit is similar to that of the feedback carburetor solenoid. In Figure 10-44 you can see the cutaway structure of the CFI injector: it avoids the complications of all the carburetor's mechanical rods and levers and float. Electrical power comes directly from the power relay, and the ECA grounds the circuit, activating the injector, for whatever pulse width (on-time) it calculates is needed for the current driving conditions. Fuel sprays from a single injector (low-pressure system, Figure 10-44) or from two injectors (high-pressure system, Figure 10-1) directly over the throttle valves. During cranking, the high-pressure CFI system pulses both injectors simultaneously in response to each PIP signal. After the engine is started, they pulse alternately at a fixed frequency. Because the fuel pressure is constant and the injector always opens the same amount (about 0.01 inch), the air/fuel mixture is

controlled entirely by the injector pulse width.

In 1988 a variation of the high-pressure CFI system was introduced on the 3.8-liter engine. Each injector has its own EGO sensor and logic circuit. A problem such as a leaking exhaust manifold on one side of the engine affects only the performance of the cylinders on that side.

Multipoint Injection (MPI). On most multipoint systems, the injectors pulse alternately in two groups. Each group is pulsed once per engine cycle during normal operation. They pulse simultaneously during cranking. In 1986 the 5-liter engine also had a sequential fuel injection system, a system in which each injector pulses individually in the firing order of the engine in coordination with its corresponding cylinder's intake stroke.

Fuel Supply System

Fuel Pump. The low-pressure CFI system uses a fuel pump mounted in the fuel tank. The fuel pump relay controls it, and the pump delivers fuel to the fuel delivery assembly—the throttle body unit, through the fuel pressure regulator.

The CFI high-pressure and MPI systems use one of two fuel delivery systems. Most use a single high-pressure, in-tank electric pump that feeds fuel at about 60 psi to the pressure regulator. Vehicles with a greater distance between the fuel tank and the engine, such as pickup trucks or vans, can use two electric fuel pumps to assure adequate fuel pressure to the injectors without the risk of vapor lock. A low-pressure "lift pump" is mounted in the tank. It pumps fuel to an in-line high-pressure pump, which increases the pressure to about 60 psi, and sends the fuel to the pressure regulator.

Either of the high-pressure pump systems generates an unregulated pressure of almost 100 psi if the fuel line is blocked. If that pressure is reached, a pressure relief switch shuts off the pump.

Carbureted EEC IV vehicles use the conventional mechanical fuel pump on the side of the engine block.

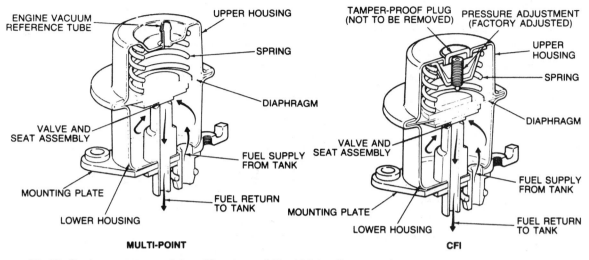

Figure 10–45 Fuel pressure regulator. *(Courtesy of Ford Motor Company.)*

Fuel Pressure Regulator.   A pressure regulator similar to those described in previous chapters controls fuel pressure to the injectors, Figure 10-45. Manifold vacuum (MAP) is routed to the MPI system pressure regulator so that it varies fuel pressure at the injectors in response to throttle position, or more properly in response to load as reflected in the intake manifold vacuum. This delivered pressure can range from a low of about 30 psi at high vacuum (low manifold pressure) conditions to as high as 40 psi at WOT (high manifold pressure). On turbocharged engines, the fuel pressure at the injectors during boost conditions can allow a delivered fuel pressure as high as 50 psi.

The CFI systems do not employ a manifold pressure-controlled regulator, and their fuel pressure stays fairly constant. The high-pressure CFI system maintains between 39 and 41 psi. The low-pressure system (on the 2.3 HSC engine) maintains about 14.5 psi at the injector tip. There is no need to modulate pressure according to load with these systems because they spray fuel ahead of the throttle, into air at ambient barometric pressure.

Fuel Pump Relay.   The fuel pump relay controls the fuel pump and is in turn controlled by the ECA, Figure 10-46. When the ignition switch is turned on, the ECA grounds the fuel pump relay coil. If the ECA does not receive a PIP signal indicating engine cranking, it will turn the fuel pump relay off after about one or two seconds. Once the engine is running, the ECA keeps the fuel pump relay energized until the ignition is turned off or until the ECA no longer receives a PIP signal.

In series with the fuel pump relay is an inertia switch, Figure 10-47. Its purpose is to disable the fuel pump if the vehicle is involved in an accident to reduce the chance of a fuel fire. The inertia switch consists of a steel ball sitting in a metal bowl. A magnet holds the ball at the bottom of the bowl. In the event of a jolt severe enough to dislodge the ball, the ball strikes a lever above it. This lever triggers an overcentering spring opening the contacts of the inertia switch and holding them open. This opens the circuit to the fuel pump and turns it off. It also pops up a reset button on the top of the inertia switch. Pushing the reset button back down reverses the overcenter switch and closes the contacts again, enabling the fuel pump to run if the ECA energizes the fuel pump relay and sends it current.

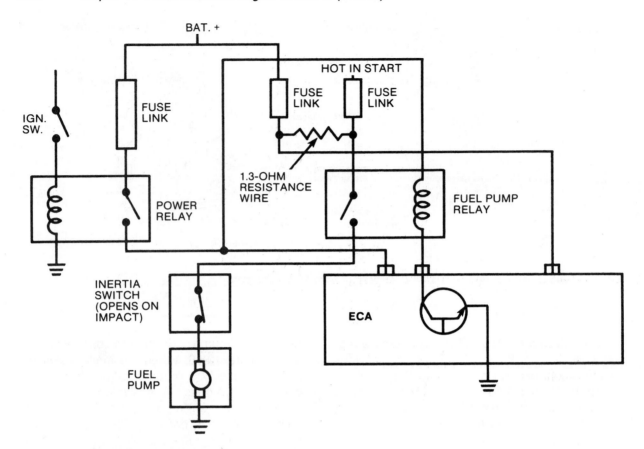

Figure 10–46 Fuel pump control circuit

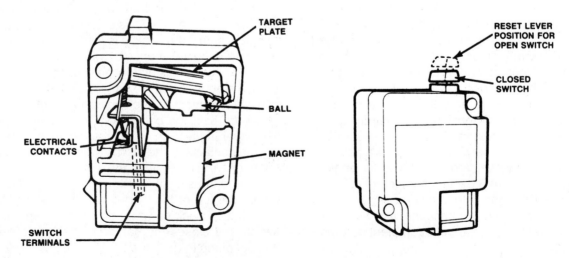

Figure 10–47 Inertia switch. *(Courtesy of Ford Motor Company.)*

Real World Problem

Anytime a fuel-injected Ford has suddenly refused to start, the technician should first check the inertia switch. Sometimes these switches have even been triggered off by another driver's parking by Braille, running into the now-disabled Ford with enough force to trigger the switch, but not enough to cause damage. Less frequently, badly worn shock absorbers on the rear axle can allow enough of a bump to trigger the inertia switch off.

This switch, located generally in the trunk, can also be damaged by moisture, so it is good practice to check to see the contacts are still clean where the harness connects. Never run a jumper wire around an inertia switch except for an in-shop test. Under no circumstances should a vehicle be released to a motorist with the inertia switch disabled.

Thick Film Integrated (TFI-IV) Ignition System

The **thick film integrated (TFI-IV)** ignition system takes its name from the TFI module bolted to the distributor housing. *Thick film* refers to the type of chip on which the module's circuit is printed. Its function is to turn on and off the ignition coil primary circuit and to control primary circuit dwell time. The spark output command from the ECA is the module's signal to open the primary circuit and fire the spark. If the spark output command fails to arrive, the TFI module will open the primary circuit in response to the PIP signal. This, of course, results in base timing only, with much less spark advance than is called for in practically all driving conditions.

Spark Output (SPOUT). The ECA receives the PIP signal from the Hall-effect switch in the distributor through the TFI module, Figure 10-48. From this signal the ECA determines engine rpm and crankshaft position. The ECA then modifies this signal (changes its spark advance timing) to achieve the best results considering the engine's speed, temperature, load, the atmospheric pressure, the EGR flow, and the air temperature. The time-modified signal is sent to the TFI module as the spark output (SPOUT) signal or command.

The SPOUT signal is a digital (on-off or square wave) signal, Figure 10-18. The leading edge of the on portion of the signal causes the TFI-IV module to open the primary ignition circuit and allows the coil's collapsing magnetic field to generate the high voltage in the secondary windings to fire the spark plug.

If the SPOUT signal does not arrive at the TFI-IV module, the module will select the base or reference signal from the PIP signal for its spark firing timing.

Computer Controlled Dwell (CCD). In early TFI-IV ignition systems, dwell (the duration of the time the coil primary circuit is energized) is controlled by the TFI-IV module. In other words, the TFI-IV module decides when to turn the coil on, and as we saw in SPOUT, the ECA tells it when to turn the coil off and fire the plug. The TFI-IV module includes circuitry that monitors engine rpm and increases dwell as engine speed increases. Beginning in 1989 for a few TFI-IV applications, the ECA instead controlled the ignition dwell. The trailing edge of the SPOUT signal tells the TFI-IV module to close the primary circuit, and the leading edge tells it to open the primary circuit and fire the plug, Figure 10-18.

The reason why the system varies the dwell with engine rpm is that all that is desired from the coil is full spark for the next plug firing. It takes a certain amount of time for the coil to fully build the magnetic field the collapse of which generates the high voltage secondary current. To keep the coil energized longer than needed is simply to use electric power as resistance heat to make the coil hot. Dwell control began much earlier than computer controls, beginning at least with the 1976 versions of electronic ignition. Even earlier, ballast resistors were used to reduce electric current through ignition primary circuits for the same reason: to keep from overheating the coil.

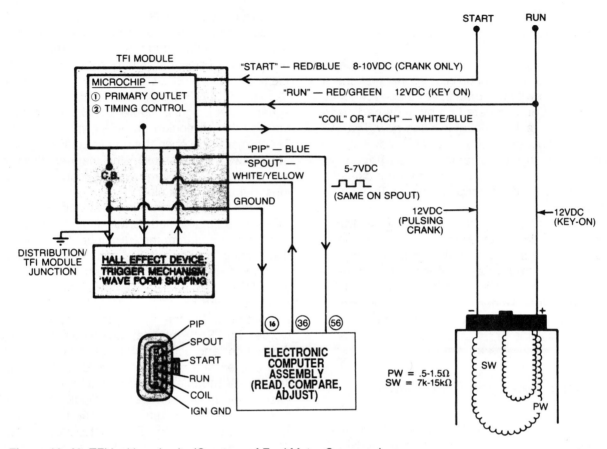

Figure 10–48 TFI ignition circuit. *(Courtesy of Ford Motor Company.)*

E-Core Coil. The TFI-IV ignition system uses an E-core coil differently shaped from the conventional coil used on prior Ford Motor Company products. The coil core resembles two capital E's turned face to face. The coil windings, molded in epoxy, are wound around the E's center horizontal bars. Because the center bars are shorter than the ones at the top and bottom, they do not quite touch each other and thus form a small air gap between them. The smaller air gap (compared to the distance between the ends of a conventional longitudinal coil core) helps the coil develop a much stronger magnetic field quickly. With no primary ballast resistor and the small core air gap, the E-coil core draws high primary circuit amperage (5.5 to 6.5 amps) and can put out as much as 50,000 volts. Even though secondary circuit amperage is very small, technicians should be very careful never to allow ignition secondary to shock them. On modern high energy ignition systems, such a shock can be more than just painful.

Octane Adjustment Rod. The early TFI-IV distributor has an octane adjustment feature. It has a rod inserted from the outside that changes its spark advance capacity, Figure 10-49. The standard rod has zero adjustment. In some cases when spark knock becomes a problem, Ford Motor Company has issued a technical bulletin authorizing the replacement of the stan-

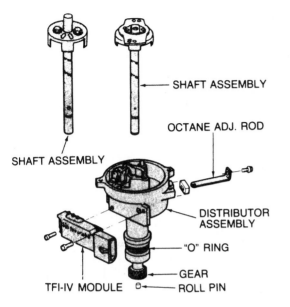

SHAFT ASSEMBLY

OCTANE ADJ. ROD

SHAFT ASSEMBLY

DISTRIBUTOR ASSEMBLY

"O" RING

GEAR

TFI-IV MODULE — ROLL PIN

Figure 10–49 Octane adjustment rod. *(Courtesy of Ford Motor Company.)*

dard rod with 3- or 6-degree spark retard rods.

Distributorless Ignition System (DIS)

3.0-Liter SHO and Supercharged 3.8-Liter Engines. In these systems, the ECA controls both dwell and timing. When the starter turns the engine, PIP and CID signals are fed to the DIS module and to the ECA. Once the DIS module receives a CID signal, it turns on coil 2 by providing a ground for the coil's primary circuit, Figure 10-19, in response to the next SPOUT trailing edge that it sees, Figure 10-50. Coil 2, Figure 10-51, builds a magnetic field while the DIS module waits for the next SPOUT leading edge. When the next SPOUT leading edge occurs, the DIS module opens coil 2's primary circuit and fires spark plugs 3 and 4.

The CID signal transition the DIS module waited for before it turned on coil 2 could have been a leading edge or a trailing edge, because either transition indicates that cylinder number 1 is at 26 degrees ATDC on either its power stroke

or its intake stroke. Following the CID transition, the DIS module selects coil 2 because coil 2 fires spark plugs 3 and 4 simultaneously, and cylinders 3 and 4 are the next cylinders at TDC, one on its compression stroke and the other on its exhaust stroke.

Having opened the primary circuit and fired coil 2, the DIS module now waits for the next trailing edge of the SPOUT signal. When that comes, the module grounds coil 3. When the leading edge of the SPOUT signal occurs next, the DIS opens the primary circuit of coil 3, firing its two plugs. Next it turns coil 1 on and off in response to the next SPOUT signal from the ECA. Obviously, the ECA controls when the cylinders fire by the time it raises the SPOUT signal voltage, and it controls ignition coil dwell by controlling how long the SPOUT signal voltage is low.

Once the DIS module has identified the number 1 cylinder, a counter circuit in the module keeps track of the coil firing sequence. The DIS module continues to monitor the CID signal to verify its cylinder counting circuit. The ECA continues to use the CID signal while the engine runs to control injector pulsing.

2.3-Liter, Dual Plug DIS. The DIS on the 2.3-liter engine works similarly to the 3.0-liter SHO and the supercharged 3.8-liter engines, but it has some unique features of its own, Figure 10-52. Because the 2.3-liter engine with DIS uses two spark plugs per cylinder (to achieve cleaner burning of the intake charge during combustion), the system has two separate coil packs: one on the right side of the engine and one on the left, Figure 10-53. The right coil pack contains coils 1 and 2. Coil 1 fires the right side spark plugs in cylinders 1 and 4, Figure 10-54. Coil 2 fires the right side spark plugs in cylinders 2 and 3. Coil 3 fires the left side spark plugs in cylinders 1 and 4, while coil 4 fires the left side spark plugs in cylinders 2 and 3.

After the engine starts, both plugs fire in each cylinder together: coils 1 and 3 fire together with coil 1 firing right plugs 1 and 4, and coil 3 firing left plugs 1 and 4. During start-up cranking, only the right coil pack fires. To prevent the left coil

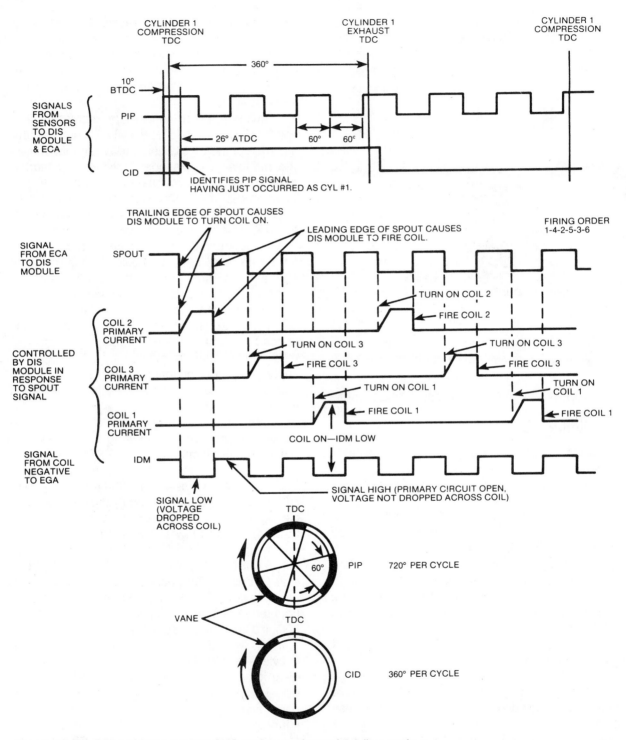

Figure 10–50 DIS coil firing 3.0-liter SHO and supercharged 3.8-liter engines

pack from working, the ECA sends a 12-volt signal to pin 6 of the DIS module, Figure 10-53. This is called the dual plug inhibit signal. With 12 volts at pin 6, the DIS module does not ground coils 3 and 4. After the ECA can tell the engine is running (from the rpm signal), it removes the 12-volt DPI signal, and the DIS module fires both coil packs. There is no spark advance during cranking.

As shown in Figures 10-50 and 10-52, when the SPOUT signal goes low, the primary coil circuit turns on; there is a large volt drop across the primary coil winding, so the voltage at the coil negative terminal, where the IDM originates, is near zero. Therefore, the IDM signal is low. When the SPOUT signal goes high, the coil primary circuit is open, and there is no volt drop across the winding; voltage at the coil negative terminal is battery voltage, and the IDM signal is high. While the 2.3-liter, dual plug engine is cranked and the system is in dual plug interrupt mode, the IDM signal is inverted, Figure 10-52. By monitoring the IDM signal, the ECA can tell whether or not

the DPI command is being responded to.

Because the 2.3-liter, dual plug engine does not use sequential fuel injection, the ECA does not need to monitor the crankshaft position in the engine cycle. So on this system, the CID signal is not sent to the ECA.

On either system, if when the engine starts cranking, the CID signal does not arrive at the DIS module, the module will randomly select the first coil to fire in response to a SPOUT signal. Once the first coil is selected, the other coils will follow in their normal sequence. If the random choice is wrong—if the computer guessed wrong when selecting a coil—the engine will not start. Each time the starter motor stops and restarts, the DIS module will make another random selection. When the random selection turns out to be the right coil, the engine will start. Since there are only two possibilities, it is unlikely that a great number of trials will be necessary on one of these systems.

If the SPOUT signal fails, the DIS module

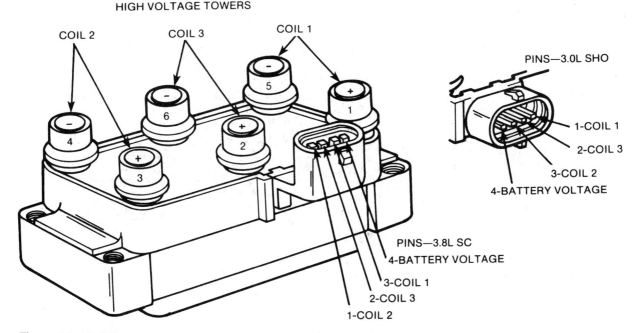

Figure 10–51 DIS coil pack, supercharged 3.8-liter and 3.0-liter SHO engines

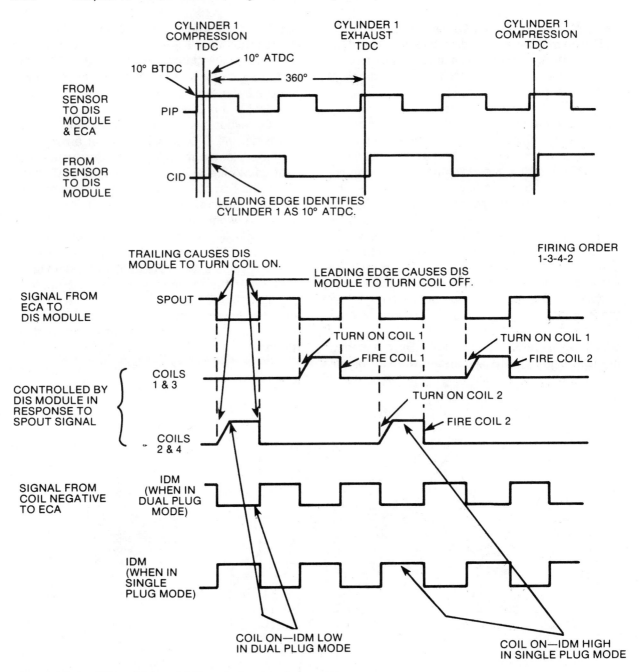

Figure 10–52 DIS coil firing, 2.3-liter, dual spark plug engine

fires the coils in response to the PIP signal. Like the TFI-IV module, the DIS module has a current limiter circuit. Primary current is limited to 5.5 amps +/- 0.5 amp.

Knock Sensor Response.   On systems using a knock sensor, the strength of the knock sensor's signal is proportional to the severity of the spark knock. When a knock signal is received by the ECA, it retards timing one-half degree per engine revolution until the knock disappears.

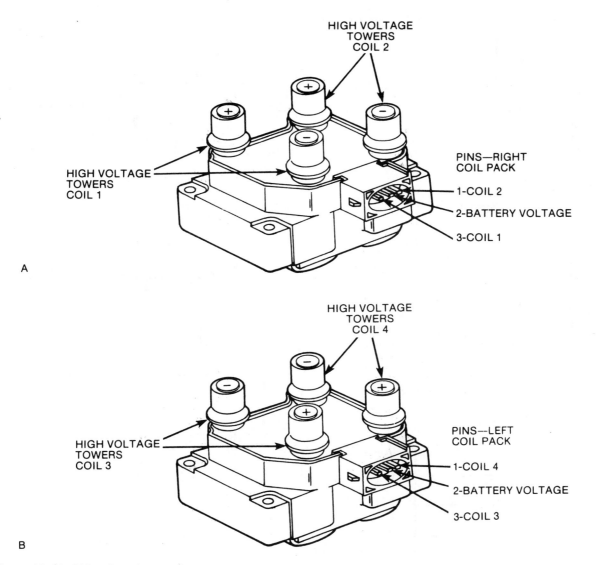

Figure 10–53 DIS coil packs, 2.3-liter dual plug engine

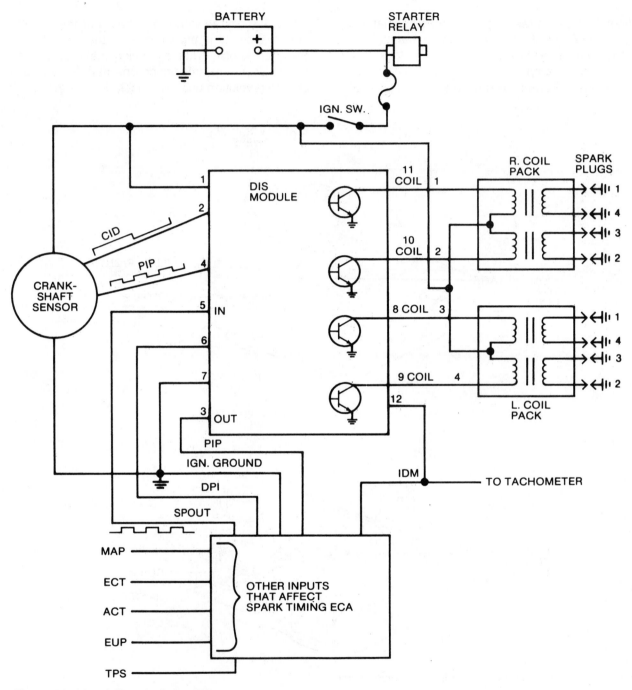

Figure 10–54 2.3-liter, dual plug DIS

Idle Speed Control

Three different systems are used for idle speed control among various EEC-IV applications:

Throttle Kicker. Throttle kicker is a vacuum-actuated device containing a diaphragm, a spring, and a plunger. It is mounted on the carburetor or CFI unit. On applications where the throttle kicker is used, normal idle speed is controlled by a conventional idle stop screw. However, when vacuum is applied to the diaphragm, the plunger extends and opens the throttle blade slightly to increase idle speed. It is sometimes referred to as a vacuum-operated throttle modulator (VOTM).

The ECA controls a solenoid that in turn controls vacuum to the throttle kicker. Part of the throttle kicker solenoid (TK solenoid) assembly is a normally closed vacuum valve, Figure 10-55. The term *normally closed* means that a device such as a switch or a valve is designed to remain closed unless it is opened by some force, in this case a solenoid. *Normally open* means just the opposite. When the solenoid energizes, the vacuum valve opens and allows vacuum to the throt-

tle kicker. The throttle kicker is activated during the following operating conditions:

- when engine temperature is below a specific temperature.
- when engine temperature is above a specific temperature.
- when the A/C compressor clutch is engaged.
- when the vehicle is above a specific altitude.

DC Motor Idle Speed Control (ISC). The ISC motor is a small, reversible electric motor. It is part of the assembly including the motor, a gear drive, and a plunger, Figure 10-56. When the motor turns in one direction, the gear drive extends the plunger; when the motor turns in the opposite direction, the gear drive retracts the plunger. The ISC motor mounts so the plunger can contact the throttle lever. The ECA controls the ISC motor and changes the polarity applied to the motor's armature to control the direction it turns, Figure 10-57. When the idle tracking switch opens (indicating throttle closed), the ECA commands the ISC motor to control idle speed. The ISC provides the correct throttle opening for cold or warm engine idle.

When the throttle opens and closes the ITS, the ECA commands the ISC motor to fully extend the plunger to function as a dashpot. In other words, when the throttle closes again during decel-

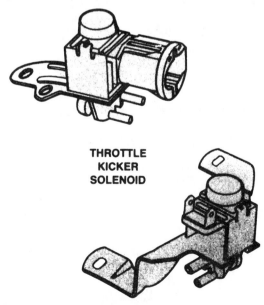

THROTTLE KICKER SOLENOID

Figure 10–55 Throttle kicker solenoid. *(Courtesy of Ford Motor Company.)*

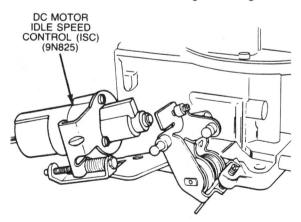

DC MOTOR IDLE SPEED CONTROL (ISC) (9N825)

Figure 10–56 ISC motor assembly. *(Courtesy of Ford Motor Company.)*

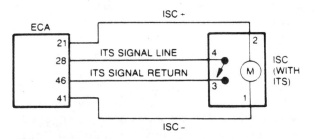

Figure 10–57 ISC motor circuit. *(Courtesy of Ford Motor Company.)*

eration, the plunger prevents the throttle from closing completely. The pressure on the plunger opens the ITS, and the motor in turn retracts the plunger to the normal idle position. This delayed throttle closing is to allow additional air into the manifold and completely evaporate fuel condensed on the manifold walls from the sudden high vacuum and temperature drop. The effect is to prevent a sudden very rich mixture on deceleration.

With the ignition off, the ECA commands the ISC motor to retract the plunger fully. This closes the throttle completely and prevents dieseling. Once the engine has completely stopped, the ECA commands the plunger to the fully extended (open valve) position in preparation for the next engine start.

Idle Air Bypass Valve Solenoid.   This idle speed control device is used on multipoint fuel injection systems. It includes an ECA-controlled solenoid operating a pintle valve, Figure 10-58. This bypass valve is so attached that when it is open, air passes from in front of the throttle butterfly blade through the pintle valve to the intake manifold side of the throttle. The ECA controls this valve by duty-cycling the solenoid. As the solenoid pulses, the valve opens. The ECA cycles the solenoid at whatever duty cycle the idle speed requires.

During a cold start, the ECA holds the idle air bypass solenoid valve at a 100% duty cycle (valve held wide open). During cranking, hot or cold, the bypass valve provides sufficient air so it is not necessary for the driver to touch the throttle. It also provides a throttle dashpot function as described above.

Thermactor Air Management

The Thermactor system is the injection system to deliver air to the exhaust manifold or the catalytic converter to aid in the reduction of HC, CO, and NO_X. During operations requiring a rich air/fuel mixture such as cold start-up or WOT, Thermactor air is dumped back into the atmosphere to avoid overheating the catalytic convert-

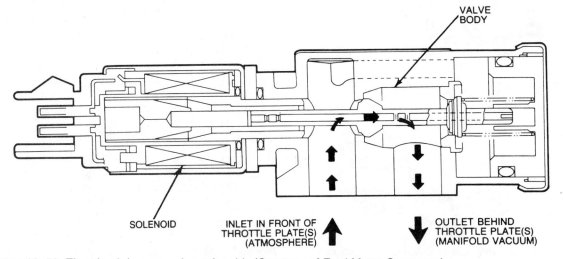

Figure 10–58 Throttle air bypass valve solenoid. *(Courtesy of Ford Motor Company.)*

er. One such condition for EEC IV and most other Ford electronic engine control systems is during engine idle. For this period, Ford tends to program the system for a rich idle mixture to maintain a good idle quality and prevent stalls.

EEC IV systems use one of two Thermactor systems. One is ECA-controlled and will be discussed here. The other is a pulse injection system using the negative pressure between exhaust pulses in the exhaust manifold (most effective and most common on four-cylinder engines) to draw air from the air cleaner past check valves into the exhaust manifold. This sys-

tem is sometimes called *Thermactor II* and is not ECA controlled.

Thermactor Air Bypass (TAB) and **Thermactor Air Divert (TAD) Valves.**   The air pump runs constantly, driven by a belt. This pump supplies air to the bypass valve, which either directs it to the divert valve or dumps it to the atmosphere, Figure 10-59. The divert valve can either direct the Thermactor air upstream to the exhaust manifold or downstream to the catalytic converter. The valves are activated by vacuum diaphragms controlled by ECA-controlled solenoids.

TAB and TAD Solenoids.   The TAB and

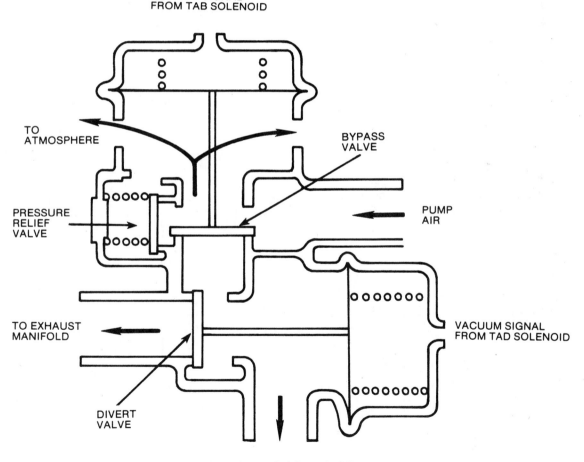

Figure 10–59 Combination TAB/TAD valves

TAD solenoids control whether or not vacuum is applied to the TAB and TAD valves. If the engine starts with a coolant temperature below 10 degrees Centigrade (50 degrees Fahrenheit), the ECA will leave the TAB solenoid circuit ungrounded. Vacuum to the bypass diaphragm is blocked, and the diaphragm spring holds the valve in the down position, bypassing pump air to the atmosphere. Once engine temperature reaches 10 degrees Centigrade, the ECA grounds the TAB solenoid circuit, Figure 10-60. Now vacuum is routed to the diaphragm, lifting the bypass valve. Injection air is now directed to the divert valve. At the same time, the ECA grounds the TAD solenoid, and it in turn directs vacuum to the divert valve diaphragm. The divert valve pulls back to close off the passage to the catalytic converter and direct the air to the exhaust manifold. Once the coolant reaches approximately 88 degrees Centigrade (190 degrees Fahrenheit), the system goes into closed loop, the ECA deactivates the TAD solenoid, and air is directed to the catalytic converter, behind the reducing bed and upstream of the oxidizing bed.

Certain operating conditions put the system back into bypass mode to protect the catalytic converter from overheating. These are:

- idle.
- failure of the EGO sensor signal to switch between rich and lean frequently enough (indicating that air/fuel mixture is not properly controlled or that the EGO sensor is cooling, or has become unresponsive with age

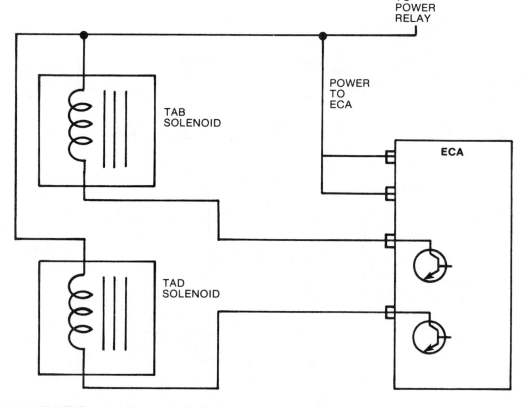

Figure 10–60 TAB/TAD solenoid control circuit

and exhaust deposits).
* acceleration or WOT operating conditions.

Canister Purge

Three types of canister systems are used on EEC IV vehicles. Some fuel injected engines (both CFI and multipoint) use a constant purge system with no ECA control. Some engines, particularly those with feedback carburetors, use an in-line, ECA-controlled, solenoid-operated vacuum valve. The third uses an ECA-controlled solenoid controlling vacuum to a ported vacuum switch (a temperature-controlled vacuum valve), which in turn controls the canister purge valve and an exhaust heat control valve.

In-line Canister Purge (CANP) Solenoid. This is a simple, normally closed, solenoid-controlled vacuum valve in the purge line connecting the charcoal canister to the intake manifold. Once the engine reaches normal operating temperature, the ECA duty-cycles the solenoid. The duty-cycle depends on the vehicle's operating conditions. This allows the canister purge to work at a controlled rate.

Canister Purge/Heat Control System. This system (very limited application) gets two-for-one from an ECA function. The ECA controls a normally closed, solenoid-operated vacuum valve. When the solenoid activates, vacuum is routed to the center port of a thermal vacuum switch, Figure 10-61. If the engine is cold, the thermal vacuum switch directs the vacuum to the top port, which connects to the heat control valve actuator. The actuator closes the heat control valve and forces exhaust gases through the crossover passage and warms the intake manifold plenum.

Once the engine warms, the wax pellet in the lower portion of the thermal vacuum switch expands and raises the valve up. It closes the top port and opens the lower port to the vacuum signal. The heat control valve opens; then the purge control valve opens and allows purging of the charcoal canister.

EGR Control

The ECA controls four slightly different EGR systems, depending on engine application and vehicle.

EGR Control (EGRC) and EGR Vent (EGRV) Solenoids. This system is a carryover from EEC III and is designed to control the amount of EGR flow (the valve can be open, closed, or anywhere in between). A position sensor (EVP) above the EGR valve tells the ECA where the EGR valve is. Two solenoids control vacuum to the EGR valve, Figure 10-62. The normally closed control solenoid allows manifold vacuum to the EGR valve when energized. The normally open vent solenoid allows atmospheric pressure into the vacuum line when not energized.

To open the EGR valve, the ECA grounds the circuits for both solenoids; the control solenoid routes vacuum to the EGR valve, and the vent solenoid blocks the vent passage. The EVP sensor reports EGR valve position to the ECA. The computer in turn manipulates the two solenoids to establish whatever EGR valve position it wants

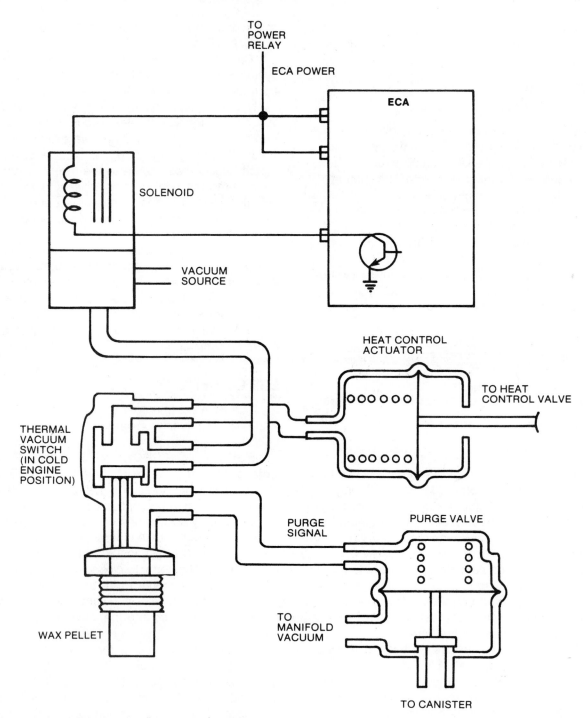

Figure 10–61 Canister purge/heat control system

for the current driving conditions. Both solenoids cycle constantly to achieve and maintain the desired EGR valve position.

EGR Shutoff Solenoid. This system uses one vacuum control solenoid operating like the vacuum control just described except it uses ported vacuum instead. When the solenoid is energized, vacuum moves the EGR valve; when the solenoid is not energized, vacuum is blocked. With this system the EGR valve may or may not have an EVP sensor. Those with the sensor work like the EGRV and EGRC solenoids combined into one, while those without the sensor are not modulated. They are either open or closed.

EGR Shutoff Solenoid with Back-pressure Transducer. This system uses one normally closed vacuum control solenoid like the system described above. The strength of the ported vacuum signal to the EGR valve is controlled, however by an exhaust backpressure variable trans-

ducer. The transducer is tied into the EGR valve's vacuum supply line and controls an air bleed into the line, Figure 10-63. Pressure signals are connected to the transducer from the exhaust system. The backpressure signal comes from a tap in the tube routing exhaust gas to the EGR valve. The control pressure signal comes from a point just downstream from the metering orifice allowing exhaust into the EGR supply pipe. These two pressures act on a valve in the transducer to position it to control the air bleed.

When exhaust pressure is low and vacuum is high, the air bleed opens and causes the vacuum signal to the EGR valve to weaken. As exhaust pressure increases and manifold vacuum decreases, the air bleed becomes more closed and strengthens the vacuum signal to the EGR valve. This way, the transducer controls how far the EGR valve opens and the consequent EGR flow. This system does not employ an EVP sensor.

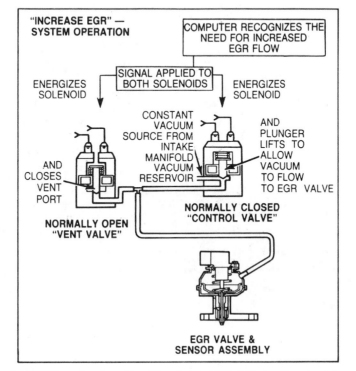

Figure 10–62 EGR control/EGR vent solenoids. *(Courtesy of Ford Motor Company.)*

EGR Vacuum Regulator (EVR)

This EGR control device, introduced in 1985, is simpler than the previous EGRC/EGRV control system. It works in a similar way, however. Like the EGRC/EGRV solenoid tandem, it allows the ECA to open the EGR valve to any degree through modulation of the signal. Since 1985, the EVR has become the standard EGR device used on most Ford engines.

The EVR is a single unit with two vacuum ports, Figure 10-64. The lower port connects to either manifold or ported vacuum depending on which engine is in the vehicle. The upper port connects to the EGR valve. The vented cover allows air to flow through the filter and down the hollow, electromagnetic core. This allows atmospheric pressure to be present above the metal disk at the bottom of the core at all times. The disk is held against the bottom of the core by a weak spring. When the engine runs and the EVR is not energized, engine vacuum creates a low pressure area below the disk. With this pressure differential, the spring cannot hold the disk against the

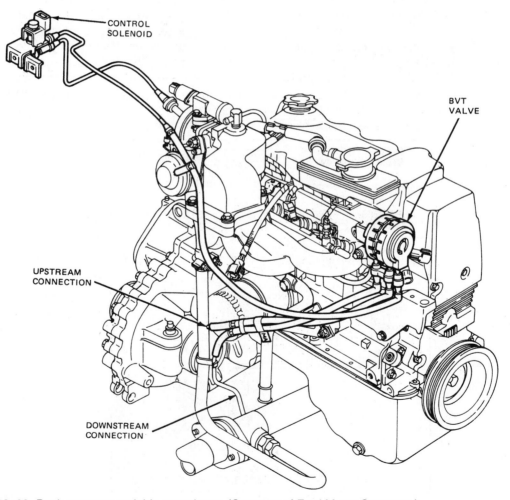

Figure 10–63 Back-pressure variable transducer. *(Courtesy of Ford Motor Company.)*

bottom of the core tightly enough to keep atmospheric pressure from filling the vacuum chamber. With atmospheric pressure in the vacuum chamber, of course, the EGR valve closes.

When the ECA wants to open the EGR valve, it grounds the EVR. The magnetic field produced in the coil pulls the metal disk tightly against the bottom of the core, sealing off the chamber below from atmospheric pressure. Engine vacuum now applies the EGR valve. By rapidly cycling the EVR on and off, the ECA can control the amount of vacuum applied to the EGR and thus how far the valve opens and the amount of gas recirculated.

The EVR is often referred to as a solenoid. But because its electromagnetic core does not move, it is really not a solenoid but rather a simple electromagnet.

Torque Converter Clutch Control

Some Ford automatic transmissions such as the A4LD, AXOD, 4EAT, and E40D employ an ECA-controlled torque converter lockup clutch. In each case, the ECA controls the converter clutch with a solenoid in the transmission, Figures 10-36 and 10-65. This solenoid moves a valve in the transmission valve body, a valve

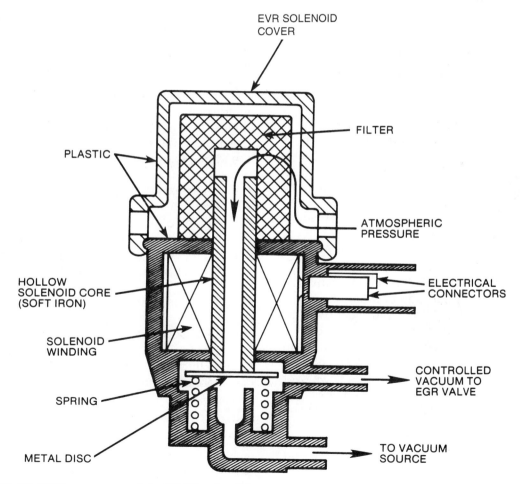

Figure 10–64 EGR vacuum regulator (EVR) solenoid

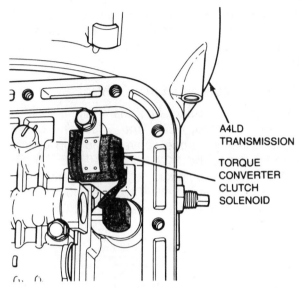

A4LD
TRANSMISSION

TORQUE
CONVERTER
CLUTCH
SOLENOID

Figure 10–65 Torque converter lockup clutch solenoid. *(Courtesy of Ford Motor Company.)*

routing transmission fluid under pressure into the torque converter and applying the lockup clutch.

The hydraulic circuitry applying the torque converter lockup clutch varies slightly between the different transmissions, but they work similarly. Figure 10-66 shows the hydraulic clutch apply circuit for the E40D transmission in the clutch-released mode. Line pressure to the converter clutch regulator valve feeds oil under pressure to the converter clutch control valve. With the converter clutch control valve in the right position as shown, line pressure oil is fed into the converter through the release passage between the transmission input shaft and the stator support. This introduces the oil into the converter on the engine side of the converter clutch plate. The pressure forces the clutch plate away from the engine side of the converter housing. The oil then flows over the clutch plate at its circumference and exits the converter through the apply passage between the stator support, and then flows to the transmission oil cooler.

When the ECA energizes the converter clutch solenoid, Figure 10-67, line pressure is applied to the right end of the converter clutch control valve. It then moves left. This reroutes the oil under pressure coming from the converter clutch regulator valve. Line pressure oil now feeds into the converter through the apply passage. With the transmission oil introduced into the converter on the transmission side of the clutch plate, the pressure must flow over the circumference of the clutch plate to get to the exhaust passage and exit the converter. This moves the clutch plate toward the engine side of the converter housing. As the friction material at the outer circumference of the clutch plate contacts the machined surface of the converter housing, it seals off the flow of oil to the exhaust passage. With no pressure on the engine side of the clutch plate and line pressure on the transmission side, the converter clutch plate is forced against the converter housing with enough force to lock it up and drive the transmission input shaft at engine speed.

There are slight variations in the torque converter clutch control strategy depending on the engine and transmission, but in general the following conditions must exist for the ECA to apply the torque converter clutch:

- engine warmed up to normal operating temperature.
- engine not under heavy load.
- part throttle cruise.
- engine within a predetermined rpm range.
- vehicle above a predetermined speed.

The solenoid will deactivate when:

- the BOO switch indicates the brakes are applied.
- the throttle position sensor indicates WOT.
- there is severe deceleration.

Turbocharger Boost Control Solenoid

Turbocharged 2.3-liter engines with EEC IV control employ an ECA-controlled boost control solenoid that allows boost pressure to rise from the normal 10 psi to as much as 15 psi. Boost pressure is controlled by a wastegate that opens at a specified pressure and allows some exhaust to bypass the exhaust turbine wheel, thus controlling the amount of boost pressure. The wastegate is controlled by a spring and diaphragm assembly (an actuator). Intake manifold pressure is routed to the actuator diaphragm. At the predetermined pressure, the diaphragm compresses the spring and moves a rod opening the wastegate.

If the ECA is satisfied with the current driving conditions, it will activate the normally closed solenoid and cycle it at 40 Hz. When the solenoid activates, it bleeds off pressure from the wastegate actuator and allows a higher intake manifold pressure before the wastegate opens.

A/C and Cooling Fan Control

Some engine applications use a separate electronic control module to control the A/C compressor clutch and the engine cooling fan (or fans). Others, typically rear-wheel-drive vehicles

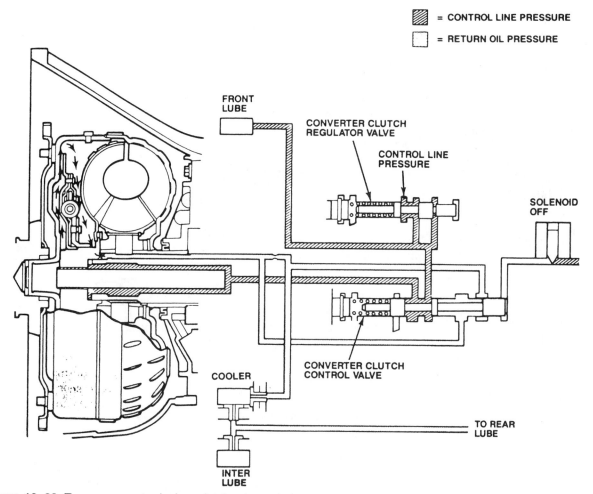

Figure 10–66 Torque converter lockup clutch released

with traditionally "north-south" mounted engines and belt-driven cooling fans, use a relay to disable the A/C clutch during WOT operation.

A/C and Cooling Fan Electronic Module. The module, Figure 10-68, is usually located in the passenger compartment. It gets input information from:

- the ECA (WOT signal).
- a coolant temperature switch (which closes at 105 degrees Centigrade [221 degrees Fahrenheit].
- the brake switch (brakes applied).
- the A/C switch (A/C on).

The module has a ground terminal and gets power from the battery through the ignition switch.

When the A/C switch is on (assuming the ignition is also on), the module provides power to the compressor clutch. During WOT operation,

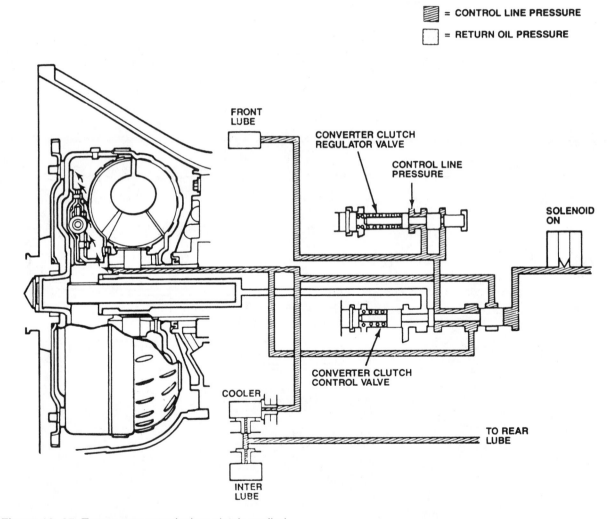

Figure 10–67 Torque converter lockup clutch applied

the ECA signals the module, which then disables the A/C clutch by deactivating it. If the WOT condition continues, the clutch stays disabled for about thirty seconds before normal A/C operation resumes. If the coolant temperature is below 105 degrees Centigrade, the module will also disable the engine cooling fan in response to the WOT signal. If the coolant switch is closed, the module will override the WOT signal from the ECA by refusing to disable the cooling fan.

If the vehicle has automatic transmission and power brakes, depressing the brake pedal sig-nals the module to disable both the A/C clutch and the cooling fan for five seconds. This is to prevent an engine stall from the load combined with the suddenly closed throttle.

Wide-Open Throttle A/C Cutoff Relay. The normally closed WOT cutout relay is in series with the A/C switch and the clutch coil. During WOT operation, the ECA grounds the WOT relay coil. The relay opens and thus disables the A/C clutch.

Inlet Air Solenoid (IAS)

A solenoid is used on some engines to con-

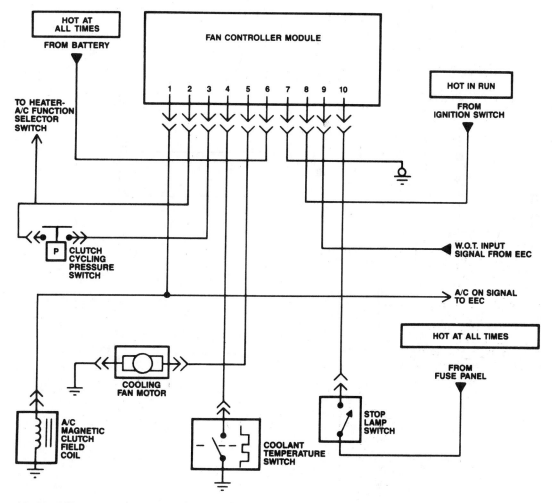

Figure 10–68 A/C and radiator fan control module. *(Courtesy of Ford Motor Company.)*

trol vacuum to the snorkel actuator controlling the air cleaner snorkel door. This function was traditionally controlled by a bimetallic vacuum valve in the air cleaner or a wax pellet thermostat. When the engine is cold, the ECA energizes the solenoid and routes vacuum to the actuator. The actuator moves the door to close the snorkel opening to ambient air. Heated air is drawn instead from a shroud around the exhaust manifold, through a duct to an opening in the bottom of the snorkel. This heated air helps evaporate fuel as it is introduced to the intake manifold.

As the engine and the air around it warm, the heated air mode is not needed. The ECA then deactivates the solenoid, vacuum to the actuator is blocked, and the door moves to open the snorkel's ambient air inlet and close the duct to the hot air.

Variable Voltage Choke

With some engines using a feedback carburetor, an ECA-controlled choke relay is used to more precisely control choke on-time, Figure 10-69. The choke cover holds an electric heating element to apply heat to the choke's bimetallic spring. When the engine starts at low temperature, the ECA

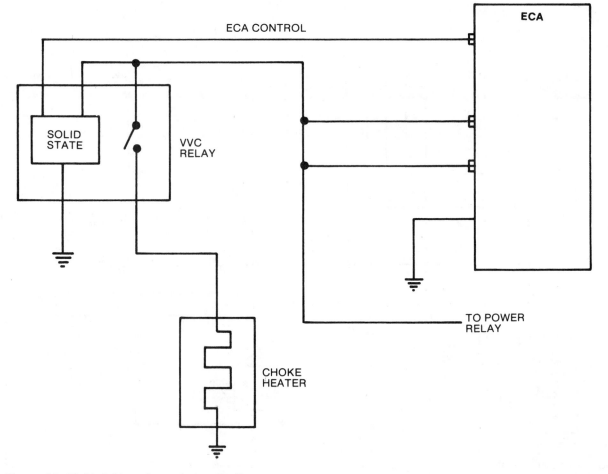

Figure 10–69 Variable voltage choke circuit

duty-cycles the relay every 2.5 seconds. When the relay contacts close, battery voltage applies to the choke heater element. When the engine reaches 27 degrees Centigrade (80 degrees Fahrenheit), the ECA holds the relay on continuously (100% duty-cycle). This allows the choke to stay on while the engine is at low temperature, but turns it off as quickly as the engine can warm up and run properly without it. This avoids the emissions problems caused by the super-rich mixture when the choke is on.

Temperature-Compensated Accelerator Pump (TCP) Solenoid

The 2.8-liter engine with a 2150A carburetor uses a TCP solenoid to control accelerator pump discharge. When the engine is below 35 degrees Centigrade (95 degrees Fahrenheit), the ECA does not ground the solenoid, and the accelerator pump functions normally. Once the engine coolant temperature reaches the specified temperature, the engine needs less fuel as the throttle is opened. The ECA energizes the TCP solenoid and routes vacuum to the accelerator pump. With vacuum applied, the pump cannot deliver as much fuel to the accelerator pump discharge nozzle.

Engine Fuel Injector Cooling Tube

In 1987 the 4.9-liter, six-cylinder in-line engine was introduced with multipoint fuel injection instead of a feedback carburetor. Because the 4.9-liter is not a cross-flow engine (the flow of gases does not enter as intake on one side and leave as exhaust on the other), the injectors are located just above the exhaust manifold. This exposes the injector and its connecting fuel line to much more heat than on most systems, particularly during the heat soak following a hot engine shutdown. Exposing the injector to this much heat can result in fuel vaporization in the injector and adjacent fuel line (vapor lock) or even seizing of the injectors. Either condition could result in hard starting, rough running after start-up, or no start at all. To prevent this, an injector cooling system was put on the 4.9-liter engine.

A tube runs almost the entire length of the engine, just above the injectors, Figure 10-70. Jets from the tube point to each of the first five injectors (injector six is not as exposed to exhaust manifold heat). A small blower blows air into the tube. The far end of the tube is closed so the cooling air is directed at the injectors. The fan is powered by an injector blower relay controlled by a module, Figure 10-71. This module is connected to a temperature-sensitive switch attached to the fuel rail. If the engine is shut down with a rail temperature of 77 degrees Centigrade (170 degrees Fahrenheit), the switch triggers the module, which turns on the fan. The module will allow the fan to run for a maximum of 15 minutes. It will turn the fan off if the temperature switch opens or if the ignition is turned on before the 15 minutes are elapsed. This system is not part of,

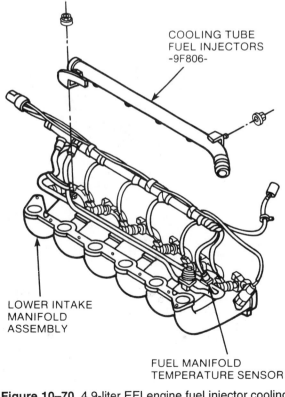

COOLING TUBE
FUEL INJECTORS
-9F806-

LOWER INTAKE
MANIFOLD
ASSEMBLY

FUEL MANIFOLD
TEMPERATURE SENSOR

Figure 10–70 4.9-liter EFI engine fuel injector cooling tube. *(Courtesy of Ford Motor Company.)*

nor controlled by, the EEC IV system. It does, as we have seen, affect start-up and driveability after heat soak.

Shift Light Indicator

Some EEC IV-controlled vehicles with manual transmissions feature a shift indicator light on the instrument panel to indicate to the operator the best time to shift to the next higher gear to optimize emissions quality. To a lesser extent,

this also indicates the best shift point for fuel economy. The ECA uses information concerning engine speed, manifold vacuum, and engine temperature to control the light. When the ECA sees an engine speed greater than 1,800 rpm with a light engine load, it grounds the indicator light circuit. The ECA will not turn on the light while the coolant temperature is low. When the transmission goes into top gear, a switch in series with the ECA and the indicator light opens to prevent the light from coming on again.

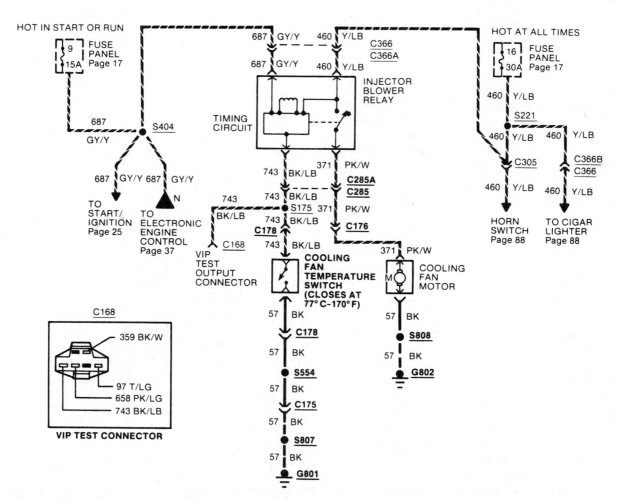

Figure 10–71 Cooling fan control circuit. *(Courtesy of Ford Motor Company.)*

VEHICLE SPEED CONTROL (VSC)

Beginning with limited applications in 1986, the cruise control system was integrated into the EEC IV system, a practice that expanded in later years. In 1988 this integration became universal for EEC IV systems. If the vehicle speed control (cruise control) is turned on, the ECA will compare vehicle speed to the commanded speed and will control two solenoids controlling vacuum to the speed control servo. The servo controls throttle position directly. These solenoids—speed control vacuum control (SCVAC) and speed control vacuum vent (SCVNT)—control vacuum to the servo just like the EGRC and EGRV solenoids (discussed earlier in this chapter) control vacuum to the EGR valve.

Top Speed Limiting

Some vehicles with the 5.0-liter engine, beginning in 1988, include a top speed limit. The ECA is programmed to control maximum vehicle speed to a value within the vehicle's original equipment tires. The ECA achieves this by reducing the fuel injectors' pulse width once the specified vehicle speed is reached.

Programmed Ride Control (PRC). Beginning in 1987, certain Fords were equipped with a PRC system. A switch on the instrument panel allows the operator to select either a plush or firm ride. The shock absorbers have adjustable damping valves, which are adjusted by rotating a small shaft extending up to the top of each shock absorber. A small electric motor mounted on top of each shock absorber turns the shaft. The motor is controlled by a PRC module. If while operating in the plush mode, certain operating conditions occur, the module will automatically adjust the shock absorbers to the firm position. Some of the information that identifies those conditions is supplied to the PRC module by the EEC IV's ECA.

Information supplied directly to the PRC module from sensors includes:

- hard braking—supplied by a pressure switch on the brake system.

- hard cornering—supplied by a sensor on the steering shaft.

Information supplied by the EEC IV ECA includes:

- more than 90% of WOT.
- speeds above 83 miles per hour.

✔ SYSTEM DIAGNOSIS & SERVICE

Malfunction Indicator Light (MIL)

Beginning in 1987, a limited number of Ford vehicles were built with a malfunction indicator light (MIL). The MIL is in the instrument panel and works much the same as the check engine light found on General Motors vehicles. Once the ECA detects a fault in one of the circuits it monitors, it turns the light on. In Self-Test, the MIL will also blink out the service codes.

The first step in diagnosing an EEC IV system problem is to verify the driveability complaint. Then go to the **Diagnostic Routine** index in the Diagnostic Routine section of the service manual and find the listing that most nearly identifies the driveability complaint. From that point on, the procedure provides a step-by-step approach to

EEC IV Diagnostic Features

One of the outstanding features of the EEC IV system is its self-diagnostic capacity. Like other systems, however, it cannot be totally self-diagnosing. The diagnostic features built into the ECA, combined with the diagnostic procedures in the service manual, will be very effective if the procedures are performed by a knowledgeable technician. The procedures seem confusing at first because of their extent and complexity, but studying and practicing the procedures make them a powerful and effective tool in diagnosing EEC IV system complaints.

finding the cause of the problem.

Pay particular attention to the pretest inspection and test preparation steps. Malfunctions in any of the engine's non-EEC IV systems can prevent the EEC IV system from functioning properly and can cause inappropriate service codes to be set.

Equipment Hookup.   All EEC IV-equipped vehicles have a Self-Test connector somewhere in the engine compartment, Figure 10-72. This connector can be connected to an analog voltmeter (Figure 10-73) or the **Star Tester** which can be purchased through a Ford dealer or from one of several other test instruments manufactured by test equipment manufacturers. The ECA displays service codes on the test instrument as a digital readout (or as needle sweeps on the analog voltmeter). A timing light should be connected at the same time. Consult the service manual or the test equipment manufacturer's instructions for specific directions on how to conduct the Self-Test.

Service Codes

The ECA monitors most of the input and output circuits. For any given set of driving conditions, it expects to see voltage values within a specified range on each monitored circuit. If the voltage is too high or too low, it will be construed as a fault (open or short), and recorded as a two-digit service code in the ECA's memory. If the fault does not recur within the next twenty ignition cycles (ignition on, then off), the ECA will erase the recorded service code. The ECA reads the service code to the technician through a special test instrument or an analog voltmeter when it is put into the Self-Test mode.

Types of Codes.   Six different types of service codes are stored in the ECA:

• *Fast codes:* These are codes used by test equipment at the vehicle assembly plant. They are used to test each vehicle on the assembly line. They contain the same information as the other service codes, but occur much too fast for most test equipment to read them. They can be seen as a slight needle quiver on an analog voltmeter or as a flicker

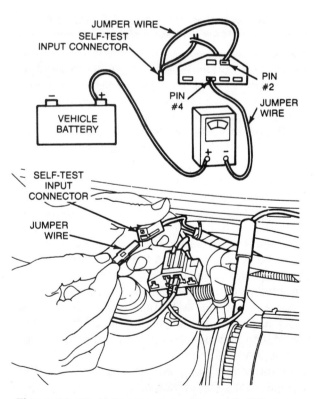

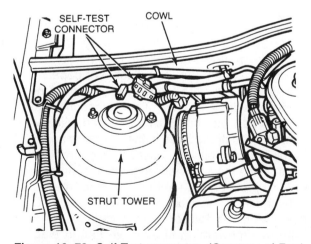

Figure 10–72 Self-Test connector. *(Courtesy of Ford Motor Company.)*

Figure 10–73 Voltmeter connection to Self-Test connector. *(Courtesy of Ford Motor Company.)*

of the LED on the harness of the Star Tester.

- *On-demand codes* (sometimes called hard codes): These codes refer to faults existing when the Self-Test is ongoing.
- *Separator code:* This code (represented by the number 10) only occurs during the key on/engine off segment of the Self-Test. It indicates to the technician that the on-demand codes are completed, and that memory codes are coming next.
- *Memory codes* (sometimes called intermittent codes or continuous codes): These codes refer to faults the ECA detected during normal vehicle operation, but that are not present during the Self-Test.
- *Dynamic response code:* This code only occurs during the engine running segment of the Self-Test and is represented also by the number 10 (the meaning of code 10 depends on the Self-Test segment in which it appears). It tells the technician that he has fifteen seconds to quickly press the accelerator pedal to the floor and release it. This allows the ECA to check the TPS, MAP, and the VAF for proper response. This is often called the "Goose Test."
- *Engine identification codes:* These codes have no useful purpose for the technician. Like the fast codes, they are intended for assembly plant tests. They indicate to the factory test equipment how many cylinders the engine has. The number 20 means four cylinders, 30 means six, and 40 means eight.

Reading Service Codes with a Voltmeter. Reading service codes on an analog voltmeter requires close attention because it is so easy to miscount when higher numerical codes display. It requires counting needle sweeps and estimating the time lapse between sweeps. A code 21 (hard code) followed by a code 10 (separator code) and then an 11 (no memory codes stored) appears as follows: the needle sweeps upscale and back. Then it immediately swings up and back again. It swings nearly full scale on a 12-volt scale. A two-second pause indicates the first digit

is complete, and the two sweeps indicate the digit is 2. After the two-second pause, one sweep occurs and is followed by a four-second pause. The single sweep indicates that the second digit is 1. The four-second pause indicates the code is complete. After the four seconds, the 21 is repeated. A six- to nine-second pause precedes the memory codes. A single sweep followed by a two-second pause and another sweep indicates a code 11. A four-second pause occurs and the code 11 is repeated.

The engine running segment of the Self-Test is presented in the same format except that a six- to twenty-second pause separates the engine identification code from the dynamic response code and a four- to fifteen-second pause separates the dynamic response code from the hard codes, Figure 10-74.

Quick-Test/Self-Test

The **Quick-Test** is a comprehensive set of procedures designed to help the technician identify problems in the EEC IV system. As presented in the Ford Service Manual, it includes preparation for testing (visual inspection and warming up the engine to operating temperature), Diagnostic Routines for symptoms without codes, and the Self-Test and Pinpoint tests to locate the exact cause of faults previously identified. The Self-Test is the part of the Quick-Test in which the ECA tests itself and its related circuits. It presents service codes for any faults it finds.

Key On/Engine Off. This is the first segment of the Self-Test. It is triggered by connecting the single-lead portion of the Self-Test connector to pin 2 of the multiple-lead part of the Self-Test connector and turning the ignition on. If a voltmeter is used to read codes, this will be done with a jumper lead. If a digital readout instrument such as the Star Tester is used, this connection is made by the button or switch as indicated in the instrument instructions. In this segment, the ECA tests first the input (sensor) circuits. Any faults detected are reported as on-demand codes. A code 11 (system pass code) indicates that no input circuit faults were found. Any hard codes identified should be repaired using the appropriate service manual before proceeding to the computed ignition timing check. After the repair is complete, repeat the test segment to verify the effectiveness of the repair.

Continuous Self-Test (Still in Key On/Engine Off Segment). The on-demand code or code 11 ("all clear"), is followed by the separator code, 10, meaning the on-demand codes are completed and memory codes (intermittent faults) are coming next. If no memory codes are stored, a code 11 will be displayed. If memory codes are displayed, they should be recorded by the technician, but no repairs should be performed at this point.

Output Cycling Test (Still in Key On/Engine Off Segment). At this point, if the throttle is pushed to WOT and released, the ECA will go into the Output Cycling Test. In this segment, the ECA will turn most of the actuators on. If the throttle is pushed to WOT and released again, it will turn the actuators off. This can be repeated as often as desired. This step, however, is not normally used unless called for by one of the Pinpoint Routines, a later part of the Quick-Test. If the Output Cycling Test is called for in one of the Pinpoint Routines and to determine whether one of the actuators is working properly, the Key On/Engine Off Test can be reentered. It is only in this segment that the Output Cycling Test can be initiated.

Computed Timing Check. This is the second segment of the Self-Test and is designed to test the ECA's ability to control ignition timing. It is initiated by having the ignition off for at least ten seconds, activating the Self-Test, and starting the engine. Any codes displayed at this time should be ignored. In this segment the ECA holds the timing at 20 degrees above base timing for two minutes. If for example, the base timing is 10 degrees BTDC, the timing will read 30 degrees BTDC. Check the timing with the timing light connected when the rest of the equipment is hooked up.

Be sure the in-line base timing connector, Figure 10-75, is reconnected correctly.

Engine Running Test. The third segment of the Self-Test tests actuators (output devices) and some sensor circuits for hard faults under

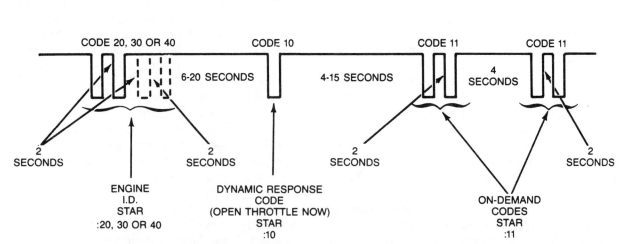

Figure parts: SELF-TEST OUTPUT CODE FORMAT ENGINE-RUNNING

Figure 10–74 Typical service code display during Engine Running segment of Self-Test. *(Courtesy of Ford Motor Company.)*

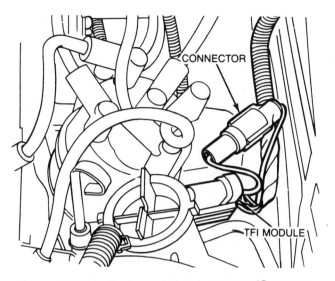

Figure 10–75 In-line base timing connector. *(Courtesy of Ford Motor Company.)*

dynamic conditions. Before going into this segment, the Self-Test should be deactivated and the engine started and run at 2,000 rpm for two minutes to warm the oxygen sensor to a point when it works normally. Turn the engine off and wait ten seconds. To activate the Engine Running segment, activate the Self-Test and restart the engine. The next display to appear is the engine identification code followed by the dynamic response code (10). The diagnostic technician now has fifteen seconds to move the throttle to WOT and release it. The ECA next issues fast codes followed by the on-demand codes. If no hard faults are observed during the engine running segment, a code 11 will display.

Continuous Monitor Test. This test, the last of the Self-Test segments, is often called the Wiggle Test. It is used to locate intermittent faults. It can be used in either of two modes: engine off or engine running. To initiate the engine off mode, turn the ignition on with the test instrument connected but the Self-Test deactivated. Begin wiggling (pulling, twisting, etc.) the wires of the sensor circuit indicated by the intermittent fault code obtained during the key on/engine off segment. Bump the sensor, manipulate any of its moving

parts, and apply heat from a heat gun to any thermistors involved. If this manipulation reproduces the intermittent open or short to which the ECA was responding when it set the code, the LED in the Star Tester harness will either flicker off, or stay off. If a voltmeter is used, the needle will pulse or swing upward and stay there. Obviously, an assistant is often useful in this test.

The engine running mode of the Wiggle Test is automatically entered two minutes after the last code displays in the engine running segment of the Self-Test if the engine is not turned off first.

Cylinder Balance Test. This is a feature of the Self-Test on the sequential injection engines. It is intended to help the technician identify a weak or noncontributing cylinder. By depressing and releasing the throttle within two minutes after the last code has been displayed in the engine running portion of the Self-Test, the system goes into the Cylinder Balance Test mode.

The ECA fixes the duty cycle rate of the idle air bypass solenoid so a stabilized idle speed is established. Engine rpm is monitored and stored in memory. Then the ECA stops pulsing the injector of the highest number cylinder. With the injector shut off, engine rpm is again measured and stored in memory. This is repeated for each cylinder with cylinder number one the last tested. The ECA then selects the lowest rpm value obtained when the injectors were selectively and sequentially shut off. It subtracts that value from the original rpm with all the injectors working. The greatest rpm drop is then multiplied by a calibrated percentage. For example, if 125 rpm was the greatest rpm drop:

$$125 \text{ rpm} \times 65\% = 81 \text{ rpm}$$

If all cylinders dropped at least 81 rpm, a code 90 displays, indicating a pass.

During the 1987 model year, the Cylinder Balance Test strategy in the ECA was expanded to allow two additional levels of the test. Level 1 might require all cylinders to drop at least 65% of what the most efficient cylinder dropped to pass the test as shown in the example. Level 2 might require all cylinders to drop at least 43% of what the most efficient cylinder dropped to pass, while level 3 might require all cylinders to drop at least 20% of what the most efficient cylinder dropped to pass. As seen in Figure 10-76, if a cylinder fails to pass level 1 but passes level 2, it is not as efficient as it should be, but it is working. If it fails levels 1 and 2 but passes level 3, it is barely working. If it fails all three levels, it is very weak or dead. If all cylinders pass level 1, they are all working efficiently and no further testing is required.

When level 1 has completed and code 90 for a pass or a code representing a cylinder number that fails has been displayed, briefly pressing the throttle part way down within two minutes will put the system into level 2 of the Cylinder Balance Test. Pressing the throttle again within two minutes after level 2 has completed will initiate level 3. After level 3 has completed, pressing the throttle again within two minutes will repeat the cylinder balance tests of level 3.

Pinpoint Procedures. This is the last part of the Quick-Test. It provides directed steps to locate the exact cause of the faults that have set the codes. If the Quick-Test steps are followed in the applicable service manual, a **Pinpoint Procedure** will be referred to for each service code requiring corrective action. The Pinpoint Procedures also provide directions that aid in finding the cause of driveability problems without accompanying service codes. The symptoms of such problems are listed with suggested things to check.

Repairs should always be made in the order in which their codes are displayed. After the repair, the Self-Test should be repeated to confirm the effectiveness of the repair. The applicable service manual should be used for each vehicle application to be sure all the Quick-Test steps match the engine, vehicle, and model year. For instance, the 1983 EFI engine does not store memory codes.

Failure Mode Effects Management (FMEM). FMEM is a strategy programmed into the ECA, beginning in mid-1986, to maintain a good engine performance if one of the most critical sensors fails. Prior to 1987, such a sensor failure would either stop the engine or put the system into Limited Operational Strategy (in which the engine runs with no spark advance and a fixed air/fuel mixture). With FMEM Strategy, the ECA will in most cases be able to replace the failed sensor's input with a plausible substitute value from another sensor. It will also set a code 98 in its Continuous Self-Test memory. Code 98, in 1987, indicates that a fault exists and that FMEM is functional.

Car	Level			Indication
	1	**2**	**3**	
	Code	Code	Code	
A	40 cyl 4 did not pass	90 all cyl passed	level 3 not needed	cylinder 4 not as efficient as it should be but is working
B	50 cyl 5 did not pass	50 cyl 5 did not pass	90 all cyl passed	cylinder 5 barely working
C	70 cyl 7 did not pass	70 cyl 7 did not pass	70 cyl 7 did not pass	cylinder 7 very weak or dead
D	90 all cyl passed	level 2 not needed	level 3 not needed	all cylinders working efficiently

Figure 10–76 Cylinder bypass test

The operator may not know a fault exists except for the following:

- idle speed may be fixed.
- the cooling fan may run all the time.
- California vehicles have a Malfunction Indicator Light on the instrument panel that will come on when the fault occurs.

All 1987 and later EEC IV vehicles employ FMEM Strategy. Figure 10-77 shows the sensors included in the FMEM Strategy and the action taken in response to each one's failure.

Self-Test Improvements. Several improvements have been made in the Self-Test over the years. Some new codes have been added, and some codes have been made more reliable. The Self-Test and the continuous monitor have the capacity to check more functions. Consequently, code numbers may change meaning from year to year. The technician must have current published information for the vehicle worked on.

Failed Sensor	Action Taken
MAP	Substitute TPS value. If TPS is faulty, use a fixed constant value. Disable Adaptive Learning. Disable EGR. Fixed idle speed.
TPS	Substitute fixed values during cranking, MAP value after engine starts.
ECT	Substitute ACT value during warm-up, fixed value after warm-up. Disable EGR. Disable Adaptive Learning. Fixed idle speed.
ACT	Substitute ECT value during warm-up, fixed value after warm-up. Disable EGR. Disable Adaptive Learning. Fixed idle speed.
EVP	No substitute. Disable EGR.

Figure 10–77 FMEM strategy

Diagnostic & Service Tips

Here are a few final suggestions when working on Ford vehicles with an EEC IV system:

- Pinpoint Procedures should not be used unless the Quick-Test directs you to do so. Each Pinpoint Procedure assumes a fault has been detected in the circuit considered. To use them without being so directed by the Quick-Test can result in unnecessary and thus ineffective repairs and the replacement of functional parts. Do not replace parts unless specific test procedures indicate they are defective.
- Do not measure voltage or connect a test light at the ECA harness connector unless specifically directed to do so by a particular test procedure. Otherwise, damage to the ECA or to sensors or actuators could result.
- Isolate both ends of a circuit and turn the ignition off before testing for shorts or continuity unless otherwise directed by a specific test procedure. This same rule applies to solenoids and switches. All circuit resistance tests must be made with the circuit electrically separated from the rest of the system, unless you are using voltage drop, resistance-by-inference techniques.
- An open circuit is any resistance greater than 5 ohms and a short is any path to ground with less than 10,000 ohms unless specifically detailed otherwise.

Adaptive Strategy and Driveability. If for some reason the continuous battery power to the ECA is removed, the adaptive strategy learning is erased. The calibration values stored in the KAM revert to the original look-up table values in the read-only memory. Because the learned values compensated for some engine wear or component inefficiency, the vehicle will possibly have driveability problems such as surging, hesitation, or poor idle until the ECA can relearn.

A parallel problem—one not always understood by inexperienced technicians—is that if a component replaced was one the adaptive strategy had compensated for, the vehicle might have driveability problems with the new, fully functional component until the ECA has an opportunity to learn the characteristics of the new or repaired part. To avoid this problem after replacement or repair of significant system components, interrupt battery power to the KAM for at least three minutes during or after such a repair. This is particularly important for major engine component replacements.

MAP/BP Relocation. Beginning in 1987, MAP and/or BP sensors were repositioned so they sit in a horizontal position with the vacuum line sloping downward to its manifold connection. This prevents fuel migration into the sensor. Fuel getting into these sensors has been responsible for many pressure sensor failures.

SUMMARY

In this chapter, we have covered the three operational strategies of the EEC IV system, the longest built of the Ford engine control systems (since 1982). We have reviewed the major sensors, for engine coolant temperature, engine speed, throttle position, manifold pressure, and residual oxygen in the exhaust. We have seen the ignition system unique to Ford, with the SPOUT and PIP signals. We saw how they work and what their input information signals to the ECA look like.

We have reviewed all the actuators and the three different kinds of fuel metering devices on EEC IV systems, the fuel jet interruption type needles, and the mixture solenoids providing air to the emulsion circuits. We saw which components most directly affected emissions and which were more concerned with driveability or fuel economy.

Finally, we looked at the major aspects of the diagnostics and repair of the EEC IV system.

▲ DIAGNOSTIC EXERCISE

What would be the effect on the system if an air pocket surrounded the engine coolant temperature sensor? Would the fuel mixture go rich or lean?

In very cold (freezing) weather a feedback carbureted car has good driveability when first started, then the engine runs rough for a relatively long time before it starts working properly again. What could be the cause of these symptoms?

If power to the computer is lost in one of the early EEC systems, can the engine stay running? Could it restart once shut off? Explain how the actuators would get their signals in your answer.

REVIEW QUESTIONS

1. Name three types of fuel-metering devices used on various EEC IV systems.
2. In what part of the ECA are the memory codes stored?
3. What type of signal does the MAP sensor produce?
4. Name at least three sensors used on EEC IV systems that are variable potentiometers.
5. Name at least three sensors used on EEC IV systems that are thermistors.
6. Name at least three sensors used on EEC IV systems that are on/off switches.
7. Name at least one actuator on an EEC IV system that is duty-cycled and one operated by pulse width.
8. What is the purpose of the IMS (inferred mileage sensor)?
9. What is the function of the PIP signal?
10. What is the function of the SPOUT signal?
11. Name three types of idle speed devices used on EEC IV systems.
12. What is the function of the TAD solenoid?

13. Although the EGR valve, Thermactor air, and the canister purge valve influence engine performance, they have a greater impact on _____.

14. What is the function of the VVC (variable voltage choke)?

15. What is the function of the IAS (inlet air solenoid)?

16. What is the function of the TCP (temperature compensated accelerator pump)?

17. List the order in which readable codes are displayed during the Engine Running segment of the Self-Test.

18. Name two points during the Self-Test when the Wiggle Test may be initiated.

19. Name two different criteria for using the Pinpoint Test.

ASE-type Questions (Actual ASE test questions rarely are so product-specific.)

20. Technician A is working on a vehicle equipped with an EEC IV system. He accidentally reverses polarity while setting up a set of jumper cables, and he fears he may have damaged the ECM. Technician B says that a diode in the power relay probably prevented damage from occurring to the ECA. Who is more likely right?
    a. A only.
    b. B only.
    c. both A and B.
    d. neither A nor B.

21. Technician A says that if the PIP signal is lost on an injected EEC IV system, the engine will not run. Technician B says that if the SPOUT signal is lost, the engine will operate on base timing. Who is correct?
    a. A only.
    b. B only.
    c. both A and B.
    d. neither A nor B.

22. Of the four steps listed here (Diagnostic Routine, Verify Complaint, Pinpoint Procedure, Self-Test), Technician A says they should be performed in the sequence in which they were just listed. Technician B says they should be performed in the following order: Verify Complaint, Self-Test, Diagnostic Routine, and Pinpoint Procedure. Who is correct?
    a. A only.
    b. B only.
    c. both A and B.
    d. neither A nor B.

23. During the key on/engine off portion of the Self-Test, Technician A says that on-demand codes are displayed followed by a separator code (10), which is followed by the memory, if any. Technician B says the on-demand codes display faults currently present and that the memory codes represent the intermittent faults. Who is correct?
    a. A only.
    b. B only.
    c. both A and B.
    d. neither A nor B.

24. Technician A says that of all the steps of the Self-Test, there are three intended to be used at the discretion of the technician or if directed to by one of the Pinpoint Procedures. Technician B says she can only recall two: the Output Cycling Test and the Wiggle Test. Who is correct?
    a. A only.
    b. B only.
    c. both A and B.
    d. neither A nor B.

Chapter 11

Recent Ford Motor Company Engine Control Systems

OBJECTIVES

In this chapter you can learn:
- ❏ the features of the PCM, the new OBD II-compliant computer used on Ford Taurus/Mercury Sable.
- ❏ the principal inputs and outputs for the new Ford system.
- ❏ modified features of some sensors and actuators previously used in Ford's computer engine controls.
- ❏ revised or updated features of the emissions controls on the system.
- ❏ an approach to the industry-standard OBD II self-diagnostic systems.

KEY TERMS

Alternate Fuel Compatibility
Freeze-Frame/Snapshot
Inertia Switch
OBD II
Standardized Diagnosis

To see the latest computerized engine controls from the Ford Motor Company, we will look at the 1996 Ford Taurus, a redesigned vehicle for the model year and one incorporating, as all vehicles must for 1996, all the details of **OBD II** diagnostics. Since the basics of this new diagnostic system are covered in Chapter 18, we will restrict mention here only to those features unique to the Ford, also referred to as EECV

ENGINE CONTROLS

Powertrain Control Module (PCM)

The main computer on the vehicle is the PCM, Figure 11-1. It controls the fuel mixture delivery and the ignition throughout the range of the engine's operational capacity. The name, of course, is OBD II mandated: it is functionally similar to that in other vehicles, but it is unique to the Ford systems. It monitors the various sensors and switches whose information is relevant to calculating the proper values for fuel injector pulse widths and firing sequence, as well as triggering actuation of the various components that perform its combustion commands.

The Ford PCM determines the gradual wear and age of the vehicle over time and any changes in altitude or barometric pressure; it can then make compensatory adjustments in its own programs to correct for whatever needs to be altered because of the changed input information.

The principal inputs to the PCM are the throttle position sensor, the mass air flow sensor, the intake air temperature sensor, the engine coolant temperature sensor, the oxygen sensor, the camshaft position sensor, the knock sensor, the power steering pressure switch, the vehicle speed sensor, and the idle air control valve (which is basically an actuator, but since the PCM remembers its position, this information should be regarded as an input).

Powertrain Control Module (PCM)

The powertrain control module (PCM)(12A650) performs the following functions:

- The PCM accepts input from various engine sensors to compute the required fuel flow rate necessary to maintain a prescribed air/fuel ratio throughout the entire engine operational range.
- Then the PCM outputs a command to the fuel injectors (9F593) to meter the appropriate quantity of fuel.
- The PCM determines and compensates for the age of the vehicle and its uniqueness, also automatically senses and compensates for changes in altitude (i.e. from sea level to mountains).

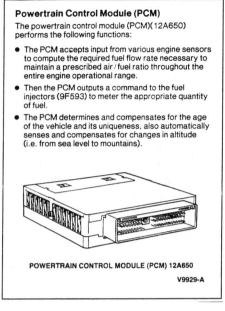

POWERTRAIN CONTROL MODULE (PCM) 12A650

V9929-A

Figure 11–1   Powertrain control module (PCM). *(Courtesy of Ford Motor Company.)*

INPUTS

Throttle Position Sensor. On the side of the throttle body is a variable potentiometer that provides a return signal to the PCM directly proportional to the position of the throttle plate, the open angle, Figure 11-2. This works in the same way TP sensors have since 1981, and the tests are still quite similar—you check for continuous variation in the resistance and return voltage throughout the throttle travel.

Mass Air Flow (MAF) Sensor. The Ford Taurus uses a hot-wire type mass air flow sensor, Figure 11-3. The hot wire in the airstream is kept at a constant temperature above the ambient air, and a hot wire sensing element measures exactly how much current is required to maintain that heat. The sensor then sends an analog voltage signal to the PCM, which corresponds to the mass of the intake air. Of course, as with all such air mass sensors, it is very important to make

Throttle Position (TP) Sensor

The throttle position sensor (TP sensor)(9B989) is:

- a potentiometer that provides a signal to the powertrain control module (PCM)(12A650). This signal is directly proportional to the throttle plate position.
- mounted to the throttle body (9E926) and is connected to the throttle plate shaft.

THROTTLE BODY SHOWN UPSIDE DOWN FOR CLARITY

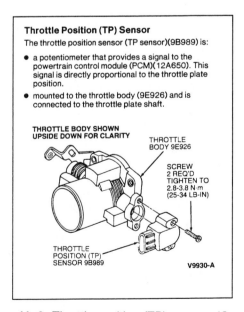

THROTTLE BODY 9E926

SCREW 2 REQ'D TIGHTEN TO 2.8-3.8 N·m (25-34 LB-IN)

THROTTLE POSITION (TP) SENSOR 9B989

V9930-A

Figure 11–2 Throttle position (TP) sensor. *(Courtesy of Ford Motor Company.)*

Mass Air Flow (MAF) Sensor

The mass air flow sensor (MAF sensor)(12B579):

- is located between the engine air cleaner (ACL)(9600) and the throttle body (9E926).
- uses a hot wire sensing element to measure the amount of air entering the engine. Air passing over the hot wire causes it to cool.
- sends an analog voltage signal to the powertrain control module (PCM)(12A650) to determine the intake air mass. The powertrain control module will then calculate the required fuel injector pulse width in order to provide the desired air/fuel ratio.

The mass air flow sensor hot wire sensing element and housing are calibrated as a unit and must be serviced as a complete assembly.

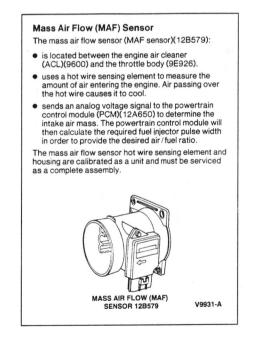

MASS AIR FLOW (MAF) SENSOR 12B579   V9931-A

Figure 11–3 Mass air flow (MAF) sensor. *(Courtesy of Ford Motor Company.)*

sure there are no downstream air leaks as any intake air that does not go through the sensor will yield a false signal to the computer. The MAF is a single service unit: it cannot be repaired as a set of parts.

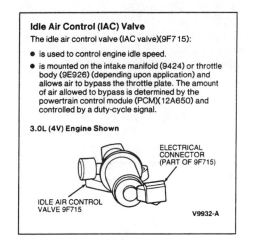

Figure 11–4 Idle air control (IAC) valve. *(Courtesy of Ford Motor Company.)*

Idle Air Control (IAC) Valve. The idle air control valve is, of course, primarily an actuator, to modulate the air bypass around the throttle plate and thus control idle speed under a variety of conditions and loads, Figure 11-4. However, since the PCM monitors where the IAC is supposed to be at a given setting and cycles it from time to time to reestablish a zero setting, it has a quasi-sensor status, too.

Intake Air Temperature (IAT) Sensor. This sensor, Figure 11-5, is a variable resistance thermistor changing in response to the temperature of the intake air charge.

Engine Coolant Temperature (ECT) Sensor. Like the intake air temperature sensor, the ECT, Figure 11-6, is also a variable resistance thermistor. It changes its resistance in response to engine coolant temperature. Both signal the PCM with a voltage corresponding to the value of the temperature measured.

Heated Oxygen Sensors (HO2S). The oxygen sensor system is different from earlier computer control systems chiefly in that there are

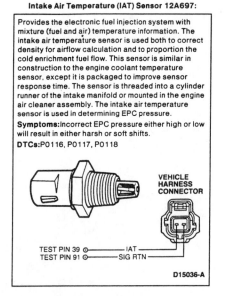

Figure 11–5 Intake air temperature (IAT) sensor. *(Courtesy of Ford Motor Company.)*

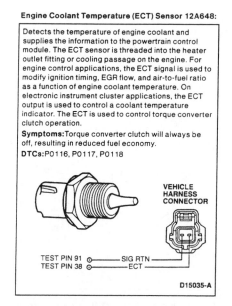

Figure 11–6 Engine coolant temperature (ECT) sensor. *(Courtesy of Ford Motor Company.)*

four of them in tandem. Two are before the catalytic converters—used as previously for control of closed-loop fuel trim—and two are *after* the catalytic converters to monitor the effectiveness of the catalytic converters, Figure 11-7. All four are electrically heated to reach operating temperature as quickly as possible. The downstream, secondary oxygen sensors should, if all is well, have a relatively flat signal. It does not matter much whether the voltage is high or low (typically they will slowly wander). What is important is that they do not reflect the kind of back and forth voltage fluctuation every second or two that the primary oxygen sensors do. If the secondary signal is a reflection of the primary sensors' signal, that means the catalytic converter is not doing anything to the exhaust gases.

Vehicle Speed Sensor (VSS).   The Ford vehicle speed sensor, Figure 11-8, uses a magnetic pick-up to send a rapid on/off pulse, the frequency of which corresponds to the vehicle's speed. The PCM uses this information, among other things, to calculate fuel mixture and ignition timing. Since the new transaxles are also in large measure electronically controlled, the VSS signal plays a role in that function.

Camshaft Position (CMP) Sensor.   On the two-valve-per-cylinder engine, the camshaft position sensor is a single Hall-effect magnetic switch activated by a vane driven by the camshaft. On the four-valve-per-cylinder engine, the camshaft position sensor is a variable reluctance sensor triggered by the high point on the left exhaust camshaft sprocket. In both cases, the sensor pro-

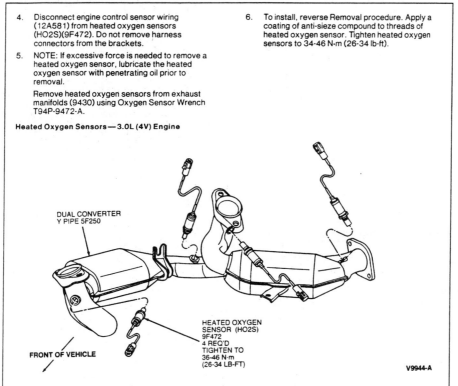

REMOVAL AND INSTALLATION (Continued)

4. Disconnect engine control sensor wiring (12A581) from heated oxygen sensors (HO2S)(9F472). Do not remove harness connectors from the brackets.

5. NOTE: If excessive force is needed to remove a heated oxygen sensor, lubricate the heated oxygen sensor with penetrating oil prior to removal.

   Remove heated oxygen sensors from exhaust manifolds (9430) using Oxygen Sensor Wrench T94P-9472-A.

Heated Oxygen Sensors—3.0L (4V) Engine

6. To install, reverse Removal procedure. Apply a coating of anti-sieze compound to threads of heated oxygen sensor. Tighten heated oxygen sensors to 34-46 N·m (26-34 lb-ft).

DUAL CONVERTER
Y PIPE 5F250

FRONT OF VEHICLE

HEATED OXYGEN
SENSOR (HO2S)
9F472
4 REQ'D
TIGHTEN TO
36-46 N·m
(26-34 LB-FT)

V9944-A

Figure 11–7 Heated oxygen sensors. *(Courtesy of Ford Motor Company.)*

the sensor provides the PCM with information used to synchronize the fuel injection sequence with the cylinder firing order.

Knock Sensor (KS).   The knock sensor, Figure 11-9, is the usual piezoelectric crystal type, which sends a voltage to the PCM if there is

detonation in the engine. This information is used to retard spark advance and eliminate the knock.

Power Steering Pressure (PSP) Switch. The power steering pressure switch, Figure 11-10, is installed in the boost side of the power steering pump. Normally it is electrically closed, but when hydraulic boost builds to the point where it could be a load on the engine, the switch opens. This provides the computer with information relevant to setting the idle speed.

Crankshaft Position (CKP) Sensor.   The crankshaft position sensor, Figure 11-11, is the main sensor for the ignition system, triggered by a 36-minus-1 tooth pulse wheel inside the engine front cover. This variable reluctance sensor generates a reversing polarity sine-wave-type signal. The information from the data stream constitutes two kinds of signal: the engine speed and the crankshaft position. The information is used to determine base timing and, with the information from other sensors that is relevant to spark timing, to calculate the amount of advance needed.

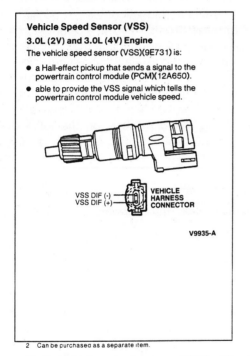

Vehicle Speed Sensor (VSS)

3.0L (2V) and 3.0L (4V) Engine

The vehicle speed sensor (VSS)(9E731) is:

- a Hall-effect pickup that sends a signal to the powertrain control module (PCM)(12A650).
- able to provide the VSS signal which tells the powertrain control module vehicle speed.

VSS DIF (-)
VSS DIF (+)
VEHICLE HARNESS CONNECTOR

V9935-A

2   Can be purchased as a separate item.

Figure 11–8 Vehicle speed sensor (VSS). *(Courtesy of Ford Motor Company.)*

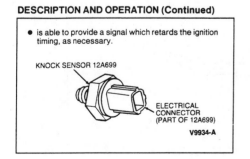

DESCRIPTION AND OPERATION (Continued)

- is able to provide a signal which retards the ignition timing, as necessary.

KNOCK SENSOR 12A699

ELECTRICAL CONNECTOR (PART OF 12A699)

V9934-A

Figure 11–9 Knock sensor. *(Courtesy of Ford Motor Company.)*

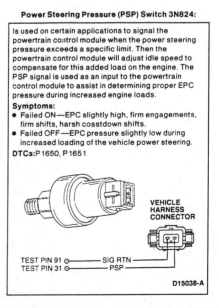

Power Steering Pressure (PSP) Switch 3N824:

Is used on certain applications to signal the powertrain control module when the power steering pressure exceeds a specific limit. Then the powertrain control module will adjust idle speed to compensate for this added load on the engine. The PSP signal is used as an input to the powertrain control module to assist in determining proper EPC pressure during increased engine loads.

Symptoms:
- Failed ON—EPC slightly high, firm engagements, firm shifts, harsh coastdown shifts.
- Failed OFF—EPC pressure slightly low during increased loading of the vehicle power steering.

DTCs: P1650, P1651

VEHICLE HARNESS CONNECTOR

TEST PIN 91 ○— SIG RTN
TEST PIN 31 ○— PSP

D15038-A

Figure 11–10 Power steering pressure (PSP) switch. *(Courtesy of Ford Motor Company.)*

OUTPUTS

Ignition Coil.   The system employs a waste spark system, with three coils in a single pack to fire six spark plugs, Figure 11-12. As with other waste spark systems, each coil fires two plugs simultaneously, one at the proper point for its upcoming power stroke and the other one (the "waste" spark) at the end of its exhaust stroke.

Fuel Pump Relay.   The fuel pump relay is part of the constant control relay module, Figure 11-13. When the ignition is first turned on, the PCM instructs the relay module to turn the pump on briefly to prime the injectors. If there is an ignition signal, the pump will stay on to maintain fuel pressure. If not, the pump is automatically shut off in a second or two. Besides the fuel pump relay, the system also employs an **inertia switch,** Figure 11-14. The purpose of this switch is to disable the fuel pump in case of an accident to prevent the possibility of a gasoline fire from the system under pressure.

Ford's inertia switches have improved in recent years, but it is not uncommon to find a car that won't start because the switch has been triggered by something less than an accident, perhaps because someone was forcibly bumped while parked. The switch contains a weight in a

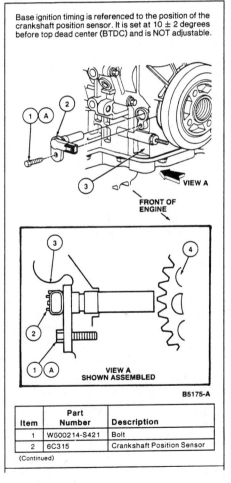

DESCRIPTION AND OPERATION (Continued)

Base ignition timing is referenced to the position of the crankshaft position sensor. It is set at 10 ± 2 degrees before top dead center (BTDC) and is NOT adjustable.

VIEW A

FRONT OF ENGINE

VIEW A
SHOWN ASSEMBLED

B5175-A

Item	Part Number	Description
1	W500214-S421	Bolt
2	6C315	Crankshaft Position Sensor

(Continued)

Figure 11–11   Crankshaft position sensor. (Courtesy of Ford Motor Company.)

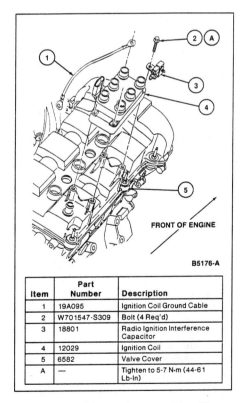

FRONT OF ENGINE

B5176-A

Item	Part Number	Description
1	19A095	Ignition Coil Ground Cable
2	W701547-S309	Bolt (4 Req'd)
3	18801	Radio Ignition Interference Capacitor
4	12029	Ignition Coil
5	6582	Valve Cover
A	—	Tighten to 5-7 N·m (44-61 Lb-In)

Figure 11–12   Ignition coil assembly. *(Courtesy of Ford Motor Company.)*

Fuel Pump Relay

Removal and Installation

NOTE: The fuel pump relay is part of the constant control relay module (CCRM) and cannot be serviced separately.

1. Disconnect battery ground cable (14301).
2. Lift up and release latch retaining constant control relay module (CCRM) to battery tray (10732).
3. Disengage four locking tangs on connector cover and remove cover.
4. Unscrew connector retaining screw and remove connector from CCRM.
5. To install, reverse Removal procedure. Tighten connector retaining screw to 1.4-2.0 N·m (12-18 lb-in).

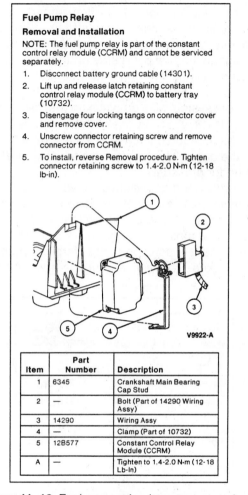

Item	Part Number	Description
1	6345	Crankshaft Main Bearing Cap Stud
2	—	Bolt (Part of 14290 Wiring Assy)
3	14290	Wiring Assy
4	—	Clamp (Part of 10732)
5	12B577	Constant Control Relay Module (CCRM)
A	—	Tighten to 1.4-2.0 N·m (12-18 Lb-In)

Figure 11–13 Fuel pump relay (constant control relay module). *(Courtesy of Ford Motor Company.)*

cone. The weight is held in place by a magnet. If a shock to the car dislodges the weight, it flips a switch that disconnects the fuel pump circuit. On the top of the switch is a reset button for just such occurrences. Since the switch is located in the trunk, it is also not uncommon to find some corrosion damage to the connector. Most experienced Ford technicians look at the inertia switch first when they find a no-start.

Fuel Injectors. The injectors used in this system, Figure 11-15, are methanol and ethanol compatible. The O-rings that seal them in the man-

REMOVAL AND INSTALLATION (Continued)

6. To install, reverse Removal procedure. Tighten screws to 1.6-2.2 N·m (14-19 lb-in).

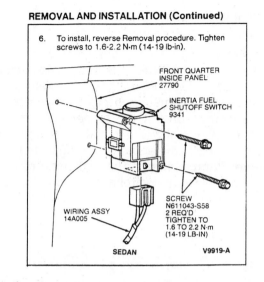

Figure 11–14 Inertia switch. *(Courtesy of Ford Motor Company.)*

ifold are likewise methanol and ethanol compatible, as should be any replacements. These injectors use a pintle designed to resist deposits from fuel additives as well. The PCM grounds each injector's circuit in turn, completing its circuit and opening the nozzle. The injection occurs during the intake stroke in the firing order of the engine.

Fuel Pressure Regulator. Also designed for **alternate fuel compatibility,** the fuel pressure regulator connects to the fuel rail (or fuel injection supply manifold, as Ford literature prefers to describe it) downstream of the injectors, Figure 11-16. It regulates the fuel pressure at the injectors through the interaction of its diaphragm-operated relief valve. One side of the diaphragm senses fuel pressure, and the other is subject to manifold pressure (vacuum). The initial fuel pressure is determined by the specially calibrated spring preload. The purpose of the arrangement is to maintain the fuel pressure drop through the nozzles to a constant pressure differential, regardless of the changes in the intake manifold pressure, and to make it simpler to calculate the fuel injection pulse width. As long as the pressure differential is constant, a fixed amount of fuel will flow for a fixed pulse width. If the pressure differential were allowed to vary, the PCM would have to compensate for different fuel quantity. Pressure is reduced from the pump's maximum outlet pressure by venting enough back to the tank to keep the rail charged at the design pressure.

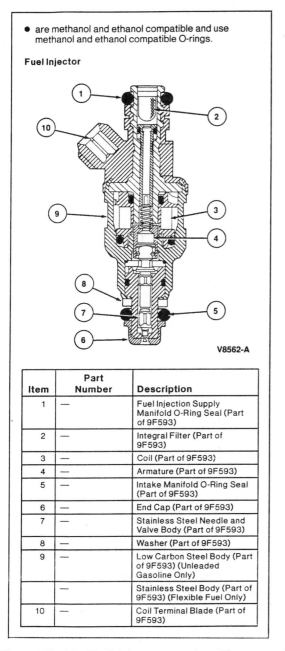

- are methanol and ethanol compatible and use methanol and ethanol compatible O-rings.

Fuel Injector

Item	Part Number	Description
1	—	Fuel Injection Supply Manifold O-Ring Seal (Part of 9F593)
2	—	Integral Filter (Part of 9F593)
3	—	Coil (Part of 9F593)
4	—	Armature (Part of 9F593)
5	—	Intake Manifold O-Ring Seal (Part of 9F593)
6	—	End Cap (Part of 9F593)
7	—	Stainless Steel Needle and Valve Body (Part of 9F593)
8	—	Washer (Part of 9F593)
9	—	Low Carbon Steel Body (Part of 9F593) (Unleaded Gasoline Only)
	—	Stainless Steel Body (Part of 9F593) (Flexible Fuel Only)
10	—	Coil Terminal Blade (Part of 9F593)

V8562-A

Figure 11–15 Fuel injector cutaway. *(Courtesy of Ford Motor Company.)*

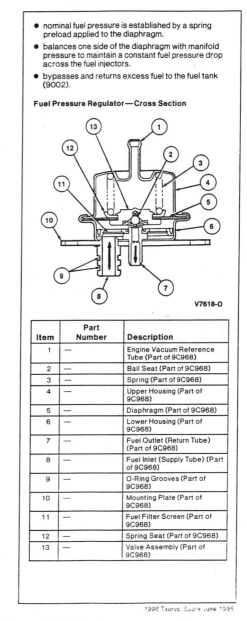

- nominal fuel pressure is established by a spring preload applied to the diaphragm.
- balances one side of the diaphragm with manifold pressure to maintain a constant fuel pressure drop across the fuel injectors.
- bypasses and returns excess fuel to the fuel tank (9002).

Fuel Pressure Regulator—Cross Section

Item	Part Number	Description
1	—	Engine Vacuum Reference Tube (Part of 9C968)
2	—	Ball Seat (Part of 9C968)
3	—	Spring (Part of 9C968)
4	—	Upper Housing (Part of 9C968)
5	—	Diaphragm (Part of 9C968)
6	—	Lower Housing (Part of 9C968)
7	—	Fuel Outlet (Return Tube) (Part of 9C968)
8	—	Fuel Inlet (Supply Tube) (Part of 9C968)
9	—	O-Ring Grooves (Part of 9C968)
10	—	Mounting Plate (Part of 9C968)
11	—	Fuel Filter Screen (Part of 9C968)
12	—	Spring Seat (Part of 9C968)
13	—	Valve Assembly (Part of 9C968)

V7618-D

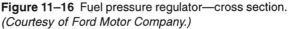
1996 Taurus. Sable June 1996

Figure 11–16 Fuel pressure regulator—cross section. *(Courtesy of Ford Motor Company.)*

EMISSIONS CONTROLS

Certain system components don't make the car go or stop, or work well or get good mileage, but they are necessary to prevent illegal toxic exhaust so that the car is allowed on the road. Some emissions components, like the PCV, work independently and have no more connection to the PCM than the glove box latch, while others are controlled as precisely as the spark advance or the fuel injection pulse width. We will look at them next.

Secondary Air Injection. While the engine is warming up to a temperature at which the system can go into closed loop and regulate the air/fuel mixture according to the responses of the oxygen sensor, the air injection pump injects air into the exhaust manifold to burn as much residual fuel (hydrocarbons) as possible to minimize undesirable emissions, Figure 11-17. On some engines, once the engine and oxygen sensor are

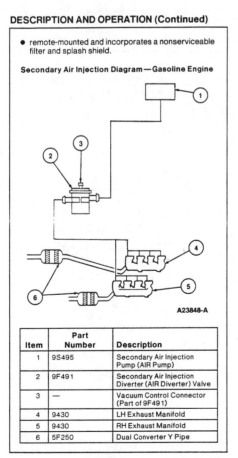

DESCRIPTION AND OPERATION (Continued)

- remote-mounted and incorporates a nonserviceable filter and splash shield.

Secondary Air Injection Diagram—Gasoline Engine

A23848-A

Item	Part Number	Description
1	9S495	Secondary Air Injection Pump (AIR Pump)
2	9F491	Secondary Air Injection Diverter (AIR Diverter) Valve
3	—	Vacuum Control Connector (Part of 9F491)
4	9430	LH Exhaust Manifold
5	9430	RH Exhaust Manifold
6	5F250	Dual Converter Y Pipe

Figure 11–17 Secondary air injection diagram. *(Courtesy of Ford Motor Company.)*

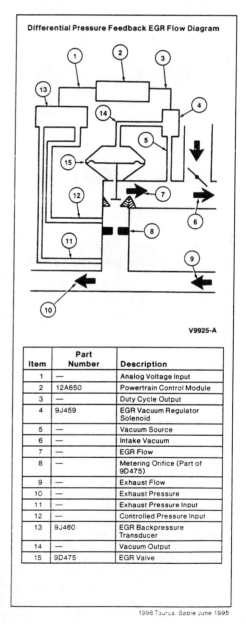

Differential Pressure Feedback EGR Flow Diagram

V9925-A

Item	Part Number	Description
1	—	Analog Voltage Input
2	12A650	Powertrain Control Module
3	—	Duty Cycle Output
4	9J459	EGR Vacuum Regulator Solenoid
5	—	Vacuum Source
6	—	Intake Vacuum
7	—	EGR Flow
8	—	Metering Orifice (Part of 9D475)
9	—	Exhaust Flow
10	—	Exhaust Pressure
11	—	Exhaust Pressure Input
12	—	Controlled Pressure Input
13	9J460	EGR Backpressure Transducer
14	—	Vacuum Output
15	9D475	EGR Valve

1996 Taurus, Sable June 1995

Figure 11–18 Differential pressure feedback EGR flow diagram. *(Courtesy of Ford Motor Company.)*

warm enough to go into closed loop and the sensor is generating its characteristic oscillating low voltage wave, the air injection switches to the catalytic converter. The switch from the exhaust manifold is the easiest way to check to see whether the system can achieve closed-loop operation.

Like other manufacturers' systems, the Ford air injection system shifts into divert mode under circumstances when the injected air could cause backfire or damage.

Item	Part Number	Description
B	—	Tighten to 45-65 N·m (33-48 Lb-Ft)
C	—	Tighten to 35-45 N·m (26-33 Lb-Ft)

EGR Backpressure Transducer

Removal and Installation

1. Disconnect engine control sensor wiring (12A581) from EGR backpressure transducer.
2. Disconnect EGR backpressure transducer hoses from EGR valve to exhaust manifold tube (9D477).
3. Remove retaining nuts or bolts (depending upon application) and EGR backpressure transducer.
4. To install, reverse Removal procedure. Tighten retaining nuts or bolts (depending upon application) to 5-7 N·m (40-61 lb-in).

EGR Backpressure Transducer, 3.0L (2V) Engine

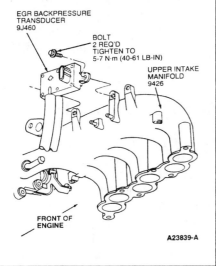

Figure 11–19 EGR backpressure transducer. *(Courtesy of Ford Motor Company.)*

Exhaust Gas Recirculation (EGR) System.

To prevent the formation of oxides of nitrogen, NO_x, which can only form at high combustion temperatures, the system meters a certain amount of now-inert exhaust gas back through the intake manifold. The effect slows the burning of the fuel and keeps combustion temperatures below the temperature at which nitrous oxides form.

The Ford system uses a pressure feedback type EGR valve, Figures 11-18 and 11-19. To enable the computer to meter the recirculating exhaust precisely, there is a backpressure transducer. This device returns a signal to the PCM corresponding to the backpressure in the exhaust system at the EGR pickup point, and this parameter is factored in with the others considered for EGR quantity calculations (engine speed, load, and temperature). Then the computer controls the EGR valve through the EGR vacuum regulator solenoid, Figure 11-20. Conditions for employment of EGR are similar to those used in previous years with previous systems.

EGR Vacuum Regulator Solenoid

1. Remove the EGR vacuum regulator solenoid as outlined.
2. Lightly blow air into the EGR vacuum regulator solenoid and verify that air does not flow.
3. Apply battery voltage and a ground to the EGR vacuum regulator solenoid at the connector.
4. Lightly blow air into the EGR vacuum regulator solenoid and verify that air does flow through the EGR vacuum regulator solenoid.
5. If air does not blow through the EGR vacuum regulator solenoid, replace the EGR vacuum regulator solenoid.
6. Install EGR vacuum regulator solenoid as outlined.

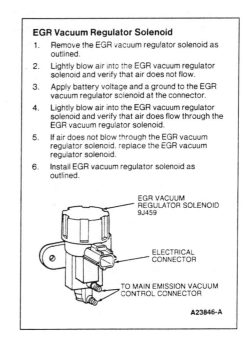

Figure 11–20 EGR vacuum regulator solenoid. *(Courtesy of Ford Motor Company.)*

✔ SYSTEM DIAGNOSIS & SERVICE

OBD II and Standardized Diagnosis. While the OBD II regulations make the computer control system on any individual car more complex, in an effort to more closely control the quality of exhaust emissions, they promise to simplify vehicular diagnosis generally **(standardizd diagnosis)**. The details of OBD II are discussed in more detail in Chapter 18, but briefly, these are the advantages:

- terms to describe components common to many vehicles have been standardized. So PCM has come to replace all the various descriptions carmakers had for their main combustion computers before.
- diagnostic trouble codes have been standardized, with enough codes listed and enough room in the "empty spaces" on the lists to add about an order of magnitude more in the future should that become necessary. If a particular engine does not have, for instance, air injection, its computer will simply not have codes for air injection stored in its memory. Different cars by different makers with the same problem will have the same trouble codes. Consequently, a given trouble code will mean exactly the same thing whether a technician is working on a Cadillac or a Toyota.
- the diagnostic connector has been standardized, with specific terminals designated for specific circuits.
- the capacities of the on-board computers will be standardized also in the sense that certain very useful tools, like the **"freeze-frame/snapshot"** features (which record practically everything that was going on when a code was set) will be present on all and accessible through all scan tools.
- the system does not force car makers to leave out any special diagnostic procedures they may develop. There are countless additional codes available, and the diagnostic connector affords a large number of specialty circuits for manufacturers to do with as they wish.

SUMMARY

In this chapter we've covered the features of the new Ford PCM, the OBD II-compliant computer and the engine controls it manages. We've looked at the new inputs and outputs of the system, and we've seen the sensors and actuators carried over or modified from the previous Ford engine management systems: the throttle position sensor, the hot-wire mass air flow sensor, the intake air temperature sensor, the engine coolant temperature sensor, the oxygen sensor, the camshaft position sensor, the crankshaft position sensor, the knock sensor, the power steering pressure switch, the vehicle speed sensor, and the idle air control valve.

We reviewed the provisions built into the system for alternative fuel capacity, using various blends of fuel that are less than 100% gasoline. We have seen the application of the standardized OBD II self-diagnostics program, including the freeze-frame/snapshot memory functions.

We've seen how the Ford system's emissions controls have been revised and updated, and we've grown more familiar with the standardized OBD II diagnostic systems.

▲ DIAGNOSTIC EXERCISE

A 1996 Ford Taurus is towed into the shop with a no-start. Spark is good, but there is no fuel pressure. Checking the circuit indicates no power to the pump. Alert to the inertia switch problem, a technician pushes the reset button on the inertia switch, but to his annoyance this does not the fix the problem. What sort of trouble codes (circuits, not alphanumeric characters) would you expect to find stored in the computer's memory? Assuming there is nothing open or shorted on the pump circuit itself, what components and component circuits could disable the pump?

REVIEW QUESTIONS

1. How does the Ford PCM detect engine wear and seasonal or altitude changes? What use does it make of this information?

2. If a car using a hot-wire air mass sensor suddenly drives into a pocket of cold air, it will take more electrical current to keep the wire at the standard temperature. How does the computer adapt to this?

NOTE: this is a trick question! Examine your assumptions!

3. The IAC (idle air control) valve functions both as an actuator and as a sensor. Explain how this is possible.

4. What should the secondary oxygen sensor's signal look like on a scope? Why?

5. The VSS (vehicle speed sensor) changes frequency as the vehicle changes speed. Does the voltage change, and what does that mean?

6. Describe how the power steering pressure switch affects the idle speed. What components are involved?

7. Under what circumstances could a waste-spark ignition system generate regular backfiring through the intake manifold? Would the system set a code?

8. Why does the system keep the pressure differential constant between fuel in the fuel rail and the air in the intake manifold?

9. Describe the different problems that might occur if the air injection system became locked in each of its three modes.

10. True or false: It is impossible to diagnose driveability problems on an OBD II car without a scan tool?

ASE-type Questions (Actual ASE tests are rarely so product-specific).

11. Technician A says OBD II means all manufacturers will use the same engine control systems. Technician B says compliance with OBD II means using the same words and the same scan tool connector. Who is right?
    a. A only.
    b. B only.
    c. both A and B.
    d. neither A nor B.

12. Technician A says the second wire in the hot-wire mass air flow sensor is a backup in case the first burns out. Technician B says the sensor's output will show a fairly constant voltage but will vary in frequency. Who is right?
    a. A only.
    b. B only.
    c. both A and B.
    d. neither A nor B.

13. Technician A says two oxygen sensors are used to make sure there is at least one reliable source of mixture feedback information for fuel trim. Technician B says one of the oxygen sensors is put behind the catalytic converter to bring it up to operating temperature more quickly. Who is right?
    a. A only.
    b. B only.
    c. both A and B.
    d. neither A nor B.

14. Technician A says the Ford inertia switch can sometimes disable the engine without the vehicle being in an accident. Technician B says it's always a good practice to reset the inertia switch anytime you have a car in the shop that is equipped with one. Who is right?
    a. A only.
    b. B only.
    c. both A and B.
    d. neither A nor B.

15. Technician A pulls a code from a Ford Taurus computer with his scan tool. The code refers him to a sensor or actuator circuit, which he carefully inspects with a volt/ohmmeter. He finds it entirely in order and he recommends replacement of the computer. Technician B says the snapshot and/or freeze-frame features should be employed first before the diagnosis is concluded. Who is right?
    a. A only.
    b. B only.
    c. both A and B.
    d. neither A nor B.

16. Technician A is working on a Ford Taurus with a rough idle, caused by a particular cylinder. Careful inspection reveals nothing wrong with the fuel delivery and ignition to that cylinder, so Technician A suggests a valve may be burned. Technician B says that wouldn't matter, since the computer has the capacity to take compensatory countermeasures for engine wear and age. Who is right?
    a. A only.
    b. B only.
    c. both A and B.
    d. neither A nor B.

17. A Ford Taurus throttle position sensor shows erratic return signal above 80 degrees of throttle opening. Since the car belongs to an elderly person who doesn't want to spend much money, Technician A suggests adjusting the throttle linkage so pushing the pedal to the floor will only open the throttle 75 degrees. Technician B says that wouldn't work, because the erratic reading above 80 degrees has already thrown off the computer's range of signal expectation for the sensor. Who is right?
    a. A only.
    b. B only.
    c. both A and B.
    d. neither A nor B.

18. A 1996 Ford Taurus has a slightly porous air inlet just downstream of the throttle body. Technician A says this will drive the mixture lean at idle and other high vacuum conditions. Technician B says this will cause the computer to set a mass air flow sensor circuit trouble code. Who is right?
    a. A only.
    b. B only.
    c. both A and B.
    d. neither A nor B.

19. A 1996 Ford Taurus is discovered to have its idle air control all the way retracted, that is, holding the bypass wide open. But idle speed is only slightly above normal, and the engine lacks power and tends to overheat. Technician A says the catalytic converter may be at fault. Technician B says the engine has simply worn out and needs to be rebuilt. Who is right?
    a. A only.
    b. B only.
    c. both A and B.
    d. neither A nor B.

20. An engine coolant temperature sensor circuit is shorted to ground. Technician A says this will make the computer think the engine is either very cold or overheated, depending on how the resistor works. Technician B says it may make no difference at all in the way the system works. Who is right?
    a. A only.
    b. B only.
    c. both A and B.
    d. neither A nor B.

Chapter 12

Chrysler's Oxygen Feedback System

OBJECTIVES

In this chapter you can learn:
- ❑ what functions the O_2 feedback system controls.
- ❑ the system's operating modes.
- ❑ under what conditions the system stays in closed loop.
- ❑ the three types of air/fuel mixture solenoids used and their control methods.
- ❑ the diagnostic procedure to determine the cause of driveability problems on models *without* on-board diagnostics.
- ❑ the diagnostic procedure to determine the cause of driveability problems on models *with* on-board diagnostics.

KEY TERMS

Inductance
Transducer

Chrysler was the first carmaker to switch from a mechanical points-and-condenser ignition system to electronic ignition, back in 1972. So the company was ready with a computer-controlled engine system in 1979 when they began systematic installation of an oxygen-sensor feedback system on many vehicles. The first system was on the California models with the slant-six engine. It combined the previous Chrysler spark advance control with an oxygen sensor that controlled a mixture solenoid. The next year the system was extended to cover four- and eight-cylinder engines as well. By 1981 it applied to federal-specification vehicles too. By that date, it also had idle speed control functions along with EGR and air injection actuation. As the system developed, the shift indicator light, the radiator coolant fan, and the vacuum solenoid controlling the car-

buretor secondary barrel on cars built for high altitudes all followed. Figure 12-1 shows a system overview.

COMBUSTION CONTROL COMPUTER

In the earliest systems, the combustion control computer was in the air cleaner, similar to the previous lean burn system's control unit. Like this unit, the early combustion control computers had problems with vibration and the variability of temperature at the air cleaner. Later versions located the computer in various more secure places in the engine compartment. Eventually, of course, Chrysler engineers would move the computer into the passenger compartment for some sys-

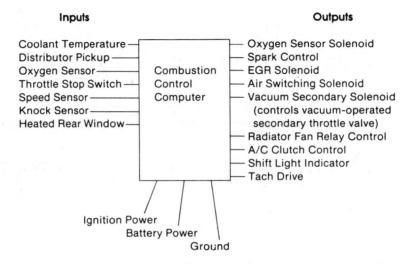

Not all inputs or outputs are used on all Oxygen Sensor Feedback applications.

Figure 12–1 Oxygen sensor feedback system overview

tems. Chrysler literature sometimes refers to the earliest controller as the "spark control computer." From 1985-on, self-diagnostics facilitate system diagnosis.

Open Loop

In open loop the computer does not use the oxygen sensor's signal, but calculates the fuel/air mixture from other sensors' inputs. The system is in open loop if any of the following conditions exist:

- the coolant is below the required operating temperature (120 degrees Fahrenheit/49 degrees Centigrade).
- the oxygen sensor is below the required operating temperature (600 degrees Fahrenheit/315 degrees Centigrade).
- the engine is at idle speed.
- the engine is accelerating the car (low manifold vacuum, relatively high load).
- the engine is decelerating the car (high manifold vacuum, relatively low load).
- the engine is restarted when hot.
- the oxygen sensor or its circuit (as well as

some other sensors' and actuators' circuits affecting mixture control) have failed.

Closed Loop

In closed loop, the computer determines fuel/air mixture from the oxygen sensor signal as well as from the other sensor signals. The system goes into closed loop only after *all* of these conditions are true:

- the coolant reaches a specified temperature (on most vehicles about 120 degrees Fahrenheit/49 degrees Centigrade).
- the oxygen sensor produces a recognizable voltage (beginning about 600 degrees Fahrenheit/315 degrees Centigrade). This voltage ranges from 0.3 to 0.9 volts. The higher the voltage, the leaner the mixture; the lower the voltage, the richer the mixture. A new oxygen sensor cycles back and forth more often than once per second; a long-used sensor begins to slow its cycles.
- a predetermined time has elapsed since the engine started.

INPUTS

Coolant Temperature

Chrysler has used three different kinds of coolant temperature sensors on different oxygen sensor feedback systems: a thermistor, a dual thermistor, and an on/off switch. See Figure 12-2.

Thermistors.   A single element thermistor was first used on six and eight cylinder engines and later on four cylinders. About 1985, some transverse engines came with dual element thermistors, Figure 12-3. A thermistor is a semiconductor that *changes* electrical resistance as its

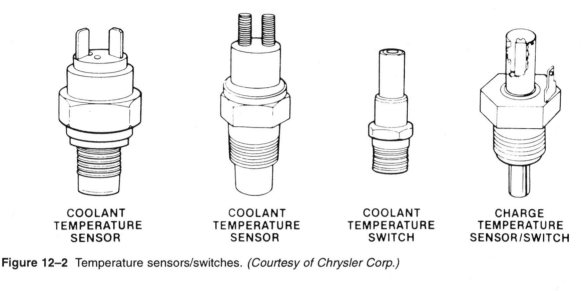

COOLANT TEMPERATURE SENSOR   COOLANT TEMPERATURE SENSOR   COOLANT TEMPERATURE SWITCH   CHARGE TEMPERATURE SENSOR/SWITCH

Figure 12–2 Temperature sensors/switches. *(Courtesy of Chrysler Corp.)*

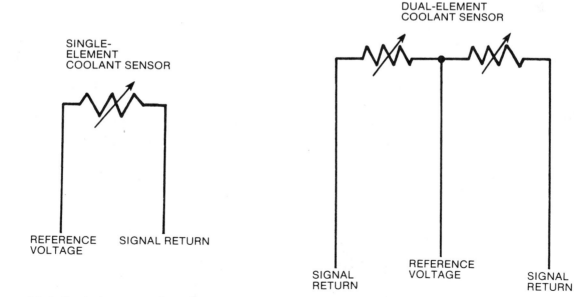

Figure 12–3 Coolant sensor schematics

temperature goes up, and the Chrysler system uses a negative temperature coefficient thermistor, which loses resistance as its temperature goes up. So the coolant temperature signal back to the computer moves back up toward the reference five-volt signal as the engine warms and the resistance goes down.

The dual element thermistor sends one signal for the fuel mixture calculation function (in general, the warmer the engine, the leaner the mixture can be with acceptable driveability, up to reasonable limits), along with spark, EGR, and canister purge. The second signal controls the radiator fan relay. Some oxygen sensor feedback systems use a separate temperature sensor in the radiator inlet tank to control the radiator fan relay.

Temperature Sensing Switch.   Most four-cylinder engines up to 1983 use a temperature sensitive switch to signal the computer once the closed loop criterion coolant temperature is reached.

Charge Temperature

Some six- and eight-cylinder engines have a temperature sensor in the intake manifold to report the air/fuel mixture temperature. Because the heated air inlet system brings the incoming air temperature up to normal operating temperature long before the coolant reaches operating temperature, some computers employ both a charge mixture temperature sensor and a coolant temperature sensor. This allows the computer to set an air/fuel mixture leaner than would be indicated by the coolant temperature alone.

Distributor Pickup

On the four-cylinder engines, engine speed and crankshaft position information comes from the Hall-effect switch in the distributor, Figure 12-4. This is an on-off, direct current signal. On the six- and eight-cylinder engines, this information comes from the dual pickup coils in the distributor, a voltage-polarity-reversing alternating current. Note that

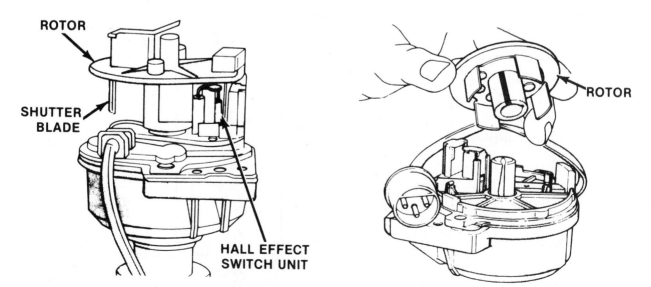

Figure 12–4 Hall-effect distributor switch. *(Courtesy of Chrysler Corp.)*

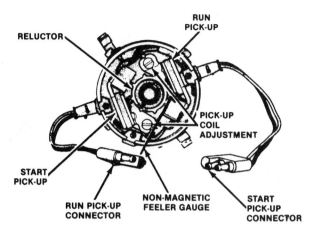

Figure 12–5 Dual pickup distributor units. *(Courtesy of Chrysler Corp.)*

the Hall-effect distributor, unlike most other computer inputs, receives an eight-volt reference signal from the computer, not the standard five-volt. Very late-model Hall-effect distributors (and some other sensors on Chrysler products) get a 9+ volt reference signal to accommodate a different kind of transistor. The inductive pickups in the six- and eight-cylinder pickups generate their own alternating current signal, Figure 12-5.

Vacuum Transducer

The computer gets intake manifold vacuum information, which corresponds inversely to engine load, from the vacuum transducer, Figure 12-6. In response to differences in intake manifold vacuum, the vacuum **transducer** changes the electrical properties of a coil through which the computer sends a reference alternating current. The intake manifold vacuum works against a spring to slide a movable metallic core in the electrical coil. As the force of the manifold vacuum increases, the metallic core is withdrawn; as the vacuum decreases, the spring drives the core back into the coil.

The computer passes an alternating current through the coil, and the alternating magnetic field produced by that current produces an **inductance,** an electromotive force produced by

the varying current in the coil. Inductance is a kind of dynamic resistance or impedance specific to alternating current circuits through a coil. The alternating current itself builds up this resistance to further current flow as a result of its fluctuating magnetic field. This inductance changes as the metallic core changes position within the coil. The deeper the core is in the coil, the lower is the inductance of the coil. The computer measures voltage drop through the coil, and that corresponds to engine load.

Oxygen Sensor

The early Chrysler feedback systems use a single wire, unheated oxygen sensor. Like all oxygen sensors, it consists of a small ceramic thimble exposed to ambient air on the inside and exhaust on the outside. A super-thin layer of platinum on each side generates a low voltage, low amperage current reflecting the difference between the oxygen in the two gases.

Throttle Stop Switch

The early Chrysler systems do not use a throttle position sensor, but instead a throttle position switch, Figure 12-7. A button switch on the throttle kicker (or dashpot) tells the computer when the dri-

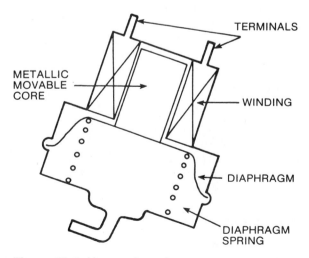

Figure 12–6 Vacuum transducer

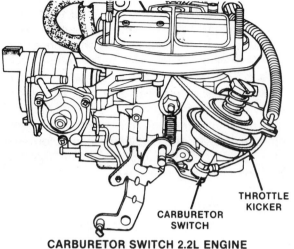

CARBURETOR SWITCH 2.2L ENGINE

Figure 12–7   Throttle stop switch. *(Courtesy or Chrysler Corp.)*

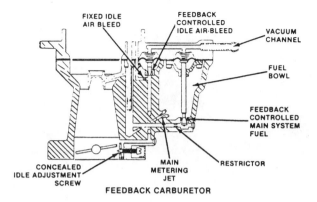

FEEDBACK CARBURETOR

Figure 12–8   Holley 6145 mixture control circuit. *(Courtesy of Chrysler Corp.)*

ver has removed his or her foot from the accelerator pedal: the circuit completes, and the signal is returned. There was one exception: in 1979 Chrysler systems used a throttle position transducer, which provides somewhat more detailed information about foot and throttle position.

Speed Sensor

An inductive pulse generator, driven by the transmission output shaft, signals the computer information about the vehicle speed. Like all inductive sensors, an increased speed of the signal is accompanied by increased voltage and frequency.

Detonation Sensor

On high-performance four-cylinder engines and on eight cylinders with four-barrel carburetors, a piezoelectric knock sensor signals the computer when frequencies characteristic of detonation occur. Again, this is an alternating current signal. This sensor is ordinarily called the knock sensor.

Heated Rear Window

Many Chrysler vehicles, particularly those with transverse four-cylinder engines, include an input from the rear window electric heater. When this heater is on, the computer can then increase the idle throttle setting to accommodate the additional load of the heavy current draw through the alternator, at just the time when the engine is producing its least torque.

OUTPUTS

Mixture Control

The O_2 feedback system has used three different mixture control solenoids on four different carburetors. Chrysler literature calls this "O_2 feedback solenoid."

Holley Carburetor, Model 6145.   This carburetor, the feedback version of the old simple one-barrel Holley, was used in 1979 and 1980 for the California models with the slant-six engine, see Figure 12-8. The system controls idle mixture through the idle air bleed volume and the main metering mixture by metering fuel, similar to the operation of the solenoid used on the Holley 6149. Air for the idle and low speed system air bleed comes through a fixed orifice plus a passage controlled by a metering rod. A vacuum diaphragm controls the metering rod. A second diaphragm controls a spring-loaded rod and "foot" to modify fuel in the main metering circuit.

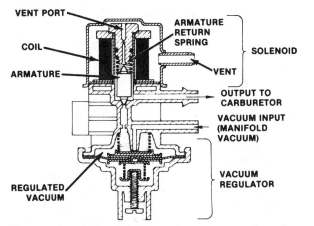

Figure 12–9 Solenoid-operated vacuum regulator for mixture control. *(Courtesy of Chrysler Corp.)*

The spring drives the rod down. If the vacuum against the diaphragm allows the rod to come down, the foot opens a valve that meters additional fuel into the main circuit.

Vacuum to the two diaphragms is simultaneous, and as vacuum increases, the mixture leans. At wide-open-throttle there is effectively no vacuum, and the mixture goes full rich. A remote, computer-controlled solenoid controls the vacuum, Figure 12-9. The lower portion of the unit is a simple vacuum regulator valve. It gets full manifold vacuum and reduces it to 5 inches of mercury (in Hg). The 5 in Hg vacuum is then applied to a valve seat higher in the assembly. A conical tip on the lower end of the solenoid armature closes this valve seat when the computer does not energize the solenoid. With the solenoid armature in this position, the regulator's output port connects to the vent port, making the output port vacuum zero. When the computer energizes the solenoid, the armature moves up, and a second conical tip on the top of the armature closes off the vent port. The lower passage opens, and the "reference" 5 in Hg vacuum is now connected to the output port. The computer duty-cycles the solenoid at whatever rate will maintain the proper fuel/air mixture, as called for by the oxygen sensor and reflected in the applied vacuum—between 0 and 5 in Hg, at the two diaphragms in

the carburetor. High vacuum leans the mixture in both idle and main metering circuits; low vacuum richens the mixture for both circuits.

Holley 6520 and Later Applications of the Holley 6145. In 1980, Chrysler applied an oxygen sensor and the two-barrel feedback Holley 6520 carburetor to four-cylinder cars. In 1981, they applied the same type of solenoid to six-cylinder engines with the Holley 6145 carburetor. This solenoid is essentially the same one used in the Holley carburetors used on some General Motors' CCC applications.

Carter BBD and Thermo-Quad. Carter carburetors on Chrysler vehicles got mixture control solenoids in 1980. Both the BBD two-barrel and the Thermo-Quad mixture solenoids work by a controlled air bleed for the idle and main metering circuits.

Air flows in the air horn above the carburetor venturi through a passage to the mixture solenoid, Figure 12-10. When the computer energizes the solenoid, air flows through the air horn passage and into the idle or main metering circuits (whichever is drawing air and fuel). Since the fuel emulsion includes more air, the mixture delivered to the cylinders goes lean. When the computer opens the mixture solenoid circuit and turns it off, the mixture goes lean. The computer's "choice," of course, follows the signal from the oxygen sensor, Figure 12-11.

Ignition Timing

The early Chrysler O_2 feedback system still follows the strategy of the still-earlier lean-burn system. The computer completes the primary ignition circuit by grounding the negative terminal. During cranking, the distributor pickup signal triggers the computer to open the primary circuit and fire the plugs at base (reference) timing.

Once the cold engine starts, the computer gradually adds spark advance, based on sensor inputs for coolant temperature, engine speed, and intake manifold vacuum. The calculations are all from its "look-up" memory, and the feedback system plays no role in spark timing (natu-

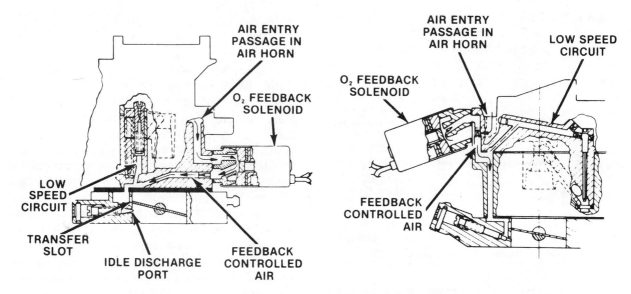

Figure 12–10 Carter carburetor mixture control, idle circuit. *(Courtesy of Chrysler Corp.)*

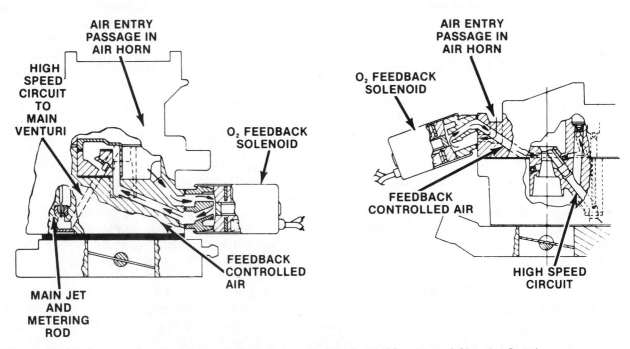

Figure 12–11 Carter carburetor mixture control, main metering circuit. *(Courtesy of Chrysler Corp.)*

rally there is *no* oxygen sensor signal when the system is cold). Once the coolant reaches a specific criterion temperature, the computer determines timing advance from manifold vacuum and engine rpm. If the throttle switch circuit closes—indicating a closed throttle—spark advance drops to reference timing.

EGR Solenoid

Only some of the early Chrysler O_2 feedback systems use the computer to control EGR. On these, the computer controls a solenoid that allows or shuts off vacuum to the EGR valve. Vacuum is shut off when:

- the engine is cold.
- the throttle is only slightly open.
- the engine has just restarted warm (and for the next minute or so).
- the throttle is wide open (approximately 80% or more). Of course, with WOT there is very little intake manifold vacuum, so even if the system malfunctioned, relatively little EGR gas would be drawn into the intake manifold.

Air Switching Solenoid

On early Chrysler O_2 feedback systems controlling the air injection system, the computer controls a solenoid that applies or disconnects the vacuum to the switch relief valve, Figure 12-12. During cold engine operation, the computer routes vacuum to the valve diaphragm and directs air from the air pump into the exhaust manifold. Once the engine gets warm enough to run in closed loop, the computer stops vacuum to the valve diaphragm and directs injected air downstream to the catalytic converter. For a brief period after each engine start, warm or cold, the computer sends vacuum to the valve to inject air into the exhaust manifold. The system includes no provision to dump air into the atmosphere except as a pressure relief.

On systems where the computer does not control the air injection system, the same strategy results from the action of a coolant temperature-controlled vacuum valve in the place of the computer-controlled solenoid. On all air injection systems, the objectives are to route air to the exhaust manifold when the engine is cold, to the catalytic converter when the system is in closed loop, and to the atmosphere when there is no need for the air injection and/or there is risk of backfire.

Vacuum Secondary

The Holley 6520 uses a progressive secondary barrel, opened by vacuum. Some of the later applications of this carburetor include a solenoid-operated vent valve on the vacuum line. When the computer opens this valve, this action locks out the carburetor secondary barrel until:

- the coolant reaches 140 degrees Fahrenheit (60 degrees Centigrade).
- vacuum drops below a specific criterion value.
- engine speed is above a specific rpm.

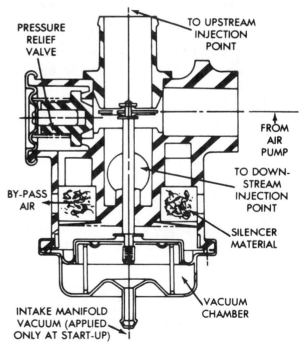

Figure 12–12 Air injection switch/relief valve. *(Courtesy of Chrysler Corp.)*

Radiator Fan Relay

On later model vehicles with transverse engines and electric fans, a radiator fan relay supplies power for both the cooling fan motor and for the air conditioning compressor clutch. Normally the circuit is open, but when either the computer or the coolant temperature switch closes it, the fan turns on to draw air through the radiator until the coolant reaches the low limit of the switch/sensor.

Shift Light Indicator

On 1985 and 1986 vehicles with manual transmissions, the computer operates an instrument light that indicates to the driver the optimum time to make upshifts. "Optimum" in this case refers to optimum emissions and fuel economy, not optimum performance or engine life.

Tachometer Drive

Some models include an electric tachometer in the instrument panel. The computer generates the tach signal from the distributor side of the ignition coil. This is essentially an on/off signal, and frequency corresponds directly to engine speed.

Other engine control features often found on feedback systems, but not necessarily controlled by the computer, are:

- a vacuum-powered throttle kicker (dashpot).
- a choke heater.
- a carburetor bowl vent solenoid.

These devices usually get their power directly from the ignition switch and are grounded directly.

✔ SYSTEM DIAGNOSIS & SERVICE

Diagnostic Procedure 1979-1984

Self-diagnostics did not appear on the Chrysler systems until 1985. Prior to that model year, diagnosis of a no-start or driveability problem followed driveability test procedures outlined in a separate booklet for each system application (by vehicle model and by build-year).

Driveability diagnosis falls into three parts:

- verification of the complaint.
- visual inspection of the vehicle.
- performance of the designated test steps.

The driveability test procedure outlines the visual inspection and the test steps. The test procedures also fall into three parts:

- no start.
- cold driveability test.
- warm driveability test.

If the engine will not start and the visual inspection is complete, begin with the first test in the no-start section. If the complaint applies only to cold driveability, begin with the first cold driveability test and likewise for a problem that occurs only when the engine is warm.

If you begin with the first test of the appropriate section and then go to the next test indicated by the results of the test just completed, the test procedure takes you through all the components and systems that could cause the driveability complaint, including systems not controlled by the computer. This, of course, is both the strength and the weakness of the early Chrysler system: granted, they thought of everything that can go wrong when they wrote the book, and going through the procedure will find everything. But it can be time-consuming if the problem is at the end of the list. So later Chryslers include self-diagnostics.

Diagnostic Procedure, 1985 and Later

In the 1985 and later vehicles, Chrysler diagnostic tests include self-diagnostics, using fault codes and other features. Failures of the most vital sensors and actuators record specific fault codes.

You need the diagnostic readout box (tool C4805 or equivalent) to put the system into one of its diagnostic modes. The 1985 system has three such modes:

- *diagnostic test mode*—The computer reads out any stored fault codes.
- *circuit actuation test mode (ATM test)*—The technician chooses an actuator circuit and activates the test. The computer then turns that circuit on and off at quarter-second intervals for five minutes while the technician checks the actuator to see that it is working properly.
- *switch test mode*—This allows the technician to verify that the computer is reading input switch signals.

If the problem is not a no-start and if the visual inspection reveals nothing amiss, the diagnostic test mode is the usual first step in the self-diagnostic driveability test procedure.

The 1986 system added a sensor test mode to the other three modes. In this mode, the computer reads out the sensor input values through the readout box.

Reading Mixture Solenoid Calibration

A voltmeter connected across the carburetor mixture solenoid (sometimes called the O_2 solenoid) will reflect the mixture commands from the computer. A voltage reading close to battery voltage indicates a lean command; a voltage reading closer to zero indicates a rich command, either because the oxygen sensor signal indicates lean or because operating conditions (such as high load or acceleration) require a rich mixture. Of course, since the actuator command is an on/off signal, the voltage reading is actually an average that reflects the duty cycle of the mixture solenoid. Hence an average reading of 7 volts would indicate approximately a 50% duty cycle. Notice also the system is set to go rich in the event the mixture control fails: this retains driveability until the vehicle can be repaired.

Real-World Problem

Overcompensation. If a saturated charcoal canister or fuel-diluted crankcase oil introduces excess fuel vapor into the intake system, the system will overcompensate. It will drive the mixture too lean while in closed loop, leaner in fact than the actuator system can deliver. This problem frequently occurs when a motorist overfills a fuel tank, pumping long after the nozzle shutoff. That practice can allow liquid fuel to fill the vapor absorption lines and eventually the charcoal canister. Likewise, a vacuum leak, for example from a leaking PCV hose, can drive the system over-rich.

SUMMARY

In this chapter, we have considered Chrysler's original computerized engine control systems, originally built in 1979. Beginning with the oxygen sensor feedback system, we learned the concepts of open and closed loop and how they govern the engine's intake air/fuel mixture, and the resulting exhaust emissions.

Among the major sensors the computer uses to calculate its outputs are the coolant and charge temperature sensor input signals. Fine-tuning of the intake air/fuel mixture follows the feedback signals from the oxygen sensor, and the spark advance is advanced as much as compatible with good driveability through the use of a knock sensor on some models.

▲ DIAGNOSTIC EXERCISE

An early Chrysler computer-controlled vehicle is brought into the shop with a stuck choke. It never opens beyond about halfway. What are the problems you might expect to find from such a condition? In particular, what kind of changes to the computer's fuel mixture delivery capacity would result? How would the computer attempt to compensate, and what would the effect of that effort be?

REVIEW QUESTIONS

1. What year did the O_2 feedback system come into widespread use?
2. List the operating modes of the O_2 feedback system.
3. List seven operating conditions that keep the system in open loop.
4. What sensor monitors engine load?
5. List five engine support functions the combustion control computer controls.
6. List the three types of mixture control solenoid used on different O_2 feedback engine applications by carburetor application.
7. For each of the mixture control solenoids by carburetor, identify the medium (air or fuel) used to control the mixture during idle and part throttle operation.

Carburetor	Idle	Off Idle
Holley 6145	_____	_____
Holley 6520	_____	_____
Carter	_____	_____

8. When the engine is cold, what inputs determine spark timing?
9. When the engine is warm, what inputs determine spark timing?
10. Under what conditions is the EGR valve closed?
11. What does the vacuum secondary solenoid do?
12. List the three parts of the driveability test procedure for vehicles without on-board diagnostics.
13. List the four parts of the driveability test procedure for vehicles with on-board self-diagnostics.

ASE-type Questions (Actual ASE tests rarely are so product-specific).

14. Technician A says O_2 feedback systems do not use a throttle position sensor. Technician B says during one model year in the early development of O_2 feedback, a type of throttle position sensor was used. Who is correct?
    a. A only.
    b. B only.
    c. both A and B.
    d. neither A nor B.
15. Technician A says there is no easy way to tell whether the feedback system is running rich or lean other than by checking for oxygen sensor codes on later model vehicles. Technician B says a voltmeter can be connected across the mixture solenoid terminals to determine whether the computer is commanding rich or lean. Who is correct?
    a. A only.
    b. B only.
    c. both A and B.
    d. neither A nor B.
16. A 1981 vehicle with an O_2 feedback system comes into the shop with a driveability problem. Technician A begins the driveability test warm procedure because the motorist reported the problem only occurs when the engine is warm. Technician B says Technician A should not start with the driveability test warm procedure. Who is correct?
    a. A only.
    b. B only.
    c. both A and B.
    d. neither A nor B.

Chapter 13

Chrysler Single-Point And Multipoint Fuel Injection Systems

OBJECTIVES

In this chapter you can learn:
- ❑ the operating modes of Chrysler's single-point and multipoint fuel injection systems.
- ❑ the function of the power module and how it works with the logic module.
- ❑ the two methods each of these systems use to control idle speed.
- ❑ how the multipoint (turbocharged) system controls spark knock.
- ❑ the three methods to control turbocharger boost.
- ❑ malfunctions in the system that stop the engine.
- ❑ the test modes to diagnose driveability problems.

KEY TERMS

Adaptive Memory
Direct Ignition/Waste Spark Ignition
Inductive Pickup/Hall-Effect Pickup
Limp-In Mode
Logic Module and Power Module
Self-Diagnostics/On-Board Diagnostics
Sensor
Sequential Fuel Injection
Single-Point/Multipoint Fuel Injection

While Chrysler produced a few Imperials in the early 1980s with an electronic fuel injection system, the first current versions of Chrysler fuel injection appeared on the 1984 2.2-liter engines. These were the **single-point** system, which Chrysler calls Electronic Fuel Injection (EFI), and the **multipoint** system, which Chrysler calls Multi-Point Injection (MPI). The 2.2-liter MPI system is often called the "turbo" system because it uses a turbocharger. Since the original engine offerings, however, other power plants have come with either the EFI or MPI system. The two systems are quite similar, so this section covers them together. The text points out variations between the two.

Before 1986 the single-point system used high fuel pressure, about 36 psi. In 1986 a low pressure (14.5 psi) injector with a ball-shaped valve replaced the previous unit with its pintle-style valve. This lower pressure injector is found on many other fuel injection systems besides Chrysler's.

LOGIC MODULE/POWER MODULE

From 1984 to 1988, two separate modules controlled the Chrysler electronic fuel injection systems. The **logic module** with the central processor was inside the passenger compartment, Figure 13-1. The **power module** works the actuators by switching the ground side circuits for the ignition coil, fuel injector or injectors, and the automatic shutdown relay.

The logic module gets most of its electric power from the power module, with a separate circuit through a fuse for memory retention. The

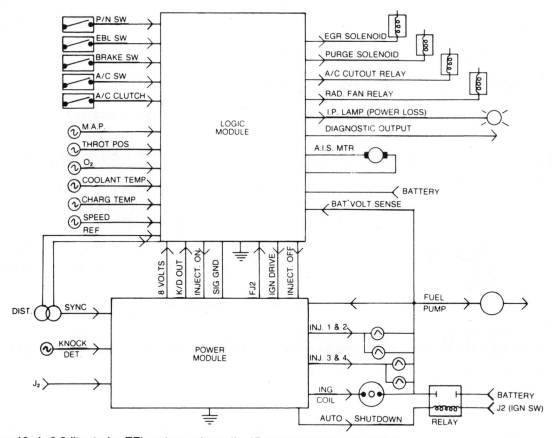

Figure 13–1 2.2-liter turbo EFI system schematic. (Courtesy of Chrysler Corp.)

logic module sends a 5-volt reference signal to its sensors and makes all the system's calculations.

Besides running the actuators, the power module also controls the charging system (as built in 1985). It does so by providing ground for the alternator's field coil, Figure 13-2. Because all of the actuator and alternator circuits are relatively high current compared to the information signal currents in the logic module, the power module is separate and is located in the engine air intake tube to carry off the heat these high current transistors convey, Figure 13-3.

The logic module signals the power module when to open and close the primary ignition circuit and when and how long to turn on the injector/injectors. The power module also provides a constant 8 volts to the logic module and to the distributor Hall-effect pickup unit.

When the ignition switch is on (run), it powers the power module through a fused circuit. Chrysler literature refers to this circuit as J2. On the printed circuit board of the power module is a rail that acts as a fuse and protects the module against reversed polarity. Should a reversed polarity occur, the fuse blows. A reserve circuit with a diode continues to supply J2 power to the power module with a slight voltage drop because of the resistance of the diode.

Single-Module Engine Control (SMEC). In mid-1987 in some 3.0-liter engine applications, the logic module and the power module were both placed as separate printed circuit boards in

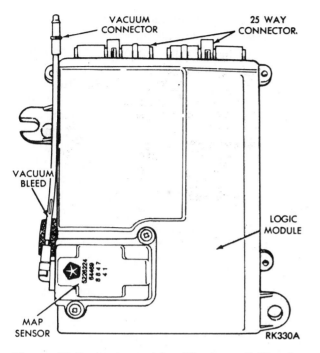

Figure 13–2 Logic module. *(Courtesy of Chrysler Corp.)*

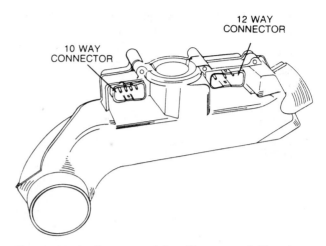

Figure 13–3 Power module. *(Courtesy of Chrysler Corp.)*

a single unit, the SMEC module. Like the earlier power module, it generates heat from the high current transistors and gets cooling air in the intake air duct. By 1988 most Chrysler vehicles used the SMEC module system.

Because most of the discussion in the rest of this section applies to either the earlier separate module systems or the later SMEC systems, the text will refer to the logic board and the power board, regardless of system unless something is specific to only one form of the Chrysler system.

The only significant technical change is the increase to between 9.2 and 9.4 volts at the logic board component of the SMEC. It uses a different kind of transistor (metal oxide semiconductor field effect transistor) on the power board requiring the higher voltage. The battery temperature sensor—used by the system to calculate the proper charging voltage—has been moved to the logic board.

Features

Automatic Shutdown (ASD) Relay.   The ignition switch energizes the ASD relay when it is in the run position. The power module energizes the ASD coil by grounding its return circuit, and the ASD relay powers the electric fuel pump, the ignition coil, and the injector(s).

The fuel pump has its own independent ground and turns on immediately. The ignition coil and injectors get power from, eventually, battery positive, but no current flows through their circuits until the power module grounds their activation circuits. From 1985 through 1987 the ASD relay was internal to the power module. In 1984 and after 1987, it is a separate relay, outside the power module, Figure 13-4. Mounting it outside saves module space and improves module reliability by the removal of the heat it generates.

WARNING: A stuck-on ASD relay will keep the fuel pump running constantly. Besides running the battery down, this could also create a significant fire hazard. Before opening any part of the fuel system on a Chrysler with recurrent battery rundown, make sure the fuel pump is not running and correct any ASD problem first.

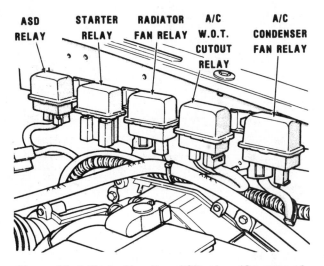

ASD RELAY STARTER RELAY RADIATOR FAN RELAY A/C W.O.T. CUTOUT RELAY A/C CONDENSER FAN RELAY

Figure 13–4 Typical location, ASD relay. *(Courtesy of Chrysler Corp.)*

Self-Diagnostics or On-board Diagnostics. **Self-diagnostics or on-board diagnostics** mean essentially the same thing. The logic board monitors its most important input and output circuits for faults and improbable readings, and records fault codes in memory when an inappropriate value appears or an actuator fails.

Adaptive Memory. **Adaptive memory** is the ability of an engine control computer to assess the success of its actuator and sensor signals, and modify its internal calculations to correct them if the desired result is not achieved. The logic board can modify some of its programmed calibrations ("maps") for fuel metering to compensate for production tolerance variations and changes in barometric pressure or vehicle altitude.

There are limits to the range of adjustment available to the adaptive memory capacity. Sometimes mechanical problems—a collapsed exhaust system, a burnt valve—throw the engine performance so far off that no countermeasure available to the computer can correct for it.

Operating Modes

Starting. When the starter cranks the engine, double injector pulses occur to richen the intake mixture. The single-point injector on the non-turbocharged engines normally pulses twice per revolution, and the multipoint injectors normally pulse in pairs, each pair spraying fuel once for each revolution. The double-pulsing at crank occurs only for a programmed time interval to avoid flooding. At this time coolant temperature alone determines the pulse width. Once the engine starts, the logic board provides fuel enrichment depending on coolant temperature and manifold pressure. The start-up enrichment "decays" to base enrichment over a programmed time interval. Once the coolant reaches the closed loop criterion temperature, of course, the feedback system controls mixture.

Primer Function. As soon as a driver turns the ignition on with the 3.0-liter engine, all six injectors pulse fuel into the intake ports. This improves starting and occurs regardless of coolant temperature, though if the coolant is warm, less fuel will spray. If the engine is cold, relatively more fuel sprays.

Open Loop. Open loop—when the oxygen sensor does not determine fuel injection pulse width—occurs when:

- the coolant is below a criterion temperature.
- the oxygen sensor is below operating temperature (approximately 600 degrees Fahrenheit/315 degrees Centigrade).
- the oxygen sensor does not switch back and forth across a switch point of about 0.45 volt.
- the vehicle is at wide-open throttle (WOT) or under similar high acceleration or load.
- when the turbocharger boost on the multipoint system reaches 1 psi or more.
- when the vehicle is decelerating or coasting with the throttle closed.
- when an engine with the single-point injector is idling.
- when there is a major sensor or actuator malfunction preventing feedback mixture control.

Closed Loop. Each system will go into closed loop only if:

- the coolant has reached the criterion temperature.
- the oxygen sensor produces a usable signal.
- a vehicle-specific time has passed since start-up.

Limp-in Mode. The logic board self-diagnostics circuits include the capacity to monitor incoming signals from the most important sensors. If one of them sends a signal out of the expected range or no signal at all, the self-diagnostics consider the sensor inoperative and shift the engine management system into **"limp-in" mode.** In this mode, the logic board substitutes a probable value for that of the failed sensor, sets a fault code and turns on the Power Loss light on the dashboard. Obviously, the logic board cannot substitute for the distributor's Hall-effect sensor signal; the engine just stops. But it can do so for coolant temperature or MAP sensors.

In limp-in mode, the engine will not develop full power nor operate with full efficiency, but it can still run. The sensors whose failure can trigger limp-in are:

- Manifold absolute pressure (MAP) sensor— The computer can create a simulated value from the inputs from the throttle position sensor and the engine speed.
- Throttle position sensor (TPS)—The computer uses the MAP sensor signal to create a substitute value.
- Coolant temperature sensor—The computer uses the charge temperature (intake manifold fuel/air mixture temperature) sensor signal as a substitute.
- Charge temperature sensor—The computer can usually manage engine performance without this information. While a fault code is set in the case of a charge temperature sensor failure, the Power Loss light does not come on.

If the distributor reference signal stops, so does the engine. There is then no way to time ignition spark.

Real World Problem

Certain mechanical problems can throw a system out of closed loop repeatedly. Suppose a car develops a series of bad oxygen sensors in a relatively short time. An experienced computer control technician will check for causes external to the system, too. This particular problem often comes when a head gasket is starting to fail and seeps coolant into the combustion chamber or exhaust. The silicone in the coolant forms a super-thin, impermeable layer on the exhaust surface of the sensor, "poisoning" it. Less frequently, coolant silicone from a burst hose or some other contaminant will get into the atmospheric side of the sensor. More rarely, silicone poisoning occurs when a technician uses older silicone-based gasket sealers.

INPUTS

Manifold Absolute Pressure (MAP) Sensor

This is a piezoresistive pressure sensor. A piezoresistor changes its electrical resistance in response to the slight bending that changes in pressure cause in it. The logic board uses its signal as a barometric pressure sensor during periods of ignition on/engine off or during certain other conditions on a few models as explained in the **Outputs** section of this chapter, Figure 13-5. The MAP sensor is usually located on the logic module.

The technician should be clear what manifold absolute pressure is. Formerly the term *intake manifold vacuum* was used. But that can be misleading (particularly after the widespread introduction of turbocharged cars). The manifold absolute pressure is the actual pressure of the air in the intake manifold. Often, of course, it is lower than ambient pressure, but it is still air pressure. If a turbocharger or supercharger increases the pressure above ambient pressure, it is still *manifold absolute pressure.* This dimension corresponds very closely to engine load.

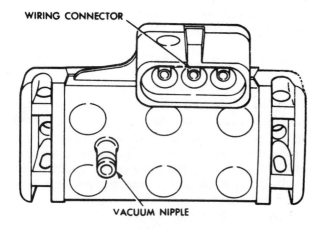

Figure 13–5 MAP sensor. *(Courtesy of Chrysler Corp.)*

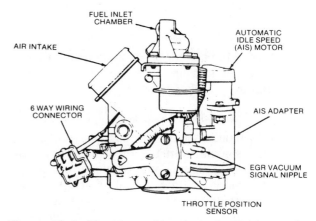

Figure 13–6 Throttle position sensor. *(Courtesy of Chrysler Corp.)*

Some earlier vehicles had a small bleed hole in the vacuum line to the MAP sensor to prevent condensation from collecting in the line. Those MAP sensors are calibrated to allow for the vacuum leak.

Real World Problem

The early MAP sensor arrangement had two problems frequently seen in repair bays. In colder climates in the winter, the system would set a code and drop into limp-in mode. But after the vehicle was in the work bay for a while, the problem mysteriously stopped without repairs. What happened was this: dirt plugged the vacuum line vent; the moisture froze and plugged the line, blocking vacuum from the sensor. The computer noticed the problem and set the code. Once the car was in the heated work bay, the ice plug melted, and the sensor worked normally.

Real World Problem

The early system used the MAP sensor as a barometric sensor when the ignition was on but the engine off, and it retained that value in memory for the duration of that trip. However, if the trip involved a significant change of altitude—up or down a sizable mountain—the engine might run very poorly until it was shut off and restarted, with no codes set because there was no sensor or actuator circuit. A driver or roadside mechanic who believed that it was bad policy to shut off an engine that was still running, however poorly, would never get the engine right until the car returned to the beginning altitude, or stalled or ran out of gas and stopped on its own.

Throttle Position Sensor (TPS)

The TPS is a rotary potentiometer—a variable resistor—at the end of the throttle shaft on the side of the throttle body, Figure 13-6. Depending on the position of the throttle, the sensor modifies the 5-volt reference signal to something less that corresponds with the throttle's current position, and returns that signal to the computer. Some engine management systems also keep track of how quickly the throttle moves, to more accurately provide acceleration enrichment when the pedal is suddenly floored, or to shut off fuel if the throttle is suddenly closed at high engine speed.

Oxygen (O₂) Sensor

The early Chrysler systems used a single-wire oxygen sensor. Some of the later systems use a three-wire sensor—the additional wires for an internal heater element.

Coolant Temperature Sensor (CTS)

The CTS is a single-element, negative temperature coefficient (meaning the resistance goes down as the temperature goes up) thermistor screwed into the thermostat housing, Figure 13-7. As it warms up, its resistance goes down. At -4 degrees Fahrenheit/-20 degrees Centigrade its resistance is 11,000 ohms, and it ranges to 800 ohms at 195 degrees Fahrenheit/90.5 degrees Centigrade, full operating temperature on these systems.

Charge Temperature Sensor (Charge Temp)

The charge temperature sensor is a single-element thermistor (again negative temperature coefficient-type) screwed into a runner of the intake manifold. It has about the same resistance

range as the coolant temperature sensor and looks similar. On early models its input contributes to cold engine enrichment calculations; otherwise it serves as a backup to the CTS. On later multipoint systems, its input helps control the air/fuel mixture when cold and boost control at all times.

Throttle Body Temperature Sensor

The low-pressure, single-point system does not include the charge temperature sensor, but instead a throttle body temperature sensor to measure temperature at the throttle body, Figure 13-8. This temperature will ordinarily be very close to intake air temperature, and this signal helps calculate the mixture for hot restarts.

Distributor Reference Pickup (REF Pickup)

A Hall-effect switch in the distributor generates the interrupted-direct-current reference pickup signal from which the logic module determines engine speed and crankshaft position. On the single-point system this same signal goes to the power board as well. Without the signal, the power board removes ground from the ASD relay,

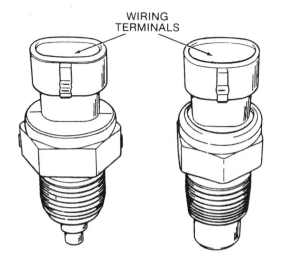

Figure 13–7 Charge and coolant temperature sensors. *(Courtesy of Chrysler Corp.)*

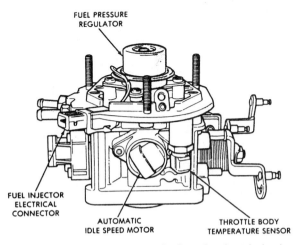

Figure 13–8 Low-pressure single-point throttle body, with throttle body temperature sensor. *(Courtesy of Chrysler Corp.)*

shutting off the ignition coil, injectors, and fuel pump. On the multipoint systems, only the logic board gets the distributor reference signal, but the logic board uses it to instruct the power board when to turn on the injectors.

A Hall-effect position sensor has specific differences from an inductive pickup, used by other manufacturers and for other purposes. The **Hall-effect sensor** generates a clean on/off signal and does not vary either in accuracy or output voltage with rpm. An **inductive sensor,** on the other hand, produces an alternating current, and the point of polarity reversal is the reference point. An inductive sensor may produce very low voltage signals at low speed, and very high voltage at higher speed. It can also change position

as the speed increases, because of the time required for the magnetic field to reverse. A Hall-effect sensor, however, does not generate its own current. It is dependent on the reference signal sent to it: the circuit is more complex and more things can go wrong.

Distributor Sync Pickup (SYNC Pickup)

Most multipoint systems use a second Hall-effect switch/sensor in the distributor, below the reference pickup, Figure 13-9. This switch has a shutter wheel with only one vane instead of four. The one vane, however, covers 180 degrees, half of the shutter wheel. The leading and trailing edges of the vane each provide a signal (on or off) to the logic and power boards. The power board uses the SYNC signal along with other information to calculate pulse width. The power board uses it to determine which pair of injectors (1 and 2 or 3 and 4) to turn on when the instruction from the logic board arrives.

Optical Distributor. A 3.0-liter V-6 engine, built by Mitsubishi, appeared on some Chrysler vehicles in 1987. This was Chrysler's first naturally-aspirated (nonturbocharged) engine with multipoint fuel injection. The computer for this engine uses a speed-density formula to calculate fuel quantity in a way similar to the way it is done on the 2.2 turbocharged engines. A speed-density

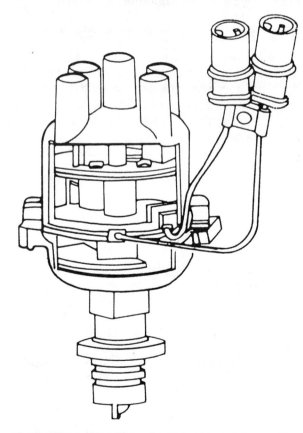

Figure 13-9 Distributor for turbocharged four-cylinder engine, with reference and sync Hall-effect pickups. *(Courtesy of Chrysler Corp.)*

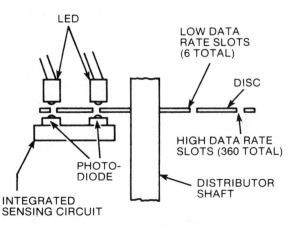

Figure 13-10 Optical distributor pickup

system does not require a vane airflow meter or a mass airflow meter to measure air.

The engine management system with this engine does not use the Hall-effect distributor, but an optical distributor with a pair of optical sensors, Figure 13-10. A disc with two sets of slits rotates with the distributor shaft. The inner set has six slits, called "low data rate" slits. The outer set has one slit for every two degrees of crankshaft rotation except for one small blank spot with no slits. This section alerts the SMEC where the number one cylinder is. The high data rate from this sensor allows the computer to track minute changes in engine speed.

Two light-emitting diodes (LEDs) are on one side of the disc, and the receptors are on the other, Figure 13-11. When one of the slits aligns with an LED, the light strikes a photodiode in the receptor, generating a small voltage applied to the base of a transistor.

We will follow the "low data rate" reference signal for the explanation, though the "high data rate" circuit works the same way: The transistor works as a ground switch for the 5-volt reference signal sent through a resistor by the SMEC computer. When the transistor turns on from the photocell pulse to its base, the voltage drops and the voltage difference across the resistor increases. When the slit in the distributor disc moves and blocks the light, the photocell shuts off; the tran-

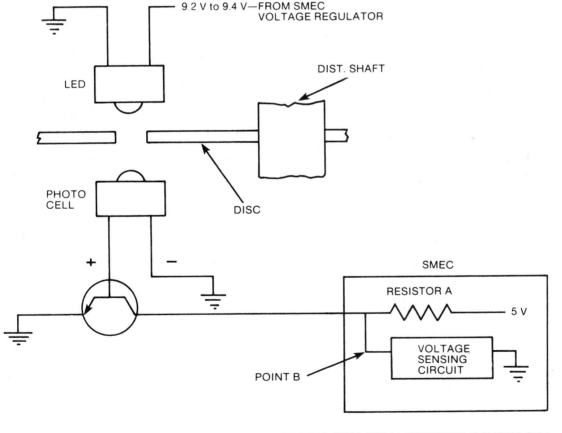

Figure 13–11 Optical distributor sensor schematic

sistor opens the reference ground circuit, and the voltage drop across the resistor disappears. Each time the voltage drops toward zero, the SMEC computer knows the next piston in the firing order has reached a specific position in its cycle. The "high data rate" sync circuit works the same way, each signal indicating the crankshaft has rotated two degrees.

Below 1,200 rpm the SMEC computer uses the sync signal to calculate spark timing and injector pulse. Above 1,200 rpm, it uses the reference signal. The reason for the switch (besides the easier task of monitoring the lower data rate at higher engine speeds) is that there are significant changes in engine speed at each power stroke, but slight differences in compression ratio, spark plug condition, and fuel atomization mean different changes for each cylinder. The high data rate signal allows the SMEC computer to monitor these changes and adjust timing and injection pulse accordingly. At higher engine speeds, there is much less variation in rpm between different cylinders' power strokes, so such detailed adjustments are not necessary.

Direct Ignition System (DIS).   In 1990, Chrysler introduced its own 3.3-liter 60-degree V-6 for some of its passenger cars and vans, Figure 13-12. This engine uses a distributorless, *direct ignition system* similar to General Motors and Ford distributorless systems. The idea of using one coil to fire a pair of plugs—one coil for every two cylinders—goes back to motorcycles, where high rpm limited available dwell time to build up the coil's field and where space for the coils was even more limited than on cars. This system is often called a **"waste spark" ignition.** Both plugs on the same coil fire simultaneously, one at the end of its compression stroke just before the power stroke; the other at the end of its exhaust stroke. Because of the low resistance in the exhaust stroke combustion chamber, the "waste" spark is lower voltage than the active one.

The electrical difference between a waste spark distributorless direct ignition system and a conventional setup is that the ignition secondary

(high voltage) windings are not directly grounded. Instead, one plug gets a negative polarity spark and the other a positive polarity. Of course, from the point of view of the air/fuel mixture, the polarity of the spark doesn't matter: any hot spark of either polarity sets the mixture burning.

The Chrysler direct ignition system, like the GM and Ford systems, uses Hall-effect crankshaft and camshaft position sensors, an ignition module and a coil pack with a coil for every 360-degree pair of cylinders (those with exactly alternating power strokes). The coil pack bolts over the ignition module.

The camshaft position sensor mounts on the timing cover and triggers off slots on the camshaft timing gear. The slots on the timing gear are coded so signals from the sensor identify which pair of companion cylinders the ignition module should fire next in response to the computer's spark timing command. This signal also determines which pair of injectors the computer pulses next. Because the camshaft sensor can identify individual cylinders (unlike the crankshaft sensor), the computer can use its signal to begin firing spark plugs and pulsing fuel injectors within the first crankshaft rotation.

The crankshaft sensor mounts on the bell housing and triggers off slots on the torque converter drive plate. There are four slots per pair of companion cylinders. The computer uses these signals to determine crankshaft position and engine speed. It uses this information in determining ignition timing, injector timing, and injector pulse width.

Vehicle Speed (Speed Sensor)

The speed sensor is a simple on/off switch that cycles eight times per speedometer cable rotation. The computer sends it a steady five-volt reference signal which it interrupts (producing an interrupted-direct-current, or for practical purposes alternating current) at a frequency corresponding to transaxle output shaft speed. The vehicle speed sensor mounts to the transaxle at the base of the speedometer cable, Figure 13-13.

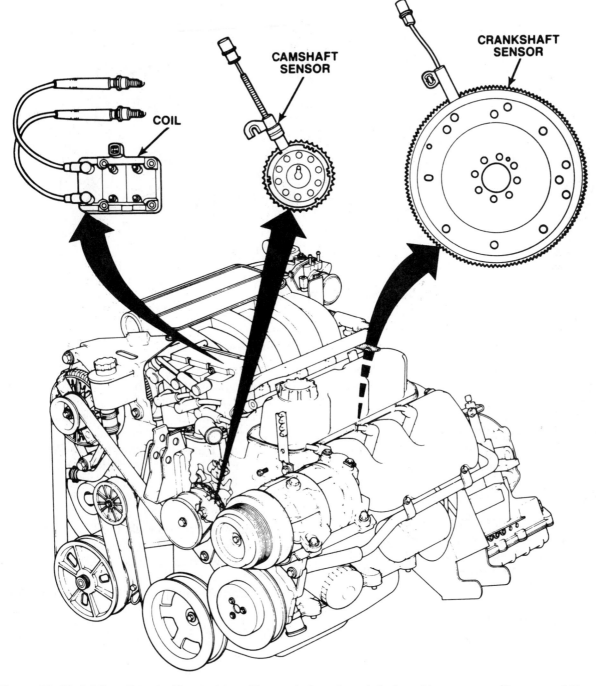

CAMSHAFT
SENSOR

CRANKSHAFT
SENSOR

COIL

Figure 13–12 3.3-liter direct ignition system with camshaft and crankshaft position sensors. *(Courtesy of Chrysler Corp.)*

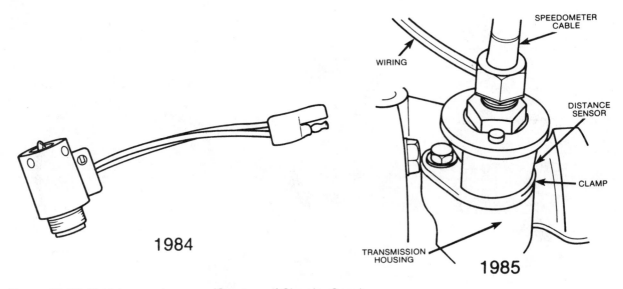

Figure 13–13 Vehicle speed sensor. *(Courtesy of Chrysler Corp.)*

Detonation (Knock) Sensor

The Chrysler knock sensor is of the piezo-electric type, bolted into the intake manifold. When excited by vibrations typical of detonation, it sends a true alternating current signal to the computer. This signal then begins countermeasures, which reduce spark advance by individual cylinders and/or reduce turbocharger boost.

Battery Temperature Sensor

A thermistor in the power board, adjacent to the battery, signals to the computer the battery's ambient temperature. The computer then uses this information to regulate alternator output and battery charging rate.

Charging Circuit Voltage

The logic board senses charging system voltage by monitoring the fuel pump power feed circuit and regulates alternator output accordingly. As we learned in the preceding paragraph, it also factors in the battery temperature to determine what the proper charging voltage should be.

Switch Inputs

Park/Neutral Switch (P/N). The P/N switch tells the logic board whether the transmission is in gear or not. Its input influences idle speed. Normal idle spark advance is canceled when the transmission is in neutral or park.

Electric Backlite (Heated Rear Window). Whenever the heated rear window switch is on, the logic board will increase the idle air flow to compensate for the additional load the heating current places on the alternator.

Brake Switch. Should the throttle position sensor/switch fail, the computer looks to the brake light switch as a signal that the throttle is closed.

Air Conditioning (A/C). Whenever the A/C is on, the A/C control circuit sends the logic board computer a signal. The logic board then increases idle intake airflow to compensate for the additional compressor and alternator load.

Air Conditioning Clutch. When the A/C system cycles the compressor clutch on and off at idle, the A/C clutch switch provides a signal for the computer to adjust the intake airflow and compensate for variations in load. Later models combine both air conditioning inputs into one signal.

Battery. One terminal of the logic board gets power directly from the battery to sustain power in the keep-alive-memory (KAM) when the ignition is off.

OUTPUTS

Injector/Injectors

Figures 13-14 and 13-15 show the low and high pressure fuel injectors used in the single-point injection systems. Each type is a solenoid-operated device.

In normal operation, the single-point injector pulses fuel twice for each engine revolution. The early multipoint fuel injection systems on four-cylinder engines pulsed one pair of injectors during the first revolution of a full cycle and the other pair during the second revolution. As explained before, this injection schedule was doubled during cranking.

Later 2.2-liter Turbo IV and 2.5-liter Turbo I systems employ **sequential fuel injection.** Each injector pulses individually in the engine's firing order. The 3.0 and 3.3-liter V-6 engines pulse injectors in sequential pairs. In each case, the logic module calculates the pulse width based on input signals from:

- CTS.
- MAP.
- RPM.
- TPS.
- oxygen sensor—in closed loop.
- charge temperature—during cold enrichment.
- speed sensor—during deceleration.
- fuel enrichment and enleanment factors programmed into the logic board.

In contrast to the more common wiring practice, the 1984 to 1987 Chrysler single-point systems have a fixed constant ground. The power module controls injection by switching power on and off (positive side switching). From 1987, the single point system used ground-side switching, making it compatible with the multipoint systems.

Electric Fuel Pump

A permanent magnet-type direct current electric motor drives the vane-type fuel pump, both submerged in fuel at the bottom of the fuel tank, Figure 13-16. One check valve functions as a pressure relief valve; the other, in the outlet port, prevents fuel from running in either direction

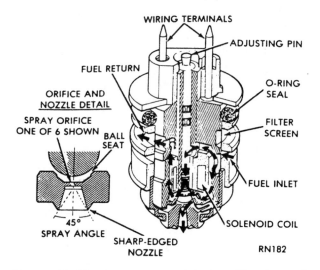

Figure 13–14 Low-pressure injector. *(Courtesy of Chrysler Corp.)*

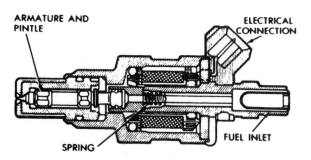

Figure 13–15 High-pressure injector. *(Courtesy of Chrysler Corp.)*

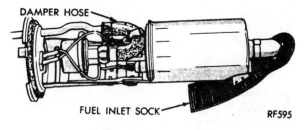

Figure 13–16   Electric fuel pump. *(Courtesy of Chrysler Corp.)*

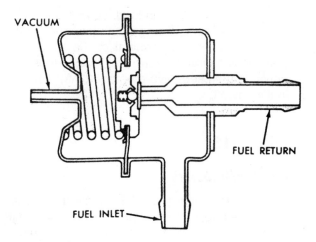

Figure 13–17 Fuel pressure regulator, multipoint-turbocharger system. *(Courtesy of Chrysler Corp.)*

when the pump is off. This keeps the line full of pressurized fuel when the engine is off and reduces fuel vapor problems.

With the ignition switch on, the power board grounds the ASD relay, powering the fuel pump. Unless there is a distributor reference signal within two seconds, the ASD will lose its ground and shut off the fuel pump. Once shut off, the ASD relay stays off until the distributor reference signal arrives at the computer, or the key is cycled off and on again.

A few vehicles, such as the Shelby GLH turbocharged Omnis, use two fuel pumps, one in the tank and another near the engine. As these vehicles sometimes use more fuel, they are more subject to vapor lock. The tandem electric pumps reduce that possibility.

Fuel Pressure Regulator.   To make air/fuel mixture calculations possible, the fuel injection system uses a fuel pressure regulator to maintain the difference between the fuel and the intake manifold constant, Figure 13-17. This is often described as a "pressure differential across the injector tip." Then the amount of fuel will correspond closely to the pulse width, with no allowances needed for pressure differences.

The different systems use different pressures and different (though similar) pressure regulators. The pre-1986 single-point system maintains a fuel pressure difference of 36 psi; the later low pressure system maintains a difference of 14.5 psi. The multipoint systems maintain a difference of 55 psi. Chapter 2 of this book explains how pressure regulators work on all systems. Fuel pressure, as

measured in the injector line on a Chrysler multipoint system, can vary from 42 to 62 psi.

On the 3.0 liter engine the fuel pressure was raised to 46 psi to avoid vapor lock problems during hot restarts.

Ignition Timing

The logic board calculates ignition timing from information reported by:

- the coolant temperature sensor.
- the distributor reference pulse (crankshaft/camshaft position sensors).
- the manifold absolute pressure sensor (MAP).
- the barometric pressure (startup MAP reading).

At warm idle, the logic board computer first uses spark advance manipulation to control normal idle speed fluctuations. If a specific amount of spark timing change does not put the idle speed at the design rpm, the logic board will use the automatic idle speed motor to change the amount of air entering the engine at idle.

Beginning in 1985, the multipoint system included a BARO-read (ambient barometric pressure) solenoid that the logic board used to briefly vent the vacuum line to the MAP sensor. During

Chrysler's Detonation Control

Chrysler put strong emphasis on performance in their turbocharged vehicles. This is reflected in their detonation control strategy. You must prevent detonation, of course, or the engine will destroy itself. There are only two ways to stop detonation on a turbocharged engine under boost: reduce the boost or retard the timing. If you reduce the boost, you lose performance; if you retard timing you raise exhaust temperature, which was already critical under boost. Chrysler's solution of retarding timing to one cylinder only raises exhaust temperature only slightly and allows the other three cylinders to keep producing at their full capacity.

the time the line was vented, the MAP sensor read ambient barometric pressure. The computer can now update this reading as often as every thirty seconds if needed, though it is very unusual for barometric pressure or altitude to change much that quickly. This vent cycle occurs at least once each time the throttle closes and engine speed is below a certain rpm. Vehicles with a turbocharger must monitor and control ignition timing very precisely, and barometric pressure significantly affects allowable spark advance.

Detonation Control. From the beginning of the turbocharged four cylinder with multipoint fuel injection, the Chrysler multipoint system has had the ability to retard the timing on just the cylinder with knock. If knock occurs, the logic board has already recorded which cylinder was just fired. It then retards the spark for just that cylinder slightly. If knock continues, the logic board will begin to reduce turbocharger boost. In 1986 a ten-second delay was built in between the time the throttle moves to wide-open throttle before the logic board responds to knock.

Wastegate Control Solenoid

Before 1985, turbocharged engines used

three types of boost control. The first was the wastegate, as described in the **Outputs** section of Chapter 5. The wastegate actuator, Figure 13-18, opens the wastegate at a boost pressure of 7.2 psi. The second method has the boost pressure sensed by the MAP sensor. If boost exceeds 7.2 psi, after snap acceleration for example, the logic board recognizes the overboost and skips fuel injection pulses until boost pressure drops to 7.2 psi. The third method uses the electronic engine speed governor: if rpm goes above 6,650, the logic board stops fuel injection until the rpm drops to 6,100.

In 1985 a fourth method was included. The logic board can operate a boost control solenoid that vents the line conveying manifold pressure to the wastegate actuator. If operating conditions are favorable, the logic board will pulse the boost control and bleed off some of the pressure. Under these circumstances the boost pressure can actually go to 10 psi before the actuator opens the wastegate. The logic board reviews these inputs for this calculation:

- barometric pressure.
- engine speed (rpm).
- coolant temperature (CTS).
- intake manifold charge temperature.
- detonation (knock sensor).

The logic board also keeps track of the engine's "detonation history," how inclined it is to knock, and under what circumstances as well as how long it has been in boost.

Revised Boost Control

Beginning in 1988, the turbo boost actuator control on the Turbo 1 engine works directly from the turbocharger itself rather than from the intake manifold pressure. This keeps the wastegate open during part-throttle operation and eliminates boost completely except at wide-open throttle. The advantages are:

- exhaust backpressure is lower.

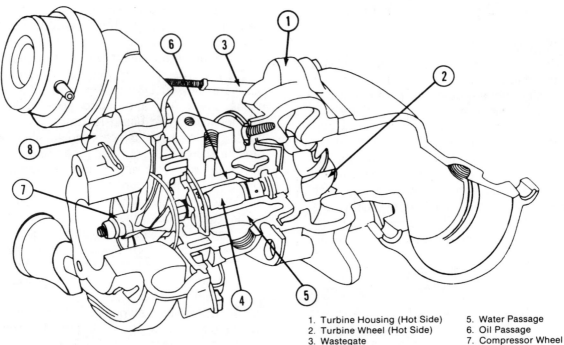

1. Turbine Housing (Hot Side)
2. Turbine Wheel (Hot Side)
3. Wastegate
4. Shaft Wheel Assembly
5. Water Passage
6. Oil Passage
7. Compressor Wheel
8. Compressor Housing

Figure 13–18 Turbocharger. *(Courtesy of Chrysler Corp.)*

• the intake air/fuel charge is cooler.
• the tendency to knock is reduced.
• part-throttle fuel economy improves.

Automatic Idle Speed (AIS) Motor

The AIS motor is a reversible electric motor, Figure 13-19. The logic board controls it based on information from the:

• TPS.
• CTS.
• speed sensor.
• P/N switch.
• brake switch.

The AIS motor moves its valve to control bypass air around the throttle plate in the throttle body. Even with the valve in its closed position,

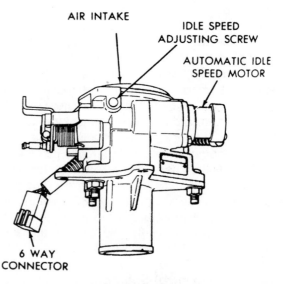

Figure 13–19 Multipoint throttle body showing AIS motor. *(Courtesy of Chrysler Corp.)*

enough air gets past the throttle plate and through the throttle body to keep the engine idling at low speed with no load. The logic board directs the AIS motor to set the valve at different positions for different operating conditions. When the vehicle is decelerating, the valve opens to prevent engine stall and to prevent condensation of fuel on the intake port walls, as well as possible backfire.

EGR Control Solenoid

This solenoid controls ported vacuum to open the EGR valve. When the logic board energizes it, vacuum stops to the EGR valve. The computer blocks vacuum when:

- the coolant temperature is below 70 degrees Fahrenheit/21 degrees Centigrade.
- the engine speed is below 1,200 rpm.
- the engine is at wide-open throttle.

When the computer de-energizes the solenoid, allowing EGR actuation, a backpressure transducer also controls the EGR valve, Figure 13-20. Some vehicle systems have a vacuum bleed in the line between the transducer and the EGR to prevent pressure buildup in the EGR valve.

EGR Valve Temperature Sensor

Chrysler installed a thermistor temperature

sensor on California versions of the 1.5-liter multipoint injection engine in the 1988 Dodge Colt. The purpose of this sensor is to detect an EGR valve that does not open when actuated. When an EGR valve opens, hot exhaust gas flows through it, and the temperature of the EGR mounting flange should go up sharply. If the SMEC computer sends an actuation signal to the EGR solenoid and does not quickly get a return signal from the EGR temperature sensor, it will store a fault code and turn on the malfunction indicator light. Because of changes in cam profile that have reduced valve overlap, Chrysler was able to eliminate the EGR valve on the turbocharged 2.2-liter engine in 1988.

Canister Purge Solenoid

The charcoal canister stores vapors from the fuel tank. So long as the coolant temperature is below 180 degrees Fahrenheit/82 degrees Centigrade, the logic board blocks the vent line to the throttle body by energizing a solenoid on the line, Figure 13-21. If the canister is not purged, it will gradually fill and will not be able to store any more vapors. If it purges too early or at other inappropriate times, it can drive the air/fuel mix-

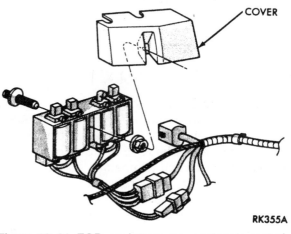

Figure 13–21 EGR, canister purge, wastegate control and BARO-read solenoids. *(Courtesy of Chrysler Corp.)*

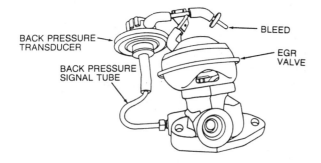

Figure 13–20 EGR valve and transducer. (Courtesy of Chrysler Corp.)

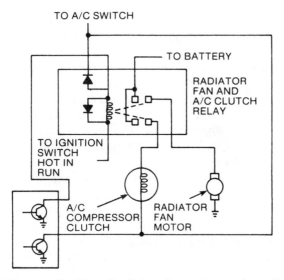

Figure 13–22 Radiator fan relay schematic. *(Courtesy of Chrysler Corp.)*

ture richer than the range of adjustment of the mixture control strategies can correct for.

Radiator Fan Relay

The radiator fan relay supplies power for both the radiator fan and the A/C compressor clutch, Figure 13-22. The logic board grounds the relay coil whenever the coolant reaches a specific temperature or when the air conditioning is turned on. Both fan and compressor clutch are controlled by grounding their relay circuits.

Charging Circuit Control

Beginning in 1985, Chrysler vehicles have alternator output regulated by the logic board working through the power board, replacing the voltage regulator. The logic board calculates the output based on the alternator's current output and on the battery temperature. Once a specific output voltage is reached, the logic board duty-cycles a transistor in the power board, grounding and completing the circuit through the alternator's field coils, Figure 13-23. The duty-cycle varies as needed to maintain the calculated out-

put voltage. This voltage depends on the battery's temperature.

A/C Cutout Relay

The A/C cutout relay is on the ground side of the A/C compressor clutch circuit, between the clutch and the A/C switch. When open, it stops the A/C clutch circuit. The logic board closes the compressor engagement circuit *except:*

- when the engine is operated at wide-open throttle.
- when engine speed is below 500 rpm.
- when the engine is cranked with the A/C switch on. The compressor cutout relay circuit remains open for ten to fifteen seconds after the engine starts.

Torque Converter Lockup Clutch

Beginning in 1988, the Torqueflite transmission came with a lockup torque converter for some vehicles. The SMEC computer activates it under specific conditions:

- When the coolant temperature is above 150 degrees Fahrenheit, *and*
- the park/neutral switch indicates the transmission is in gear, *and*
- the brake switch indicates the brakes are not applied, *and*
- the throttle angle, as reported by the TPS, is above a certain minimum.

The converter clutch works hydraulically, controlled by the computer. The actuation solenoid mounts on the valve body transfer plate and receives third clutch oil when the transmission goes into third gear. If the computer activates the solenoid by grounding its circuit while third clutch oil pressure is applied, the torque converter locks the crankshaft to the input shaft. If not, the open solenoid bleeds the oil pressure off, and the torque converter clutch releases. This works in a way very similar to the GM and Ford systems.

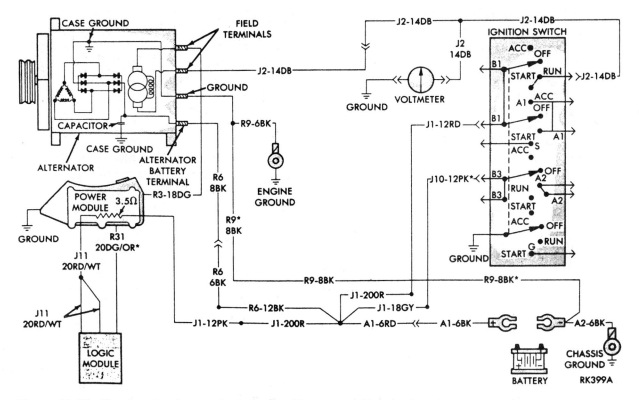

Figure 13–23 Charging circuit control schematic. *(Courtesy of Chrysler Corp.)*

Power Loss Lamp

The Chrysler Power Loss lamp serves the same function as the Check Engine lamp on GMs. If the computer has encountered a fault in any of its sensors, actuators, circuits or internally in itself or the power board, it stores a code and (for most faults) turns on the Power Loss lamp and sets the system in limp-in mode. The lamp also comes on when the key first turns the ignition on as a bulb check. If the fault disappears within one driving trip, the lamp stays on until the next start-up. Then the lamp will be off and the system will come out of limp-in, but the code will stay in the computer's memory for the next thirty key off/on cycles without that problem. In the 1988 and later models, if a sensor fails, the computer generates its own substitute value for that

sensor's readings and lights the lamp. If the sensor comes back into the proper range, however, the computer goes back into normal operation without the need for ignition cycling to exit limp-in mode.

Vehicle Maximum Speed Governor

Also beginning in 1988, vehicles with 3.0-liter and 2.2-liter Turbo I engines have a vehicle maximum speed governor in the computer's program. Because of the speed rating of the tires installed on the vehicle, the computer shuts off all fuel if the vehicle exceeds 118 mph.

SMEC-Controlled Cruise Control

Most of the Chrysler SMEC-computer-controlled systems include the cruise control in the engine management system. The system is similar to the Ford system described in Chapter 10 of this book. One additional feature of the Chrysler variation is a third solenoid called a dump solenoid. The ignition switch powers the dump solenoid through a fuse, the cruise control on/off switch, and the brake switch. Unlike most computer actuators, the dump solenoid has a fixed ground. Either turning the cruise control switch off or stepping on the brake will interrupt power to the dump solenoid, immediately bleeding off vacuum to the servo and letting the throttle return spring pull the linkage to idle.

✔ SYSTEM DIAGNOSIS & SERVICE

A thorough diagnostic procedure is published by Chrysler Corporation for each model year and application. This driveability test procedure should be in the service manual you use on any particular car.

On-Board Diagnostics

Besides directing the combustion activity of the engine, the logic board also monitors specific input and output circuits for faults (voltage values out of range or that don't change appropriately). If a fault is detected, the computer will store a two digit code in its diagnostic memory. These codes and much other diagnostic information can be obtained by putting the system in one of its diagnostic modes using a diagnostic readout box (tool C-4805 or its equivalent). On some systems the fault codes can be read out and the switches tested with the power loss lamp itself.

Test Modes

Diagnostic Test Mode.   Connect the readout box to the diagnostic test terminal. Early terminals are under the hood, later ones may be found in the passenger compartment. Turn the ignition on-off-on-off-on in five seconds or less. The readout box then displays any stored codes.

When it is finished with everything stored in its memory, it displays code 55.

Switch Test Mode.   Once the preceding test is complete and code 55 shows, make sure that all system input switches are turned off. Then turning on any of the input switches will cause the displayed code 55 to change as long as the logic board gets the incoming signal. Turning the switch off makes the code 55 reappear.

Circuit Actuator Test Mode (ATM).   After the diagnostic test mode is completed and code 55 shows in the readout box, press the ATM button on the readout box until the desired ATM code number is displayed. When the button is released, the ATM will begin. The logic board cycles the designated actuator on and off at two-second intervals for five minutes or until you turn the ignition off. This provides ample time to examine the actuator directly and be sure it is working properly. Each ATM code is a two-digit number representing each of the actuator circuits this mode can test. They are listed in the service manual with the fault codes.

Sensor Test Mode.   Beginning in 1986, this test mode allows choosing an individual sensor, and the readout box will display the return voltage signal that the sensor is sending the logic board at that moment.

Engine Running Test Mode.   This test checks a number of sensors and actuators in use. Connect the readout box to the diagnostic test connector and start the engine. Observe the readout box display. Once the oxygen sensor is hot enough, the display will switch between 0 (lean) and 1 (rich). The readout box can also command the idle speed motor to increase idle speed. On turbocharged models, striking the intake manifold near the knock sensor with a tool as explained in the service manual will make the number 8 appear alongside the switching 0-1. The number 8, of course, indicates that ignition timing has been retarded.

Test Mode 10.   The random access memory (RAM) of the SMEC computer is less volatile than that of the logic module. Removing battery power from it, one of the procedures used to clear the memory for logic module-equipped vehicles, may not erase the stored fault codes for some time. So for these systems, there is another sensor test, mode 10. When the technician selects test mode 10, all fault codes are erased. Needless to say, fault codes should not be erased until after they have been written down for later diagnosis.

The SMEC system also has a few additional fault codes and some changes from previous diagnostic procedures. For example, some of the harness connectors are hard to probe with test instruments, so as a result the revised diagnostic procedures may require disconnection of the harness and the performance of open circuit tests.

Testing Without the Readout Box. Cycling the ignition switch on and off three times in five seconds triggers the logic module to flash out codes through the power loss lamp if the readout box is not connected. It flashes on and off to indicate the code number. For example, two flashes separated by a short pause of about one-half second, followed by a pause of about two seconds and then three flashes, separated again with a short pause indicates code 23. Once the codes are flashed out, the system goes into switch test mode. As long as it receives signals from the switches, the logic module will flash the power loss lamp on and off in response to the input switches being correspondingly cycled. The other test modes, however, cannot be performed without the readout box.

Driveability Test Procedure

The driveability test procedure is designed to use the test modes just described and to help diagnose engine performance complaints not caused by internal computer problems. It should be used whenever driveability problems are present. The first step, however, is to verify the problem. Ordinarily, the technician should drive the vehicle under the same conditions in which the owner discovers the problem.

The second step is a visual inspection. The

driveability test in the Chrysler literature provides a guide for this, specific to the vehicle diagnosed. If no problem is found during visual inspection, the most appropriate driveability test section should be used. If the engine does not start, obviously, use the no-start section. If the problem occurs only when the engine is cold, use the cold driveability section and so on. Once a driveability test section is begun, start with the beginning of the test and follow each step to the end.

WARNING: Because of the high fuel pressure in many of these systems, it is very important to relieve the pressure before opening the fuel system. Serious eye and fire hazards are present if a fitting is loosened while the system is still pressurized. In addition, use only the correct part numbers for fuel system hose and hose connections when making repairs.

SUMMARY

In this chapter, we have covered single and multipoint fuel injection systems used by Chrysler. We have seen the relative roles of the logic and power modules in controlling combustion in the engine under all driving conditions. The logic module contains the memory and calculation equipment for the system, while the power module handles the actuation work. Likewise, we reviewed the way the computer controls the idle speed under various conditions, how it controls ignition timing, and how it prevents knock. The text explained how the Chrysler system controls turbocharger boost to optimize both power and engine life, first retarding spark to a specific cylinder, and only should that fail, to end knock by reducing turbocharger boost. The new incorporation of control of the charging system output voltage was discussed. This accommodates battery condition when very cold, and can also smooth idle by reducing charge at that low-torque engine speed. We also reviewed the sensors such as the battery temperature sensor and the way the computer uses the voltage in the fuel pump circuit to calculate the alternator field strength. We've seen the role of the automatic shutdown relay in powering the fuel pump, and we've discussed limp-in mode, self diagnostics, and how to understand and repair any problems discovered.

▲ DIAGNOSTIC EXERCISE

1. Sometimes noncomputer-related problems can cause difficulties in computer-controlled systems. What kinds of problems might arise from a cracked alternator bracket or a bad water pump bearing? What effect might leaking valve stems have on the fuel mixture controls?

2. Describe the problems that would be caused by a high resistance battery negative connection to the engine block and the kinds of tests you would use to determine this was the problem.

REVIEW QUESTIONS

1. Name four different operating modes of the Chrysler single-point and multipoint fuel injection systems.
2. What is the source of the logic board's electrical power?
3. Name three actuators the power board controls according to instructions from the logic board.
4. What is the function of the automatic shutdown relay?
5. What purpose does adaptive memory serve?
6. Why is there a vacuum bleed hole in the MAP sensor line on some early Chrysler systems?
7. What two distributor signals does the logic board get on turbocharged systems?

8. Name four sensor inputs affecting ignition timing.
9. Name two functions the radiator fan relay controls.
10. Describe the spark knock control strategy for turbocharged engines.
11. Describe the four strategies used to control boost pressure on turbocharged engines.
12. Name four inputs whose failure throws the system into limp-in mode.
13. Name an input whose failure causes the engine to stop.
14. What turns on the power loss lamp?
15. Name four test modes used for driveability problem diagnosis.
16. What are the first two steps of the driveability diagnosis procedure?

ASE-type Questions (Actual ASE test questions are rarely so product-specific.)

17. Technician A says the logic board will first control idle speed with the AIS motor. Technician B says the logic board will first use spark timing to control idle speed. Who is correct?
  a. A only.
  b. B only.
  c. both A and B.
  d. neither A nor B

18. Technician A says on late-model Chrysler fuel-injected engines the logic and power boards control charging circuit voltage. Technician B says the charging system on those vehicles no longer has its own voltage regulator. Who is correct?
  a. A only.
  b. B only.
  c. both A and B.
  d. neither A nor B

19. Technician A says the engine running test mode of the Chrysler diagnostic sequence allows the technician to test individual actuator circuits for proper operation. Technician B says the diagnostic test mode of the Chrysler

diagnostic sequence allows the technician to test individual sensor circuits for their proper voltage signal. Who is correct?

a. A only.
b. B only.
c. both A and B.
d. neither A nor B.

20. A car with Chrysler's EFI system comes into the shop with a complaint of poor idle during warm-up. What driveability test section should be used to diagnose this problem?

a. driveability test cold.
b. driveability test warm.
c. both A and B.
d. neither A nor B.

Chapter 14

Chrysler Multiplexing And Computer Developments

OBJECTIVES

In this chapter you can learn:
- ❏ about the introduction of multiplexing, and what that means for operation, diagnosis and repair.
- ❏ how Chrysler has tied the vehicle electronics together on a data bus network, controlling many components with a single harness.
- ❏ how the system communicates with its various elements using binary code transmitted in specific ways at a certain baud rate. How the Chrysler data collision avoidance system works.
- ❏ what a range-switching sensor is, and why Chrysler engineers have introduced it for certain purposes.
- ❏ what a stepper motor is, and why it is used on the idle speed control system.
- ❏ about the new types of crankshaft and camshaft position sensors.
- ❏ about changes in the fuel injection system, capitalizing on increased capacities of the computer control system.

KEY TERMS

Baud
Binary Code
Data Bus Network
Multiplexing
Range-Switching Temperature Sensors
Stepper Motor
Voltage Spike

Once carmakers put a fast, powerful computer in a car to control complicated engine combustion under *all* ordinary driving conditions, it was not long before they decided to make the computer control more and more of the vehicle's systems. After all, most of the time the computer is so far ahead of the engine it has nothing to do but go along for the ride. Some computers even shut themselves down almost completely once they have finished calculating the next power stroke's mixture and spark timing, and set an "alarm" to rouse themselves for the next calculation. The greater reliability and lower cost of semiconductor-based controllers (compared to electrical and mechanical devices) makes this choice inevitable.

POWERTRAIN CONTROL MODULE

By the mid-1990s, the engine control computer had so many additional jobs beyond combustion that it became the PCM, the powertrain control module.

The reason carmakers needed more powerful, faster computers was to meet more stringent federal emissions and fuel economy standards. For example, to fine-tune the fuel/air mixture and spark timing even further, beginning in 1992 Chrysler changed to range-switching temperature sensors for coolant and air temperature, Figure 14-1.

Range-Switching Temperature Sensor. The purpose of the change is to telescope the sensitivity of the sensor in the area of greatest importance: around operating temperature. It also accommodates the problem that the thermistor's reaction to changes in temperature is not linear, that is, there is more change in the cooler (lower) temperatures than at the upper end. Unfortunately, for engine management purposes, we're more interested in knowing about the higher temperatures in fine detail for air/fuel mixture control.

The sensor works like this: Below 125 degrees, there is a fixed 10,000-ohm resistor in series with the sensor's thermistor. At about 125 degrees, the PCM turns on a 1,000-ohm resistor

in parallel with the 10,000-ohm fixed unit. This toggles the resistance to 909 ohms.

This spreads out the resistance more over the range the computer is looking for. Of course, what it is actually measuring is the voltage drop across the resistor pair. This range shift means, of course, that the computer must know that the change that occurs around 125 degrees is not a sudden cooling of the engine but the resistance switch. This "expectation" is programmed into the PCM's memory.

When the second resistor is switched into parallel, the total fixed resistance is now considerably lower than that of the thermistor. There is a wider range of voltage drop available to be monitored and greater accuracy obtainable.

Idle Speed Stepper Motor.   Chrysler has also changed to a more precise **stepper motor** to control idle speed. The stepper motor rotates a

Temp. of	Sensor Resistance	Temp. Change	Resistance Change	OHM Change Per Degree	Volt Drop Across Sensor	
− 20	156,667 Ω				4.7V	
		10°	50,388Ω	5038.8		
− 10	106,279 Ω				4.57V	WITH
40	25,714 Ω				3.6V	10,000 Ω
		10°	6,302Ω	630.2		FIXED
50	19,412 Ω				3.3V	RESISTANCE
110	4,577 Ω				1.57V	
		10°	1,244	124.4		
120	3,333 □				1.25V	

SENSOR CIRCUIT SHIFT (909 Ω FIXED RESISTANCE RATHER THAN 10,000 Ω)

140	2,338 Ω				3.6V	
		10°	406Ω	40.6		
150	1,932 Ω				3.4V	WITH
200	839 Ω				2.4V	909 Ω
		10°	125Ω	12.5		FIXED
210	714 Ω				2.2V	RESISTANCE
240	435 Ω				1.62V	
		10°	64.5Ω	6.45		
250	371 Ω				1.45V	

Figure 14-1 Temperature, resistance, and voltage drop change with range-switching temperature sensor

precise amount for each pulse the computer sends it, so by tracking the number of pulses, the computer knows the position of the air bypass valve.

THE MULTIPLEXING SYSTEM

Data Bus Networks

Data bus networks, or **multiplexing,** may be the most significant expansion of the role of computers since they first appeared on cars. This wiring method effectively makes the entire electrical system of the vehicle one computer, Figure 14-2. The technique also allows a reduction in the number of sensors. Where previously, for example, the computer may have used the information from one coolant temperature sensor, the radiator fan switch may have worked from another, redundant sensor/switch.

Data Bus/Binary Code. Each sensor is wired directly to the controller that makes princi-

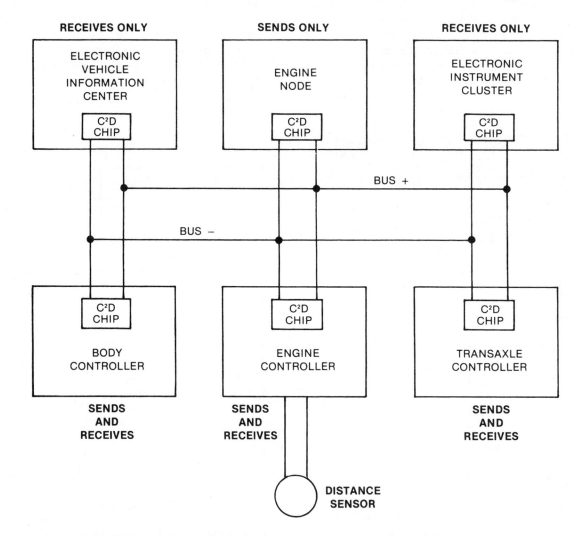

Figure 14-2 Data bus network schematic

ple use of its information. That controller also sends the information, in **binary code,** through a data bus—a single circuit connected to all the car's various microprocessors. Each device on the network includes a code-reading device called a C^2D chip that reads and sends messages onto the wire. Some devices can only send or only receive, depending on their purpose.

Each receiver gets all messages on the network. The C^2D chip compares the coded message to its memory list to see whether the information is relevant to its own function.

Chrysler Collision Detection (C^2D) System. Each sending device can broadcast information at any time the network is free. To resolve the problem of two devices attempting to send simultaneously, Chrysler systems employ the C^2D system. Each message starts with an ID code, a byte (an eight-digit binary number) that conveys the message's priority, content type, and size in bytes, Figure 14-3. The message ID byte with the most zeroes to the left of the first 1 has the highest priority and gets broadcast first. In 1992, Chrysler multiplexing systems transferred information at a rate of 7812.5 baud (7812.5 digital bits per second).

Different vehicles may have different components wired to their "internal internet," but on many Chrysler products these include:

- body control computer.
- engine controller (SBEC).
- vehicle theft alarm (VTA).
- electronic vehicle information center (EVIC).

- engine node (a module in the front of the vehicle with sensors providing air temperature, and direction of travel).
- electronic transmission controller (EATX).
- mechanical instrument cluster (MIC).
- traveler.
- overhead console.

Magnetic Interference. The data bus itself consists of two wires twisted together to reduce magnetic interference and the generation of false information by nearby electric components. Since the wires are twisted together, any magnetic induction affects both wires at the same time, canceling out most of the problem.

FUEL SYSTEM

Pressure Changes. By 1996 Chrysler fuel injection systems had changed further in two specific ways: fuel pressure is now held constant at approximately 49 psi by a mechanical regulator independent of the PCM and of intake manifold vacuum. Second, the fuel injection plumbing is one way: there is no return line (except for the pressure relief valve at the pump, which is, as before, submerged in fuel in the tank).

Diagnostic & Service Tip.

These changes have service caution relevance. Make sure you relieve the fuel pressure before opening any part of the fuel system. The higher pressure can spray more fuel farther and harder. Also, it may be necessary to remove the fuel filler cap when working on the fuel system. The pressure it retains can be enough to push fuel through the lines, with the attendant fire risk should that line be open. Keep in mind that should the system be in any actuation test mode that energizes the ASD relay, there will be full fuel delivery for as long as seven minutes. That can just about empty a fuel tank.

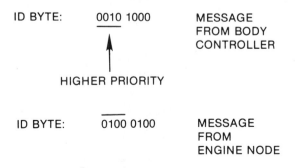

ID BYTE: 0010 1000   MESSAGE
                           FROM BODY
                           CONTROLLER

             ↑

HIGHER PRIORITY

ID BYTE: 0100 0100   MESSAGE
                           FROM
                           ENGINE NODE

Figure 14-3 C^2D chip priority code

✔ SYSTEM DIAGNOSIS & SERVICE

The combination of more powerful microprocessors, higher speed, more connections and more memory, means the system can detect and report on many more events in the vehicle. The diagnostic trouble code memory can store everything from DTC 12 (battery disconnected from PCM within the last 50 key-on cycles) or DTC 34 (open or short in the cruise control solenoid circuit) to DTC 13 (no change in MAP sensor reading when the engine starts). The newer PCMs can even modify injection pulse width to correct for low charging system voltage (a lower voltage can't open the pintle as quickly, so the delivered amount is slightly reduced). Besides the previous functions of the ASD relay, it now sends a signal to the computer to indicate that it is energized, and it powers the oxygen sensor heaters.

Camshaft Position Signal. An example of the indirect way the computer works with engine controls can be seen in the signal from the camshaft.

The notches on the camshaft do not correspond to TDC for any of the cylinders, and two pair of the 180-degree-apart position notches are identical while one is nonexistent, Figure 14-4. But the computer can recognize from the single three-pulse signal for cylinder 4 (in the firing order) or from no signal at all combined with the crankshaft position signal exactly where the cylinders are in their two-revolution engine cycle. The computer does not need to distinguish cylinder 2 from 5 nor 3 from 6. Once it has identified 1 and 4, it knows where all the pistons and valves are for the combustion sequence because the crankshaft throws do not change relative position.

2.4-Liter Camshaft Position Sensor. The camshaft position sensor for the 2.4-liter engine works even more subtly. At the end of the camshaft, held by offset alignment tabs, is an eccentric-**polarity** disc "target magnet," Figure 14-5. The north and south poles of this disc magnet cover different degrees of arc on each side of

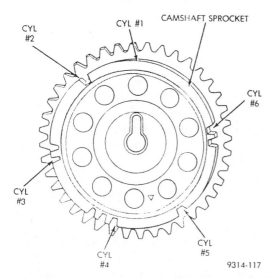

Figure 14-4 Camshaft sprocket with cylinder identification notches. *(Courtesy of Chrysler Corp.)*

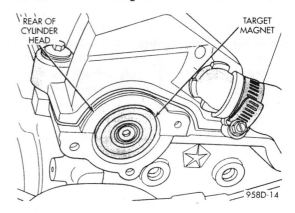

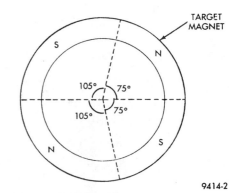

Figure 14-5 Target magnet camshaft position sensor. (Courtesy of Chrysler Corp.)

the disc. As the target magnet rotates, the camshaft sensor reacts to the change of polarity. Its output switches from 5.0 volts to 0.3 volt as the disk rotates from north to south, and the dif-

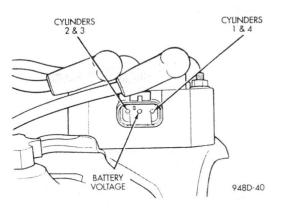

Figure 14-6 Four-cylinder ignition coil primary circuit terminals. *(Courtesy of Chrysler Corp.)*

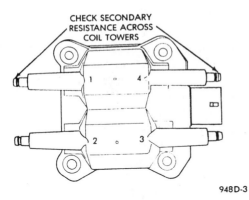

Figure 14-7 Four-cylinder ignition coil secondary circuit terminals. *(Courtesy of Chrysler Corp.)*

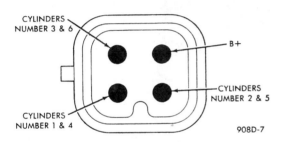

Figure 14-8 Six-cylinder ignition coil primary circuit terminals. *(Courtesy of Chrysler Corp.)*

ference of polarity arc identifies the camshaft position.

Ignition Coils. Because of the waste spark ignition secondary, the ignition coil for the four-cylinder engines requires only three primary circuit inputs (Figures 14-6 and 14-7) and for the six-cylinder only four (Figure 14-8). Check coils for resistance only when cool; a hot coil will not afford a reliable continuity and resistance check.

Resistance through the primary circuit of each coil should be between 0.45 and 0.65 ohm at 70 to 80 degrees Fahrenheit. Resistance through the secondary circuits should be 7,000 to 15,800 ohms. See Figures 14-7 and 14-9.

Powertrain Control Module. Sometimes it's necessary to remove the PCM, Figure 14-10. Several precautions will prevent any electrical damage from such things as **voltage spikes,** and accidental groundings of terminals. First, disconnect *both* battery cables, negative cable first. Then unscrew the PDC (power distribution center) from the bracket. Take out the battery heat shield and the battery, then move the PDC reward for access to the PCM. Then remove the 40-pin connector, squeezing the tabs and pulling the connector rearward. The PCM then unbolts for removal.

Vehicle Theft Alarm System. Many late-model Chryslers include an extensive VTA system, which can trigger the alarm and disable the

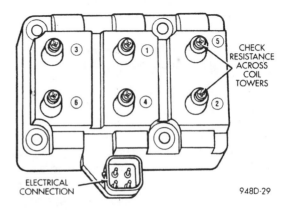

Figure 14-9 Six-cylinder ignition coil secondary circuit terminals. *(Courtesy of Chrysler Corp.)*

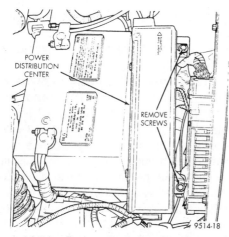

Figure 14-10   Removal of the Chrysler powertrain control module. *(Courtesy of Chrysler Corp.)*

engine once the system is armed. It requires 16 consecutive seconds to arm itself after the vehicle is locked and all the doors are closed. There are two semi-independent sections of the system, engine compartment, and passenger compartment. Either can remain effective should the other be disabled, and separate DTCs are set for failures in them. Any of the following will trigger the alarm and disable the ignition and fuel systems:

- opening any door.
- opening the hood.
- turning the ignition to ON or unlock (the ignition can be turned to the accessory without triggering the alarm).

Diagnosis of VTA failures requires a specific Body Diagnostic Procedures manual.

Real World Problem

A motorist brings in a car with a peculiar driveability problem. Inquiry discovers that the problem started when he and a friend installed a new, more powerful CD player in the vehicle. Any time the onset of a driveability problem corresponds to installation of a component—even if it has no obvious connection with engine control—check to see whether the routing of the power cables is too close to the data bus or other components of the computer control system. Check particularly to see that the data bus wires have not been unwound or rerouted. Powerful magnetic pulses from speakers and other accessories can generate spurious signals in sensor and actuator wires. Ordinarily, this can be remedied by rerouting the offending cables to some harmless location.

SUMMARY

In this chapter, we have covered multiplexing and what that means for operation, diagnosis and repair. As we have seen, multiplexing is the interconnection of multiple microprocessors, sensors and actuators on a single double-strand connecting circuit.

We've learned about the data bus network, enabling a single computer to control many different actuators and monitor many different sensors over that single harness. We've covered the binary code that transmits information throughout the system, the encoding system that identifies the source and relative importance of each packet of information, and the data collision avoidance system to keep the flow usable.

We've seen the new range-switching temperature sensors that telescope the sensor's range at the area of maximum interest for engine management purposes, and the stepper motor used for controlling idle speed. We've learned about the new crankshaft and camshaft position sensors as well as changes in the fuel injection and ignition systems. These features afford greater capacities of the more powerful computers now employed by Chrysler.

▲ DIAGNOSTIC EXERCISE

When a technician follows a late-model Chrysler's coolant temperature sensor return signal as the engine warms up, the signal smoothly and gradually reflects the change in the engine's temperature. But the computer has set a code indicating a coolant temperature sensor fault. Assuming the signal is getting to the computer and that the computer itself is working properly, what could be the cause of this code?

REVIEW QUESTIONS

1. What advantage does multiplexing have over a wiring harness with separate circuits for each component on the system? What disadvantages?
2. What is gained by telescoping the temperature sensor's information signal by range switching? What can the engine control system do more accurately?
3. What might happen if the data bus wires were unwound or routed near another circuit that carried high current or high voltage?
4. The binary code signals transmitted over the data bus are encoded in such a way as to insure that only one signal is transmitted at one time, the Chrysler Collision Detection system. Explain the other purposes of signal encoding.
5. Explain how the camshaft position sensor on the 2.4-liter engine works. Include the concept of magnetic polarity in your answer.
6. Under what temperature conditions should you check an ignition coil? Cold, during warm-up or hot? What will happen if you choose the wrong temperature?
7. What are the precautions that should be observed before removing the PCM, and what is the purpose of these precautions?
8. Can a defect in the vehicle theft alarm system prevent the engine from starting?
9. On the latest Chrysler port fuel injection systems, fuel pressure is held constant regardless of manifold absolute pressure. What does that tell you about the computer's fuel metering strategy?

ASE-type Questions (Actual ASE tests are rarely so product-specific).

10. Technician A says the more powerful, faster computers used in newer cars were needed to meet tighter emissions requirements. Technician B says the more powerful computers are used to perform many tasks unrelated to emissions control. Who is correct?
    a. A only.
    b. B only.
    c. both A and B.
    d. neither A nor B.
11. Technician A says the new range-switching temperature sensors use one range for the lower engine temperatures, and another for the higher temperatures. Technician B says both sensors are used for the full range of engine temperatures. Who is correct?
    a. A only.
    b. B only.
    c. both A and B.
    d. neither A nor B.
12. Technician A says as the temperature of the engine coolant goes up, the resistance through the sensor goes up. Technician B says the computer "expects" a change of reading at about 125 degrees Fahrenheit. Who is correct?
    a. A only.
    b. B only.
    c. both A and B.
    d. neither A nor B.
13. Technician A says the stepper motor that controls the idle air bypass can be set to specific positions by the computer. Technician B says the computer uses a specific stepper motor setting when it wants a specific idle speed. Who is correct?
    a. A only.
    b. B only.
    c. both A and B.
    d. neither A nor B.
14. Technician A says the data bus wire is wound to prevent electrical interference from other components. Technician B says the data bus carries the current to open the fuel injectors when the computer decides that is what to do. Who is correct?

a. A only.
b. B only.
c. both A and B.
d. neither A nor B.

15. Technician A says all information on the data bus network is sent to every component connected to it. Technician B says not every component can use each piece of information that is transmitted. Who is correct?
    a. A only.
    b. B only.
    c. both A and B.
    d. neither A nor B.

16. Technician A says the Chrysler computer system can sense low charging system voltage and correct fuel injection pulse width for any low output. Technician B says a low charging system voltage can't open the fuel injector as fast as the correct one. Who is correct?
    a. A only.
    b. B only.
    c. both A and B.
    d. neither A nor B.

17. Technician A says that with a waste spark system, the computer doesn't need to know anything about camshaft position. Technician B says fuel injection sequencing depends on the camshaft position signal. Who is correct?
    a. A only.
    b. B only.

c. both A and B.
d. neither A nor B.

18. Technician A says you should check a Chrysler waste spark ignition coil for resistance when it is hot so you can catch any running problems. Technician B says the six-cylinder coil has only four primary circuit connections. Who is correct?
    a. A only.
    b. B only.
    c. both A and B.
    d. neither A nor B.

19. Technician A says you should remove both battery cables before you remove or disconnect the PCM. Technician B says you can feel any voltage spike in a circuit. Who is correct?
    a. A only.
    b. B only.
    c. both A and B.
    d. neither A nor B.

20. Technician A says the fuel pressure regulator on a late-model Chrysler is defective because fuel pressure does not change with MAP signal. Technician B says the new system uses a constant fuel pressure, regardless of other system parameters. Who is correct?
    a. A only.
    b. B only.
    c. both A and B.
    d. neither A nor B.

Chapter 15

European (Bosch) Engine Control Systems

OBJECTIVES

In this chapter you can learn:
- ❑ the two basic kinds of Bosch Control Systems.
- ❑ the distinction between the Motronic systems and the earlier versions.
- ❑ to identify the operating conditions which cause different operational modes.
- ❑ how the control unit determines engine load.
- ❑ the major input sources and output actuators.
- ❑ the controlled functions that are standard Motronic features.
- ❑ the major diagnostic steps of Motronic systems.

KEY TERMS

Continuous Injection
Lambda
Motronic
Pulsed Injection

The Robert Bosch Corporation does not build cars or engines, but components and control systems. Carmakers buy these specially-engineered components and control systems from the Bosch Corporation to use on their own vehicles. An early pioneer in fuel injection systems, the Bosch Corporation makes many of the products found on most German and Scandinavian cars, as well as those on many American and Japanese vehicles. Bosch also licenses several of its systems for manufacture by other firms, so an understanding of the Bosch systems in turn provides an understanding of many systems built by other sources as well. A modern English car with a Lucas control system, for example, is actually controlled by Bosch designed components and controllers.

SYSTEM OVERVIEW

The Bosch company has been involved in fuel injection since the 1920s, when they began with diesel injectors. Their experience with gasoline injection goes back to military vehicles in the late 1930s. One of the earliest of their systems for passenger cars was the direct, diesel-style injection system used on a few of the gasoline-fueled Mercedes-Benz SLs. That system was a mechanical forced injection system very similar to diesel injection systems used on trucks.

All the Bosch engine management systems are computerized developments of earlier fuel injection systems, mostly mechanical. While our focus in this book is on computer-controlled systems, it would be difficult to understand what the

Bosch engineers were doing without some explanation of the earlier mechanical systems on which their computers work.

While Bosch has built a version of throttle body injection for a limited number of European manufacturers, all of the systems used on vehicles imported to the United States have been multipoint injection systems. The advantages of multipoint are the same for Bosch as for any other injection system builder, and these advantages have been described in Chapter 3.

Bosch systems work in one of two basic ways: pulsed injection systems and continuous systems. D-Jetronic and L-Jetronic systems are pulsed, Figure 15-1; K-Jetronic systems are continuous (also referred to as Continuous Injection System, CIS), Figure 15-2. D comes from the German word for (intake manifold) pressure, "Druck," the major sensor in that system. L comes from the German word for air, "Luft," the airflow sensor or meter being that system's major sensor. K comes from the German word for continuous, "Kontinuerlich," because the fuel flows continuously from that system's injectors as long as the engine is running. When the letter "E" is added, this indicates the system is electronic, that is, controlled by a computer. The term **Motronic** applies to the more fully developed engine management systems, covering spark timing, fuel injection, and some other vehicle functions. In general, a Motronic system uses a single computer to manage all engine functions,

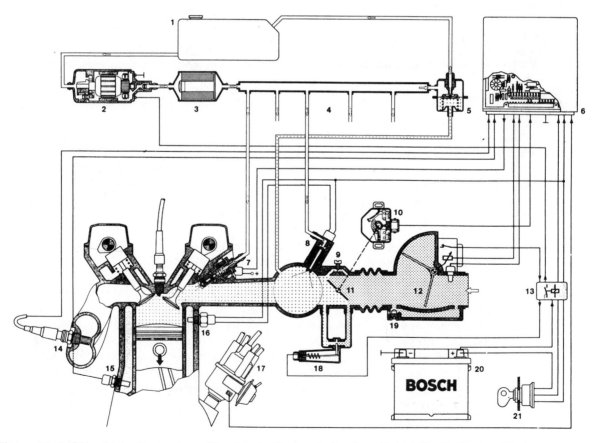

Figure 15–1 Pulsed injection system. *(Reprinted with permission from Robert Bosch Corporation.)*

while the earlier systems used independent controllers, typically one for ignition and another for fuel injection.

The **pulsed injection** systems work either sequentially or in clustered groups, varying fuel delivery by pulse width just as domestic systems do. The continuous systems spray fuel from port injectors whenever the engine is running, varying the amount sprayed to correspond with the amount of air entering the engine. Both types of systems have been elaborated into computer-controlled systems, with improved driveability, fuel economy, and emissions quality. Once warmed up to operating temperature and driving

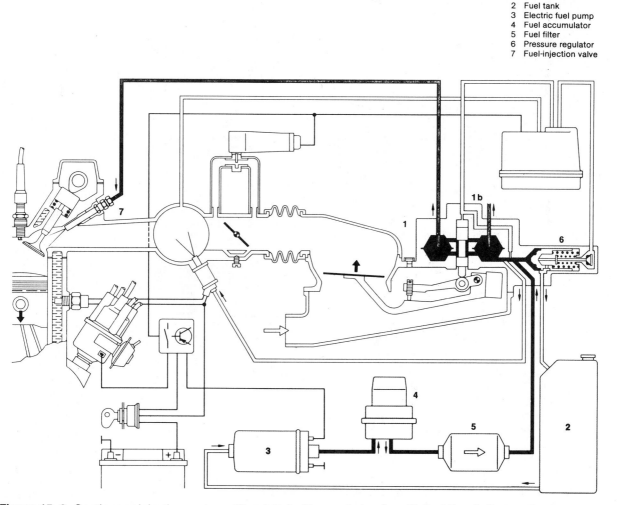

Schematic diagram of
the K-Jetronic.
Functional area:
Fuel supply
1   Mixture control unit
1 b Fuel distributor
2   Fuel tank
3   Electric fuel pump
4   Fuel accumulator
5   Fuel filter
6   Pressure regulator
7   Fuel-injection valve

Figure 15–2 Continuous injection system. *(Reprinted with permission from Robert Bosch Corporation.)*

Functional diagram taking as an example the L-Jetronic equipped with Lambda closed-loop control.

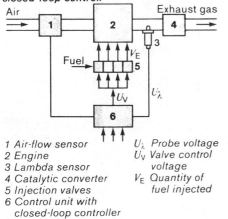

1 Air-flow sensor
2 Engine
3 Lambda sensor
4 Catalytic converter
5 Injection valves
6 Control unit with closed-loop controller

U_λ Probe voltage
U_V Valve control voltage
V_E Quantity of fuel injected

Figure 15–3 Closed loop schematic. *(Reprinted with permission from Robert Bosch Corporation.)*

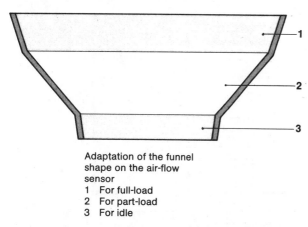

Figure 15–4 K-Jetronic air cone. *(Reprinted with permission from Robert Bosch Corporation.)*

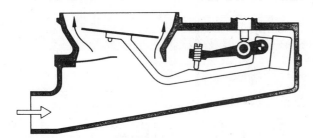

Adaptation of the funnel shape on the air-flow sensor
1 For full-load
2 For part-load
3 For idle

Figure 15–5 K-Jetronic air cone taper for different air flows. *(Reprinted with permission from Robert Bosch Corporation.)*

normally, the system controls the air/fuel mixture by a closed loop feedback process, Figure 15-3.

The **continuous-injection** (K- and KE-Jetronic) systems are the most common of the Bosch systems, used on virtually all European makes at different times. Incoming air lifts or lowers (depending on the vehicle manufacturer) a plate suspended on a lever in a specially shaped cone, Figures 15-4 and 15-5. The movement of the plate (in the original mechanical systems) completely determines the amount of fuel injected through the fuel distributor. In the later, computerized systems the control unit can vary the

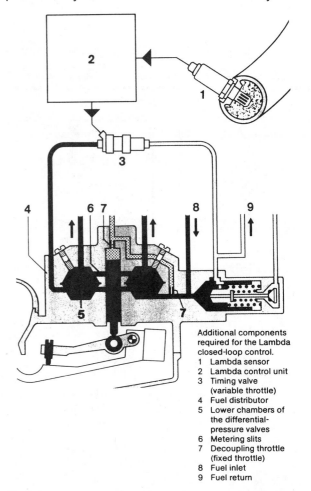

Additional components required for the Lambda closed-loop control.
1 Lambda sensor
2 Lambda control unit
3 Timing valve (variable throttle)
4 Fuel distributor
5 Lower chambers of the differential-pressure valves
6 Metering slits
7 Decoupling throttle (fixed throttle)
8 Fuel inlet
9 Fuel return

Figure 15–6 Air sensor plate and fuel plunger. *(Reprinted with permission from Robert Bosch Corporation.)*

mixture by varying the pressure differential between the upper and lower parts of the fuel distributor in response to the signals from sensors, thus controlling the delivered air/fuel mixture, Figure 15-6. The control unit also employs programming for enrichment under cold engine, acceleration, and some other driving conditions.

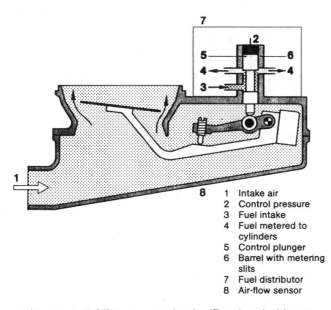

1	Intake air
2	Control pressure
3	Fuel intake
4	Fuel metered to cylinders
5	Control plunger
6	Barrel with metering slits
7	Fuel distributor
8	Air-flow sensor

Figure 15–7 Mixture control unit. *(Reprinted with permission from Robert Bosch Corporation.)*

Mixture Control Unit

The major fuel and air component of the continuous injection systems is the mixture control unit, Figure 15-7. This combines the airflow sensor plate and the fuel distributor. As the incoming air lifts (or on some models lowers) the plate, it moves a lever directly connected to the fuel distributor. The lever lifts a precisely machined plunger in a barrel with special slots to meter the fuel to each cylinder, Figure 15-8. The farther the plunger is raised, the more fuel passes the plunger, through the barrel and into the cylinders, corresponding to the increased amount of air moving past the air plate.

The original K-Jetronic systems worked mechanically to adjust the air/fuel mixture. Once the system became computer controlled to adjust the mixture in response to signals from sensors, the fuel distributor was modified to include a pressure actuator, a device to vary the pressure difference between the upper (delivery) and lower (control and return) halves of the fuel distributor, Figure 15-9.

The control unit does this by varying the current (amps) flowing through the pressure actuator. This then controls fuel bypass and the pressure difference and delivery rate. The current passes

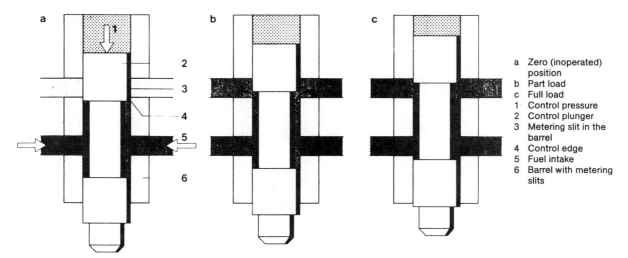

a	Zero (inoperated) position
b	Part load
c	Full load
1	Control pressure
2	Control plunger
3	Metering slit in the barrel
4	Control edge
5	Fuel intake
6	Barrel with metering slits

Figure 15–8 Fuel plunger. *(Reprinted with permission from Robert Bosch Corporation.)*

through an electromagnetic coil that raises a restrictor baffle in the fuel return line, determining the control pressure. Electromagnetism is directly proportional to the current, regardless of the volt- age. The pressure difference between the two halves of the fuel distributor flexes a thin metal diaphragm which allows fuel to pass into the injec- tor lines. The more the diaphragm flexes down,

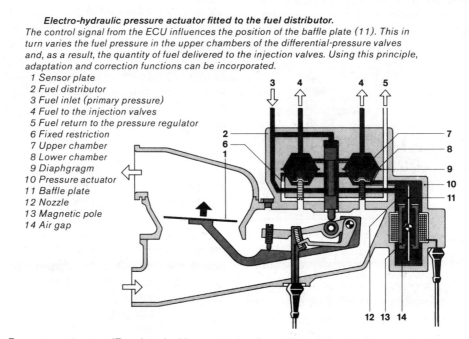

Electro-hydraulic pressure actuator fitted to the fuel distributor.
The control signal from the ECU influences the position of the baffle plate (11). This in turn varies the fuel pressure in the upper chambers of the differential-pressure valves and, as a result, the quantity of fuel delivered to the injection valves. Using this principle, adaptation and correction functions can be incorporated.

1 Sensor plate
2 Fuel distributor
3 Fuel inlet (primary pressure)
4 Fuel to the injection valves
5 Fuel return to the pressure regulator
6 Fixed restriction
7 Upper chamber
8 Lower chamber
9 Diaphgragm
10 Pressure actuator
11 Baffle plate
12 Nozzle
13 Magnetic pole
14 Air gap

Figure 15–9 Pressure actuator. *(Reprinted with permission from Robert Bosch Corporation.)*

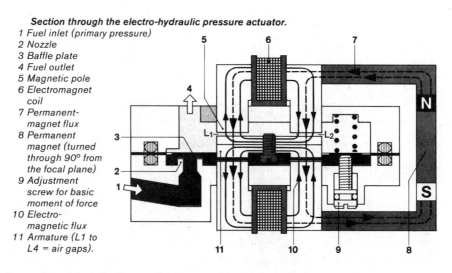

Section through the electro-hydraulic pressure actuator.

1 Fuel inlet (primary pressure)
2 Nozzle
3 Baffle plate
4 Fuel outlet
5 Magnetic pole
6 Electromagnet coil
7 Permanent-magnet flux
8 Permanent magnet (turned through 90° from the focal plane)
9 Adjustment screw for basic moment of force
10 Electro-magnetic flux
11 Armature (L1 to L4 = air gaps).

Figure 15–10 Operation of fuel distributor diaphragm. *(Reprinted with permission from Robert Bosch Corporation.)*

the more fuel is injected and vice versa, Figure 15-10. By returning more fuel to the tank, the actuator reduces the pressure difference, flexes the diaphragm up, and reduces fuel delivery, Figure 15-10. By restricting return fuel flow to the tank, the actuator increases the pressure difference, flexes the diaphragm down, and increases fuel delivery.

Bosch continuous injection systems are often called mechanical fuel injection systems, but it should be noted that they are primarily hydraulic in the way they work.

Most continuous injection systems are installed separately from the engine, on a bracket in the intake air line. On certain V-form engines, notably Mercedes-Benz and Volvo, the control unit is mounted directly on the engine, much like a carburetor.

The injectors in a continuous system spray whenever the fuel pump is operating (powered as on domestic systems through a relay). On many systems, the injectors include a special vibrating needle in the tip which aids in atomizing the fuel. Often the action of these vibrating needles audible as a hiss with the hood open and the engine at idle. This is normal and does not require any corrective action.

PULSED SYSTEMS

The pulsed-injection systems (D-Jetronic, L-Jetronic, and LH-Jetronic) vary the air/fuel mixture by varying pulse width, just as with domestic systems. As we'll see, the sensors and actuators work very similarly.

The earliest of the modern Bosch systems was the D-Jetronic. Named for the German word for pressure, D-Jetronic uses intake manifold pressure as the principle input for the calculation of how much fuel to inject. While Bosch left this system years ago after tests and experience showed that air mass measurements were much more accurate for such a calculation, the intake manifold pressure information is nonetheless used on many systems as we have seen in earlier chapters.

The D-Jetronic systems are also unique in their employment of analog computers. Check Chapters 2 of this book for more information about analog versus digital computers. A few of the early L-Jetronic systems also used an analog computer. These systems work more slowly and by variable voltages rather than by faster, digital systems. D-Jetronic systems only control fuel injection and have no control over ignition. Most of them do receive a signal from the distributor (from a second set of contact points in early models) to time the fuel spray.

Service & Diagnostic Tip

Since Bosch D-Jetronic systems calculate fuel delivery entirely by intake manifold pressure, a vacuum leak on these systems does not cause a bad idle from a lean mixture (unless there is a large vacuum leak to just one cylinder). However, a vacuum leak does cause a high idle speed in a way that can be puzzling to technicians unfamiliar with the system. Checking for vacuum leaks is done the same way as with other fuel induction systems.

The pulsed injection Bosch systems do not employ the airflow plate or fuel distributor from the K-Jetronic systems. Instead they measure the incoming air volume (or in later systems, mass), and the control unit calculates the proper pulse width under all engine conditions and opens the injectors to deliver the correct amount of fuel. The injectors for these systems are identical to those in domestic fuel injection/engine management systems. The most important sensor for this operation (after the engine is started and warmed up) is the airflow meter. While this is similar to the airflow meter used on some Ford systems, it is a Bosch-engineered component, so we will describe it in some detail here.

Airflow Meter

The airflow meter is basically a lightly-sprung swinging door in the airstream, sharing a pivot shaft with a potentiometer that modifies the return voltage signal to the computer, Figure 15-11.

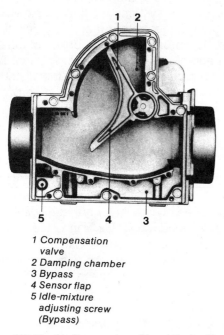

1 Compensation
   valve
2 Damping chamber
3 Bypass
4 Sensor flap
5 Idle-mixture
   adjusting screw
   (Bypass)

Figure 15–11 Air flow meter. *(Reprinted with permission from Robert Bosch Corporation.)*

Paired with the airflow meter door is a second, equal-sized door in a blind passage. The second door serves two functions: On four-cylinder engines, the individual cylinder pulses can be so distinct that they create pulses in the intake manifold, moving the sensor back and forth with each pulse. This is particularly likely at low engine speeds. The second door serves as a damper to prevent this sensor door pulsing. If the door were allowed to pulse, obviously, the computer would receive inaccurate signals about the amount of air taken into the engine. On many of the swinging-door airflow meters there is an additional door or pop-out plug in the center of the flap to allow a backfire to occur without destroying the expensive airflow meter when the air door would slam shut.

A later version of the L-Jetronic system, called the LH-Jetronic, uses a hot-wire air mass sensor, that works just like hot-wire air mass sensors used in domestic engine management systems, Figure 15-12.

MOTRONIC

The Motronic system is Bosch's first multi-function engine management system. In addition to controlling the air/fuel mixture, it controls

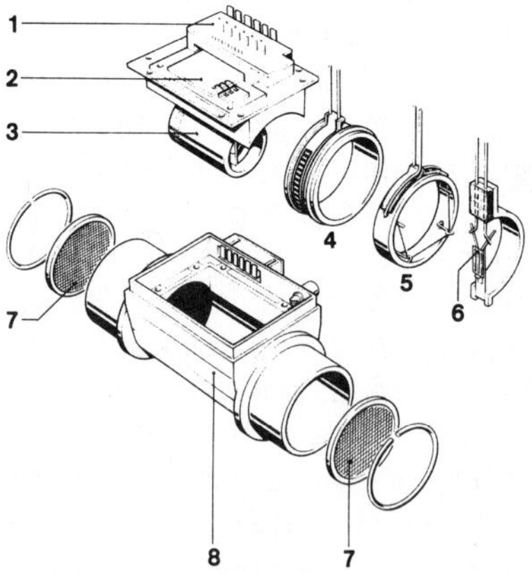

1 Printed board
2 Hybrid circuit
   In addition to the
   resistors of the
   bridge circuit, it also
   contains the control
   circuit for
   maintaining a constant
   temperature and the
   self-cleaning circuit.
3 Inner tube
4 Precision resistor
5 Hot-wire element
6 Temperature-
   compensation
   resistor
7 Guard
8 Housing

Figure 15–12 Air mass sensor. *(Reprinted with permission from Robert Bosch Corporation.)*

ignition timing and dwell. The system may also be designed to control idle speed, EGR operation, evaporative emissions, automotive transmission, and a turbocharger, depending on vehicle application

CONTROL UNIT

The Bosch Motronic system is similar to the L-Jetronic system and like its predecessor controls fuel metering with pulse width modification. The Motronic's computer is a digital unit with the ability to control additional functions, such as spark timing. An overview of the Motronic system is provided in Figure 15-13.

The control unit's mounting brackets also serve to dissipate heat from the power transistors that drive the ignition and injectors, Figure 15-14. A thirty-five pin connector connects to the vehicle's system harness.

Bosch and Imports

Of the imported European cars with a comprehensive computerized engine control system, most use a Bosch system. It first appeared on selected BMWs. Most of the systems used on Japanese cars imported into the United States are built using Bosch patents. Practically all the gasoline engine multipoint injection systems for domestic vehicles use Bosch components or patents.

Main Relay

A main relay, Figure 15-15, similar in operation to the power relay used on Ford EEC systems (see Chapter 10), powers the control unit as soon as the ignition is turned on. This relay also includes a diode to protect the control unit against voltage surges and accidental polarity reversals.

Inputs	Control Unit	Outputs
Airflow Meter		Fuel Injection
Engine Speed		Fuel Pump Relay
Crankshaft Position		Ignition Timing
Throttle Position		Dwell Control
Coolant Temperature		RPM Limit
Air Temperature		Peak Coil Current Cutoff
Lambda Sensor (oxygen sensor)		
Knock Sensor	Main Relay	The following functions are
Altitude Sensor		system options (available to the
Starter Signal		vehicle manufacturer):
Battery Voltage		
		Rotary Idle Adjuster
		Turbo Boost Control
		EGR Control
		Canister Purge
		Transmission Control
		Start-Stop Control

Noncomputer-Controlled Functions: Fuel Pressure Regulation, Fuel Pressure Pulsation Damper, Cold-Start Injector and Thermo-Time Switch, and Auxiliary Air Device (used in place of Rotary Idle Adjuster)

Figure 15–13 Overview of Motronic system

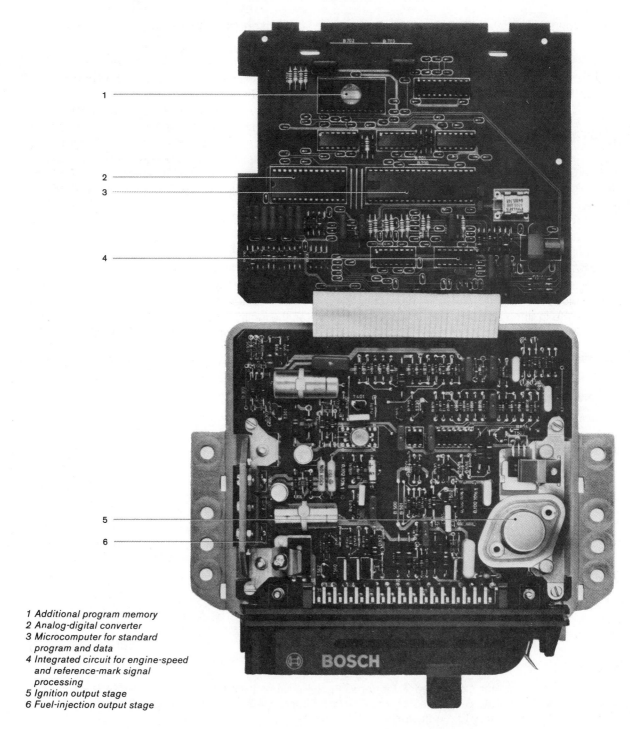

1 Additional program memory
2 Analog-digital converter
3 Microcomputer for standard
   program and data
4 Integrated circuit for engine-speed
   and reference-mark signal
   processing
5 Ignition output stage
6 Fuel-injection output stage

Figure 15–14 Control unit. *(Reprinted with permission from Robert Bosch Corporation.)*

Figure 15–15 Main relay. *(Reprinted with permission from Robert Bosch Corporation.)*

On-Board Diagnostics

The first few model year applications of Motronic do not feature any form of self-diagnosis except if a fault is detected in the closed-loop operating circuit, a calculated pulse width aimed at maintaining an air/fuel mixture near stoichiometric will be used. Later versions, particularly those imported after the implementation of OBD II regulations, include extensive self-diagnostics.

OPERATING MODES

The control unit is programmed with different operational strategies as driving conditions change. Most of the differences in strategy occur during open-loop mode.

Cranking

Two fuel-metering programs can be employed during engine cranking. One is based on cranking speed, the other on temperature. At lower crank speeds the fuel quantity injected stays constant regardless of airflow fluctuations (input from the airflow sensor is not reliable in these conditions because of pulsing caused by individual cylinder intake strokes). At higher cranking speeds, air intake diminishes slightly as a result of lower volumetric efficiency, so fuel quantity injected is also reduced. During cold cranking the control unit adds an enrichment program in addition to the speed-dependent program. Some versions of the Bosch systems include a cold-start injector in the intake manifold, Figure 15-16, while some merely add injection pulses at the port injectors. Ignition timing is also adjusted depending on cranking speed and coolant temperature. The cold-start injector system is very similar to that used on several American vehicles: it sprays a priming load of fuel into the cold intake manifold upstream of the main injectors. The extra distance the fuel must travel to reach the cylinders affords slightly more time for the gasoline to vaporize, making startup easier.

Service & Diagnostic Tips

Sometimes when a vehicle has been used in a warm climate for a season, a certain amount of moisture can collect in the fuel line to the cold-start injector. This moisture can then allow rust to form since the injector does not have the occasion to turn on and flush out the contamination. When the cold weather returns in the fall, the injector turns on, but is plugged with rust and the car won't start. At this point, you will have to service the cold start injector by cleaning or replacing it.

On applications without a cold-start injector, the control unit pulses the injectors several times per crankshaft revolution during cranking instead of once per engine cycle, the way it normally

The Motronic system.

1 Fuel tank
2 Electric fuel pump
3 Fuel filter
4 Fuel distributor
5 Pressure regulator
6 Pulsation damper
7 Control unit
8 Ignition coil
9 High-tension
   distributor
10 Spark plug
11 Injection valve
12 Cold-start valve
13 Idle-speed adjusting
    screw
14 Throttle valve
15 Throttle-valve switch
16 Air-flow sensor
17 Air-temperature
    sensor
18 Lambda sensor
19 Thermo-time switch
20 Engine-temperature
    sensor
21 Auxiliary-air device
22 Idle-mixture adjusting
    screw
23 Reference-mark sensor
24 Engine-speed sensor
25 Battery
26 Ignition-starting
    switch
27 Main relay
28 Pump relay

Atmospheric pressure

Manifold pressure

Exhaust

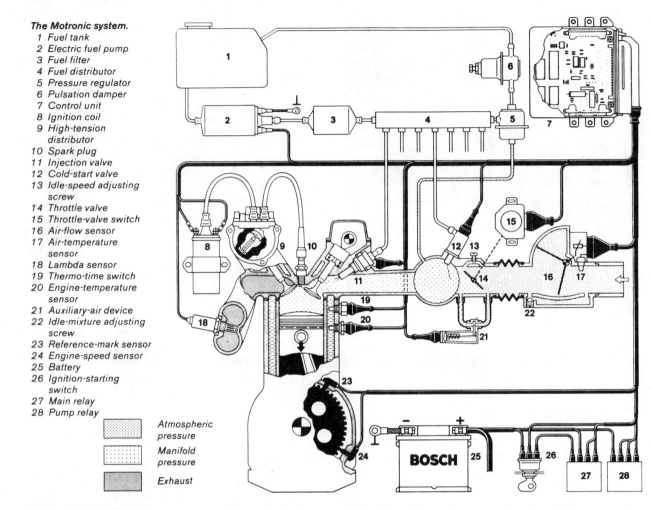

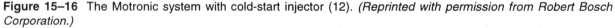

Figure 15–16 The Motronic system with cold-start injector (12). *(Reprinted with permission from Robert Bosch Corporation.)*

works. This should both enhance fuel evaporation and avoid flooding the spark plugs with wet fuel. The enrichment decays to zero over a specific and small number of engine revolutions. The number depends on the engine coolant temperature when cranking begins. If the rpm reaches a preset value before the specified number of revolutions has occurred, the cold-cranking enrichment program stops anyway. The preset rpm value also depends on the engine coolant temperature when cranking begins.

After Start-up

As an engine starts cold, the intake port surfaces and combustion chamber walls, ceiling, and piston top are also cold. Fuel can condense on these surfaces, leading to poor combustion. During the brief time it takes for these areas to warm, the control unit initiates a post-start enrichment program to maintain a good idle quality and improve throttle response. Once this begins, the enrichment also decays within a brief time to zero. This period of time depends on coolant temperature at the time of start-up crank. The control unit also advances the ignition timing relatively more with the engine cold than it will in other circumstances. All of this is to provide the most efficient engine power with the least possible amount of fuel and exhaust emissions. The advanced spark also helps warm the combustion chamber temperature as quickly as possible.

Warm-up

As the engine warms up, and the post-start period has expired, additional enrichment is based on a combination of engine coolant temperature and load. An idle speed increase is applied to prevent stalling and improve driveability and to hasten warm-up. The idle speed increase occurs during the post-start mode and continues into the warm-up mode. As an added means of bringing the oxygen sensor and catalytic converter to operating temperature as quickly as possible, some versions of the Bosch Motronic system retard ignition timing during warm-up to make the exhaust gas somewhat hotter.

Acceleration

Whenever the driver quickly opens the throttle from an earlier fixed position, the control unit will briefly enrich the mixture. This works to serve the same purpose as the accelerator pump in a carburetor and solves two problems: 1) that air accelerates into the combustion chamber faster than the fuel might, and 2) that it is harder to get the fuel to vaporize in higher pressure intake air. The degree of accelerator enrichment is affected by the engine coolant temperature.

Wide-open Throttle (WOT)

During wide-open throttle operation, the control unit commands a fixed enriched air/fuel mixture. To avoid fuel quantity miscalculations from airflow sensor fluctuations, at this speed the fuel quantity is calculated based on engine speed.

The objectives of the mixture strategy at wide-open throttle is to produce maximum engine torque without allowing detonation and physical damage. Emissions quality is not specifically optimized during wide-open throttle operation because WOT use is ordinarily restricted to driving conditions where safety considerations (passing, perhaps) outweigh the disadvantages of momentarily adverse emissions.

Deceleration (Overrun)

When the vehicle is decelerated, the control tapers the fuel delivery to zero and retards the ignition timing. Retarding the timing provides better engine braking and reduces the emission of hydrocarbons in the event there is still residual fuel on the intake port walls.

If the engine speed falls below a specified rpm or if the driver opens the throttle, the control unit returns fuel injection and spark advance to the appropriate level. This occurs gradually over a programmed number of engine revolutions rather than suddenly, for driveability and a smooth transition. On some Bosch systems, there is no fuel cutoff during deceleration, and on these a fuel-enrichment program is used to avoid engine bucking and misfire as well as high hydrocarbon emissions from misfires.

Closed Loop

The Bosch Motronic system, like other engine control systems, goes into closed loop once the engine coolant temperature and the oxygen sensor reach normal operating temperature. Bosch and European manufacturers' technical literature frequently refers to the oxygen sensor as a "lambda sensor." While the oxygen sensors work the same way, a different description is used for the optimal air/fuel ratio. Until recently, European engineers identified the ideal ratio by the Greek letter **lambda.** In this country the ideal mixture was always called the "stoichiometric" ratio. Both terms mean the same thing and both oxygen sensors work the same.

INPUTS

Airflow Meter

The Bosch airflow meter used on the Motronic system is the same as described in the pulsed systems section of this chapter. The shaft of the sensor moves the wiper of a potentiometer designed to provide accurate voltage signals to the control unit, Figure 15-17. This unit is designed to provide consistent return signals in spite of aging and rapid changes in temperature.

The control unit can calculate the pulse width for each cylinder by comparing the information from crankshaft position and speed with airflow meter data. The injection then occurs during the appropriate cylinder's intake stroke.

Engine Speed Sensor

The control unit senses engine speed using a pulse-generating device that uses the flywheel teeth as the pulse-triggering device, Figure 15-18.

Reference (Crankshaft Position) Sensor

The crankshaft position or reference sensor works like the engine speed sensor except that its trigger is usually a single pin on the flywheel, Figure 15-19. This sensor provides the control unit with information about crankshaft position.

Throttle Valve Switch

The throttle valve switch contains two sets of electrical contact points, Figure 15-20. Each set has a separate circuit and receives a reference

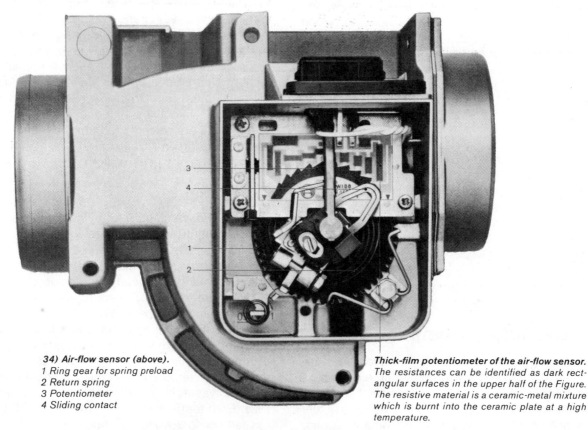

34) Air-flow sensor (above).
1 Ring gear for spring preload
2 Return spring
3 Potentiometer
4 Sliding contact

Thick-film potentiometer of the air-flow sensor.
The resistances can be identified as dark rectangular surfaces in the upper half of the Figure. The resistive material is a ceramic-metal mixture which is burnt into the ceramic plate at a high temperature.

Figure 15–17 Air flow meter. *(Reprinted with permission from Robert Bosch Corporation.)*

voltage from the control unit. This circuit enables the control unit to determine if each of the switches is open or closed. When the throttle is closed, the idle contacts close, completing the circuit for that switch. During part-throttle operation (all normal driving modes) both switches are electrically open. At wide-open throttle the full load contacts

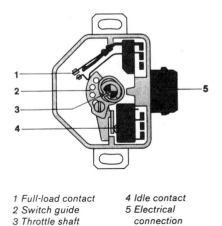

1 Full-load contact   4 Idle contact
2 Switch guide   5 Electrical
3 Throttle shaft   connection

Figure 15–20 Throttle valve switch. *(Reprinted with permission from Robert Bosch Corporation.)*

1 Permanent magnet   5 Winding
2 Housing   6 Flywheel
3 Engine block   ring gear
4 Soft iron core   7 Reference mark

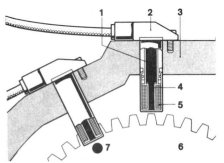

Figure 15–18 Engine speed and reference mark sensors. *(Reprinted with permission from Robert Bosch Corporation.)*

Figure 15–19 Crankshaft and reference sensors. *(Reprinted with permission from Robert Bosch Corporation.)*

close, completing that switch's circuit. The control unit can thus determine three driving conditions from the one unit. The throttle valve switch (containing both switches) is attached to the end of the throttle shaft.

Coolant Temperature Sensor

The coolant temperature sensor is a thermistor bolted into the water jacket near the thermostat housing. Its operation is similar to those on domestic vehicles.

Air Temperature Sensor

The intake air temperature sensor is a thermistor mounted in the intake opening of the airflow sensor, Figure 15-21. Its signal indicates the temperature of the intake air.

Oxygen Sensor (Lambda Sensor)

As previously explained, the earlier Bosch systems use oxygen sensors that work like familiar domestic units, but the literature describes them as "lambda" sensors, using the common German term for air/fuel stoichiometry.

1 Electrical connection, 2 insulation tube, 3 connector, 4 NTC resistor, 5 housing, 6 rivet pin, 7 securing flange. Arrow denotes the direction of intake air.

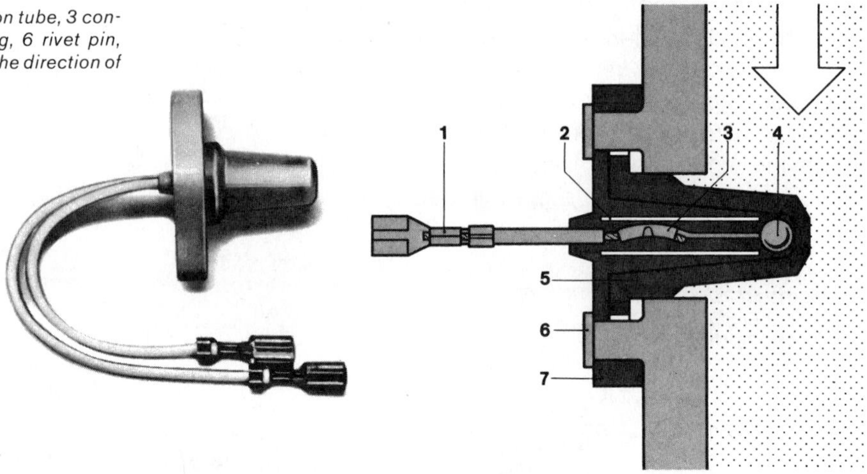

Figure 15–21 Air temperature sensor. *(Reprinted with permission from Robert Bosch Corporation.)*

Knock Sensor

The Bosch Motronic system uses a piezo-electric knock sensor, Figure 15-22. The sensor generates a voltage signal in response to most engine vibration frequencies, but the signal is strongest at frequencies resonant with detonation knock frequencies. At a vibration between 5 and 10 kHz, the control unit recognizes knock and begins retarding the spark. The knock sensor is located on the block between two cylinders. On some engines, particularly V-type engines, two knock sensors are used. Combined with the information the computer has about crankshaft position, this can identify the individual cylinder with knock, to retard the ignition timing to that cylinder selectively (a feature only on later versions of the Motronic system).

Lambda and Air/Fuel Ratio

German engineers have long used the Greek letter lambda (L) to represent a ratio reflecting the air/fuel ratio. The relationship lambda represents is:

Lambda = actual inducted air quantity ÷ theoretic air requirement

The theoretical air requirement is 14.7 units (by weight). If the actual air quantity used was 14.7, then lambda = 1. Using this method of expression, an air/fuel ratio richer than 14.7:1 is expressed as a number less than 1. For example, 0.9 is equivalent to 13.23:1 air/fuel ratio (13.23/14.7 = 0.9). An air/fuel ratio leaner than 14.7:1 is expressed as a number greater than 1. For example, 1.03 is equivalent to 15.14:1 air/fuel ratio (15.14/14.7 = 1.03).

Altitude Sensor

The altitude-sensing device or barometric pressure sensor is an aneroid unit attached to the wiper of a potentiometer similar to the one described in the **Inputs** section of the introductory Ford chapter, Chapter 9. On most Bosch-equipped vehicles, it is located in the air cleaner.

Starter Signal

The cranking control circuit provides a signal to the control unit when the starter is being cranked.

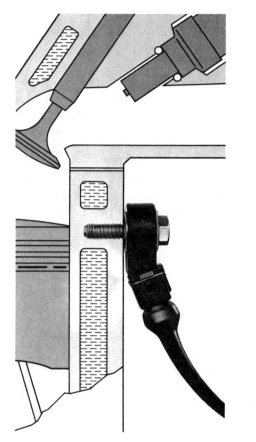

Figure 15–22 Knock sensor. *(Reprinted with permission from Robert Bosch Corporation.)*

Battery Voltage

A system voltage input from the vehicle's electrical system is provided to the control unit so it can monitor battery and charging system voltage.

OUTPUTS

Generally Motronic systems control the operation of the fuel injectors, the engine speed limiter, the fuel pump relay, ignition timing, and dwell. Specific vehicle manufacturers may use other features designed for the particular car.

Injectors

The injectors used on Bosch systems are like those used in domestic built multipoint fuel injection systems. In fact, domestic manufacturers have historically used Bosch injectors for their multipoint injection systems.

Real World Problem

A common source of vacuum leaks, or "false air" as the Bosch literature sometimes calls it (because it does not go through the airflow meter), is around the injectors. The seal is usually a single O-ring. With heat and age, these O-rings can crack and cause a vacuum leak.

It is fairly common for an injector O-ring to leak after the injector is removed and reinstalled, even if it was not leaking air before, because the removal and replacement disturbed the seal. While it is not always necessary to replace O-ring seals when reinstalling an injector, they should be inspected carefully and replaced if necessary. Engine oil is generally regarded as the best O-ring installation lubricant, though at least one manufacturer favors 80-weight gear lubricant. Failure to inspect, and if necessary replace, the injector O-ring can allow a technician to build another new problem into a problem car.

NOTE: Air shrouded injectors are used by Volkswagen and several other carmakers on some of their models. Idle air does not pass through the regular intake channels but through a special idle air passage that leads to collars around individual injectors. This specially routed idle air serves to keep the airflow high where air meets fuel and improves vaporization of the fuel. This idle air shroud plumbing, however, constitutes another source of potential vacuum leaks. On early Volkswagen models, a plastic elbow near the idle air controller often developed cracks or splits which were very hard to spot.

The Bosch injectors are solenoid-operated, pintle-type valves. In open-loop operation, the control unit calculates pulse width based on engine load and engine speed; engine load is calculated from airflow and engine speed information. Figure 15-23 shows a three-dimensional graph or "map" correlating load, speed and air/fuel ratio for a given engine application. Air fuel ratio is expressed as lambda, explained earlier, but the translation into more familiar domestic stoichiometric terms is straightforward.

This map is stored digitally in the control unit's read-only memory. It functions as the look-up table to determine pulse width. The arrows indicate increased load, speed, and richness. Each line intersection represents a relative value of the desired air/fuel ratio.

The control unit can, however, modify the pulse width value represented by each line intersection in response to the following information:

- engine temperature.
- throttle position as indicated by the throttle switch. The control unit also employs a special mixture richening pulse width command at warm WOT to prevent combustion detonation.

- air density. The airflow sensor input must be corrected at higher elevations to allow for the thinner air. This correction is made with input from the air temperature sensor and, on the vehicles so equipped, the altitude sensor.
- battery voltage. Just as on domestic fuel injection systems, the amount of fuel that flows through the injector is affected by the injectors response time, which is directly reflective of charging system voltage. The control unit can detect this voltage and modify the pulse width within a certain range to correct the fuel delivery.

Engine Governor. For each engine with a Bosch control system, there is a specific maximum allowable engine speed. If the engine speed exceeds that limit by more than 80 rpm, the control unit will shut off fuel injection until the engine falls 80 rpm below the limit, Figure 15-24. This "redline" limit is to prevent mechanical damage to the engine from over-revving. The system uses fuel cutoff rather than spark override to keep from dumping excessive hydrocarbons into the atmosphere.

Please note, however, that the engine gover-

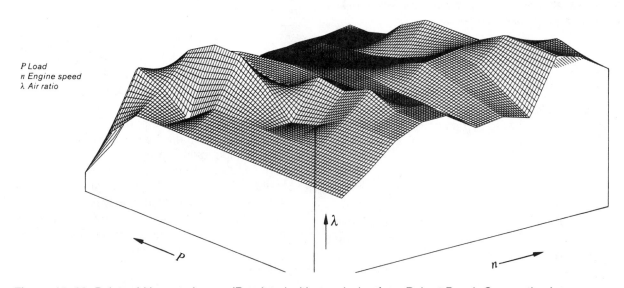

P Load
n Engine speed
λ Air ratio

Figure 15–23 Pulse width control map. *(Reprinted with permission from Robert Bosch Corporation.)*

nor function can only protect the engine from mechanical damage caused by powered over-revving. If a driver downshifts into a lower gear at high speed, the governor can shut off the fuel, but driven by vehicle inertia, the engine could still

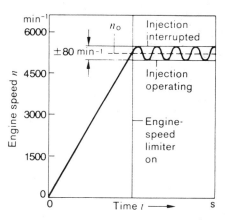

Limiting maximum engine speed n_0 by suppression of fuel-injection pulses.

Figure 15–24 Engine speed governor. *(Reprinted with permission from Robert Bosch Corporation.)*

spin fast enough to break something, probably a valve or connecting rod.

Fuel Pump Relay

As on most fuel injection systems, the fuel pump is indirectly controlled on Motronic systems through a fuel pump relay. Power is available at the relay whenever the ignition switch is turned on. The control unit then grounds or ungrounds the relay's coil to activate or shut off the fuel pump. If the engine does not turn at a minimum specified speed, the control unit will shut off the relay to reduce the risk of fire. As on many other fuel injection systems, the high pressure fuel pump is located inside the fuel tank, submerged in fuel, Figure 15-25. Earlier systems employed a fuel pump just outside the tank, on the line to the engine compartment and just ahead of the fuel filter (some have the filter in the engine compartment).

Fuel Pressure Regulator. The Bosch fuel pressure regulator, like most others, uses intake manifold absolute pressure to maintain a constant pressure differential between the fuel in

1 *Electric fuel pump*
2 *Fuel filter*
3 *Fuel distributor tube*
4 *Injection valve*
5 *Cold-start valve*
6 *Pressure regulator*
7 *Pulsation damper*

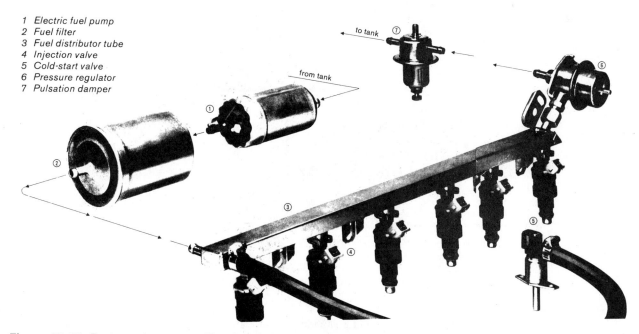

Figure 15–25 Fuel supply system. *(Reprinted with permission from Robert Bosch Corporation.)*

the fuel rail and the air in the intake manifold, Figure 15-26.

Fuel Pressure Pulsation Damper. The pulsation damper, also in Figure 15-26, works to eliminate the noise of fuel pulses from the pump. While it looks much like the pressure regulator, it is not connected to manifold vacuum and it functions as a surge accumulator to absorb the pressure pulses from the opening and closing of the injectors. The pulsation damper is in the return line downstream from the pressure regulator. The higher pressure Bosch systems, as well as those on four-cylinder engines, are more likely than the low pressure systems or those on five-plus cylinder engines to use pressure dampers because the effects of the pulses are greater in them. Higher pressures are often preferred by carmakers because the fuel is more finely atomized under higher pressure, other things being equal.

Ignition Timing

The control unit operates the ignition coil by ground switching the primary windings, Figure 15-27. Like fuel injection pulse width, ignition timing is based on engine load and speed. Figure 15-28 shows a three dimensional map of spark advance for one engine. Arrows point in the directions of increased load, speed, and spark timing advance. The line intersections represent the spark advance for all combinations of engine speed and load. These values are stored digitally as a look-up table in the control unit's read-only memory. The control unit reviews sensor inputs and recalculates spark timing anew between each cylinder firing. The control unit can modify the values in the read-only map in response to inputs from:

- coolant temperature sensor.

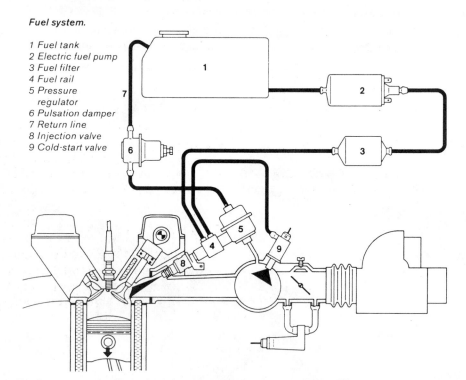

Fuel system.

1 Fuel tank
2 Electric fuel pump
3 Fuel filter
4 Fuel rail
5 Pressure regulator
6 Pulsation damper
7 Return line
8 Injection valve
9 Cold-start valve

Figure 15–26 Fuel pressure regulator (right) and pulsation damper (left). *(Reprinted with permission from Robert Bosch Corporation.)*

Motronic ignition subsystem.

1 Ignition switch	*5 Plug connectors*
2 Ignition coil	*6 Spark-plugs*
3 High-tension distributor	*7 Control unit*
4 Spark-plug leads	*8 Battery*

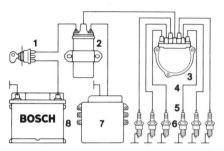

Figure 15–27 Ignition circuit. *(Reprinted with permission from Robert Bosch Corporation.)*

- intake air temperature sensor.
- throttle position switch (during WOT operation, timing is set for maximum torque with retard applied only to prevent detonation).
- altitude sensor (if present).

The control unit modifies the spark advance according to the information from either the coolant temperature sensor or the intake air temperature sensor. This modification, of course, depends on the driving conditions. At starting crank, timing depends entirely on engine speed and coolant temperature. At low cranking speeds and low temperatures, the timing is close to TDC. At higher cranking speeds and with warmer engines, the timing is only slightly advanced. With high cranking speed even at low temperature, timing is more advanced. Each of these combinations improves starting and post start-up driveability.

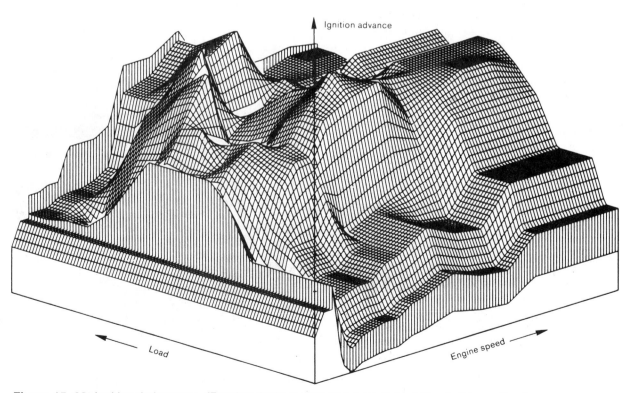

Figure 15–28 Ignition timing map. *(Reprinted with permission from Robert Bosch Corporation.)*

Once the engine starts cold, timing is advanced significantly for a brief period to reduce the need for fuel enrichment. Once the engine is warm enough to reduce the likelihood of stalling, the ignition timing is retarded on some vehicles to speed up engine warm-up, even at the cost of lower delivered engine torque. Increased exhaust temperature helps warm the oxygen sensor and catalytic converter more quickly.

On earlier systems, during warm idle, ignition timing is used to control idle speed. With no additional load, transmission in neutral, and air conditioning off, the timing can be retarded slightly and still maintain the desired idle speed. This keeps combustion chamber surfaces hotter to reduce the production of unburned hydrocarbons. If there is a load such as the transmission going into gear, the idle speed will begin to slow. The control unit will respond by advancing ignition timing enough to achieve more torque and bring the idle speed back to the desired speed. This saves opening the throttle and reduces fuel consumption. On applications with the rotary idle adjuster, idle air will increase if advancing the spark timing alone is not enough to keep the idle speed at the desired setting.

At wide-open throttle, normal or higher engine coolant and/or intake air temperature cause the control unit to reduce spark timing at those load combinations where there is highest probability of spark knock. If at a specific load there is acceleration short of WOT, the control unit retards the existing spark advance and then gradually begins to restore it. This reduces both spark knock and the production of oxides of nitrogen.

Spark Knock Control

On earlier systems the spark knock control was separate from the Motronic unit. Later it was incorporated into the control unit. The later versions, controlled by the computer, allow a more detailed control of spark advance. When the knock sensor indicates a detonation signal, the control unit recognizes the knock and immediately reduces spark advance until the knock disap-

pears. Then it gradually readvances timing toward the original value until the knock returns.

Dwell Control

The Motronic control unit manages ignition coil dwell in the same way as the General Motors HEI system. As engine speed increases, dwell increases to insure complete magnetic saturation of the coil. The Bosch system also considers charging system voltage when calculating dwell.

Coil Peak Current Cutoff

If the ignition coil remains on with the engine stopped, both the ignition coil and the driver (transistor) in the control unit may overheat or burn out. To avoid this problem, the control unit automatically turns off the ignition coil anytime engine rpm drops below a specified speed, usually about 30 rpm, well below a successful starting crank speed.

Figure 15–29 Idle air bypass device. *(Reprinted with permission from Robert Bosch Corporation.)*

Idle Speed Control

Bosch Motronic systems use two different methods to control idle speed.

Auxiliary Air Device. On some applications, warm idle speed is controlled by an idle air bypass device in the idle air bypass passage, Figure 15-29. After a cold start and during warm-up, idle speed is increased by the air routed through an auxiliary air valve. A bimetallic strip inside the valve holds the valve open at low temperature, Figure 15-30. With the passage open, additional air bypasses the throttle blade. This air has, however, passed the airflow sensor so the computer knows the correct amount of fuel to inject. In addition, once the ignition is turned on, voltage is applied to a heating element wrapped around the bimetallic strip. As the strip warms up, it gradually closes the valve and blocks the flow of auxiliary air, slowing the idle as the engine warms. Once the engine is completely up to operating temperature, its heat is enough to keep the valve closed.

Rotary Idle Control Valve. Other applications use the rotary idle control valve, controlled directly by the computer, Figure 15-31. This valve controls engine idle speed regardless of engine coolant temperature. At the end of an armature shaft, a rotary valve or slider valve can move in the idle air bypass, Figure 15-32. The DC electric motor has a permanent magnetic field and two armature windings. By controlling the voltage applied to the two windings, the control unit can incrementally rotate the armature and thus position the rotary valve to control idle air bypass. The control unit's memory has a programmed idle speed that it wants to maintain for any given engine coolant temperature. It first uses ignition timing to adjust idle speed; it then uses the rotary idle adjuster to bring idle speed to the desired rpm if the timing adjustments have been insufficient.

Turbo Boost Control

Bosch Motronic systems use a combination of spark retard and boost pressure reduction to control spark knock on turbocharger engines. Particularly on smaller engines, Bosch engineers feel reducing boost alone puts unnecessary limits on engine performance. They also find that just using spark retard to control knock is a poor choice because just retarding ignition timing raises exhaust temperature. A temperature rise can be threatening to the already high temperature turbocharger turbine. Using a combination of some of both measures is the strategy they pre-

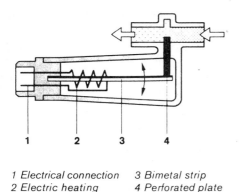

1 Electrical connection   3 Bimetal strip
2 Electric heating   4 Perforated plate

Figure 15–30 Auxiliary air bypass device schematic. *(Reprinted with permission from Robert Bosch Corporation.)*

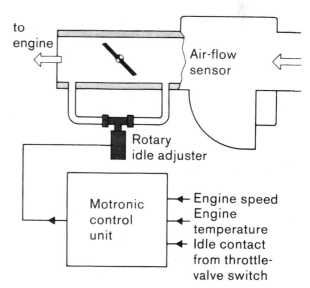

Figure 15–31 Rotary idle adjuster circuit. *(Reprinted with permission from Robert Bosch Corporation.)*

fer. Spark retard is the fastest measure (provided the knock is caused by spark and not by preignition): it can be applied to the very next power stroke after the detection of knock. It takes several intake strokes even after the wastegate opens before the manifold pressure begins to come down because the turbocharger must slow and the intake air must start to cool. As boost pressure does come down, however, the Motronic system starts to readvance the ignition timing to retain as much engine torque as possible. Later systems include the capacity to retard ignition timing to individual cylinders, as described in earlier chapters on domestic systems.

EGR Valve Control

Early European Bosch-equipped vehicles did not use EGR systems to control the production of oxides of nitrogen (NO_x) at high combustion temperatures. Instead they used the reduction feature of the catalytic converter and the residual exhaust retained by the valve overlap. This method, however, does not allow the precise controls available with an electronically controlled EGR valve, so later versions employ EGR systems functionally similar to those used on most domestic-built systems.

Evaporative Emissions Control

Motronic system vehicles use an evaporative emission system similar to those on domestic vehicles. Fuel vapors from the fuel tank are stored in a charcoal canister, Figure 15-33. Once the system goes into closed loop, the control unit activates a solenoid, opening a valve in the purge line connecting the canister to the intake manifold. As usual, precise mixture control then follows from the oxygen sensor signal.

Electronic Transmission Control

Later versions of the Motronic system include sensors to indicate the transmission state. A vehicle speed sensor, transmission gear selection sensor, and kickdown sensor provide the control unit with this information. On some systems the control unit determines the transmission's hydraulic pressure and shift points (solenoid valves replacing traditional spool-type shift valves), Figure 15-34. Shift point look-up tables (one for economy, one for performance, one for manual shifting) stored in the control unit's read only memory provide optimum transmission shifting according to the programming look-up chart selected by the driver.

Smooth shifts with reduced clutch slip and

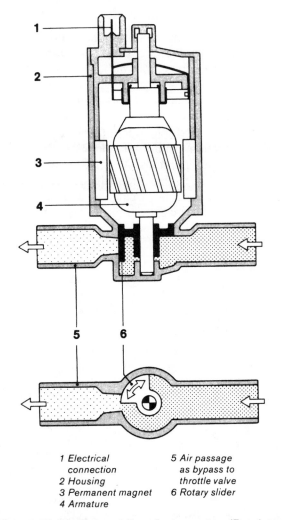

1 Electrical connection
2 Housing
3 Permanent magnet
4 Armature
5 Air passage as bypass to throttle valve
6 Rotary slider

Figure 15–32 Rotary idle adjuster motor. (Reprinted with permission from Robert Bosch Corporation.)

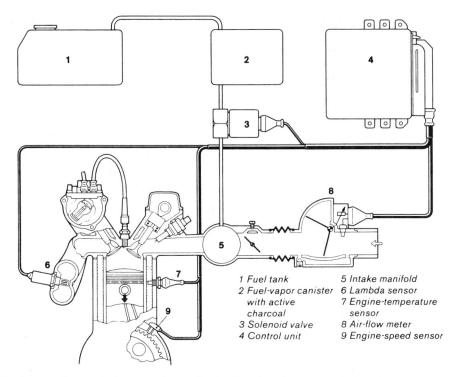

Figure 15–33 Evaporative emissions control system. *(Reprinted with permission from Robert Bosch Corporation.)*

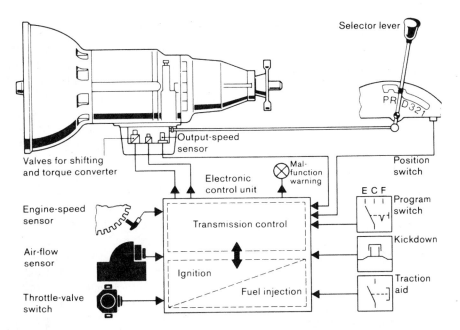

Figure 15–34 Electronic transmission control. *(Reprinted with permission from Robert Bosch Corporation.*

wear are achieved, especially during WOT operation, by momentarily retarding ignition timing to reduce engine torque output during the shift. Motronic transmission control can thus improve fuel economy, shift quality, transmission torque capacity and expected transmission life.

Start-Stop Control

Another optional feature that can be added to a Motronic system is the start-stop control. This feature is on some European cars, and may become available on domestic cars. This applies only to vehicles with manual transmissions for safety reasons. This feature helps conserve fuel and reduce emissions.

The start-stop control uses a vehicle speed sensor, a clutch pedal position sensing switch, and a separate control module. If vehicle speed falls below about 1 mph with the clutch pedal depressed, the module signals the control unit, which shuts off the fuel. As long as the driver keeps the clutch pedal down, the module will activate the cranking circuit and fuel injection will resume as soon as the driver presses the accelerator. This system does not work until the engine is warm.

✔ SYSTEM DIAGNOSIS & SERVICE

The vehicle manufacturer establishes the diagnostic procedures for each vehicle, so procedures vary somewhat from one make to another. The diagnostic procedures here are representative of Bosch Motronic systems.

Troubleshooting Guide

The troubleshooting guide consists of a list of driveability complaints such as: Will Not Start Cold, Erratic Idle During Warm-up, and Backfires. Each complaint is accompanied by a list of possible causes. The possible causes should be checked either in the order of easiest to check, or of most likely to cause the problem, based on the technician's familiarity with the vehicle and the

Variable Displacement System

Another feature Bosch uses on some vehicles is a variable displacement system not unlike Cadillac's modulated displacement system. The system includes:

- deactivating cylinders by shutting off the injectors to selected cylinders.
- having the intake and exhaust valves of the deactivated cylinders continue functioning.
- providing intake and exhaust manifold valving that allows the exhaust from working cylinders to be passed through the deactivated cylinders to keep them up to normal operating temperature.

problem. It is important that all components or systems identified in the possible cause list for the complaints that are not parts of the Motronic system are working properly before proceeding to the tests for the Motronic system.

Motronic Test Section

The service manual is likely to have a section that contains test procedures for each component of the Motronic system. The tests in this section should not be conducted until the engine is at normal operating temperature. In testing components and their circuits, remember that connections cause more problems than components.

SUMMARY

In this chapter we have learned about the history of Bosch fuel injection systems; we have considered the two mechanically distinct kinds of Bosch engine management systems, pulsed and constant fuel injection. We have seen the differences between the K-Jetronic (constant flow from the injectors), D-Jetronic (with pulsed injectors controlled principally by the intake manifold pressure), L-Jetronic (with fuel delivery determined by an air-door sensor) and finally Motronic (a complete engine management system, controlling mixture as well as ignition timing and other functions).

We have reviewed the difference between today's Motronic systems and the predecessor systems. The different operating conditions and how they work in the different operational modes were covered. We have seen the several ways the control unit calculates engine load. We have considered the major input sources and output actuations, as well as the major functions. Finally, we have looked at the diagnostic approaches.

▲ DIAGNOSTIC EXERCISE

A Volkswagen with a Motronic system and an automatic transmission stalls whenever it is put into a forward gear. The car works normally in reverse, and will idle indefinitely without a problem. There is no connection between the engine management system and the transmission except for the neutral start switch. Why does the engine stall in forward gears?

REVIEW QUESTIONS

1. Name seven operating conditions that cause a Motronic system to select a specific strategy for fuel and/or ignition timing control.
2. How does the Motronic system protect its control unit against reversed polarity?
3. What is the function of the airflow sensor?
4. What are the sources of engine speed and crankshaft position input?
5. List seven control unit-controlled functions that are standard features of the Motronic system.
6. Name at least six optional control unit-controlled features.
7. What does the term *lambda* mean as used in automotive application?
8. Name at least two functions affected by battery voltage.

9. What two measures are used to control spark knock on turbocharger applications?
10. What does the troubleshooting guide do?
11. What does the test section do?

ASE-type Questions (Actual ASE test questions are rarely so product specific).

12. Technician A says Motronic's pulse width during open-loop operation is determined based on data concerning engine speed and throttle position. Technician B says that Motronic's pulse width determination in open-loop is based on engine speed and load. Who is correct?
    a. A only.
    b. B only.
    c. both A and B.
    d. neither A nor B.
13. Technician A says that Motronic determines engine load from engine speed and throttle position. Technician B says that Motronic determines engine load from engine speed and airflow. Who is correct?
    a. A only.
    b. B only.
    c. both A and B.
    d. neither A nor B.
14. Technician A says that Motronic controls idle speed with ignition timing and idle air flow. Technician B says there are two different methods of controlling idle airflow. Who is correct?
    a. A only.
    b. B only.
    c. both A and B.
    d. neither A nor B.
15. Technician A says that Motronic limits maximum engine speed by shutting off the injectors if the specified speed is exceeded. Technician B says that some Motronic engine applications limit engine speed by shutting off the ignition instead of the injectors. Who is correct?
    a. A only.
    b. B only.

c. both A and B.

d. neither A nor B.

16. A K-Jetronic system runs rich. Technician A measures the voltage at the pressure actuator and finds that it is constant regardless of engine speed or load. He suggests that the computer is faulty. Technician B says leaking injectors could cause the problem. Who is correct?

    a. A only.

    b. B only.

    c. both A and B.

    d. neither A nor B.

17. A Motronic-controlled system has an excessively high warm idle speed but no other symptoms. Technician A says to check for a vacuum leak between the throttle and the valves. Technician B says to check for a leak between the air flow sensor and the throttle. Who is correct?

    a. A only.

    b. B only.

    c. both A and B.

    d. neither A nor B.

18. Shortly after school resumes in September, a teacher finds her Motronic-equipped car is hard to start when cold. Technician A says it's time for a tuneup. Technician B says to check the electrical circuit to the cold-start injector. Who is correct?

    a. A only.

    b. B only.

    c. both A and B.

    d. neither A nor B.

19. A motorist who drives his car very hard complains that the engine misses and vibrates at high rpm. Technician A says the ignition coil is suspect. Technician B says the fuel filter is clogged. Who is correct?

    a. A only.

    b. B only.

    c. both A and B.

    d. neither A nor B.

20. An early Motronic car has backfired while cranking. Now it won't start. Technician A says the injectors may be damaged. Technician B says to inspect the airflow meter. Who is correct?

    a. A only.

    b. B only.

    c. both A and B.

    d. neither A nor B.

Chapter 16

Asian Computer Control Systems

OBJECTIVES

In this chapter you can learn:
- ❏ the basics of the Nissan engine control system.
- ❏ the basics of the Toyota engine control system.
- ❏ how each system can be diagnosed using OBD II diagnostic procedures.
- ❏ characteristic real world problems of each system.

KEY TERMS

Ceramic Zirconia Oxygen Sensor
EGR Temperature Sensor
EVAP System Pressure Sensor
Flow Bench
Front and Rear Oxygen Sensors
Fuel Temperature Sensor
Fuel Trim Compensation Factor
Hot-Film Air Mass Sensor
Two-Trip Malfunction Detection

In this chapter, we will concentrate on Toyota and Nissan computer control systems. While these are not the only Asian cars imported to North America, they are representative, and common; and their control systems are typical of many of the other Asian manufacturers' products. Because of the widespread use of shared components by Japanese manufacturers, an understanding of one company's systems often provides an understanding of the systems used by others. Further, while there are Asian vehicles manufactured in Korea, these control systems are basically Japanese and the overwhelming number of such imports come unmodified (or simplified) from Japanese cars, so this is where our concentration shall be. In each case, we consider only late model vehicles, not the earlier and simpler systems. Many earlier vehicles used almost off-the-shelf Bosch controls or those built under Bosch license. Early feedback carburetor systems work similarly to domestic computer-controlled carburetor systems.

Real World Problem

In the last years that feedback carburetors were used on many Japanese cars, the carburetors became much more complicated in order to consistently deliver a stoichiometric air/fuel mixture to the cylinders. Most feedback carburetors were precisely adjusted on specialized equipment called **flow benches** at the factory to meet very narrow tolerances. If you see an adjustment screw marked with paint on one of these carburetors, do not break the paint loose or attempt to adjust the carburetor using this screw. It is very difficult to get the adjustment right with a flow bench and almost impossible without one. In addition, the flow and angle specifications are not published. Never touch a painted adjustment screw on a feedback carburetor, or you will almost certainly not be able to get the unit to work properly again.

NISSAN: GENERAL SYSTEM DESCRIPTION

Typical late-model Nissans use a **hot-film air mass sensor** as the principal input to the computer to determine the fuel pulse width and spark advance. The computer, of course, controls the fuel mixture and ignition based on information stored in its memory as well as inputs from all its other sensors. For a general overview of the entire system, see Figure 16-1.

Nissans use both sequential injection and injections pulsed together, or simultaneous injection. The sequential mode injects fuel once every complete engine cycle (two crankshaft revolutions) during the corresponding cylinder's intake stroke. The simultaneous injection mode sprays all the injectors at once, twice per complete engine cycle. The simultaneous injection mode is used when the engine is starting or if the engine is in fail/safe mode or if the camshaft position sensor signal is lost and the engine is running on the crankshaft position sensor signal alone.

The Nissan control system does not set the ignition timing consistently at the edge of knock and use the knock sensor signal as a routine control input. Instead, the spark advance map is scaled to avoid knock under all but the extremes of load, engine temperature, and low fuel octane equivalency.

Besides normal conditions, the computer also modifies ignition timing under special circumstances:

- during starting.
- during warm-up.
- during idle.
- if the engine is overheated.
- during acceleration.

The Nissan computer control system employs special fuel enrichment measures under many of the same specific circumstances:

- during warm-up.
- during cranking and starting.
- during acceleration.
- when the engine is overheated.
- during high load and high vehicle speed conditions.

The system employs special lean fuel mixtures (including complete fuel shutoff in some cases) under other circumstances:

- during deceleration.
- during low-load, high vehicle speed conditions.

Deceleration fuel cutoff will stop as the engine falls below a speed of 2,200 rpm with no load. Fuel cutoff does not begin unless the unloaded engine is above approximately 2,700 rpm.

Mixture Ratio Self-Learning.   The feedback control system employs a self-learning ability, similar to systems used on domestic cars. Like them, the Nissan system monitors the success of its pulse-width modification measures to keep the air/fuel mixture stoichiometric by input from the oxygen sensor. The measures used are called "fuel trim." The computer can modify its working copy of the hard-wired mixture-to-pulse-width map depending on the results it sees from the oxygen sensor. It uses both a short-term and a long-term **fuel trim compensation factor,** similar to GM's integrator and block learn. The short-term factor corrects for temporary circumstances requiring fuel mixture adjustment; the long-term factor comes into play for gradual engine component wear and changes in the environment, such as altitude or seasonal climactic change.

Air Conditioning Cutout.   To enhance engine performance and/or to preserve air conditioner components, the air conditioning compressor is automatically disengaged regardless of the control settings if:

- the throttle is fully open.
- during start-up cranking.
- at high engine speeds (to protect the compressor).

Evaporative Emissions System.   The Nissan evaporative emissions system like most

ENGINE AND EMISSION CONTROL OVERALL SYSTEM

System Chart

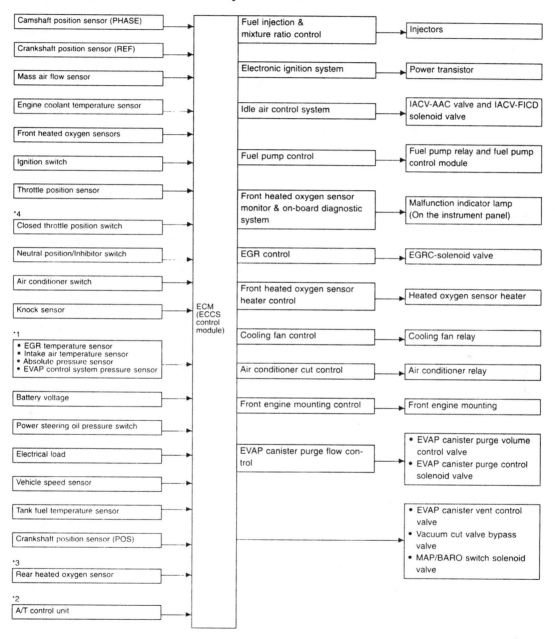

*1: These sensors are not directly used to control the engine system. They are used only for the on-board diagnosis.
*2: The DTC related to A/T will be sent to ECM.
*3: This sensor is not used to control the engine system under normal conditions.
*4: This switch will operate in place of the throttle position sensor to control EVAP parts if the sensor malfunctions.

Figure 16–1 Nissan system overview. *(Courtesy of Nissan Motors.)*

other carmakers, uses a charcoal canister that stores fuel vapors from the tank to prevent their escape into the atmosphere. The system depends on the vapor pressure/vacuum relief valve in the fuel filler cap, and the vapor recovery shutter in the filler tube. The computer activates the EVAP canister purge volume control valve to operate the system under normal driving conditions. The fuel mixture pulse-width adjustments are made in response to oxygen sensor information since there is no way for the computer to know how much fuel vapor, if any, is coming through the vapor canister.

INPUTS

Camshaft Position Sensor (PHASE). The camshaft position sensor, Figure 16-2, is in the front engine cover facing the camshaft sprocket.

This information distinguishes which cylinder is at the firing position. The camshaft position sensor is a coil-and-magnet pulse generator: it consists of a permanent magnet wound in an electrical coil. When the gap between the camshaft sprocket and the magnet changes, the magnetic field changes and generates a voltage signal in the sensor output line. The computer uses the camshaft position signal to sequence the fuel injectors. At the beginning of the start-up cranking period the camshaft position signal has not yet occurred, and all the injectors are simultaneously pulsed for engine start-up.

Camshaft position sensors can be checked for shorts and opens with an ohmmeter, as well as for residual magnetism at the tip of the sensor with a strip of ferrous metal or a screwdriver tip.

Crankshaft Position Sensor (REF). The REF crankshaft position sensor, Figure 16-3, is on the upper oil pan facing the crankshaft acces-

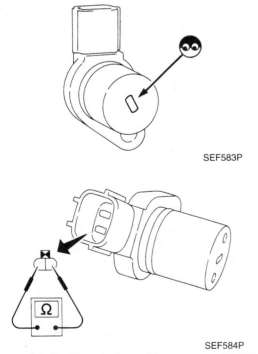

SEF583P

SEF584P

Figure 16–2 Camshaft position sensor (PHASE). *(Courtesy of Nissan Motors.)*

SEF585P

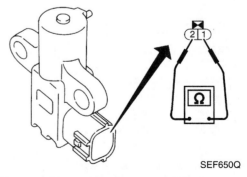

SEF650Q

Figure 16–3 Crankshaft position sensor (REF). *(Courtesy of Nissan Motors.)*

sory belt pulley. It signals the computer when the TDC position for individual cylinders occurs. This sensor is a coil-and-magnet inductive signal generator producing a signal that varies in voltage with engine speed. Resistance through its coil should range between 470 and 570 ohms at room temperature.

Mass Airflow Sensor. The hot film air mass sensor is located in the intake air duct between the air filter and the throttle, Figure 16-4. It should indicate a signal voltage of below 1 volt with the ignition on and the engine off, and should show between 1.0 and 1.7 volts at idle. With the engine running at higher speeds and under load, the signal voltage can rise to as much as 4.0 volts. Because of the sensitivity of the sensor to disturbances in the air flow, a frequent check should be done to make sure there is no dust or foreign material on or near the sensor.

Engine Coolant Temperature Sensor. As with other sensors, the engine coolant temperature sensor modifies a reference signal from the ECM to correspond to the monitored parameter, Figure 16-5. The resistance of the thermistor in the sensor reduces as the engine temperature increases. Information from the engine coolant temperature sensor plays a significant role in engine management, because a richer mixture is needed for cooler temperatures to make sure there is enough fuel vaporized to burn. At operating temperatures, the system switches to a closed-loop control system, using the signal from the oxygen sensor. The engine coolant temperature sensor, however, is critical to the determination of when the system goes into closed loop.

Front Heated Oxygen Sensors. Nissans use **ceramic zirconia oxygen sensors,** Figure 16-6. While the operation is similar to other types of sensors, the amplitude of the signal is slightly larger, with the signal switching from about 1 volt to almost zero as the mixture cycles from lean to rich and back across the stoichiometric ratio. The computer gets the sensor's output signal and constantly readjusts the injector pulse width to correct the air/fuel mixture and keep the signal cycling. On V-type engines, two front oxygen sensors are used, one on each bank; and the computer controls the mixture delivered to each

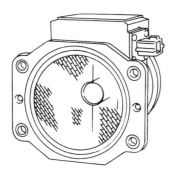

Figure 16–4 Mass airflow sensor. *(Courtesy of Nissan Motors.)*

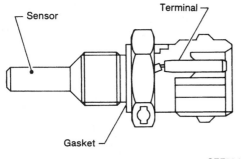

SEF594K

Figure 16–5 Engine coolant temperature sensor. *(Courtesy of Nissan Motors.)*

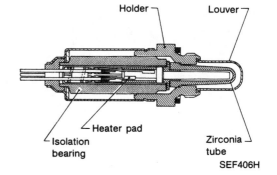

SEF406H

Figure 16–6 Heated oxygen sensor. *(Courtesy of Nissan Motors.)*

delivered to each bank independently of that delivered to the other. The oxygen sensors are heated to bring the system to closed loop as quickly as possible and to prevent the system from falling out of closed loop should the vehicle idle for a long time.

Diagnostic & Service Tip

Very frequently the problem with an oxygen sensor is not that it has failed completely, but that it has begun to respond too slowly to changes in the residual exhaust oxygen. Once it does so, the computer will not be able to properly cycle the injector pulse width to keep the mixture stoichiometric. There is no specific number of cycles per unit of time that an oxygen sensor should register, but the newer systems should work more quickly than the older. If in doubt, monitor the signal cycling of a "known-good" sensor and compare it to one you suspect.

Real World Problem

Oxygen sensors, particularly zirconia-based oxygen sensors, are relatively brittle and delicate components because of the ceramic element. If an oxygen sensor is dropped much more than a foot to a hard surface such as a concrete shop floor, it will probably sustain an invisible internal crack and will have to be replaced. An oxygen sensor is much more delicate than a spark plug as well as more expensive, so they should always be handled carefully.

Ignition Switch. When the ignition switch is turned on, the computer becomes active and sends the reference signals to all the sensors that receive one. It also begins monitoring all the return signals and prepares to energize the appropriate actuators.

Throttle Position Sensor. The computer learns about accelerator pedal movement and position from the throttle position sensor, Figure 16-7. As with many other linear measurement

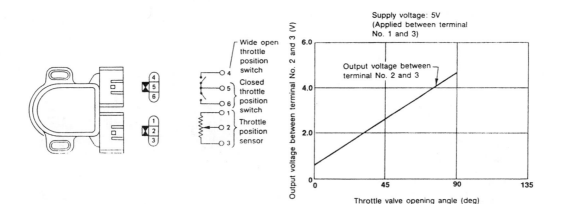

MEC693BA.

Figure 16–7 Throttle position sensor. *(Courtesy of Nissan Motors.)*

sensors, it receives a reference voltage from the computer, routes it through a variable resistor, and returns a corresponding voltage to the computer. Throttle position signals range from about 0.5 volt to about 4.6 volts from closed throttle to wide-open throttle.

Closed Throttle Position Switch. The closed-throttle position switch, Figure 16-8, tells the computer when the driver has removed his or her foot from the accelerator pedal. The computer then knows to control the idle speed depending on information from its other sensors. The closed-throttle switch is either on or off.

Park/Neutral Switch. The park/neutral switch allows the starter circuit to complete only with the transmission in one of these two gears. It also sends a signal to the computer when the vehicle is in gear. This signal is used for idle control, spark timing, and sometimes fuel mixture.

Air Conditioner Switch. When the air conditioning is switched on, the computer receives a signal which it uses to determine proper idle speed.

Knock Sensor. The Nissan knock sensor is a piezoelectric crystal type, sending a voltage signal to the computer in the event it sustains vibrations at a frequency characteristic of detonation. If the computer receives that signal, it retards ignition timing until the knock disappears. As already mentioned, the Nissan spark advance map is designed to prevent knock except under unusual fuel quality, load, or temperature conditions.

The knock sensor can be electrically inspected for shorts or opens. Nissan literature calls for an ohmmeter capable of measuring up to 10 megohms. Like oxygen sensors, knock sensors are brittle and should not be reused if they have been dropped on a hard surface.

EGR Temperature Sensor. The **EGR temperature sensor,** Figure 16-9, reports changes in the EGR passage temperature. While this information does not directly control engine management functions, it is used for emissions diagnosis. If the computer sees no signal or the wrong signal from the EGR temperature sensor, this indicates that insufficient exhaust is recirculating through the EGR system, leading to excessive production of NO_x in the exhaust. The sensor is a thermistor, changing its resistance and thus the return signal with temperature changes.

Intake Air Temperature Sensor. The intake air temperature sensor, Figure 16-10, mounts in the air cleaner housing and transmits a voltage signal corresponding to the intake air temperature. The sensor uses a thermistor that is responsive to changes in temperature. As temperature rises, the resistance of the thermistor goes down.

Absolute Pressure Sensor. The Nissan absolute pressure sensor, Figure 16-11, connects to the MAP/BARO switch solenoid through a hose. It detects both ambient pressure and intake manifold pressure. As the pressure rises, the voltage signal rises.

Throttle position switch
built into throttle
position sensor

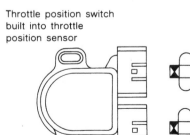

SEF402Q

Figure 16–8 Closed throttle position switch. *(Courtesy of Nissan Motors.)*

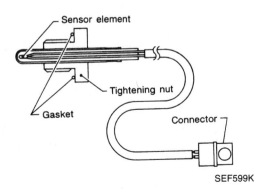

SEF599K

Figure 16–9 EGR temperature sensor. *(Courtesy of Nissan Motors.)*

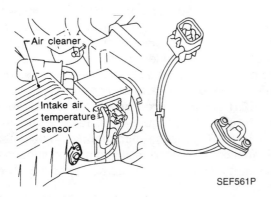

SEF561P

Figure 16–10 Intake air temperature sensor. *(Courtesy of Nissan Motors.)*

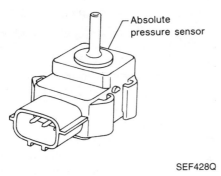

SEF428Q

Figure 16–11 Absolute pressure sensor. *(Courtesy of Nissan Motors.)*

EVAP Control System Pressure Sensor. To monitor the effectiveness of the EVAP control system, the computer receives a signal from the **EVAP control system pressure sensor,** Figure 16-12. While this does not directly affect engine control functions, it does play a role in on-board emissions diagnosis. This sensor will indicate if there is a leak in the vapor canister system.

NOTE: This sensor cannot distinguish a leak in the system from a loose fuel filler cap, which will trigger the same trouble code and light the malfunction indicator lamp (MIL). Check for a properly-functioning fuel filler cap first, including a pressure test of the cap, whenever this sensor indicates a problem.

Battery Voltage/Electrical Load. The computer constantly monitors the voltage received from the battery and charging system and will set the check engine light if it falls below proper voltages. This signal is used to modify the injector pulse width and the ignition coil dwell if system voltage falls below the standard charging system voltage.

Power Steering Oil Pressure Switch. The power steering oil pressure switch, Figure 16-13, is on the power steering high pressure tube. When a power steering load occurs, it triggers this switch, signaling to the computer to increase the idle speed setting. This switch conducts battery voltage and can be checked for continuity with an ohmmeter.

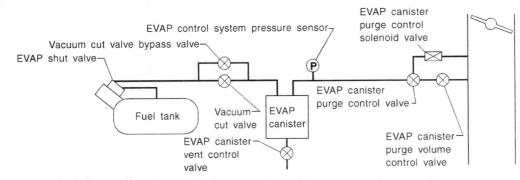

SEF782Q

Figure 16–12 Evaporative emission system diagram. *(Courtesy of Nissan Motors.)*

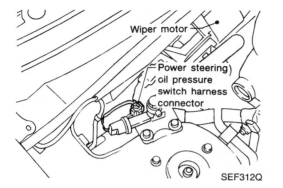

Figure 16–13 Power steering oil pressure switch. *(Courtesy of Nissan Motors.)*

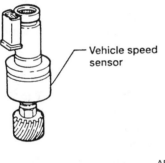

Figure 16–14 Vehicle speed sensor. *(Courtesy of Nissan Motors.)*

Vehicle Speed Sensor. The vehicle speed sensor, Figure 16-14, is in the transaxle on the output shaft. Its pulse generator sends a signal corresponding to vehicle speed to the speedometer, which conveys the signal to the computer. This information is used for both fuel pulse width and ignition timing calculations.

Tank Fuel Temperature Sensor. The system uses a thermistor-type **fuel temperature sensor** to monitor the temperature of the fuel in the tank. The sensor is accessible under the rear seat, Figure 16-15. The fuel temperature sensor receives a reference voltage from the computer and returns to it a voltage signal corresponding to the temperature of the fuel, from 3.5 volts when the fuel is at 68 degrees Fahrenheit to 2.2 volts when the fuel reaches 122 degrees. Information from this sensor indicates changes in fuel viscosity and its inclination to vapor lock.

Crankshaft Position Sensor (POS). The crankshaft position sensor, Figure 16-16, plays a familiar role in spark advance timing, fuel injec-

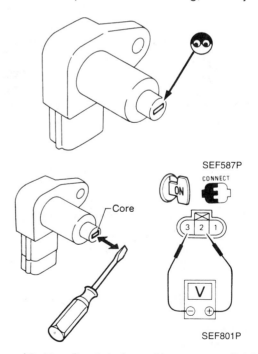

Under the rear seat

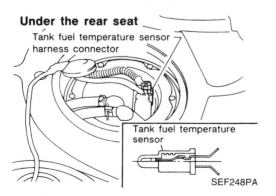

Figure 16–15 Tank fuel temperature sensor. *(Courtesy of Nissan Motors.)*

Figure 16–16 Crankshaft position sensor (POS). *(Courtesy of Nissan Motors.)*

tion timing and in the newest vehicles emissions monitoring. It is located on the flywheel end of the oil pan and signals one degree changes in the flywheel position to the computer. In these later cars, the crankshaft position sensor is the principal source of information about cylinder misfire. When a cylinder misfires, whether from spark, fuel, or mechanical problems, the engine slows slightly at the point when that cylinder's power stroke should occur. At idle, this slowing is perceptible as rough running, but the crankshaft position sensor can report the miss to the computer even at higher speeds.

The emissions-related reason for tracking cylinder miss is that if the cylinder does not fire, it does not consume the fuel that entered with the intake stroke. The first consequence would be a considerable overheating of the catalytic converter from all the increased fuel and oxygen. The second consequence would be a failed catalytic converter, and the third would be the dumping of raw hydrocarbons into the air after the catalytic converter could no longer burn them up.

Once the misfiring cylinder is identified by the crankshaft position sensor, the computer can then shut off the fuel injector to that cylinder. This is not, of course, a complete countermeasure, because the oxygen is still pumped through the cylinder and into the exhaust. This means the oxygen sensor's reports will be inaccurate (overlean) and the computer will drive the system rich. So this is one of the few circumstances when the computer turns on the check engine light the first time the problem occurs, without waiting for it to recur on a subsequent trip.

Rear Heated Oxygen Sensor. Nissans that conform to the OBD II regulations include a rear heated oxygen sensor. This sensor works in the same way as the front sensor or sensors, but is used primarily to monitor the effectiveness of the catalytic converter. As long as the catalytic converter is working properly, the rear heated oxygen sensor's signal should remain almost flat or only slowly cycling over its output range. The computer tracks the signal output and will set a code should the signal mirror that of the front sensor. If

the front sensor fails, the computer will make use of the rear sensor to control air/fuel mixture as closely as possible.

Automatic Transmission Diagnostic Communication Line. There is a communications link between the automatic transmission control unit and the engine control computer which conveys information related to transmission malfunctions. Any such information erased from the transmission unit must be separately erased from the engine control computer.

OUTPUTS

Fuel Injectors. The multiport fuel injectors are Bosch-type, used on many vehicles. Activated by a constant power source grounded through the computer, each injector opens by lifting its pintle off the seat and allowing fuel to spray through the nozzle for a fixed amount of time, calculated by the computer on the basis of all the system inputs.

Ignition Coil Power Transistor. Late model Nissans use one coil per cylinder in a direct ignition system, Figure 16-17. Because of an internal diode to prevent flashback on the secondary circuit, you can only measure resistance between terminals 1 and 2. If a bad coil is suspected, moving the coil from one cylinder to another to see whether the cylinder miss follows is the accepted technique. No internal repairs are possible; replacement is the only service.

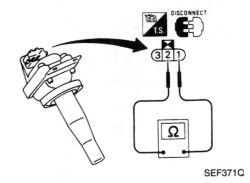

SEF371Q

Figure 16–17 Ignition coil resistance test. *(Courtesy of Nissan Motors.)*

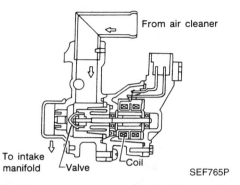

Figure 16–18 Idle air control valve (IACV)—auxiliary air control valve (AAC). *(Courtesy of Nissan Motors.)*

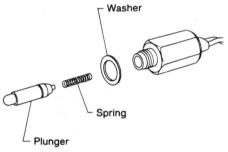

Figure 16–19 IACD-FICD solenoid valve. *(Courtesy of Nissan Motors.)*

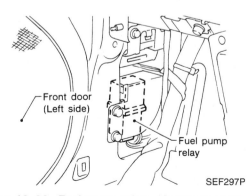

Figure 16–20 Fuel pump relay. *(Courtesy of Nissan Motors.)*

IACV-AAC Valve and IACV-FICD Solenoid Valve. The idle air control valve—auxiliary air control (IACV-AAC) valve, controls throttle bypass air that determines the engine's idle speed, Figure 16-18. The computer activates this component according to inputs from the engine coolant temperature sensor, electric load sensor, air conditioning switch, and power steering pressure switch. Once removed from the engine, the operation of this valve can be checked by cycling the ignition switch on and off to observe whether or not the valve shaft moves smoothly forward and backward according to the ignition switch position. Removal and reinstallation of the valve should only be done with the ignition switch off.

The IACV-FICD solenoid combines with the IACV-AAC valve to constitute the idle air adjusting unit. The IACV-FICD solenoid is strictly an on-off solenoid moving a plunger in the air passage, Figure 16-19. It can be checked separately from the vehicle for smooth function.

Fuel Pump Relay and Fuel Pump Control Module. The fuel pump relay is activated for one second after the ignition switch is first turned on, to pressurize the fuel system before starting the engine, Figure 16-20. The computer will leave the relay energized as long as it receives input from the crankshaft position sensor (POS) at the flywheel. The fuel pump control module adjusts voltage supplied to the fuel pump to control the amount of fuel flow, Figure 16-21. It reduces the voltage to approximately 9.5 volts except during:

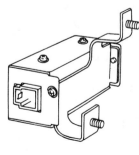

Figure 16–21 Fuel pump control module. *(Courtesy of Nissan Motors.)*

- engine start-up cranking.
- when the engine coolant is below 45 degrees Fahrenheit.
- within 30 seconds after start-up with a warm engine.
- under high load and high speed conditions.

Malfunction Indicator Lamp. The control unit turns on the malfunction indicator lamp (MIL) when it detects a fault in one of the engine's control or activator circuits. Most malfunctions will have to occur in two successive trips before the computer will turn on the MIL, but certain malfunctions that represent either failure of the engine to run properly or a risk of damage to a component will trigger the MIL the first time the problem is encountered. See the service manual to determine which problems fall into which category for specific models and years.

EGR Solenoid Valve. The EGR solenoid works in response to signals from the ECM. When the ECM grounds the circuit, the coil in the solenoid is energized, and a plunger then moves to cut the vacuum signal from the throttle body to the EGR valve. When the ECM sends an off signal, vacuum passes through the solenoid valve to the EGR valve. The system is thus designed to be fail-safe with the EGR valve responding to manifold vacuum.

Function of the EGR valve is monitored by the EGR temperature sensor, Figure 16-22. The computer can distinguish between EGR flow when there should be none and no EGR flow when there should be some by this temperature sensor, and it sets different codes depending on which condition it detects.

The EGR valve can also be checked manually. If it doesn't move smoothly, it may have mechanical problems.

Fuel Pump Pressure. Fuel pump pressure with zero vacuum at the MAP sensor, should be approximately 43 psi. At idle, the fuel pressure should read approximately 34 psi.

Fast Idle Cam. Instead of a stepper motor, some Nissans use a fast idle cam on the throttle linkage. The fast idle cam works by a thermostatically heated spring.

Oxygen Sensor Heaters. To insure that the engine control system can go into closed loop as soon as possible, both the **front and rear oxygen sensors** include heater circuits to get them to operating temperature as quickly as possible. The computer will store different trouble codes for a failure of the heater circuit from a failure of the oxygen sensor output circuit.

Cooling Fan Relay. The engine control computer operates the radiator cooling fans through a relay depending on the engine coolant temperature, vehicle speed, and whether the air conditioner is turned on or not. There are two different fan speeds available to the computer. There are also differences in the fan speed maps for cars built for the California market, though the rest of the cooling system is the same.

Air Conditioner Relay. When the air conditioner is turned on, the a/c relay provides the engine control computer a signal which is used to modify the control of the idle speed.

Front Engine Mount Control. To keep the engine vibration to a minimum, the engine control system also electrically varies the dampening effect of the front engine mount. The mount is in a soft setting when the engine is idling, and in a firm setting when the engine is doing work moving the vehicle under load.

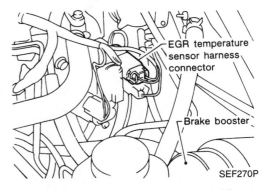

Figure 16–22 EGR temperature sensor. *(Courtesy of Nissan Motors.)*

EVAP System Controls. The late model Nissans use a more complex evaporative control system, storing fuel vapors in a charcoal canister as has been used for many years, but with a more complex set of controls and monitoring measures. Refer to Figure 16-12 under the **Inputs** section for a complete system diagram. The basic actuator is the EVAP canister purge control valve, Figure 16-23. When the computer sends an off signal to the control solenoid valve, the vacuum between the intake manifold and the EVAP canister purge control valve is shut off. The EVAP canister purge volume control valve controls the passage of vapors from the EVAP canister purge control valve to the intake manifold,

Figure 16-24. The system includes, as was described under the **Inputs** section in this chapter, an EVAP control system pressure sensor, Figure 16-12, intended to detect leaks in the system or a faulty, loose, or missing fuel filler cap. While the EVAP control system differs somewhat from one year to the next and from one model to another, fundamentally they all work the same way, storing fuel vapors in the charcoal canister until the engine, running in closed loop, can vent the canister into the intake manifold and burn the stored vapors. Corrections to the fuel mixture required by the variable concentration of fuel vapors entering the engine fall within the ability of the oxygen sensor feedback control system.

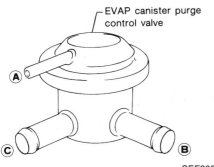

Figure 16–23 EVAP canister purge control valve. *(Courtesy of Nissan Motors.)*

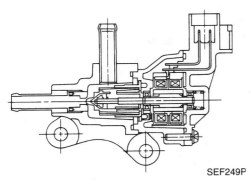

Figure 16–24 EVAP canister purge volume control valve. *(Courtesy of Nissan Motors.)*

Diagnostic & Service Tip

With one coil per cylinder buried in the valve cover on late model vehicles, it is somewhat more difficult to get spark advance and engine speed information without special Nissan tools. Nonetheless, you can remove the coil from the number one cylinder, attach an extension spark plug cable, and make the measurements with conventional equipment. Many shops use a similar technique on the GM/Oldsmobile Quad-4 engine.

✔ SYSTEM DIAGNOSIS & SERVICE

Late model vehicles include the OBD II standardized diagnostic system. In addition, they have retained the earlier Nissan self-diagnostics. This older system uses a screwdriver-actuated switch in the computer itself (ordinarily under the front passenger seat, but behind the instrument panel near the glove box on newer vehicles), and displays trouble codes either by flashes of the check engine light or, on the earliest systems, by flashing lights on the computer itself. The older systems, of course, are simpler, rely on fewer inputs and outputs, and provide less information for diagnosis. Check the shop manual for the specific information corresponding to the model and year of the vehicle you are working on.

Nissans use a **two-trip malfunction detection** logic; that is, for most component circuit failures the computer will store a fault on the first trip in which it is detected and will turn on the check engine light if it recurs on the next trip (a trip is defined as an engine start-up, followed by enough driving to bring the engine to operating feedback control system condition). There are several different ways to erase stored faults in Nissan systems, including changing the computer screwdriver switch from Test Mode II to Mode I or using the scan tool, so read the manual for the specific car before you perform these tests. Disconnecting the battery will erase the codes (as well as the self-learning data), but it can take as long as twenty four hours of disconnection before the energy stored in the capacitors discharges fully.

TOYOTA: GENERAL SYSTEM DESCRIPTION

As an example of a Toyota computerized engine control system, we will look at the 1MZ-FE V6 engine in the late model Toyota Camry.

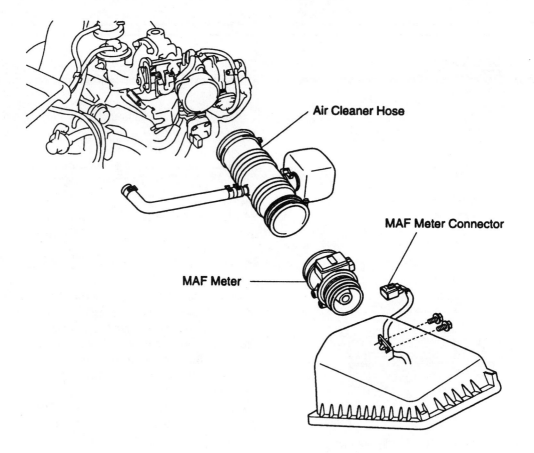

Figure 16–25 Mass air flow sensor (meter)

Other systems are similar, with appropriate modifications for engine type and vehicle style.

The Toyota system uses a mass airflow sensor to determine how much fuel to inject at a given engine condition. The six fuel injectors are sequentially operated in the firing order of the engine. Spark is delivered through individual coils at each spark plug, similar to the Nissan system described earlier in this chapter.

INPUTS

Mass Air Flow Sensor. The mass airflow sensor provides the engine control computer with a fluctuating voltage signal corresponding to the amount of air passing through the air intake system, Figure 16-25. With the sensor jumpered as explained in a shop manual, its output signal can be monitored by a voltmeter for function. Check the shop manual tables for resistance checks across sensor terminals.

Heated Oxygen Sensors. The Toyota system uses front and rear oxygen sensors, Figure 16-26, in the same way as the Nissan system described earlier in this chapter. On V-6 engines, there are two front oxygen sensors, one for each bank. The front oxygen sensors generate the signals the computer uses for closed loop fuel mixture control, and the rear sensor is used to determine the effectiveness of the catalytic converter. Should either front sensor fail, the rear sensor signal is used to help maintain fuel trim. The oxygen sensors all include a heater circuit to bring them up to operating temperature as quickly as possible and run the engine in closed loop.

Throttle Position Sensor. The throttle position sensor, Figure 16-27, provides the computer with information about where the driver's foot has positioned the accelerator pedal. On the Toyota system, the throttle position sensor is adjustable using a feeler gauge between the throttle stop screw and stop lever and an ohmmeter on the terminals as shown in the shop manual. A defective throttle position sensor can lead to the wrong idle speed and poor driveability at other speeds.

Engine Coolant Temperature Sensor. The engine coolant temperature sensor (ECT), Figure 16-28, varies its internal resistance inversely with temperature; that is, the warmer it gets, the lower its resistance goes. It threads into the water jack-

Bank 1 Sensor 1

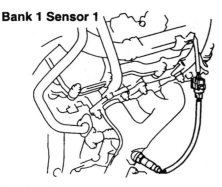

Bank 2 Sensor 1

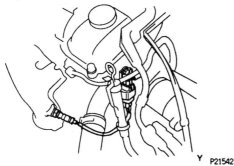

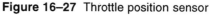

Figure 16-26 Heated oxygen sensor

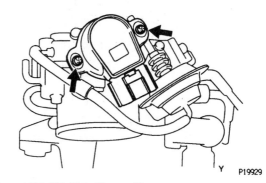

Figure 16-27 Throttle position sensor

et near the thermostat outlet and can be checked with an ohmmeter. At normal shop temperature, it should indicate approximately 2,000 ohms. The computer uses information from the coolant temperature sensor to calculate fuel mixture, spark advance, and ignition timing. Its information also plays a role in determining when to go into closed loop, when to engage the EGR system, and when to open the passages to the evaporative emissions charcoal canister.

Knock Sensors.   V-6 engines use two knock sensors, one on each bank, Figure 16-29; in-line engines use one. The knock sensor, as on other systems, is a piezoelectric design that, when vibrated at a frequency characteristic of knock, sends a signal to the computer, which then retards spark advance until the knock disappears. Spark advance is then readvanced until

the knock reappears. As with most knock sensors, function can be checked by tapping on the engine block adjacent to the knock sensors with a tool while observing the spark advance (warm engine, closed loop conditions). Knock sensors are relatively delicate and should never be dropped. If dropped to a hard surface, they will frequently sustain damage.

EGR Temperature Sensor.   To determine whether its EGR system actuation commands are effective, the computer monitors information from the EGR temperature sensor, Figure 16-30. When the EGR passage opens, hot exhaust gas travels through the channel and the temperature sensor should reflect an increased temperature. This information does not enter into direct control of the engine, but does play an important role in the self-diagnostics of the system's emissions controls.

Camshaft Position Sensor.   The latest Toyota V-6 engines use a direct ignition system with one coil per cylinder rather than a distributor, and hence they require a camshaft position sensor, Figure 16-31. This sensor is on one of the camshaft sprockets on the accessory drive side of the engine. The purpose of this sensor is to allow the computer to determine which cylinder is approaching its power stroke firing position, information it uses for both spark firing and for fuel injection sequencing. The camshaft position sensor is a coil-and-magnet inductive sensor, generating a current that alternates as the camshaft turns. Electrical resistance should range from a cold value of 835

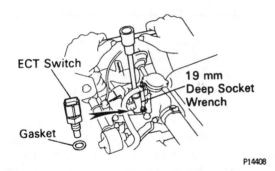

Figure 16-28 Engine coolant temperature sensor

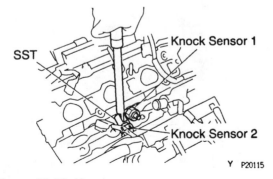

Figure 16-29 Knock sensors

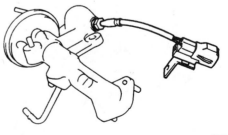

Figure 16-30 EGR gas temperature sensor

ohms to a hot resistance of 1,645.

Crankshaft Position Sensor. The crankshaft position sensor, Figure 16-32, is adjacent to the crankshaft pulley. It provides the computer

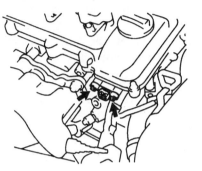

Figure 16–31 Camshaft position sensor

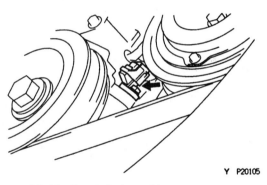

Figure 16–32 Crankshaft position sensor

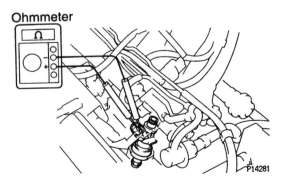

Figure 16–33 Fuel injector resistance check

with information about the position and speed of the crankshaft, information employed for fuel injection, spark timing, and idle control as well as functions restricted to on-board diagnostics. Resistance for the magnet-and-coil crankshaft position sensor should range from a cold of 1,630 to a hot of 3,225 ohms.

OUTPUTS

Fuel Injectors. The fuel injectors, Figure 16-33, are Bosch-design units, one for each cylinder. The computer sprays them sequentially in the firing order of the engine during the corresponding cylinder's intake stroke. At room temperature, each injector's coil should have approximately 13.8 ohms resistance. Injectors are also subject to mechanical restrictions and blockages, so they should be separately inspected for flow and spray pattern if there is reason to suspect they are not working properly.

Ignition Coils. The latest V-6 engines use a direct ignition system with one coil and driver transistor per pair of cylinders (a waste spark system), Figure 16-34. Ignition coil resistances should be 0.70 to 1.10 ohms through the primary circuit and 10.8 to 17.5 kilohms through the secondary circuit, ranging from cold to hot temperatures. With individual coils of this design, the most time-efficient test for a bad coil is often to merely shift the coil from one cylinder to another

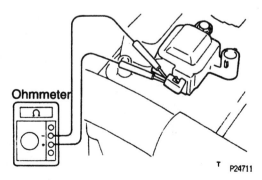

Figure 16–34 Ignition coil resistance check

to see whether the cylinder miss follows the coil or not. If it does not, of course, diagnosis should focus on fuel or mechanical aspects of the affected cylinder.

Fuel Pump Circuit. The fuel pump on many Toyota systems is under a plate beneath the rear seat, Figure 16-35. The computer energizes a main fuel injection relay, Figure 16-36, which in turn activates the fuel pump. As with most systems, the fuel pump comes on for approximately one second after the ignition key is turned on to prime and pressurize the injectors and fuel rail and shuts off if no camshaft position signal is received after the priming second.

The fuel from the pump fills the fuel rail, and the pressure is controlled by the fuel pressure regulator, Figure 16-37. The regulator bolts into the end of the fuel rail and reduces pressure corresponding to intake manifold vacuum, sending unused fuel back to the tank. The fuel pressure

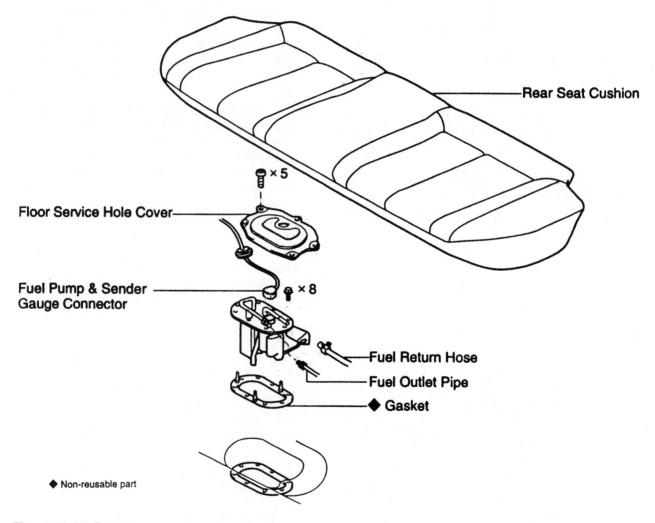

Rear Seat Cushion

× 5

Floor Service Hole Cover

Fuel Pump & Sender Gauge Connector

× 8

Fuel Return Hose

Fuel Outlet Pipe

◆ Gasket

◆ Non-reusable part

Figure 16–35 Fuel pump

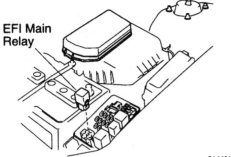

EFI Main Relay

P14401

Figure 16–36 EFI (fuel pump) main relay

regulator, of course, is not directly controlled by the computer, but responds to the pressure from the fuel pump and the partial vacuum in the intake manifold.

Idle Air Control System. Figure 16-38 shows the idle air control (IAC) valve. It bolts to the throttle body and provides a bypass channel for air to pass the throttle blade for idle speed control, depending on the engine coolant temperature, electric, and other accessory loads and whether the car is in a drive gear or not. The IAC valve is easily removed from the throttle body and can be

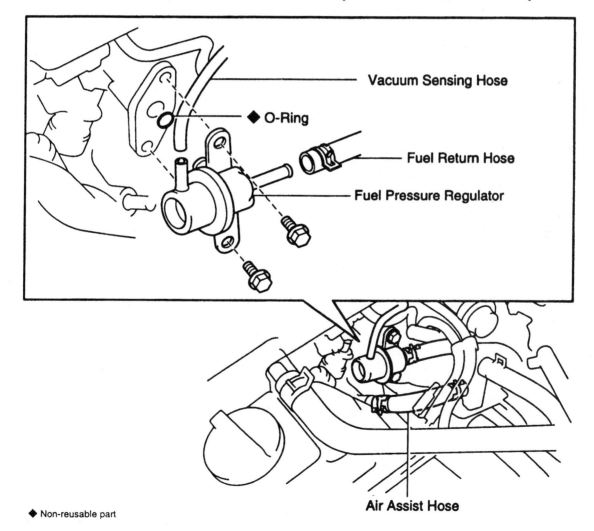

Vacuum Sensing Hose

◆ O-Ring

Fuel Return Hose

Fuel Pressure Regulator

Air Assist Hose

◆ Non-reusable part

Figure 16–37 Fuel pressure regulator

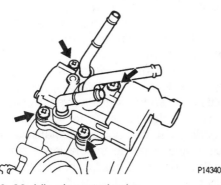

Figure 16–38 Idle air control valve

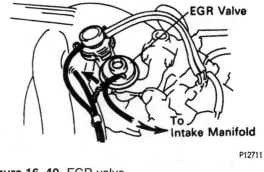

Figure 16–40 EGR valve

inspected by jumpering the connections to see whether the valve opens and closes properly.

Diagnostic & Service Tip

On the Toyota IAC valve and on others of similar design, it is almost always a better service practice to renew the gasket between the valve and the throttle body any time it is removed. Failure to do so can allow "false air" to pass around the gasket and into the intake manifold, making it harder or impossible for the computer to correctly determine the idle speed. This kind of problem will ordinarily not set a code, so solving it will take more time than the repair is worth.

EGR System. The exhaust gas recirculation system on late model Toyotas uses a vacuum modulator, Figure 16-39, to activate EGR valve, Figure 16-40. The EGR system should not be in operation at all below an engine coolant temperature of 131 degrees Fahrenheit. The EGR vacuum modulator can be opened and cleaned/inspected for holding vacuum. With the engine warm, check that the VSV and modulator pass vacuum along to operate the EGR valve itself and that it opens when actuated. Exact connections to perform these checks can be found in the appropriate shop manual.

EVAP Control System. As with most manufacturers, Toyota uses a charcoal canister, Figure 16-41, to store fuel vapors and burn them later when the engine is running. With the newer OBD II-compliant vehicles, any detected

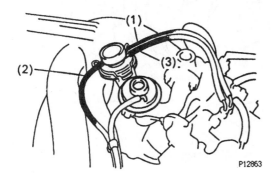

Figure 16–39 EGR vacuum modulator

Figure 16–41 EVAP system charcoal canister

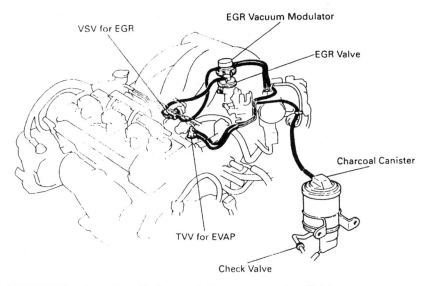

Figure 16–42 EGR/EVAP schematic with thermostatic vacuum valve (TVV)

vapor leaks—including a leaking or missing fuel filler cap—can set a code and turn on the check engine light. The vapor venting does not occur until the thermostatic vacuum valve, Figure 16-42, has reached a temperature of between 104 and 138 degrees Fahrenheit. The system also includes a check valve, Figure 16-43, which should pass air from the yellow port to the black port but not backwards. Be sure to reinstall the check valve with the black port facing the purge port side.

✔ SYSTEM DIAGNOSIS & SERVICE

OBD II Diagnostics. With the latest model Toyotas, the OBD II scanners that can be used on other vehicles can be used on these as well, using the standardized diagnostic link connector in the lower left side of the instrument panel, Figure 16-44. Each vehicle also employs a standard set of diagnostic trouble codes (DTCs) that are common to all manufacturers. More information about the OBD II program is available in Chapter 18 of this book.

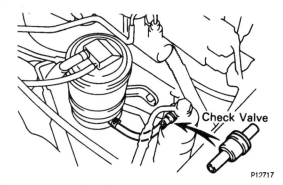

Figure 16–43 EVAP system check valve

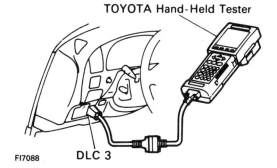

Figure 16–44 OBD II diagnosis with scan tool

Toyotas use a two-trip fault recording program, as do many other manufacturers. By recording a trouble code in memory the first time it occurs, but not turning on the malfunction indicator light (check engine light) until it recurs in the next successive trip, this reduces the chances of having an excessive number of problems that cause motorists concern. Two-trip fault recording programs also retain the sensitivity to system malfunctions that is needed to keep the vehicle's emissions performance at its desired state. Almost all component or signal failures that can affect the

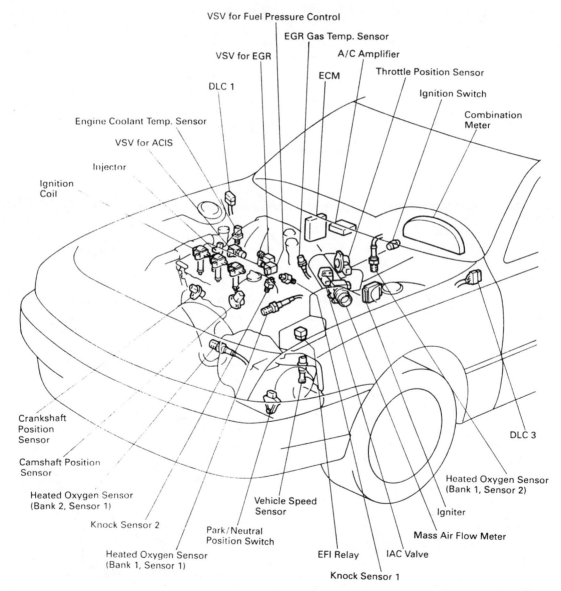

Figure 16–45 Component locations

vehicle's emissions or fuel economy performance will be reflected in the information available on the scan tool. Certain crucial component signals, which would either cause damage to the system or prevent the vehicle from operating, trigger a trouble code on their first occasion.

If the trouble code indicates a problem on a given circuit, remember that it is the entire circuit that has failed, not just the specific component. While finding and checking the component itself is a reasonable first step, Figure 16-45, the problem may actually be in the harness or in the computer itself.

SUMMARY

To focus on Asian engine control systems in this chapter, we have studied on late model Nissan and Toyota vehicles. We saw how the earlier feedback systems employed complex carburetors that had to be carefully set at the factory to deliver a proper mixture and retain driveability.

The chapter considered the different injection patterns (simultaneous and sequential) employed for start-up, patterns to avoid the use of a cold-start injector, and patterns to optimize emissions during the engine's dirtiest phase: start-up cranking. The different fuel enrichment strategies as well as spark advance map alterations were considered. We saw how the self-learning capacities allowed the systems to adjust fuel mixture and spark timing from recorded experience. And we considered the more elaborate fuel canister vapor absorption and storage systems. We learned the use of an EGR temperature sensor to monitor the effectiveness of the bypass valve's circuit. We learned of the interconnections with the transmission.

Both the Nissans and the Toyotas use a nonwaste-spark direct ignition with one coil for each cylinder. And both use the new tandem oxygen sensors to provide information about the effectiveness of the catalytic converter.

We have also seen how each can be diagnosed using standardized OBD II scan tools, as well as noted the more mundane problems that can occur even with high-tech systems such as these.

▲ DIAGNOSTIC EXERCISE

A customer brings in a late-model Nissan with the check engine light on. There are no driveability complaints, either from the customer or noticeable to a technician on a test drive. When checked with the scan tool at the OBD II diagnostic link connector, the computer indicates a problem in the charcoal canister purge system. What is the most likely cause of this problem? How long should the repair take if the most likely cause is the real one?

REVIEW QUESTIONS

1. Hot-film air mass sensors work quite similarly to hot-wire sensors. Explain the difference and why a manufacturer might favor one or the other.
2. One could say the Nissan engine control system has a knock sensor but makes much less use of it than domestic systems do. What is meant, and why?
3. Explain the Nissan strategy for sequential and simultaneous fuel injections. What is the purpose for the two methods? What is the connection with the camshaft position sensor?
4. What would be the effect of a failure of the Nissan system's deceleration cutoff measures?
5. What driveability symptoms would occur if there were foreign material stuck to the hot-film air mass sensor?
6. What happens to an oxygen sensor signal if there is a crack in the ceramic substrate? What happens to the signal if deposits accumulate on the sensor's inside electrode?

7. If a closed throttle position switch were to fail open, what would be the symptoms? What symptoms would occur if it fails closed?

8. True or false: If the EGR temperature sensor does not show a change of temperature when the valve is actuated, the computer will disable the EGR system and set a code.

9. What is the purpose of the fuel temperature sensor? Would there be driveability symptoms if it failed?

10. What are the differences in the ignition circuits between the direct ignition circuits described in this chapter, and waste-spark direct ignition systems described in other chapters? What would be the principal differences in ignition problems you might find?

11. Why does the Toyota program wait for certain faults to repeat again on a second trip before they are recorded in memory with a dash light indication?

ASE-type Questions (Actual ASE test questions will rarely be so product specific.

12. A late-model Nissan has a cylinder miss. Technician A says a quick test to see whether it is spark or fuel is to check the crankcase oil for any residual fuel. Technician B says a burned valve might have allowed damage to the catalytic converter. Who is correct?
   a. A only.
   b. B only.
   c. both A and B.
   d. neither A nor B.

13. An ignition coil has failed in a late-model Nissan. Technician A says the quad driver in the computer may also be damaged. Technician B says to check the spark plug as well before replacing the coil. Who is correct?
   a. A only.
   b. B only.
   c. both A and B.
   d. neither A nor B.

14. A V-6 Toyota has three separate oxygen sensors. Technician A says this allows the system to set the fuel mixture correctly even if one or the other of the sensors is working, as long as two remain. Technician B says only the front oxygen sensors include a heating coil. Who is correct?
   a. A only.
   b. B only.
   c. both A and B.
   d. neither A nor B.

15. Technician A says a late-model Nissan with a defective camshaft position sensor will not be able to sequence the fuel injection. Technician B says the vehicle could still run in limp-in mode. Who is correct?
   a. A only.
   b. B only.
   c. both A and B.
   d. neither A nor B.

16. Checking the fuel injectors on a late-model Nissan reveals all the injectors firing together at starting crank. Technician A says this means the system is in a failure mode and should have turned on the check engine light. Technician B says a defective intake air temperature sensor could cause this problem. Who is correct?
   a. A only.
   b. B only.
   c. both A and B.
   d. neither A nor B.

17. Some late model Nissans use two crankshaft position sensors, one at the crankshaft accessory drive pulley and another at the flywheel. Technician A says the vehicle cannot start without signals from both sensors. Technician B says the sensors are redundant to allow the vehicle to operate in an emergency. Who is correct?
   a. A only.
   b. B only.
   c. both A and B.
   d. neither A nor B.

18. A late model Toyota has poor performance and fuel economy complaints and a variety

of trouble codes. Inspection reveals the air filter element is missing. Technician A says general wear to the engine from ingested dirt could be the cause of all the problems. Technician B says contamination on the mass airflow sensor alone could explain everything. Who is correct?

a. A only.
b. B only.
c. both A and B.
d. neither A nor B.

19.   A coolant temperature sensor on a late-model Toyota is open-circuited. Technician A says this would mean the vehicle would never reach closed loop. Technician B says the vehicle would always run in closed loop. Who is correct?

a. A only.
b. B only.
c. both A and B.
d. neither A nor B.

20. A late-model Toyota has an excessively high idle speed. Technician A says air leaking into the system downstream of the mass airflow sensor could be the cause. Technician B says a faulty park/neutral switch could be the problem. Who is correct?

a. A only.
b. B only.
c. both A and B.
d. neither A nor B.

21. A late-model Toyota stalls immediately at any speed if the air conditioning is turned on. Technician A says a bad air conditioner clutch may be responsible. Technician B says a bad air conditioner switch sensor may be the cause. Who is correct?

a. A only.
b. B only.
c. both A and B.
d. neither A nor B.

Chapter 17

Electronically Controlled Diesel Engine Systems

OBJECTIVES

In this chapter you can learn:
- the fundamentals of the General Motors 6.5-liter diesel engine control system.
- the fundamentals of the Ford/Navistar/International 7.3-liter diesel engine control system.
- the fundamentals of the Volkswagen 1.9-liter TDI diesel engine system.
- basic principles of how diesel engines are similar to and different from gasoline-fueled engines.

KEY TERMS

Annulus Ring
Compression Ignition (CI)
Direct Injection
Glow Plugs
Injection Pressure
Injection Timing Stepper Motor
Lift Pump
Prechamber
Pumping Losses
Ricardo Chamber
Spark Ignition (SI)
Transfer Pump
Unthrottled
Vacuum Pump

Rudolph Diesel invented the compression-ignition engine named for him almost 100 years ago, and for special applications requiring exceptional durability or exceptional fuel economy or other circumstances, the hardy oil-fueled, fuel-injected engine has been the powerplant of choice.

DIESEL AND GASOLINE ENGINES

Since all the other engines considered in this book are gasoline-fueled, following are the main differences between a gasoline engine and a diesel engine, as well as the features that make a diesel more suitable to specific applications (and less suitable to others).

First, there is no electric coil spark system in a diesel engine. Combustion of the diesel fuel occurs entirely because of the temperature of the air into which it is sprayed. A gasoline engine is called **spark ignition (SI)**; a diesel engine is called **compression ignition (CI).**

Second, there is no throttle in a diesel engine. The diesel engine draws as much air as possible into each cylinder for each engine cycle, regardless of load or the position of the accelerator pedal (more properly called the fuel pedal on a diesel engine). Power output is determined entirely by the amount of fuel injected for each cylinder event. Because of its unthrottled intake, a diesel engine "breathes" much more air than an equivalent gasoline-fueled engine except when the latter is at wide-open throttle. The diesel engine, however, ordinarily does not burn all of its oxygen; it burns only what it has fuel for and just pumps the rest through. Mixture control is absolutely important for a gasoline-fueled engine; it makes little or no difference for a diesel.

Third, in a diesel engine, the fuel oil injection is sprayed directly into the combustion chamber (or into an adjacent portion of that chamber, either a **Ricardo chamber** or a **prechamber**). The fuel begins to burn as soon as injection begins. No fuel is introduced into the intake air stream during its passage through the air intake, filter, tubing, or manifold runners. Nothing but air passes the intake valve.

Fourth, compression ratios in diesel engines are much higher than in gasoline-fueled engines, typically between 17 to 1 and 24 to 1 (gas engine compression ratios are usually 8 or 9 to 1).

The higher compression ratio is one of the reasons for a diesel engine's greater efficiency: more air per cylinder event allows more torque per power stroke. In addition, diesel fuel has about 30% more energy per gallon than gasoline, and more diesel fuel can be produced from a barrel of crude oil than any other kind of fuel. For certain applications, specifically (but not only) marine, diesel fuel has the considerable advantage of being much less flammable than gasoline and highly resistant to evaporation and the formation of vapors.

Finally, diesel engines have much lower **pumping losses** since they do not develop a manifold vacuum. A gasoline engine running at any setting less than wide-open throttle must pull every piston on the intake stroke down against partial vacuum through the intake manifold. These "pumping losses" can mean less horsepower available at idle or part-throttle in gasoline-powered engines.

A diesel engine, therefore, is ordinarily much more economical to operate at partial throttle settings where most driving is done. A diesel can idle using almost no fuel for very long periods of time, though its increased efficiency can allow the engine to cool and carbon to form in the combustion chambers if that is done regularly. This is the

reason many bus and truck companies use a high idle when parked with the engine unloaded: there is virtually no increase in fuel economy, but the combustion chambers are kept hot and clean.

A diesel engine does not generally have problems with either detonation (knock) or overheating. Since there is no fuel in the combustion chamber prior to fuel injection, there is nothing to ignite or detonate. If there is knock, it is either from a mistimed injection pump, fuel of the wrong cetane rating (equivalent to octane for gasoline), or a leaking injector. On diesel engines with very considerable piston ring wear, there is occasionally the phenomenon of the "runaway" diesel, which starts to run on engine lubricating oil leaking around the piston rings. Under such a condition each combustion event will have knock, and the only way to stop the engine is to choke off the intake air, plug the exhaust, or stop the engine with the drivetrain in gear using the vehicle's brakes (not always an option with automatic transmissions).

Diesels are very resistant to overheating because of their thermal efficiency. Except for vehicles designed to move very slowly, such as garbage trucks, most diesel engines can use a much smaller radiator and cooling system than an equivalently powered gasoline-fueled engine needs. A clogged radiator or burst hose, of course, will still allow overheating.

Real World Problems: Diesel Smoke

A major incentive for the development of electronically controlled diesel engines has been the elimination of diesel smoke. Due to the excess oxygen present in diesel combustion chambers, these engines actually have less toxic emissions than equivalently powered gasoline engines. However, diesel emissions are visible and thus more objectionable to people.

Even with the electronic controls, the three kinds of diesel smoke will still provide evidence to the technician about what is wrong with the system.

Blue smoke means the same as with gasoline-powered engines: too much engine lubricating oil is getting around either the piston rings or the valves and is burned in the combustion chambers. Major engine repair is the fix.

Black smoke indicates a lack of oxygen getting into the combustion chambers. The most common cause of this is a clogged intake air filter, though anything that restricts the flow of air into the engine would have the same effect. A major objective of the electronic controls is to provide the computer with some way to estimate air flow, so the system can reduce the amount of fuel injected. Thus, on a properly functioning electronic diesel, a clogged air filter will result in a noticeable drop in engine power before it results in black smoke. A leaking injector, however, can cause black smoke with any kind of diesel.

White smoke is fuel, unburned and blown through the exhaust. It is a result of insufficient heat in the combustion chamber to ignite the fuel. The most common cause of white smoke is an engine that is too cold, one or more defective **glow plugs** (electric resistance heaters extending into the combustion chamber), or a thermostat stuck open. An engine that is blowing white smoke is also contaminating its lubricating oil with fuel, and the reduced viscosity of the oil can cause rapid wear of engine parts.

There are two kinds of diesel combustion chambers used on modern engines, prechamber (or Ricardo chamber) and **direct injection.** The prechamber is a small, separate and removable chamber recessed in the cylinder head above the piston. The injector and the glow plug are in the prechamber. The advantages of the prechamber design are easier starting, and somewhat smoother and quieter operation. The disadvantage of the prechamber design is reduced fuel economy because of more combustion chamber surface to conduct heat into the water jacket, and the work (and thus friction) needed to force the

air into, and the combustion gases out of, the prechamber.

The direct injection diesel is about 15% more efficient than prechamber designs, chiefly because of the elimination of the two factors just mentioned. Direct-injection diesels are most common on larger commercial vehicles, but the small 1.9-liter Volkswagen diesel described in this chapter is a direct-injection design, as is the medium-size 7.3-liter International/Ford. Ordinarily direct-injection diesels have the combustion chamber recessed into the top of the piston rather than into the cylinder head. The fuel injection is to the center of the piston, adjacent to the glow plug.

Because a diesel engine uses the heat of the rapidly compressed intake air to bring the temperature up to ignition temperature, many diesels require some auxiliary source of heat when started cold. For all diesel engines, there is a temperature below which it is physically impossible for them to start without auxiliary heat, usually around zero degrees Fahrenheit.

The most common of these auxiliary heat sources is the glow plug. A glow plug is nothing more than an electric heater shaped into a rod and extending into the combustion chamber. Most systems use some sort of coolant temperature sensor to determine how long the glow plugs must be powered to reach a suitable fuel ignition temperature, and this period can range up to as much as a minute or more if the temperature is very low. Other engines, notably the Cummins engine used in certain Dodge pickups, use an electric heating grid in the intake manifold to heat the incoming air.

In either case, this explains why the diesel requires a much larger battery to start than an equivalently powerful gasoline engine. Between the considerable power drain of the glow plugs or heat grid and the need to use a much more powerful starter motor to create the much higher compression pressures (for all diesel engines, there is a certain combination of combustion chamber temperature and cranking rpm that must be exceeded if the engine is to start), diesels often have multiple batteries to get started. Sometimes these batteries are wired with a series-parallel switch to insure maximum cranking speed. Others use compressed air pneumatic turbine starters, which are relatively immune to temperature problems.

Once running, most diesels can keep running even with a complete electrical system failure as long as they don't run out of fuel, lubricating oil pressure, or coolant.

At least this was true with purely mechanical diesels prior to those we are considering in this chapter. Complete electrical failure, of course, is extremely rare and disables most engine monitoring equipment, so a driver would do well to stop for repairs even with the older-type engines unless there is a genuine emergency.

Given all the things that older, purely mechanical diesel engines could do, two questions arise: why do gasoline engines still exist, and what is there to control with a computer?

Gasoline engines still exist primarily because diesels do not have the advantages of lightness and moderate manufacturing cost that gasoline engines do. The typical diesel engine is built much more robustly (and thus more expensively) than the typical gasoline engine.

Also, a diesel engine can often weigh almost twice as much and cost three times as much as a gasoline engine for a given power output. It will get better fuel economy and last longer.

The principal reasons for computer controls of diesel engines is to improve emissions and fuel economy. While the most important feature of gasoline engine emissions control is the air/fuel mixture, a diesel engine always runs with more than enough air (unless as we saw, the air filter has become clogged with dirt). What influences the diesel's emissions more than anything else is the exact point of injection timing. In addition, on some older mechanical-injection engines, fuel delivery could go quickly to maximum when the driver's foot pushed the fuel pedal to the floor, but it took slightly longer for the air induction system to reach higher volumetric efficiency levels.

GENERAL MOTORS 6.5-LITER DIESEL: SYSTEM OVERVIEW

Like gasoline-fueled new vehicles, the 6.5-liter GM diesel is now OBD II-compliant, making scan tool DTC-based diagnostics available for the first time. Of course, diesel systems are fundamentally different, so you will need the proper shop manual to find the appropriate codes, which are, of course, fewer and simpler than the hundreds available for gasoline-fueled vehicles. They are also, of course, different.

The 6.5-liter GM diesel is a V-8 type engine, with pushrods operating the valves through rocker arms. Its compression ratio is 21.3 to 1. Some applications of the engine use a turbocharger, particularly those for heavier vehicles. Turbocharger boost is ordinarily controlled between 2 and 8 psi at peak torque. Most (but not all) call for a diesel catalytic converter on the exhaust system. Sometimes, a diesel catalytic converter has been described as more of a ceramic burn-off filter, but for practical purposes, it means the same thing as it does on gasoline-fueled vehicles. Because of the diesel engine's **unthrottled** free breathing, of course there is no need for an oxygen sensor feedback system. The amount of oxygen in the exhaust corresponds to nothing about the completeness of the fuel burning.

The GM 6.5-liter diesel engine uses a high-swirl precombustion chamber to mix the air and fuel as completely as possible. Like most diesels, its maximum torque output comes at a fairly low engine speed, 1,700 rpm, allowing a long power stroke for the most complete fuel burning. Maximum horsepower occurs at about double the maximum torque speed.

Prechamber Operation. Intake and exhaust strokes on a prechamber diesel are identical to those on gasoline-fueled engines, but we will look briefly at the compression and power strokes for the differences. In Figure 17-1, the compression stroke, you can see how the greatly compressed and heated air is forced into the small, precisely machined prechamber. The turbulence of the air in the prechamber promotes atomization of the fuel.

During the power stroke, Figure 17-2, fuel injection has begun and the mixture is burning in the prechamber. The expanding gas blows out of the prechamber and into the main combustion chamber, forcing the piston down and turning the crankshaft. Notice there is almost no room between the piston top and the cylinder head except in the small flame channel recess just below the prechamber. This narrow space above the piston sometimes allows carbon buildup, which can make audible hammering when the engine is started cold, before the block can warm

up and expand from the heat. Since fuel injection continues for a brief period of the power stroke after the beginning of combustion, you can see why **injection pressure** must be much higher than for gasoline-fueled injectors. Typically a diesel injector sprays fuel at several thousand psi pressure.

WARNING: Because diesel injection pressures are so high, never allow the injector to spray anywhere close to your skin, either from the pump on the engine or with an injection pressure tester. On this and all other diesel injection systems, the pressure is high enough to easily blow the fuel through the skin and tissue. This kind of accident can result in severe blood poisoning and may even require amputation of the affected finger or hand.

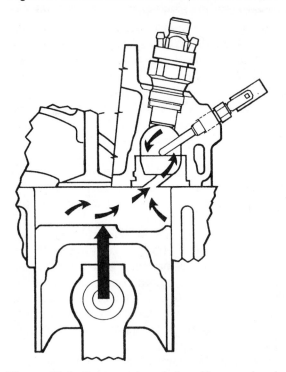

Figure 17–1 Compression stroke with a prechamber diesel. *(Courtesy of General Motors Corporation, Service Technology Group.)*

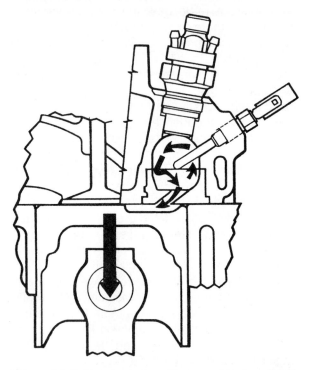

Figure 17–2 Power stroke with a prechamber diesel. *(Courtesy of General Motors Corporation, Service Technology Group.)*

The 6.5-liter GM diesel uses a computer, Figure 17-3, to control various functions of engine management. The computer is ordinarily located in the passenger compartment of the vehicle. The computer controls the following system functions:

- fuel quantity.
- injection timing.
- glow plugs.
- turbocharger boost.
- EGR valve.
- cruise control.
- transmission control.
- self-diagnostics.

The inputs and outputs of the system are not tied one-to-one, but are interrelated as with gasoline-fueled systems. Hence the coolant temperature sensor and the accelerator pedal position sensor both can affect fuel volume. Load and engine speed both affect how the fuel injection timing is determined.

INPUTS

Engine Coolant Temperature Sensor. The engine coolant temperature sensor is the familiar type, with a resistance that goes down as the temperature goes up. This variable resistance modifies the five-volt reference voltage into a lower signal voltage returned to the computer. The sensor threads into the water jacket next to the thermostat outlet. Some 6.5-diesels use a dual-thermostat system to better control coolant flow, but this does not affect the way the coolant temperature sensor works or mounts.

Intake Air Temperature Sensor. The intake air temperature sensor is similarly a variable resistor, losing resistance as the air temperature goes up. It bolts into the air horn just upstream of the intake manifold.

EGR Control Pressure/BARO Sensor. The EGR control pressure/BARO sensor, Figure 17-4, provides the computer with a barometric reading when the EGR system is off, and information about the pressure in the EGR actuation vacuum lines when it is on.

Turbocharger Boost Sensor. On vehicles equipped with a turbocharger, the turbocharger boost sensor mounts to the top of the air intake horn, Figure 17-5. This provides the computer with information about how much fuel can be injected and how to actuate the wastegate controls.

Accelerator Pedal (Fuel Pedal) Position Sensor. The fuel pedal position sensor, Figure 17-6, is similar to the throttle position sensor in

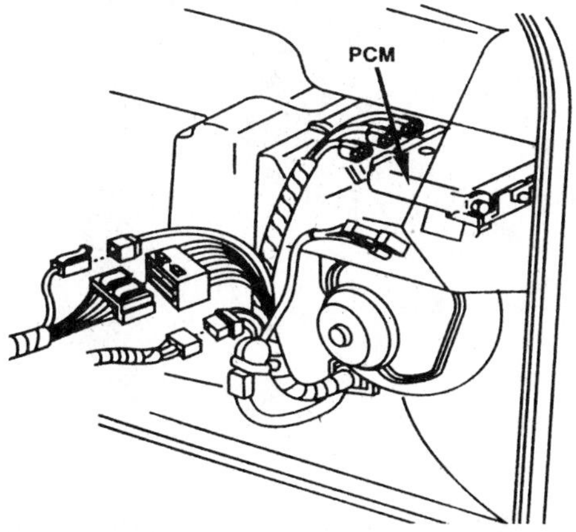

Figure 17–3 Computer location. *(Courtesy of General Motors Corporation, Service Technology Group.)*

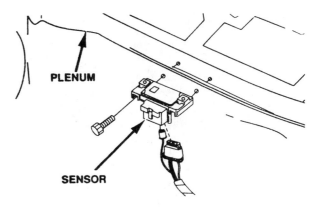

Figure 17–4 EGR control pressure/BARO sensor. *(Courtesy of General Motors Corporation, Service Technology Group.)*

gasoline-fueled engines. On the diesel engine, however, it mounts to the fuel pedal lever, inside the passenger compartment. More importantly, it replaces the mechanical linkage previously used to convey the driver's intentions to the injection pump. This system is a true drive-by-wire arrangement. The fuel pedal position sensor is actually three independent variable resistors in one box. Each one gets a five-volt reference signal from the computer and modifies that according to the fuel pedal position. This sensor means the system is basically a drive-by-wire system, and it has deep fail-safe modes. If there is a problem with the accelerator pedal position sensor, the following results occur:

- if one sensor fails, the computer stores a trouble code, but does not turn the check engine light on, and the vehicle operates normally.

- if two sensors fail, the computer turns on the check engine light and stores a code. The vehicle will operate at noticeably reduced power.

- if all three sensors fail, the check engine light comes on and the trouble codes are stored. In addition, the engine will run only at idle speed. Due to the considerable low speed torque of a diesel engine, the vehicle may be able to move along the road, even up hills, but slowly.

Optical Sensor. The optical sensor, Figure 17-7, is located in the fuel pump, and is the most important component in the diesel engine system. The optical sensor is responsible not only for the high pressure injection of the fuel, but also for the timing of the injection—the most significant variable in diesel engine performance, both in fuel economy and in emissions quality.

The sensor consists of two optical receivers separated from their emitters by a thin disk. The disk has two sets of notches, the outside one with 512 notches, and the inside one with eight. The inner notches include one larger notch to identify cylinder position. Since the fuel pump turns at camshaft rather than crankshaft speed, the two signals from this provide information about the

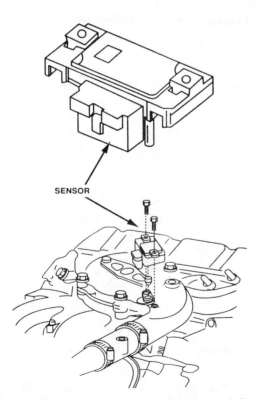

Figure 17–5 Boost sensor. *(Courtesy of General Motors Corporation, Service Technology Group.)*

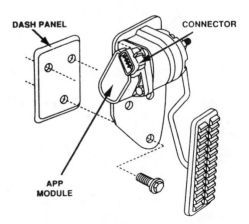

Figure 17–6 Accelerator pedal position sensor. *(Courtesy of General Motors Corporation, Service Technology Group.)*

position of the crankshaft, pistons, and camshaft.

The computer sends the optical sensor a five-volt reference signal, which is then modified by the movement of the notched disk in the fuel pump, and returned as two information signals—a high resolution signal from the 512-notch ring, and a camshaft position signal. If either signal is out of calibration, the computer sets a trouble code.

Diagnostic & Service Tip.

Many shops with extensive GM diesel experience report more problems with the optical sensor than with any other part of the GM diesel system. Checking for proper operation is an important test to perform early in any troubleshooting procedure on these engines. The sensors inside the injector pump may be damaged by any gasoline coming into contact with them.

Fuel Temperature Sensor. Diesel fuel viscosity is affected more by temperature than is gasoline, so the system includes a fuel temperature sensor, Figure 17-7. This sensor is part of the optical sensor.

Crankshaft Position Sensor. The crankshaft position sensor is in the front engine cover, Figure 17-8. This sensor is a Hall-effect signal generator, receiving a five-volt reference from the computer. As long as no reluctor vane lines up with the sensor tip, the sensor's magnetic field passes through the Hall-effect switch and turns it off, leaving the reference signal at five volts. As the reluctor vane lines up, the magnetic field moves to pass through the vane instead, turning on the Hall-effect switch and dropping the reference voltage to near zero. Erratic signals, shorts and opens in the sensor circuit are all recorded by the computer as trouble codes.

Vehicle Speed Sensor. The vehicle speed sensor is the same type unit as on gasoline-

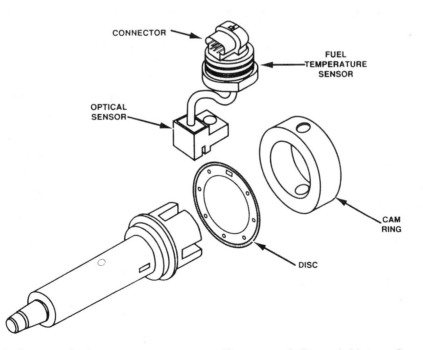

Figure 17–7 Optical sensor/fuel temperature sensor. *(Courtesy of General Motors Corporation, Service Technology Group.)*

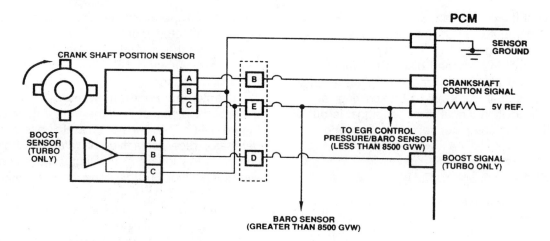

Figure 17–8 Crankshaft position sensor. *(Courtesy of General Motors Corporation, Service Technology Group.)*

fueled vehicles. Like those systems, late models use a signal buffer and may include transfer case low range signal switches and separate transmission input and output speed sensors for four-wheel-drive units.

Cruise Control System.   Driver controls for the diesel cruise control are identical to those for gasoline engines. Yet since the diesel engine's power comes entirely from the amount of fuel injected, the actuator works differently, moving linkage to increase or reduce fuel delivery instead of opening or closing a throttle. The system includes the usual disabling functions if the clutch or brake pedal is moved. Some systems employ a vacuum pump to work a vacuum servo. The pump is needed because the engine does not inherently develop any useful vacuum.

A/C Signal.   When the air conditioning is turned on, the computer receives a signal that it uses for idle speed control. At full power settings, the computer may also disengage the compressor clutch.

Self-Diagnostic Command.   Since the 6.5-liter GM diesel meets OBD II requirements, it can be diagnosed with a scan tool through the standard connector. While codes are different for a diesel engine, the procedure for calling them up is the same. Check a shop manual for an expla-

nation of each of the stored trouble codes.

Glow Plug Feedback Signal.   Since this diesel engine uses glow plugs for auxiliary heat when starting a cold engine, the computer receives a feedback signal from the glow plugs when they are turned on. The computer will calculate from several inputs, most importantly the engine coolant temperature and engine speed, how long to turn the glow plugs on and when to turn them off. They usually continue to heat from the time the ignition switch is first turned on until shortly after the engine starts and is running smoothly.

Injection Pulse Width Feedback Signal.   The computer controls fuel injection through the fuel solenoid driver (see the following **Outputs** section). The fuel solenoid driver also returns a pulse-width modulated signal back to the computer indicating when the injector plunger closes. This provides the computer with information about how much fuel was injected in this combustion cycle. The computer employs a calibrated injection pump-mounted resistor to determine fuel delivery rates, the value of which is retained in the computer's memory. If the computer memory has been erased or the computer itself replaced, it will relearn this resistance in the next running cycle.

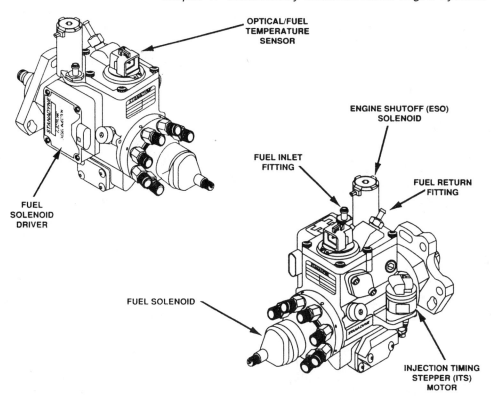

Figure 17–9 Electronic fuel injection pump. *(Courtesy of General Motors Corporation, Service Technology Group.)*

If the injection pulse width feedback signals are out of range, the computer stores the corresponding trouble code in its memory for recall with a scan tool.

OUTPUTS

Fuel Injection Pump.   The heart of any diesel is the high pressure fuel injection pump, Figure 17-9. The fuel injection pump forces a very fine spray of fuel into the combustion chamber to initiate and continue the power stroke. This pressure may be as high as 3,000 psi, in order to properly atomize the fuel and to overcome the combustion pressure.

The injection pump is actually a series of pumps. Fuel from the frame-mounted electric **lift pump** first arrives at the **transfer pump,** a vane-type pump that varies the pressure it delivers according to engine speed. At idle, transfer pump outlet pressure is 20 to 30 psi. At high engine speed, the pressure may be over 100 psi. A relief valve will open if the pressure reaches 125 psi. The pressure regulator also includes a viscosity compensating port. Pressure relief fuel returns to the transfer pump inlet passage.

Injection.   From the transfer pump, fuel travels to the charging **annulus ring.** This component has eight slots, one per cylinder. From it, fuel moves to two places: the pump housing, and the pumping plungers of the rotor.

The metering of the fuel into the pumping plungers is the most important part of understanding how the 6.5-liter diesel works. Fuel enters from two places: the inlet ports in the annulus rotor, and the control valve. As the rotor spins, the annulus slots move into and out of alignment with the inlet slots. When they are aligned, fuel under transfer pump pressure enters the chamber. When the slots are not aligned, fuel flow stops. The second source of fuel is through the control valve. When the fuel solenoid is off, the valve opens, and fuel can run through the center of the pump shaft into the plunger chamber. The fuel that enters the plunger chamber has been metered for injection delivery.

The actual injection is driven by the plungers. They ride on low friction shoe-and-roller assemblies in an eight-lobed cam ring. When the shoe-and-roller assemblies are in the valleys between the lobes, the plungers are fully open. When the shoe-and-roller assemblies ride up the lobe, the plungers move inward and force the fuel through the injector lines to the injectors, Figure 17-10. This fuel metering occurs eight times for each revolution of the pump rotor, for each complete engine cycle of two crankshaft rotations.

The computer controls the amount of fuel injected by determining when to turn the fuel solenoid on and off. The longer it is on, the more fuel is injected; the shorter, the less. At warm idle, the solenoid is on for only a brief moment, enough to inject a small droplet of fuel. At high power settings, the solenoid remains on for most of the rotor motion, injecting a maximum amount of fuel. After the injectors have sprayed the amount of fuel the computer has determined is correct for the current conditions, it disengages the solenoid and spills the fuel back into the fill/spill chamber.

These steps can be seen graphically in the following Figures 17-11 to 17-14.

Idle Quality.   One of the important advantages of the electronic control of injection volume is on improvement in idle quality. On mechanical diesel injection engines, it is difficult to manufacture all the combustion chambers to have exactly the same compression ratio. In addition, variations in the amount of fuel injected are greatest at idle speeds because of manufacturing tolerances in the injection pump. The slight variations in compression ratio and idle fuel delivery cause a much rougher idle on a mechanically injected diesel than

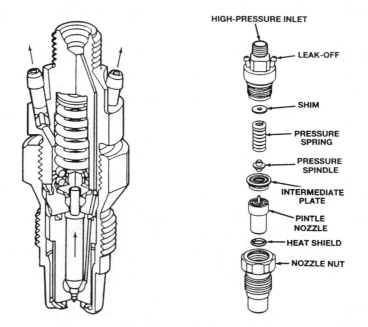

Figure 17-10 Diesel fuel injector. *(Courtesy of General Motors Corporation, Service Technology Group.)*

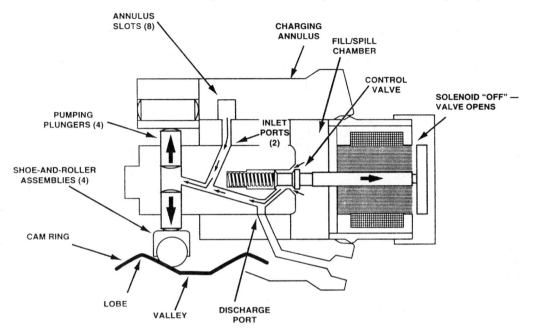

Figure 17–11 Fuel injection metering sequence. *(Courtesy of General Motors Corporation, Service Technology Group.)*

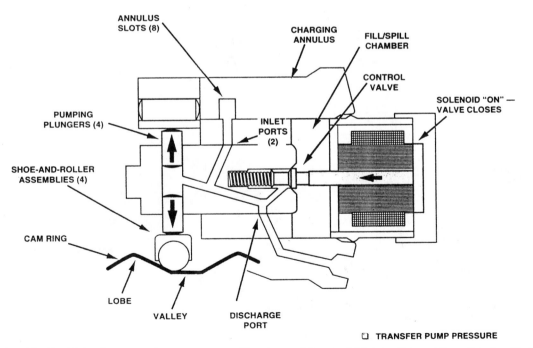

Figure 17–12 Fuel injection metering sequence. *(Courtesy of General Motors Corporation, Service Technology Group.)*

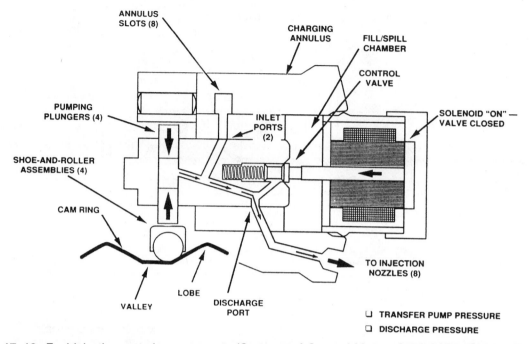

ANNULUS SLOTS (8)

CHARGING ANNULUS

FILL/SPILL CHAMBER

CONTROL VALVE

PUMPING PLUNGERS (4)

INLET PORTS (2)

SOLENOID "ON" — VALVE CLOSED

SHOE-AND-ROLLER ASSEMBLIES (4)

CAM RING

TO INJECTION NOZZLES (8)

LOBE

DISCHARGE PORT

VALLEY

□ TRANSFER PUMP PRESSURE
□ DISCHARGE PRESSURE

Figure 17–13 Fuel injection metering sequence. *(Courtesy of General Motors Corporation, Service Technology Group.)*

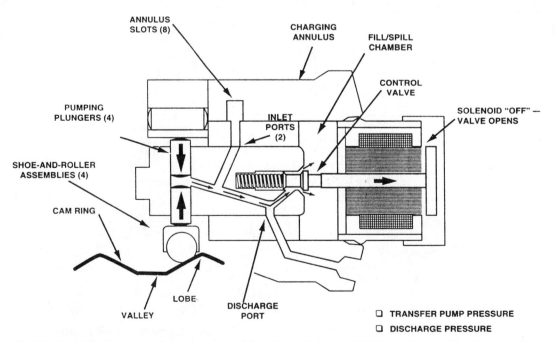

ANNULUS SLOTS (8)

CHARGING ANNULUS

FILL/SPILL CHAMBER

CONTROL VALVE

PUMPING PLUNGERS (4)

INLET PORTS (2)

SOLENOID "OFF" — VALVE OPENS

SHOE-AND-ROLLER ASSEMBLIES (4)

CAM RING

VALLEY

LOBE

DISCHARGE PORT

□ TRANSFER PUMP PRESSURE
□ DISCHARGE PRESSURE

Figure 17–14 Fuel injection metering sequence. *(Courtesy of General Motors Corporation, Service Technology Group.)*

on a gasoline engine. With electronic controls the computer can monitor idle speed variations and inject different amounts of fuel to each cylinder to make the idle speed much smoother.

Injection Timing. Injection volume is just one of the parameters controlled. The other major one is injection timing. Just as with gasoline engines, combustion must start earlier in the compression stroke as the engine turns faster, and there are different optimal injection timing points with different loads.

This system uses an **injection timing stepper motor,** Figure 17-15, to achieve this timing control. The stepper motor uses a sliding piston that rotates the injection cam ring with a pin. As the piston moves in its bore, it turns the cam ring to the position determined by the computer. The stepper motor does not actually move the piston; instead it moves a servo valve plunger that uses transfer pump pressure to move the advance piston itself, Figure 17-16. When the computer wants more injection advance, the stepper motor retracts in steps. This moves its pivot arm, and the control lever moves away from the servo valve. Spring pressure moves the valve away from the advance passage, and pressurized fuel enters the passage and pushes the piston in the advance direction. When the computer wishes to

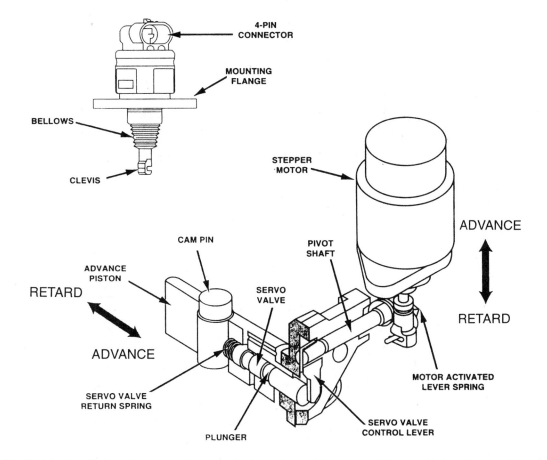

Figure 17–15 Injection timing stepper motor and advance piston. *(Courtesy of General Motors Corporation, Service Technology Group.)*

retard injection timing, it extends the stepper motor arm, pushes the control lever in, and reverses the hydraulic force on the advance piston. The layout is similar to the movement of a spool valve in an automatic transmission or a power steering system.

Engine Shutoff (ESO) Solenoid. The only way to stop a running diesel engine other than stalling it is to shut off either air or fuel. The computer controls the engine shutoff solenoid mounted on the fuel pump. As long as this solenoid is actuated, fuel can flow to the injectors. When it is shut off, the fuel passage is blocked and the engine stops immediately.

Injection Timing Stepper Motor. On the right side of the injection pump is the injection timing stepper (ITS) motor, Figure 17-17. The motor contains two coils controlling injection pump timing through four circuits, a high and low position for each coil. The computer will, of course, set a code for any problem with the ITS circuits. The range of timing is very limited, however, and if the computer, the timing gears, the front engine cover, the crankshaft position sensor, or anything else affects injection timing, a special procedure must be followed to mechani-

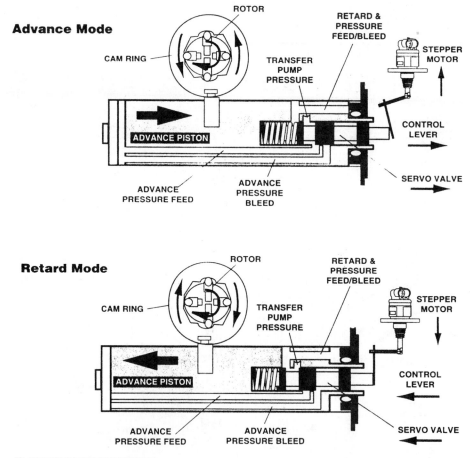

Figure 17–16 Advance piston operation. *(Courtesy of General Motors Corporation, Service Technology Group.)*

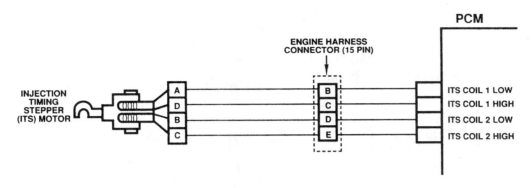

Figure 17–17 Injection timing stepper motor operation. *(Courtesy of General Motors Corporation, Service Technology Group.)*

cally set the injection pump within one degree of accuracy. This procedure is outlined in the appropriate shop manual. The point of the procedure is to allow the computer to "relearn" exactly where TDC occurs. Some GM literature refers to this as "TDC offset recovery."

Glow Plug Control.   It is frequently hard to start a diesel engine cold because of the relatively slow crank provided by the starter and the cold combustion chamber walls which may not allow

the compressed air to reach the ignition flash point of the injected fuel. The 6.5-liter GM diesel uses eight glow plugs to assist in providing this initial heat, Figure 17-18.

The computer controls the activation of the glow plugs based on information from the engine coolant temperature sensor, the crankshaft position sensor (for engine rpm at crank and just after start-up), and the feedback signal from the glow plugs themselves. It activates the high-current

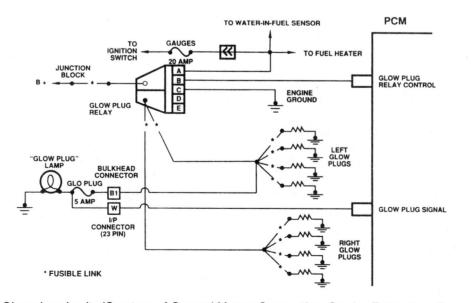

Figure 17–18 Glow plug circuit. *(Courtesy of General Motors Corporation, Service Technology Group.)*

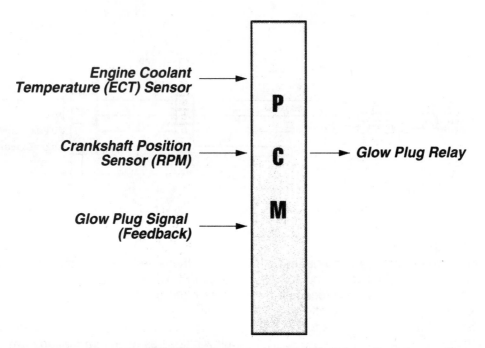

Figure 17–19 Glow plug relay schematic. *(Courtesy of General Motors Corporation, Service Technology Group.)*

glow plugs through a special high-current relay, the glow plug relay, Figure 17-19. The computer also operates the glow plug lamp on the dashboard that informs the driver when to begin engine cranking. The wait period varies depending on the temperature signal.

CAUTION: According to GM, never install a jumper across the high-current terminals of the glow plug relay, or immediate damage to the glow plugs can result. Glow plugs can be tested for continuity to determine if one has burned out, but an amperage measurement is not practical because most designs use less current as they get warmer, so the results are not relevant. On earlier (1994 and 1995) models, an amperage draw of 14 amps per glow plug, or 55 amps per engine bank, is normal.

EGR System.   Diesel engines, just like gasoline engines, can produce undesirable NO_x

gas in their combustion chambers if combustion temperatures go much above 2,500 degrees. As in gasoline engines, exhaust gas is metered back through the intake system to provide an inert gas that will keep the peak temperatures below the

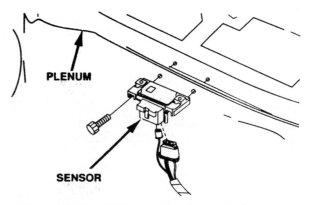

Figure 17–20 EGR control pressure/BARO sensor. *(Courtesy of General Motors Corporation, Service Technology Group.)*

NO_x formation levels. Inputs for these calculations come from the accelerator pedal position sensors, the crankshaft position sensor, the EGR control pressure/BARO sensor and the engine coolant temperature sensor.

The EGR control pressure/BARO sensor mounts on the cowl, and monitors the absolute pressure in the EGR vacuum line, Figure 17-20. The computer uses this input for a barometric pressure reading when the EGR system is off, just after the key is turned on but before the engine starts, for example. If the EGR control pressure sensor reports a vacuum too high or too low, the computer will shut off the EGR system and set the appropriate trouble code.

Actual metering of recirculated exhaust gas is done by the EGR valve, controlled by the EGR solenoid and EGR vent solenoid, Figures 17-21, 17-22, and 17-23. The computer activates these solenoids by providing a ground path for their voltage, applied through the ignition switch. The EGR solenoid cycles to apply a desired amount of vacuum to open the EGR valve. The EGR vent solenoid dumps any vacuum in the line to close the EGR valve during EGR-off conditions. Failure of either solenoid

will trigger a corresponding trouble code and turn on the check engine light.

Boost Control System. The system uses a wastegate to control the boost provided by the turbocharger in accord with information the computer gets from the accelerator pedal position sensors, the boost sensor, and engine speed information from the crankshaft position sensor (Figure 17-24).

Not all applications of the 6.5-liter GM diesel use a turbocharger, and when it is not installed, this information/actuation circuit is not part of the system. The boost sensor is very similar to a MAP/BARO sensor on a gasoline-fueled engine: it is a resistor that changes following the sensed pressure. As pressure increases, resistance goes down. The computer uses this information to calculate how much fuel can be delivered to the engine.

The engine uses a special **vacuum pump** to actuate the wastegate, as well as the other vacuum-operated devices, the EGR solenoid, and EGR vent solenoid. A special vacuum pump is needed on a diesel engine because the engine does not develop vacuum behind a throttle, as do gasoline-fueled engines. The computer tracks

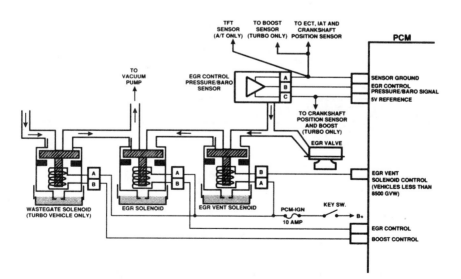

Figure 17–21 EGR valve circuit operation. *(Courtesy of General Motors Corporation, Service Technology Group.)*

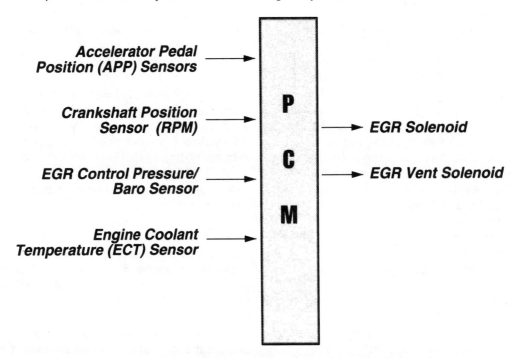

Figure 17–22 EGR control inputs and outputs. *(Courtesy of General Motors Corporation, Service Technology Group.)*

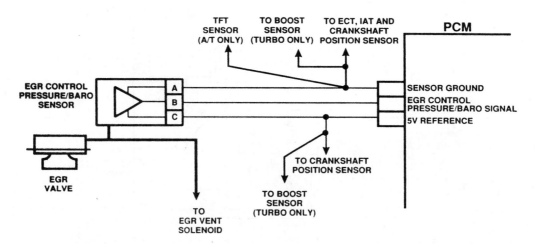

Figure 17–23 EGR control pressure/BARO sensor circuit operation. *(Courtesy of General Motors Corporation, Service Technology Group.)*

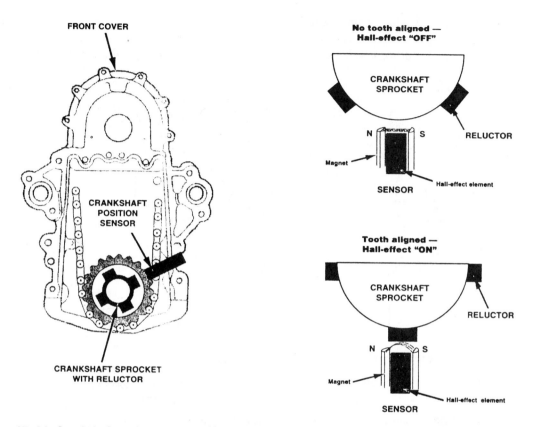

Figure 17–24 Crankshaft position sensor. *(Courtesy of General Motors Corporation, Service Technology Group.)*

the intake manifold pressure by the pressure sensor and cycles the wastegate solenoid to prevent overboost, Figure 17–25.

Transmission Controls. Automatic transmissions mated to the 6.5-liter GM diesel are also controlled by the engine computer. There are slight differences depending on which transmission is installed and on whether the vehicle is four- or two-wheel-drive. The torque converter lockup works the same way as on gasoline-fueled engines. On many late-model diesel vehicles, the work of the transmission control module, as in a gasoline-fueled vehicle, is directed entirely by the engine control computer. There is a significant difference between two- and four-wheel-drive in that the four-wheel-drives include a transmission output shaft speed sensor to enable the computer to track differences between transfer case settings. In the two-wheel-drive vehicle, only the single vehicle speed sensor is required. As with many other newer GM vehicles, there is a vehicle speed sensor buffer module, essentially a VSS signal interpreter which provides vehicle speed information to the computer, the antilock brake system, and the speedometer odometer.

The fundamental differences between the transmission controls for the diesel engine and those for the gasoline-fueled engine consist in the diesel's greater torque at lower speeds. This changes the shift points to keep the engine at a more favorable rpm. Because of this greater low speed torque, the diesel can also stay in a given gear longer than an equivalent gasoline-fueled engine.

Lift Pump. While the Diesel engine mechanically drives the injection pump, there is another pump to prime the injection pump. The **lift pump,**

Figure 17-26, is very similar to fuel pumps on gasoline engines and provides relatively low (15 psi) pressure to purge any air from the injection

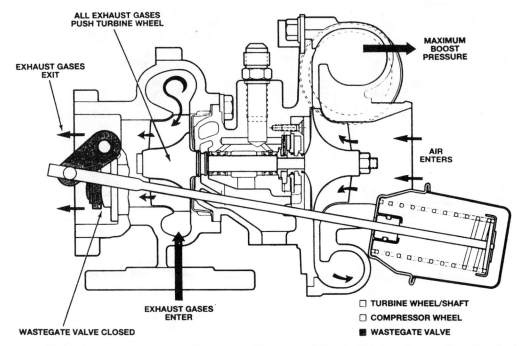

Figure 17–25 Turbocharger wastegate and actuator. *(Courtesy of General Motors Corporation, Service Technology Group.)*

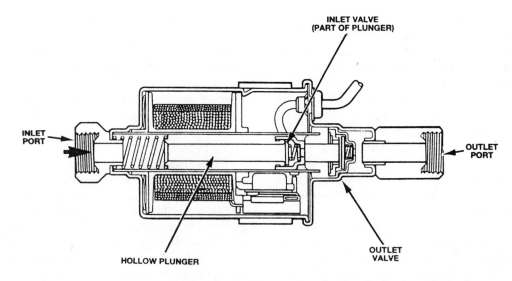

Figure 17–26 Lift pump. *(Courtesy of General Motors Corporation, Service Technology Group.)*

pump circuits. The lift pump is turned on by the ignition switch and kept on by a switch circuit through the oil pressure sender. Late model vehicles use a computer-controlled fuel pump relay, and the oil pressure sender is redundant.

FORD/INTERNATIONAL/NAVISTAR 7.3 DIRECT INJECTION DIESEL: SYSTEM OVERVIEW

Certain Ford trucks are available with the Navistar/International 7.3-liter direct injection diesel engine. This engine is built for Ford by International and is used on a number of other medium-duty trucks as well. Like the GM diesel described earlier, it is a compression-ignition engine. We will not repeat the information about the differences between gasoline and diesel engines here, but will assume the reader understands them.

The Ford/International engine is a direct injection design; that is, there is no prechamber above the cylinder head. Instead there is a small cavity in the piston top into which the fuel is directly sprayed. The glow plugs also extend into this cavity. As is typical for many direct injection diesels, the compression ratio is somewhat lower than for a prechamber design. This engine has a compression ratio of approximately 18 to 1.

This system is unique among current electronically controlled diesels because it does not use an injection pump. Instead injection is done electronically by the individual injectors. It is also unique in the injection pressure developed; under conditions of high power output, injection can reach as high as 21,000 psi! The precaution mentioned earlier about not allowing injected fuel to come anywhere near people is ten times more pronounced with these injectors.

INPUTS

The Ford/International system uses eight

basic sensors on which to base the computer's calculations. These are:

- fuel pedal position sensor.
- camshaft position sensor.
- injection control pressure sensor.
- boost pressure sensor.
- oil temperature sensor.
- oil pressure sensor.
- coolant temperature sensor.
- ambient air temperature sensor.
- barometric pressure sensor.
- exhaust backpressure sensor.

Most of these sensors are tied to the OBD II diagnostic program in Ford vehicles (not applicable, however to medium-duty International trucks). When the computer detects an absent or out-of-range signal, it will set a code, as described in the appropriate shop manual.

For a system schematic, see Figure 17-27.

OUTPUTS

The engine drives a seven-piston fixed displacement axial pump, pressurizing the engine oil to a pressure between 450 and 2,750 psi during normal operation. This is not an injection pump; it merely provides the individual injectors with highly pressurized engine oil. The computer selects the desired pressure from its various inputs, particularly engine coolant temperature and engine speed. The colder the engine or the greater the load and speed, the higher the pressure selected will be. The desired pressure is obtained by electronic control of the rail pressure control valve (RPCV), Figure 17-28.

The computer then actuates the injectors by commands to the injector drive module (IDM), basically a heavy-duty solid-state relay bank.

Electronic Injectors. The key to the system is in the electronic/hydraulic injectors, Figure 17-29. These work by a combination of electronic control and hydraulic power multiplication.

Fuel enters the injectors at the lower end in

International HEUI* Fuel System Operation
*Hydraulically Activated Electronically Controlled Unit Injection

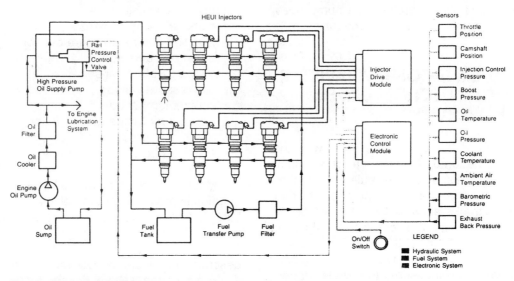

T444E Shown, I-6 engines are the same except only 6 injectors.

Figure 17–27 System schematic. *(Courtesy of Navistar International, Engine Division.)*

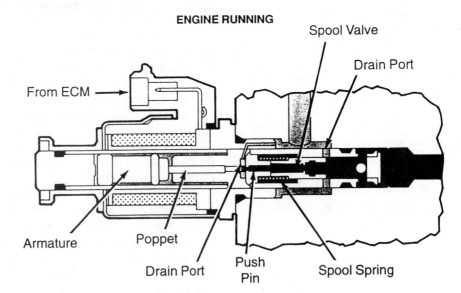

Figure 17–28 Rail pressure control valve. *(Courtesy of Navistar International, Engine Division.)*

the injector tip charging area. The high pressure engine oil enters the injector at the top in the intensifier piston space. Fundamentally, the way the injection pressure is built is this: a small amount of fuel to be injected waits in the space under the intensifier piston. When the computer actuates the injector, the drive module triggers the electronic solenoid at the top of the injector, raising the poppet valve. This simultaneously closes the drain passage and opens the passage to the rail, pressurized as described previously.

Since the diameter of the top of the intensifier piston is seven times larger than the diameter of the lower fuel delivery piston, the pressure is multiplied by the same factor. So if 500 psi is applied to the top of the piston, 3,500 psi is applied to the injector tip. If 3,000 psi is applied to the top, 21,000 psi is applied to the tip.

Since the only thing activating these injectors is the computer's command, variations in delivery volume and injection advance timing are simply and entirely a matter of internal computer software. No moving parts have to shift or adjust or rotate except for the computer's sensors.

When the computer deactivates the solenoid, the springs return the intensifier piston and the nozzle needle valve to their rest positions, refilling each chamber with fuel for the next injection.

VOLKSWAGEN 1.9-LITER TDI DIESEL: SYSTEM OVERVIEW

Volkswagen has built small fuel-efficient diesel cars for many years, but the new 1.9-liter turbocharged direct-injection diesel is their most advanced compression ignition product yet, Figure 17-30.

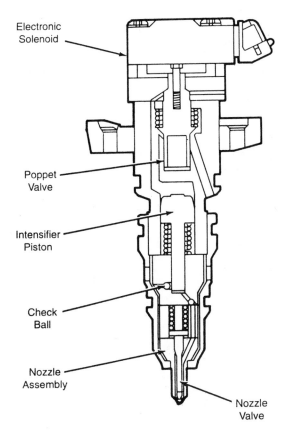

Figure 17–29 Electronic/hydraulic injector. *(Courtesy of Navistar International Transportation Corp.)*

Figure 17–30 VW 1.9-liter TDI. *(Courtesy of Volkswagen of America, Inc.)*

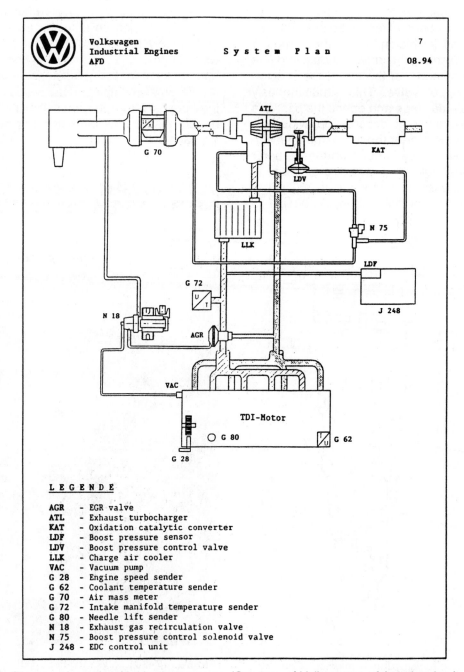

Figure 17–31 System schematic with air mass sensor. *(Courtesy of Volkswagen of America, Inc.)*

This small four-cylinder engine is a fully-electronic direct-injection design (all their previous diesels used prechambers). Very similar to the GM in basic design (except for its direct injection), the system is a drive-by-wire system with injection quantity and advance determined by the computer from driver commands and sensor inputs. The car includes a diesel catalytic converter to burn off residual hydrocarbons left in the exhaust. The catalytic converter, as is universal on diesels so equipped, does not reduce NO_x, but only the hydrocarbons.

INPUTS

Besides the usual sensors for engine coolant temperature, boost pressure, and intake air temperature, the VW TDI system uses a hot-wire air mass sensor, just as used on cars, to measure exactly how much air is entering the engine, Figure 17-31. The other unusual sensor is a needle lift sensor on cylinder 3, which indicates to the computer exactly what the achieved timing advance is and enables it to correct for this as needed.

SUMMARY

In this chapter, we have covered the basics of the General Motors 6.5-liter electronically controlled diesel engine, the Ford/Navistar/International 7.3-liter electronically controlled diesel and the Volkswagen 1.9-liter TDI engine. We have also considered the basic differences between diesel engines and gasoline-fueled engines, as well as diagnostic approaches to the former.

Each of these engines includes a novel drive-by-wire accelerator system, with no mechanical linkage between the accelerator pedal and the engine. Instead power demand information gets to the computer through variable resistance potentiometers, similar to a throttle position sensor on a gasoline engine.

In each engine, the computer controls the injection quantity and injection timing from the injectors. In the GM and VW engines, the computer acts on solenoids and a stepper motor at the injection pump; in the International/Ford system, the injection is entirely electrical with the computer determining pressure, quantity (through pulse width), and timing for each cylinder for each power stroke.

Because each of these systems is installed in vehicles required to be OBD II-compliant, diagnosis is possible using the standardized scan tool, a DVOM, and for some sensors, a lab scope. Codes are stored in the same way as for gasoline-fueled engines, except the codes are specific to diesel engines instead of gasoline.

▲ DIAGNOSTIC EXERCISE

Diesel engine problems fall into three general categories: no-starts, rough running, and smoke, black or white. Describe what differences there would be in diagnosing these problems with the new electronically controlled diesels covered in this chapter.

REVIEW QUESTIONS

1. On the GM 6.5-liter diesel engine, what would be the effect on an engine running at steady cruise speed if the alternator failed and the electrical system voltage slowly went down as the battery discharged? What causes this effect?

2. In each of the engine systems covered in this chapter, could the engine run if the crankshaft position sensor failed? If so, would there be any difference in performance? If not, what prevents fuel injection?

3. On the Ford/International electronic diesel engine, one cylinder is dead. Cranking indicates identical compression pressure on each cylinder, and there is no air in the fuel

system. How should you check the injector?

4. An inattentive motorist has run his Volkswagen 1.9- liter TDI diesel car out of fuel. Now it won't start. What should be done to diagnose and repair this problem?

ASE-type Questions (Actual ASE test questions are rarely so product specific.)

5. An electronically controlled diesel engine will run, but does not respond to the fuel pedal (accelerator pedal) at all. Technician A says the accelerator pedal position sensor may have failed. Technician B says the computer itself may be defective. Who is correct?
   a. A only.
   b. B only.
   c. both A and B.
   d. neither A nor B.

6. A diesel-fueled truck was inadvertently filled with gasoline and started. It ran for a few minutes while the gasoline circulated into the fuel delivery system, but then shut off. Technician A says many sensors and actuators in the injection pump fluid circuit may have been damaged. Technician B says to drain the tank of gasoline, refill it with diesel fuel and run it until it runs smoothly. Who is correct?
   a. A only.
   b. B only.
   c. both A and B.
   d. neither A nor B.

7. An electronically controlled diesel has a battery with a low charge. Technician A says to use jumper cables to another vehicle and start it normally. Technician B says the vehicle's battery must be first recharged com-

pletely and then the vehicle started from it, with no jumper. Who is correct?
   a. A only.
   b. B only.
   c. both A and B.
   d. neither A nor B.

8. An electronically controlled diesel has smoke in the exhaust. What would you check if:
   a. the engine exhaust had black smoke.
   b. the engine exhaust had blue smoke.
   c. the engine exhaust had white smoke.
   d. the engine exhaust had white smoke on cold days right after start-up.

9. A GM 6.5-liter electronically controlled diesel has various driveability complaints. Technician A notices an OBD II diagnostic link connector under the dashboard and wants to try his scan tool, but Technician B says it won't be of any help since all the diesel's sensors and actuators are different from those on a gasoline-fueled vehicle. Who is correct?
   a. A only.
   b. B only.
   c. both A and B.
   d. neither A nor B.

10. Technician A unplugs all the injectors to avoid pressure problems on a Ford/International 7.3-liter electronically controlled diesel engine while preparing for a diesel compression test. Technician B says the pump will still develop high pressure. Who is correct?
    a. A only.
    b. B only.
    c. both A and B.
    d. neither A nor B.

Chapter 18

OBD II Self-Diagnostics

OBJECTIVES

In this chapter you can learn:
- ❑ the reasons for the new OBD II program.
- ❑ major aspects of the OBD II program.
- ❑ standardized features of all manufacturers' OBD II program.
- ❑ representative strategies used to monitor engine management system malfunctions.
- ❑ basics of the use of scan tools and lab scopes for diagnosis.

KEY TERMS

Misfire Detection
OBD II
Scan Tool
Standardized DLC 16-Pin Diagnostic Connector

With the 1996 models, many car manufacturers built their first **OBD II**-compliant vehicles. The **OBD II** program is intended to standardize the diagnosis of emissions and driveability-related problems on all new cars sold in the United States. The new standardized self-diagnostic systems are on all cars for the 1996 model year (with waivers for some carmakers with unique technical difficulties).

WHY OBD II?

The OBD II system (and its predictable successors) should be the basis for driveability and emissions diagnosis for many years to come. It makes diagnostic tools, codes and procedures similar, regardless of manufacturer or country of origin. The system was originally crafted to allow plenty of room for growth and the incorporation of many additional subsystems. While there are some differences among carmakers and among

models (because of different system control components), the purpose of the program is to make the diagnosis of emissions and driveability problems simple and uniform in the future; it will no longer be necessary to learn entirely new systems for each manufacturer.

The chemistry of gasoline combustion, the mechanics of a four-cycle engine and the emissions control strategies that have proved successful are the same for all carmakers. These facts plus federal law should make emissions and driveability diagnosis both more successful and easier to learn in the future.

Much of this information involves technical changes introduced gradually, since emissions concerns first began to shape combustion control measures and since in any given year most carmakers' changes are largely enhancements or refocusings of systems introduced previously. Some subsystems described here started as early as 1963 (PCVs); others appear on only a few 1996 models; some will remain unique to specific vehicles. The OBD II system affords the options to do all of this.

WHAT WILL OBD II DO?

The plan behind OBD II is that any automotive service technician can diagnose any vehicle built according to the standard using the same diagnostic tools. As of yet, the system is too new to tell whether that promise will be fulfilled completely. Meanwhile, carmakers can introduce special diagnostic tools or capacities for their own systems, so long as standard scan tools, along with digital volt/ohmmeters and oscilloscopes can analyze the system. These dealer tools can, of course, have additional capacities beyond the designated OBD II functions. One of the mandated capacities of OBD II systems, for example, is "freeze frame," the ability of the system to record data from all its sensors and actuators at a time when the system turns on the malfunction indicator light (MIL). General Motors expands this capacity to include "failure records," which does the same thing as freeze frame, but includes any fault stored in the computer's memory, not just those related to emissions component circuit failures.

The goal of OBD II is to monitor the effectiveness of the major emission controls and to turn on the malfunction indicator light (MIL) and store a diagnostic trouble code (DTC) whenever the effectiveness deteriorates to a point where research indicates the emission level reaches 1.5 times the allowable standard for that gas and that vehicle, based on the federal test procedure.

Besides enhancements to the computer's capacities, the program requires some additional sensor hardware to monitor the emissions performance closely enough to fulfill the tighter constraints and beyond merely keeping track of component failures. In most cases this hardware consists of an additional heated oxygen sensor down the exhaust stream from the catalytic converter, upgrading specific connectors and components to last the mandated 100,000 miles or 10 years, in some cases a more precise crankshaft or camshaft position sensor (to detect misfires more accurately) and the new standardized 16-pin data link connector (DLC), Figure 18-1.

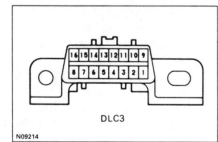

DLC3

N09214

Terminal No.	Connection	Voltage or Resistance	Condition
2	Bus ⊕ Line	Pulse generation	During transmission
4	Chassis Ground	↔ Body Ground 1 Ω or less	Always
5	Signal Ground	↔ Body Ground 1 Ω or less	Always
16	Battery Positive	↔ Body Ground 9 ~ 14 V	Always

Figure 18-1 OBD II regulations require a standardized, universal 16-pin DLC (Data Link Connector) for all 1996 cars sold in this country.

STANDARDIZATION

Besides the closer monitoring of emissions performance, the other major change of OBD II is the standardization of diagnosis. While not all vehicles use identical systems, there is much overlap in the types of systems used (catalytic converters, oxygen sensor feedback, etc.), and the OBD II program is designed to reduce the confusion between one system and another by mandating not only the standard diagnostic link, but also the specific codes and the descriptions of components in manufacturers' literature. These standards and standard descriptions and trouble codes were all prepared by the Society of Automotive Engineers (SAE) to achieve the following:

- common terms and acronyms (SAE standard J1930).
- common data link connector (DLC) and location (SAE standard J1962).
- common diagnostic test modes (SAE standard J2190).
- common scan tools (SAE standard J1979).

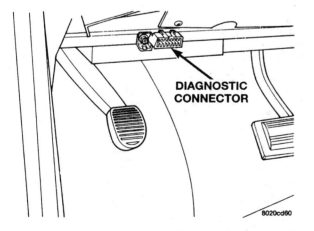

Figure 18-2 Not only is the DLC's configuration standardized, but so is its position—it must be between the left end of the instrument panel and a position no more than 300mm to the right of center. This will make finding it easy for the technician. *(Courtesy of Chrysler Corporation.)*

- common diagnostic trouble codes (SAE standard J2012).
- common protocol standard (SAE standard J1850).

Common Terms. All vehicle manufacturers will have to employ common names and abbreviations for components serving similar purposes. For example, the sensor reporting crankshaft position and speed information to the computer will be called a crankshaft position sensor by each manufacturer, and it will be abbreviated CKP. The computers will be described as PCMs. Most manufacturers began using these terms for their 1993 model year vehicles.

Common Data Link Connector (DLC). Each vehicle will have a standard shaped and sized **16-pin diagnostic connector,** using the same pins for the same information and located somewhere between the left end of the instrument panel and a position 300 millimeters to the right of the center (Figure 18-2). Notice that while certain terminals of the connector are designated for specific purposes, others are left free for the manufacturer to use as desired or are not employed on current model vehicles.

Common Diagnostic Test Modes. These test modes are common to all OBD II vehicles and all will be accessed using an OBD II scan tool. Each mode is described in the following paragraphs.

Mode 1. Parameter Identification (PID) mode allows access to certain data values, analog and digital inputs and outputs, calculated values and system status information. Throughout the service literature, there will be references to PID values. Some of the PID references are from a generic OBD II PID list that all scan tools must be able to reference. If a referenced nongeneric PID is not on this list, it can be accessed with the manufacturer-specific scan tool or equivalent. If a generic scan tool is used for nongeneric OBD II PIDs, a string of characters (a hexadecimal number in some cases) may have to be entered. The necessary numbers will be supplied by the scan tool maker. Later model and updated generic scan tools will have the capacity to directly access all or selected manufacturers' codes.

Mode 2. Freeze Frame Data Access mode permits access to emission-related data values from specific generic PIDs. These values represent the operating conditions at the time the fault was recognized and logged into memory as a DTC. Once a DTC and set of freeze frame data are stored in the computer's memory, they will stay in memory even when additional emission-related DTCs are stored. The number of such sets of freeze frame data that can be stored is, of course, limited. On General Motors vehicles, for example, the number of such sets is five for the 1996 model year.

There is one type of failure that is an exception to that rule: misfire. Fuel system misfires will overwrite any other type of data and are not themselves overwritten. They can be removed only with the scan tool. When the scan tool is used to erase a DTC, it automatically erases all the freeze frame data associated with that DTC event.

Mode 3. This mode permits scan tools to obtain stored DTCs. The information is transmitted from the car computer to the scan tool following an OBD II Mode 3 request. Either the DTC, its descriptive text or both will display on the scanner. The specific menu access techniques to emission-related DTCs is left up to the scan tool manufacturer, but such data should be relatively simple to extract. As the number and sophistication of scan tools increases, independently supplied scanners will be able to access and interpret the manufacturer-specific codes as well. Complete printed lists of DTCs for all manufacturers are available.

Mode 4. The PCM (powertrain control module) reset mode allows the scan tool to clear all emission-related diagnostic information from its memory. Once the PCM has been reset, the PCM stores an inspection maintenance readiness code until all the OBD II system monitors or components have been tested to satisfy an OBD II trip cycle without any other faults occurring. Quite specific conditions must be met before a given engine start and vehicle movement constitutes a "trip" for the OBD II system, which will be described somewhat later in this chapter.

Mode 5. The oxygen sensor monitoring test result indicates the on-board sensor fault limits and the actual oxygen sensor voltage outputs during the test cycle. The test cycle includes specific operating conditions that must be met (engine temperature, load, speed, etc.) to complete the test. This information helps determine the effectiveness of the exhaust catalytic converter. Here are some, but not all, of the available tests and test identification numbers:

Test ID	Test Description	Units
01	Rich to lean sensor threshold voltage for test cycle	VOLTS
02	Lean to rich sensor threshold voltage for test cycle	VOLTS
07	Minimum sensor voltage for test cycle	VOLTS
08	Maximum sensor voltage for test cycle	VOLTS

Mode 6. The output state mode (OTM) allows a technician to activate and deactivate the system's actuators on command and through the scan tool. When the output state mode is engaged, the actuators can be controlled without affecting the radiator cooling fans. The low and high speed radiator cooling fans are turned on separately, without energizing other output components.

Common Scan Tools. For OBD II, a **scan tool** must access and interpret emission-related diagnostic trouble codes regardless of the vehicle make or model. Most brand-specific and better quality aftermarket scan tools can also access additional information regarding driveability problems and other systems controlled or monitored by the PCM.

The scan tool includes a harness that mates with the standardized 16-pin connector (Figure 18-3).

Common Diagnostic Trouble Codes. SAE J2012 determines a five-character alphanumeric code in which each character has a specific meaning. The first character is the prefix letter indicating the range of the function:

P = Powertrain
B = Body
C = Chassis

The second character (and the first number) indicates whether the DTC to follow is a standard SAE code or one specific to the manufacturer:

0 = SAE
1 = Manufacturer

The difference is that a DTC with 0 as the second character means the same thing regardless of the make or model of the vehicle, while a DTC with 1 as the second character has its meaning defined by the manufacturer of that vehicle, so its meaning may be different from exactly the same DTC on another make and model of car.

The third character of a powertrain DTC (one beginning with P) indicates the system subgroup:

0 = total system
1 and 2 = fuel/air control

3 = ignition system/misfire
4 = auxiliary emission controls
5 = idle/speed control
6 = PCM and inputs/outputs
7 = transmission
8 = non-EEC powertrain

The fourth and fifth characters identify the specific fault detected. Some OBD II vehicles will still flash a two-digit code through the malfunction indicator lamp, as previously. Typically these codes are not numerically identical with the last two characters of the OBD II (SAE J2012) code, but correspond to the manufacturer's older code tables.

Standard Protocol. A "protocol" in computer language is merely an agreed-upon digital code the computer uses to communicate with the scan tool. Compliance with OBD II means each manufacturer uses the same multiplexing language between the PCM and its sensors and actuators and with the diagnostic information sent to and received from the scan tool through the DLC.

MONITORING CONDITIONS

OBD II standards require the engine management system to detect faults, set DTCs, turn on or off the malfunction indicator light, or erase

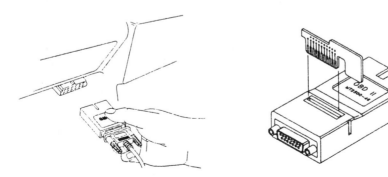

Figure 18-3 The connector of a generic scan tool will fit the universal DLC. This aftermarket unit uses plug-in "personality keys" to help take advantage of each car maker's unique, proprietary diagnostics, those that go beyond OBD II. *(Courtesy of Snap-on Tools Company, Copyright owner.)*

the DTCs for each monitored circuit according to very specific sets of operating conditions. The following definitions help identify the conditions determined so far:

Warm-up Cycle. OBD II standards define a warm-up cycle as a period of vehicle operation, after the engine was turned off, in which coolant temperature rises by at least 40 degrees Centigrade and reaches at least 160 degrees Centigrade. Most OBD II DTCs are automatically erased after 40 warm-up cycles if the failure is not detected again after the MIL is turned off. See the chart, Figure 18-4. Some manufacturers retain erased DTCs in a 'flagged' condition; forgiven, as it were, but not forgotten. This can be

useful if the technician notices a pattern of component failure, all of which might be related to a single intermittent cause like low fuel pressure.

Drive Cycle. A "drive cycle" contrasted with an "OBD II drive cycle" consists of an engine start and vehicle operation that brings the vehicle into closed loop and includes whatever specific operating conditions are necessary either to initiate and complete a specific OBD II monitoring sequence or to verify a symptom or confirm a successful repair. A monitoring sequence is an operating strategy designed to test the operation of a specific system, function or component. For examples, the computer can open and close the EGR valve during deceleration and monitor the

Monitor	Monitor Type (When it Completes)	Number of Malfunctions on Separate Drive Cycles to Set Pending DTC	Number of Separate Consecutive Drive Cycles to Light MIL and Store DTC	Number and Type of Drive Cycles with No Malfunction to Erase Pending DTC	Number and Type of Drive Cycles with No Malfunction to Turn MIL OFF	Number of Warm-Ups to Erase DTC after MIL is Extinguished
Catalyst Efficiency	Once per OBD II Drive Cycle	1	3	1	3 OBD II Drive Cycle	40
Misfire Type A	Continuous		1		3 (Similar Conditions)	40
Misfire Type B/C	Continuous	1	2	1	3 (Similar Conditions	40
Fuel System	Continuous	1	2	1	3 (Similar Conditions)	40
Oxygen Sensor	Once per Trip	1	2	1 Trip	3 Trips	40
Exhaust Gas Recirculation	Once per Trip	1	2	1 Trip	3 Trips	40
Comprehensive Component	Continuous (When Conditions Allow)	1	2	1 Trip	3 Trips	40

Figure 18-4 OBD II DTC/MIL function chart

MAP sensor input to see whether the EGR is working, or during cruise the computer can open and close the canister purge valve while observing the oxygen sensor signal, thus testing both components.

OBD II Trip. The OBD II trip, often shortened just to "trip," consists of an engine start following an engine off period, with enough vehicle travel to allow the following OBD II monitoring sequences to complete their tests, as shown in Figure 18-5:

- Misfire, fuel system and comprehensive system components. These are checked contin-uously throughout the trip.
- EGR. This test requires a series of idle speed operation, acceleration and deceleration to satisfy conditions needed for completion. It is performed once per trip.
- HO2S. This test requires a steady speed for about 20 seconds at speeds between 20 and 45 mph after warmup to complete. This test is also performed once per trip.

All OBD II scan tools include a readiness function showing all the monitoring sequences on the vehicle and the status of each: complete or incomplete. If vehicle travel time, operating con-

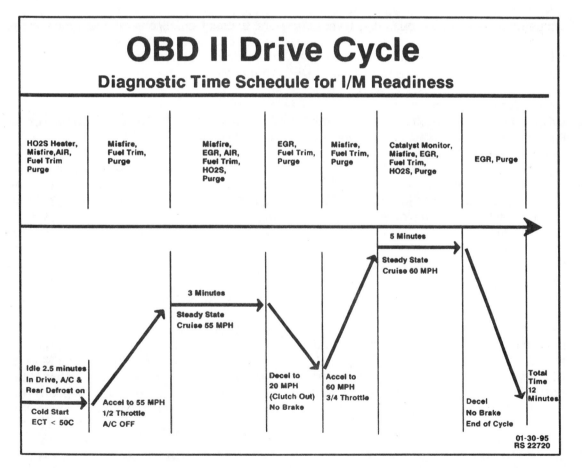

Figure 18-5 The car must be driven according to this schedule in order to complete the self-tests shown. *(Courtesy of General Motors Corporation, Service Technologies Group.).*

ditions or other parameters were insufficient for a monitoring sequence to complete the test, the scan tool will indicate which monitoring sequence has not completed the test.

OBD II Drive Cycle. This specific set of driving conditions will satisfy all requirements for all OBD II monitoring sequences (Figure 18-5). The driving instructions are these:

1. Start the engine. Do not turn the engine off for the remainder of the drive cycle.
2. Drive the vehicle in any convenient manner for at least four minutes. Any drive mode, including idle, is acceptable for step 2 and step 3.
3. Continue driving until the engine coolant temperature reaches an operating level of 180 degrees Fahrenheit or greater.
4. Idle the engine for 45 seconds.
5. Use approximately 1/4 open throttle to accelerate from a standstill to 45 mph over a period of about 10 seconds.
6. Drive between 30 and 40 mph at a steady throttle position for at least one minute.
7. Drive for at least four minutes at speeds between 20 and 45 mph. If traffic conditions cause slowing below 20 mph, extend the time by an equal amount to insure the four minutes total. Do not employ a WOT throttle at any time.
8. Decelerate and idle for at least 10 seconds.
9. Accelerate to 55 mph at 1/2 throttle (the elapsed time should be about 10 seconds).
10. Cruise between 40 and 65 mph, maintaining a steady throttle position for at least 80 seconds.
11. Bring the vehicle down to idle speed standstill.
12. Check with the scan tool for on-board diagnostic system readiness test results and any stored DTCs.

Similar Conditions. Once the malfunction indicator light has been turned on for misfire or a fuel system fault, the vehicle must go through three consecutive drive cycles including operating conditions similar to those existing at the time the fault was first detected to turn the malfunction indicator light off. Similar conditions mean:

- engine speed within 375 rpm of the DTC flagged condition.
- engine load within 10% of the same.
- engine temperature the same (either cold or warmed-up).

Note that with OBD II it is no longer necessary for a circuit to fail twice (or three times for a catalytic converter fault) to set a DTC and turn on the MIL. Notice also that achieving operating conditions similar to those when the DTC was first set could take some time, particularly if they are unusual conditions. If the problem initially appeared only at WOT and high rpm, these conditions may not be met again until the next time Granny borrows the car.

SETTING DTCS AND TURNING ON THE MIL

When an emissions-related fault is detected for the first time, a DTC relative to that fault is stored as a pending or intermittent code. Different manufacturers describe this preliminary code differently. Ford calls it a "pending" code; Chrysler calls it a "maturing" code; and General Motors calls it a "failed last time" code. Under whatever name, during the next drive cycle the pending fault will be erased if the monitoring sequence that first detected the fault is repeated and the same fault does not recur. If the fault does recur on the second drive cycle, then the DTC is stored (now as a "history" code in General Motors language), and the computer turns on the MIL.

There are two current exceptions to this arrangement. In the first, the misfire monitor can store a DTC and start flashing the MIL in response to its first detection of a type A misfire (a type A misfire could overheat and damage the three-way catalytic converter). In the second exception, the catalyst monitoring sequence must detect a fault in three OBD II drive cycles before storing a DTC and turning the MIL on.

Different faults must fit different criteria for turning the MIL on or off. If the MIL was turned on by a monitoring sequence for the heated oxygen sensor, the EGR or the total system, it will turn off after three consecutive OBD II trips during which the problem does not recur under the expected circumstances. If a misfire or fuel mixture monitoring sequence detected the fault, it will be turned off after three repeated OBD II trips with the right conditions met but the problem not appearing. Similarly, a MIL illumination that resulted from a fault detected in the catalytic converter effectiveness monitoring sequence will also be automatically turned off after three OBD II trips with the proper conditions met but no recurrence of the fault. Turning the MIL off, however, does not automatically erase the stored DTC.

Erasing a DTC requires 40 warm-up cycles without the problem recurring. These 40 cycles begin only after the MIL is turned off. DTCs can also be erased by a technician using the scan tool.

DIAGNOSTIC MANAGEMENT SOFTWARE

Each PCM includes diagnostic management software to organize the complex testing procedures. The terms used for this diagnostic management software varies by manufacturer. Ford calls theirs the "diagnostic executive," while Chrysler calls the same thing the "task manager."

Each monitoring sequence performs its tests under a unique set of operating conditions, involving specific temperature, engine speed, load, throttle position and time duration from start-up. These conditions must be met for the test to be completed. The diagnostic management software determines the sequence in which the tests will be run, whether the proper conditions have been met for each test and whether the duration of the test was long enough. If not, the test is aborted, and that trip is not counted relative to that test. The managing software waits for the next opportunity to run the appropriate monitoring sequence.

For an example, let's say an EGR monitoring sequence has detected a fault on the previous trip, as shown in Figure 18-5. A pending DTC has been set, and the diagnostic management software is waiting for the next trip to confirm the fault (and store the DTC and light the MIL) or to confirm correct functioning of the system and erase the pending DTC. During the next period of vehicle operation, the software waits for engine data indicating the correct temperature and idle speed, with over four minutes since start-up. If these conditions have been met, it begins the EGR monitoring sequence, testing the EGR system. Suppose there has been enough idle time and the correct amount of acceleration, followed by a steady throttle position at 34 mph for well over a minute. Then let's assume five minutes of driving at speeds from 23 to 44 mph, but no WOT operation, followed by unbraked deceleration and idle for 15 seconds. Then the ignition is turned off.

But a trip (the minimum for an EGR monitoring sequence test) has not occurred yet because there was no 1/2 throttle acceleration to 55 mph. The EGR system has neither passed nor failed the second test; the diagnostic management software is still waiting for the next trip so the test can be run again complete. This sequence was enough of a drive cycle, however, to enable the misfire and fuel system monitoring sequences to complete their tests.

Sometimes the diagnostic management software may cancel a test because of an additional problem with significant connections to the test circuit. For example, if the computer already knows the oxygen sensor is not working, it will not conduct a monitoring sequence test of the catalytic converter because the results would be inconclusive. That test will be postponed until the oxygen sensor circuit problem is corrected and the correction is recognized by the computer.

Sometimes the computer will not run a given test because it conflicts with some other test currently underway. For example, if the computer is running the EGR monitoring sequence, it will not run the catalytic converter monitoring sequence at the same time. When the EGR test is underway, the variations in EGR gas temporarily cause conditions in the catalytic converter that are not representative of normal converter operation.

Sometimes the diagnostic management software will run a test, but suspend it until another monitoring sequence has been completed and its results have been evaluated. For another example, the catalytic converter monitoring sequence might be delayed until the oxygen sensor test sequence has been successfully completed.

Freeze Frame Data. Besides storing detected DTCs, the diagnostic management software keeps a running track record of all the relevant engine parameters for a given circuit. If a fault is detected and recorded, that information is stored as a "snapshot." This data is used by the diagnostic management software for comparison and identification of similar operating conditions when they recur. The data is also available to the diagnostic technician for further information about what might be amiss in the system. This information can be accessed with the scan tool. Freeze frame information typically includes:

- the DTC involved
- engine rpm
- engine load
- fuel trim (short and long term)
- engine coolant temperature
- MAP and/or MAF values
- operating mode (open or closed loop)
- vehicle speed.

The freeze frame data storage capacity for OBD II is only required to store one freeze frame for a DTC, though some manufacturers have chosen to install greater capacity (five in GM's case). On the basic system, freeze frame data is stored only for the one that occurred first, unless the later DTC is a misfire or fuel system fault. In that case, the diagnostic management software replaces the stored data from the lower priority DTC with the freeze frame data relative to the misfire or fuel system DTC.

MONITORING SEQUENCES

As mentioned above, a monitoring sequence (sometimes called simply a "monitor") is an operating strategy the computer uses to check the operation of a specific circuit, system, function or component. Some of the specific monitoring sequences commonly used are described in this section.

Catalyst Efficiency Monitoring Sequence

As the engine's fuel system cycles from slightly lean to slightly rich in response to the oxygen sensor signal, an effective three-way catalytic converter (TWC) stores some oxygen during the lean cycles and uses that oxygen to oxidize (burn) the excess hydrocarbons during rich periods. (Of course, an air injection system, if present, supplies additional oxygen.) Because of this phenomenon, the percent of oxygen in exhaust discharged from the converter tends to be almost constant despite variations in the fluctuating amount entering. An effective converter will even this out, but one that has begun to lose its capacity to store oxygen will not be able to do so; therefore its downstream oxygen signal will begin to fluctuate with the oxygen content entering, just as the upstream sensor does (Figure 18-6).

Catalytic converter effectiveness is tested once per OBD II drive cycle, and this test is conducted by comparing the readings of the downstream (rear), secondary oxygen sensor with the signals from those in front of the converter. The downstream, secondary oxygen sensor is sometimes called the "catalyst monitor sensor" (CMS). As the PCM commands the fuel mixture to be rich and lean across the stoichiometric notch, the upstream oxygen sensors should produce a rising and falling voltage signal in response to the residual oxygen after the combustion chambers.

The downstream sensor (CMS) should produce a signal of much lower frequency and amplitude, as shown in Figure 18-6.

To prevent crack damage to the ceramic element of the CMS, some manufacturers delay turning on the heater for a period of time, usually until the engine coolant temperature sensor indicates a warmed-up engine. This allows the water condensation in the exhaust system to evaporate. To prevent mixups, the downstream sensor

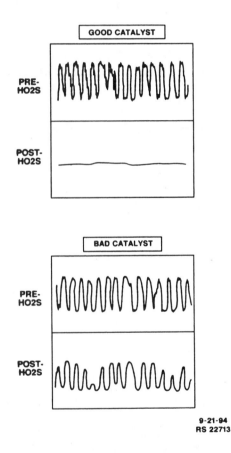

Figure 18-6 There will be a great difference between the voltage signals of the two oxygen sensors if the catalytic converter is operating efficiently, but the signal from the rear O2 sensor will start to fluctuate along with the front sensor if the catalyst is going bad. *(Courtesy of General Motors Corporation, Service Technologies Group.).*

has a different harness connector than the upstream sensor, though operation of each sensor is identical.

Misfire Monitoring Sequence

Any time a cylinder misfires, the raw fuel and air are pumped from that cylinder into the exhaust and through the catalytic converter. With this much oxygen and fuel dumped into and burned in the catalytic converter, it gets very hot. The ceramic or aluminum-oxide honeycomb begins to melt into a solid mass, ceasing emissions functions and plugging the exhaust with a molten lump. Emissions performance and driveability both suffer dramatically.

Misfires are therefore monitored continually by measuring the contribution of each cylinder to engine speed. This is referred to as **misfire detection.** As the engine runs, the crankshaft speed is not actually constant, but changes slightly as each cylinder delivers torque during its power stroke. In between power strokes, the crankshaft actually slows down by a few rpm, only to accelerate back up with the next cylinder's power stroke. By using a very high data rate crankshaft position sensor, manufacturers can provide the computer with a means to track these slight variations in engine rpm. When it does not detect the appropriate acceleration for a given cylinder (identified by the crankshaft position sensor, compared in some cases to data from the camshaft position sensor), the computer knows this cylinder has misfired.

Even on engines that are running perfectly, combustion is not perfect, and there are occasional misfires. Most OBD II systems allow a random misfire rate of about 2% before a misfire is flagged as a fault.

The OBD II misfire monitoring sequence includes an adaptive feature compensating for variations in engine characteristics caused by manufacturing tolerances and component wear. It also has the adaptive capacity to allow for vibration at different engine speeds and loads. When an individual cylinder's contribution to engine speed falls below a certain threshold, however, the misfire monitoring sequence calculates the vibration, tolerance and load factors before setting a misfire DTC.

There are two different misfire thresholds: type A misfires and type B/C misfires (Figure 18-7).

Type A Misfire Monitoring Sequence. If during 200 revolutions of the crankshaft, the misfire monitoring sequence detects a misfire rate

that would cause the catalytic converter temperature to reach 1600 degrees Fahrenheit or above, the MIL will begin to flash, and the system defaults to open loop to prevent the fuel control system from commanding a greater pulse width (because of the extra oxygen in the exhaust stream from the misfired cylinder). Once the engine is out of the operating range where the high misfire occurs, the MIL stops flashing, but

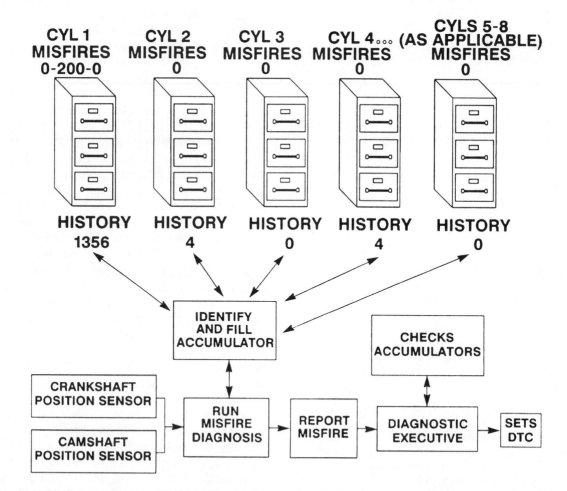

Figure 18-7 Misfire counters are similar to files kept on each cylinder. Current and historical misfire counters are maintained, and the Diagnostic Executive reviews this information before setting a DTC. *(Courtesy of General Motors Corporation, Service Technologies Group.).*

stays on constantly. A DTC is immediately set. Many manufacturers whose products use sequenced fuel injection have programmed the PCM to turn off one or two injectors in the faulty cylinders, to keep from pumping fuel into the exhaust. As a safety exception, if the engine is under load, as when passing or climbing a hill, the PCM does not deny fuel to the misfiring cylinder or cylinders when that capacity is available.

Type B/C Misfire Monitoring Sequence. If during 1,000 revolutions of the crankshaft the misfire monitoring sequence detects a misfire rate of 2 to 3 percent, it sets a pending DTC. At the same time, all the operating conditions at the time are recorded as a freeze frame. If the same pattern is repeated on the next drive cycle, the diagnostic management software will turn the MIL on.

Fuel System Monitoring Sequence

Though this is one of the highest-priority monitoring sequences, it is also one of the simplest. Whenever the system is running in closed loop, the fuel system monitoring sequence continuously watches short-term fuel trim and long-term fuel trim (as GM's former integrator and block learn are now called).

If a problem such as a vacuum leak, air flow restriction or incorrect fuel pressure causes the adaptive fuel control to make changes exceeding a predetermined limit in the short- or long-term fuel trim, the fuel system monitoring sequence reports a failure, and a pending DTC is set. On the next drive cycle, if the failure does not reappear, the pending code is erased; if it does appear again, a DTC is set and the MIL is turned on.

Oxygen Sensor Monitoring Sequence

The monitoring sequence tests upstream and downstream oxygen sensors separately, testing each once per drive cycle. Once the diagnostic management software identifies the correct engine operating conditions, the computer pulses the injectors at a fixed duty-cycle rate. The oxygen sensor monitoring sequence checks the frequency of the oxygen sensor signal to see that it produces a signal corresponding to the cycle rate of the injectors. The frequency is high enough that a slow-responding oxygen sensor will not be able to keep up and will exhibit a reduced amplitude signal as well. To pass the oxygen sensor monitoring sequence, the oxygen sensor must generate a voltage output greater than 0.67 volts, switch across 0.45 volts a minimum number of times during a 120 second period, and demonstrate a rapid voltage rise and fall.

Because of the way the catalytic converter works, the rear oxygen sensor, sometimes called the catalyst monitor sensor (CMS), should normally produce a low amplitude and fairly low frequency signal. During CMS testing, the computer forces the air/fuel ratio from rich to lean to force the CMS to produce higher amplitudes. If a rich air/fuel condition is momentarily sustained going into the combustion chambers, available stored oxygen in the catalytic converter is consumed, and the converter will contain less oxygen. In consequence, the CMS signal should go higher in voltage. If a lean condition is then momentarily sustained, the converter gets saturated with oxygen and the CMS signal should drop in voltage.

Certain automotive manufacturers use gold-plated pins and sockets for the oxygen sensor harness connector to obtain increased reliability, to meet the extended period of emissions warranty required by the law.

EGR System Monitor

Different manufacturers use different methods to obtain EGR system feedback, to confirm that the system is working under engine conditions in which it is needed. One of the simplest methods is the one used by Chrysler.

In normal operation, when the EGR valve opens, the pulse width to the injectors is somewhat reduced, to compensate for the oxygen displaced by the inert exhaust gas. Chrysler's strategy is to select an operating condition that meets the test criteria, including the criterion that the EGR valve should open. Then, without any other change in the pulse width command to the injectors, the EGR valve is disabled and closed. This should cause the air/fuel ratio to go lean, since there is in effect more oxygen in the mixture. Monitoring the oxygen sensor feedback signal will thus indicate the proper (or improper) function of the EGR valve.

Comprehensive Component Monitoring Sequence (CCM)

Remaining inputs and outputs affecting emissions may not be individually tested by a monitoring sequence. They are instead checked by the CCM. In many cases, monitoring these components is done in the same way it was on earlier OBD I systems (Figure 18-8).

Analog inputs are checked for open circuits, shorts and information signals that are out of range by monitoring the analog to digital converter input voltages. Typical examples are:

- intake air temperature (IAT)
- engine coolant temperature (ECT)
- throttle position (TP)
- manual level position (MLP)
- mass air flow (MAF) and manifold absolute pressure (MAP).

Digital and frequency input signals are checked by plausibility. This is done by using other sensor values and calculations to see whether a given sensor's reading is approximately what would be expected for the existing conditions. For example, the diagnostic management software compares the CKP signal to the CMP signal. Some examples of these interrelated signals are:

- crankshaft position (CKP)
- camshaft position (CMP)
- ignition diagnostic monitor (IDM)
- vehicle speed (VSS)
- output shaft speed (OSS).

Some inputs are constantly monitored, others can only be checked when they are actuated

so a resulting change in conditions can be observed. All the PCM outputs are tested while the vehicle is operating. Some are checked continuously; others require system actuation to observe changes of state.

A defective idle air control (IAC) system can cause either incorrect idle speed or a rough idle, either of which can also produce increased or fluctuating emissions performance. The IAC is a

California Air Resources Board (CARB) OBD II Comprehensive Component Monitoring List of Components Intended to Illuminate MIL

Important: Not all vehicles have these components.

Components Intended to Illuminate MIL
Transmission Range (TR) Mode Pressure Switch
Transmission Turbine Speed Sensor (HI/LO)
Transmission Vehicle Speed Sensor (HI/LO)
Transmission Vehicle Speed Sensor (HI/LO)
Ignition Sensor (Cam Sync, Diag)
Ignition Sensor Hi Res (7x)
Knock Sensor (KS)
Engine Coolant Temperature (ECT) Sensor
Intake Air Temperature (IAT) Sensor
Throttle Position (TP) Sensor A, B
Manifold Absolute Pressure (MAP) Sensor
Mass Air Flow (MAF) Sensor
Automatic Transmission Temperature Sensor
Transmission Torque Converter Clutch (TCC) Control Solenoid
Transmission TCC Enable Solenoid
Transmission Shift Solenoid A
Transmission Shift Solenoid B
Transmission 3/2 Shift Solenoid
Ignition Control (IC) System
Idle Air Control (IAC) Coil
Evaporative Emission Purge Vacuum Switch
Evaporative Emission Canister Purge (EVAP Canister Purge)

Figure 18-8 By CARB (California Air Resources Board) regulations, Comprehensive Component Monitoring will illuminate the MIL if there is a failure in any of these components. *(Courtesy of General Motors Corporation, Service Technologies Group.).*

closed-loop feedback to the IAC solenoid if the idle speed reported to the computer by the crankshaft position sensor (CKP) is not correct. The comprehensive component monitoring sequence checks to see whether the corrections made exceed predetermined limits.

Other outputs and actuators are checked for open and short circuits by monitoring the voltage in the actuator's driver circuit. The actuators are energized in almost every case by switching the driver circuit to ground, which reduces the voltage to nearly zero. Whenever the actuator is not energized, the voltage in the circuit should be at the charging system voltage. The following are actuators typically monitored by OBD II systems:

- idle air control coil
- EVAP canister purge vacuum switch
- fan control (high speed)
- heated oxygen sensor heater (HO2S)
- catalytic converter monitoring oxygen sensor (CMS)
- wide-open throttle air conditioning cutout (WAC)
- electronic pressure control solenoid (EPC)
- shift solenoid 1 (SS1)
- shift solenoid 2 (SS2)
- torque converter clutch (TCC)

Faults in the last four of these actuator circuits will most frequently result in the computer's turning on the transmission control indicator light (TCIL) rather than the MIL, if there is such a light on the vehicle.

Control of the two oxygen sensor circuits, the primary upstream mixture control sensor and the downstream catalytic converter monitoring sensor, is different for different manufacturers. Ford, for example, uses the PCM to actuate the heater circuits. In that case it is easy to monitor whether the circuit is working by monitoring the circuit voltage.

General Motors handles the testing differently. In these vehicles, the PCM feeds a bias voltage of approximately 0.45 volts to the HO2S signal terminal. When the oxygen sensor reaches operating temperature and starts to generate a signal, the bias voltage is turned off, and the system works normally. When the ignition is turned on, it feeds battery voltage to the front oxygen sensor heater circuit, which has a fixed ground. After a cold start the PCM measures how long it takes for the forward oxygen sensor to start generating signals. The sensor, of course, reaches operating temperature faster with the heater circuit energized. If the PCM determines from its memory that it took too long to start generating mixture feedback signals, a pending DTC is set. The amount of time allowed for the sensor to start producing signals depends on the ECT and IAT temperature signals.

Chrysler uses yet another strategy to test the feedback control front oxygen sensor heater circuit. The heater is powered directly by the automatic shutdown relay (ASD), controlled by the PCM. After the ignition is shut off, the PCM uses the battery temperature sensor to sense ambient temperature. It then waits for a specific time, based on the ambient temperature, for the oxygen sensor to cool down long enough to stop generating any signal. After that time, the PCM energizes the ASD relay. If the heater brings the oxygen sensor back to operating temperature, it resumes generating a signal, even though the engine is not running. If the sensor does not produce a signal within a predetermined amount of time, a pending DTC is set.

Evaporative Fuel System Integrity

To further guard against the possibility of volatile hydrocarbons leaking from the evaporative emissions system, OBD II requirements call for a detection system that can detect a leak equal to a 0.040-in. opening in the system (Figure 18-9). While manufacturers use a variety of means to check this, the most prevalent system seems to be to equip the system with a vacuum solenoid that can, as the test is run between ignition on and engine cranking, use the MAP sensor to check for residual vapor pressure caused by the fuel vapors. Note that such a sys-

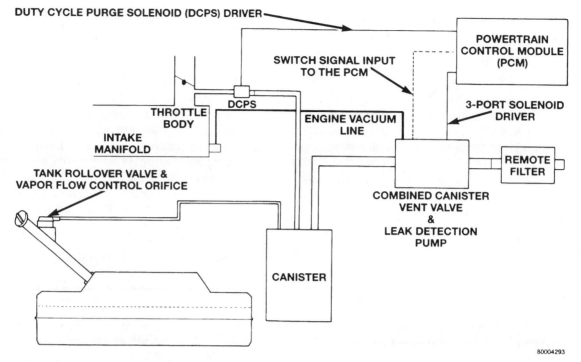

DUTY CYCLE PURGE SOLENOID (DCPS) DRIVER

SWITCH SIGNAL INPUT TO THE PCM

POWERTRAIN CONTROL MODULE (PCM)

DCPS

THROTTLE BODY

INTAKE MANIFOLD

ENGINE VACUUM LINE

3-PORT SOLENOID DRIVER

TANK ROLLOVER VALVE & VAPOR FLOW CONTROL ORIFICE

REMOTE FILTER

COMBINED CANISTER VENT VALVE & LEAK DETECTION PUMP

CANISTER

80004293

Figure 18-9 A complex network of components and monitor sequence is required to assure the integrity of the evaporative system. *(Courtesy of Chrysler Corporation.)*

tem, and other proposed systems, will be thrown off if the fuel tank cap is loose or leaks pressure when it is latched. A DTC is set in the usual way.

Secondary Air (AIR) Monitoring Sequence

Manufacturers are permitted to use a variety of methods to monitor the air injection system, provided it detects failures of the designated 1.5 times increase in emissions. The most common means is to use the upstream or downstream oxygen sensor to check on the amount of oxygen in the exhaust while commanding the air into the different parts of the exhaust system.

While most vehicles built to the OBD II standards in 1996 do not employ an air injection system, those that do will be able to monitor its function with the downstream oxygen sensor.

Chlorofluorocarbon (CFC) Monitoring

While the OBD II standards required the monitoring of CFCs, other federal laws prohibit the production of the product, so virtually no vehicles will have that type of refrigerant. Without the CFCs, there is no requirement to monitor them.

DOMESTIC CAR MAKER CODES

Ever since computerized engine management systems became popular in the early 1980's, the domestic carmakers have been using their own on-board diagnostics, which included proprietary fault or trouble codes. These can be made to flash out on the MIL, then one can look the code number up on a chart for the cause of the problem. Since the 1980's, with a proper scan tool, you have been able to access the data stream of both GM and Chrysler Corp. vehicles (Ford started making this information available to the technician much later).

With the advent of OBD II, DTCs have become standardized, but there are also make-specific codes (designated by a "1" after the first letter; generic codes have a "0" here).

It is still possible, in some cases, to access the old MIL codes as well. On a Chrysler product, for instance, you turn the ignition key on-off-on-off-on within a span of five seconds, then count the flashes of the MIL. A code 14, for example, tells you that the MAP sensor voltage is too low, which corresponds to generic/universal OBD II code P0107.

See the table (Figure 18-18) at the end of this chapter for Chrysler DTCs and what they mean, both those available through a generic scan tool and those that can be flashed out on the MIL.

EUROPEAN APPROACH

Robert Bosch is the major European authority on the subject of computerized engine management systems, and its most highly-evolved systems are in the Motronic family, used by Mercedes-Benz, Volkswagen, Volvo, and others. Motronic has always meant a combination of fuel and spark management since it was first introduced in 1979. Now it has additional functions and numerical series designations like software, Motronic 4.3 is upgrading to 4.4 as of this writing.

The way the European carmakers adapt Motronic to satisfy OBD II regulations is interesting. In Mercedes-Benz C-Class cars, for instance, the Motronic control module (M-B calls the system HFM-SFI for Hot Film engine Management-Sequential Fuel Injection) is next to a proprietary diagnostic connector that is to be used with the company's own scan tool, which is used by dealership technicians. OBD II regulations are satisfied by a separate dedicated diagnostic module con-nected to Motronic through a CAN (Controller Area Network) bus, as shown in Figure 18-10. This fulfills the legal requirements and sends data to a universal/generic DLC (Data Link Connector) under the dash. The OBD II module's only outputs are to the canister purge switchover valve and the MIL. The CAN bus also allows information to flow to and from the electronic accelerator or cruise/idle speed control module.

In Motronic, just as in every other computerized engine management system regardless of country of origin, there is a comprehensive list of sensors, or "operating-data acquisition" devices, including a chassis accelerometer to differentiate potholes from misfiring. This is different from the approach of other makers, such as GM, which uses signals from the ABS wheel speed sensors to indicate rough road surfaces.

The plausibility check of the air mass meter is a good example of the self-tests. The computer makes continuous calculations from throttle angle and rpm to arrive at a probable injection duration (similar to the operation of a speed-density EFI system). Then, it compares this value to that being obtained from the MAF sensor, and stores a DTC (Diagnostic Trouble Code) if there is a large enough discrepancy.

A complex test procedure is required for the evaporative emissions system. The following is Volvo's means of performing it:

1. The canister valve is closed and the EVAP valve is inhibited, which should mean the tank is sealed and the pressure inside is stable. If pressure falls, a DTC for the EVAP valve is stored.

2. The canister valve is opened, venting the system. The EVAP valve starts to cycle and fresh air is drawn through the canister. Pressure in the tank should begin to fall slowly. A rapid drop will set a DTC for the canister shut-off valve.

3. With the canister valve closed, the EVAP continues to cycle, which should make tank pressure fall quickly. If not, the DTC will be

**HARD WIRED INPUT SIGNALS
FROM HFM TO OBD 2 :**

- CRANK SHAFT POSITION
- CAMSHAFT POSITION
- BEFORE CATALYST O2 SENSOR
- AFTER CATALYST O2 SENSOR

FAULTS DETECTED BY OBD 2:

- MANIFOLD ABSOLUTE PRESSURE SENSOR
- BOTH O2 SENSORS
- MISFIRES
- CATALYST EFFICIENCY
- PURGE SWITCH OVER VALVE
- CAN
- BATTERY VOLTAGE

**FAULTS DETECTED BY OBD 2 VIA
LOGIC CHAINS:**

- CAM ADVANCE UNIT MECHANICAL
- EGR AND SWITCH OVER VALVE MECHANICAL
- AIR INJECTION MECHANICAL
- PURGE SWITCH OVER VALVE MECHANICAL
- TRANSMISSION UPSHIFT DELAY SYSTEM

HARD WIRED INPUTS TO OBD 2 :

- MANIFOLD ABSOLUTE PRESSURE SENSOR

OUTPUTS OF OBD 2 :

- PURGE SWITCHOVER VALVE
- CHECK ENGINE LAMP

NS9/1

OBD II

CAN

CAN

CAN

HFM

E GAS

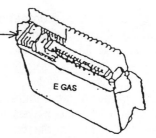

**FAULTS FROM HFM
SENT TO OBD 2
VIA CAN:**

- AIR MASS
- COOLANT TEMP
- O2 SENSOR HEATER
- FUEL TRIM
- FUEL INJECTOR
- KNOCK SENSORS
- CRANK POSITION SENSOR
- CAM POSITION SENSOR
- CAM ADVANCE UNIT ELECTRICAL
- EGR SWITCH OVER VALVE ELECTRICAL
- AIR INJECTION RELAY ENERGIZING SIDE
- PURGE VALVE ELECTRICAL
- VEHICLE SPEED SIGNAL
- CLOSED THROTTLE POSITION
- TRANS UPSHIFT SWITCH OVER VALVE ELECTRICAL
- INTAKE RESONANCE VALVE ELECTRICAL
- FULL LOAD CONTACT

**FAULTS FROM ELECTRONIC
ACCELERATOR TO OBD 2 :**

IDLE MALFUNCTION

Figure 18-10 In this Mercedes Benz HFM-SFI/Bosch Motronic computerized engine management system, a separate OBD II module is used to meet regulations. The CAN is a Controller Area Network bus. *(Courtesy of Mercedes-Benz, North America, Inc.)*

for substantial leakage.

4. The EVAP valve is now closed, which should leave a vacuum in the tank. If this falls slowly, a small leak DTC will be recorded.

To speed up the completion of trips during diagnosis and service, use this driving schedule from Volvo:

- With the coolant below 77 degrees Fahrenheit. and the A/C off, start the engine and put the transmission in gear.
- Accelerate gently to 1,500-2,000 rpm, then drive for five minutes at this speed.
- Idle the engine in gear for 70 seconds.
- Drive for six minutes at the above speed.
- Idle the engine in gear for 40 seconds
- Drive for five minutes at 1,500-2,000 rpm again.

ASIAN EXAMPLE: TOYOTA

The first car maker on the street with a fully OBD II-compliant engine control system was Toyota, as early as 1994. This was a considerable engineering accomplishment, especially in light of the fact that some other import carmakers had stated early on in discussions about the OBD II situation that they would have to use add-on modules to meet the standards (similar to the way Mercedes-Benz/Robert Bosch have handled it).

Toyota's OBD II strategies closely resemble those of our domestic carmakers (Figure 18-11). Just as with every other vehicle manufacturer, however, Toyota uses some proprietary diagnostics connected with its own DTCs, as shown in Figure 18-12.

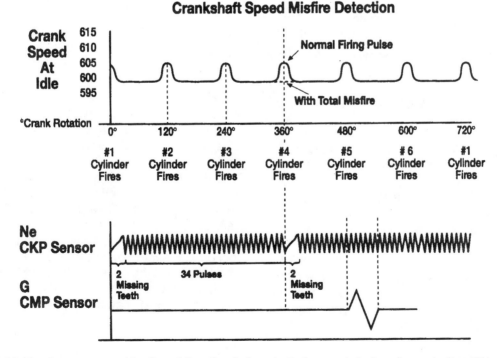

Figure 18-11 Toyota uses a combination of the signals from both the crankshaft and camshaft position sensors to detect the variations in engine rpm that may indicate misfire.

DTC CHART (Manufacturer Controlled)

DTC No. (See Page)	Detection Item	Trouble Area	MIL*	Memory
P1300 (EG-266)	Igniter Circuit Malfunction	• Open or short in IGF or IGT circuit from igniter to ECM • Igniter • ECM	○	○
P1335 (EG-269)	Crankshaft Position Sensor Circuit Malfunction (during engine running)	• Open or short in crankshaft position sensor circuit • Crankshaft position sensor • ECM	−	○
P1500 (EG-270)	Starter Signal Circuit Malfunction	• Open or short in starter signal circuit • Open or short in ignition switch or starter relay circuit • ECM	−	○
P1600 (EG-271)	ECM BATT Malfunction	• Open in back up power source circuit • ECM	○	○
P1780 (EG-273)	Park/Neutral Position Switch Malfunction	• Short in park/neutral position switch circuit • Park/neutral position switch • ECM	○	○

*:   − .... MIL does not light up
     ○ ... MIL lights up

Figure 18-12 This chart shows some of the DTCs that are specific to Toyota.

DIAGNOSTIC EQUIPMENT

Scan Tools. OBD II performs continuous tests to make sure everything that affects emissions is working properly, and provides universal DTCs to help you in troubleshooting.

But you still need diagnostic equipment. The most prominent example of this is the scan tool, a microprocessor-based hand-held test instrument that gives you the capability of accessing the engine management data stream (Figure 18-13). In other words, it allows you to tap into the computer system through a lead that plugs into the DLC (Data Link Connector). You can then carefully consider this information to see if it makes sense when compared to the symptoms that caused the car to be brought in for service.

OBD II regulations state that DTC's and a large number of engine management sensor signals, computer commands, etc. must be readable on a universal generic scan tool, and that all cars must carry a universal 16-pin DLC under the dash.

A scan tool will give you a great deal of important troubleshooting information, but many technicians do not make use of its full capabilities, using it only to pull codes.

That is unfortunate because the data a scan tool provides will in most cases be exactly what is needed to find the cause of the trouble. Here is a partial list of the information available on GM cars through a typical scan tool:

• DTCs, both current and historical
• Oxygen sensor signal in millivolts

- Rear oxygen sensor signal in millivolts
- Loop status
- Rich/lean flag
- Power enrichment on/off
- Oxygen sensor cross-counts

- Injector pulse width in milliseconds
- Fuel trim cell
- Fuel trim index
- Short-term fuel trim, and average
- Long-term fuel trim, and average

Data Cable — used for testing all vehicles.

Optional Communication Cable — used to connect the scanner to the Snap-on MT1670 *Scribe* printer or the MT1765 *Counselor XL* oscilloscope.

Quick ID Button Allows vehicle identification without connection to vehicle power

Controls — These three controls provide all scanner operation.

Thumbwheel

Yes (Y) button

No (N) button

LED Indicators

Monitor common on/off engine control functions. See cartridge label and LED menu in CUSTOM SETUP program for LED descriptions.

Vehicle Test Cartridges

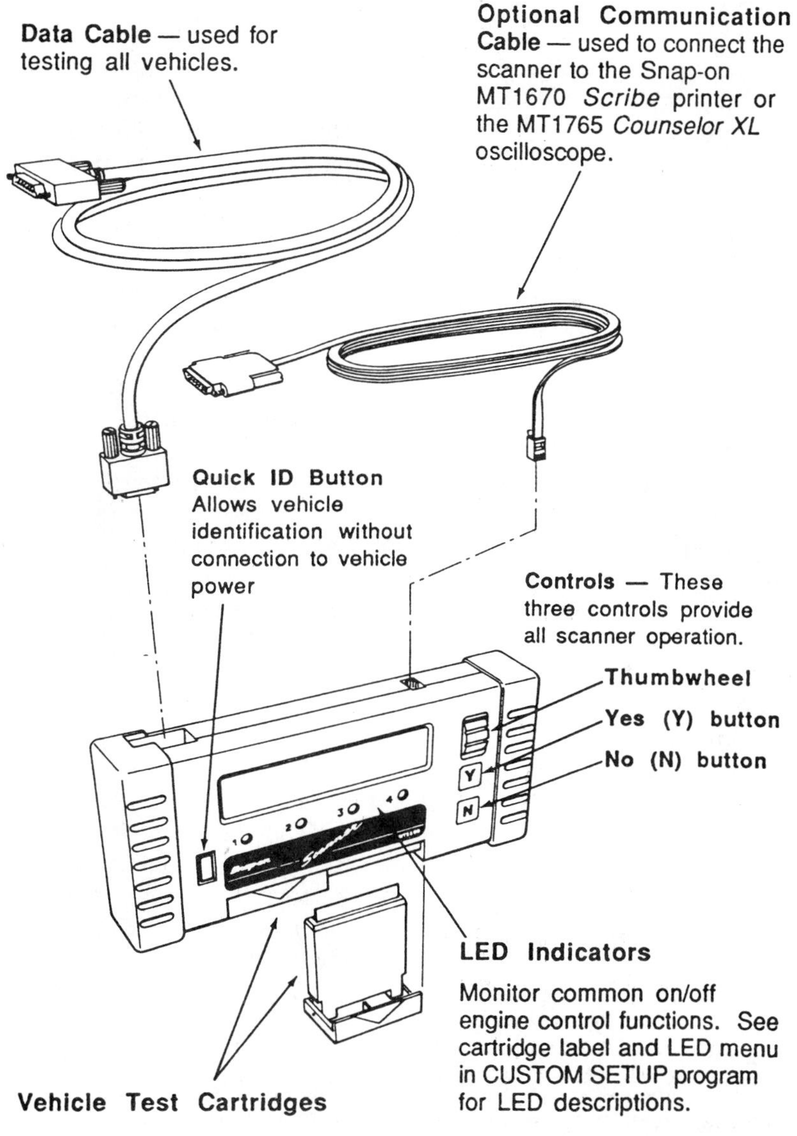

Figure 18-13 This aftermarket scan tool features plug-in modules for various makes and models. The data cable can accommodate both pre-OBD II and OBD II cars by means of adapters. *(Courtesy of Snap-on Tools Company, Copyright owner.)*

- Engine rpm
- Desired idle rpm
- Coolant temperature
- Intake air temperature
- MAP signal
- Barometric pressure signal
- Throttle position volts
- Throttle angle as a percentage
- Calculated air flow
- Low octane fuel spark modifier
- Spark advance in degrees
- Knock retard in degrees
- Knock signal (yes/no)
- Open/closed loop indication
- Converter high-temperature condition (yes/no)
- Air control solenoid (port/atmosphere)
- EGR desired position as a percentage
- EGR actual position as a percentage
- EGR pintle position in volts
- EGR duty cycle percentage
- EGR auto-zero (inactive/active)
- Idle air control position
- Wastegate position
- Park/neutral position
- PRNDL switch position
- Commanded gear
- Brake switch on/off
- Mph/kph
- TCC/shift light on/off
- 4th gear switch on/off
- A/C request (yes/no)
- A/C clutch on/off
- Battery voltage
- Fuel pump volts
- Intake tune valve on/off
- Purge duty cycle percentage
- Purge learn memory
- PROM ID
- Time from start

As mentioned above, this is the kind of information that will help you cure the majority of driveability/emissions problems, regardless of whether or not they have set a code.

Another extremely important feature is that you can read all that important data simply by plugging the scan tool into the DLC and pushing buttons or scrolling through menus. In fact, most technicians who have taken the time to learn the simple basics of scan tool operation typically use it before they even consider hooking up any other type of diagnostic equipment. It is usually the first tool deployed in driveability/emissions diagnosis.

A scan tool may also allow you to activate components normally controlled by the PCM. In Chrysler Corporation's Circuit Actuation Test Mode, for instance, the company's DRB (Diagnostic Read-out Box) scan tool (or after-market equivalent) can actuate:

- All ignition coils individually
- All fuel injectors individually
- Idle air control motor
- Radiator fan control module
- A/C clutch relay
- Auto shutdown relay
- Duty cycle EVAP purge solenoid
- S/C servo solenoids
- Generator field
- Torque converter clutch solenoid
- EGR solenoid
- Fuel system test
- Speed control vacuum solenoid
- Speed control vent solenoid
- Fuel pump relay
- All other solenoids/relays
- Set rpm in 100 rpm increments from 900 to 2,000

This test procedure can be very helpful because if a component functions as it is supposed to (you may hear or feel a click, see fuel spray, etc.), you can be fairly sure its wiring and driver circuit are okay.

Certain computer system functions are only accessible by the carmakers' own scan tools; for example, bi-directional control of the braking system and the ability to force transmission shifts. Forcing abnormal system operation can cause system damage or even uncontrolled braking. The manufacturers have taken the position that the ability to control these functions should be left to dealership personnel.

Of course, scan tools designed for use on 1996 and newer cars also fulfill the requirements for OBD II tests. Besides engine data such as listed above, there is a category called Specific Engine Data. In GM, for example, this includes:

- information specific to EGR and EVAP system diagnosis
- data required to verify that the EGR and EVAP systems are operating properly
- information specific to the diagnosis of misfire
- information specific to the HO2S 1 and HO2S 2 sensors
- data required to verify the proper operation of both oxygen sensors

Another category is DTC Data. This includes Freeze Frame data, which is information gathered at the moment a DTC is set. From this, the technician can re-create the conditions that were present at the time. Also in DTC Data are Failure Records, which contain the information present at the time a diagnostic test was failed, but this data is not necessarily associated with MIL activity.

An especially powerful troubleshooting combination is the use of a scan tool along with a four-gas (HC, CO, O_2 and CO_2) or five-gas (add NO_x) infrared exhaust analyzer. This will allow you to compare sensor signal or computer command information with actual tailpipe emissions to see if the logical result of these readings is indeed present.

Lab Scopes. No matter how well your scan tool works and how well you understand what it is telling you, there will be occasions when it simply will not let you "see" the problem. Perhaps the "glitch" occurs so quickly that the scan tool's display cannot show it, or maybe the OBD system simply is not programmed to recognize this discrepancy.

In these cases, you may find the lab scope (also called the Digital Storage Oscilloscope, or DSO, or simply the digital scope) to be very helpful. This device, whether console or hand-held, will never become obsolete because it reads the val-

ues that every electronic device, past, present, and future, must work with. The basic principles of electricity and electronics do not change even though car models and generations of computerized engine management systems do (Figure 18-14).

The productive use of the lab scope requires a considerable learning effort on the part of the technician. You should have a thorough knowledge and comfortable understanding of the principles of electronics in general, and engine management systems in particular. Given that, you can check individual components and circuits regardless of design or country of origin, and be quite confident in your findings. Take any TPS as a simple example, sweep it with your lab scope properly attached and you'll know for sure whether it is good or bad.

The live analog oscilloscope has been used for ignition troubleshooting for many decades. It is an entirely different type of device from the modern lab scope, and is simply not useful for CEC diagnostics.

Reduced to essentials, a lab scope is a visual voltmeter. The waveforms it displays would be impossible to see with any other type of equipment. A DMM (Digital Multi-Meter) or scan tool shows you values interpreted to fit the display

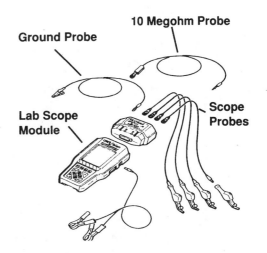

Figure 18-14 The modern hand-held lab scope is powerful, yet convenient to use. *(Courtesy of OTC, Div. of SPX Corp.)*

method, whereas a scope shows what is actually happening in the real world.

You can use a lab scope to check anything you might have been testing with a DMM, including sensor outputs (Figure 18-15), ground circuit voltage drops, battery voltage checks, etc. You will fix more cars with it than with a DMM because you will see the quality of the signal rather than just the quantity, and you will learn to recognize normal patterns.

The way a lab scope acquires and displays signals is what gives it its troubleshooting power, and also what distinguishes it from the traditional live analog ignition scope. Analog scopes need a repetitive signal, which they display in real time (in other words, at the actual moment when these electrical events occur).

Although a lab scope is a digital device, input signals are still analog. The method used by a digital scope to process and display such signals is known as sampling. Samples of the input signal are taken, then converted into digital values, which are stored in memory. These are then strung together as a trace on the screen. So, what a lab scope is showing you is not a live signal, but a representation of a signal that took place a fraction of a second before.

Since the sampling rate is typically millions of samples per second, every important detail of a signal or event is displayed (Figure 18-16). This speed will allow you to see and identify any momentary signal irregularities that might be causing trouble in the way the engine runs. You will also be able to observe the waveforms whenever you want, repeatedly, if necessary, because they are stored in memory.

A typical modern lab scope will have dual or multiple trace capabilities (Figure 18-16), which means you can input and view two or more separate signals at the same time.

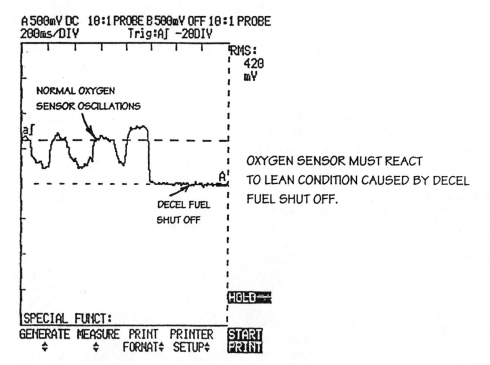

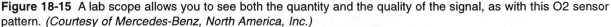

Figure 18-15 A lab scope allows you to see both the quantity and the quality of the signal, as with this O2 sensor pattern. *(Courtesy of Mercedes-Benz, North America, Inc.)*

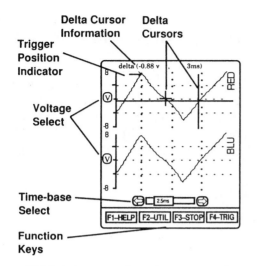

Figure 18-16 Lab scopes have two or more traces, so you can compare signals, that of an oxygen sensor to injector pulse width, for example. *(Courtesy of OTC, Div. of SPX Corp.)*

This will allow you to compare signals, and see how one affects the other. For example, you could put oxygen sensor voltage on Channel A and fuel injector pulse on Channel B, then watch to see if pulse width responds to changes in the O_2 signal.

So, think of the lab scope as a high-speed visual voltmeter (Figure 18-17) that provides a clear, graphic look at signal quality. With it, you will be able to catch momentary glitches, spikes, noise, and waveform shapes that are unusual for the component being tested.

SUMMARY

In this chapter we've covered the why's and the how's of OBD II, and shown that it will make diagnosis much more similar for all vehicles. We have seen how the program is standardized for

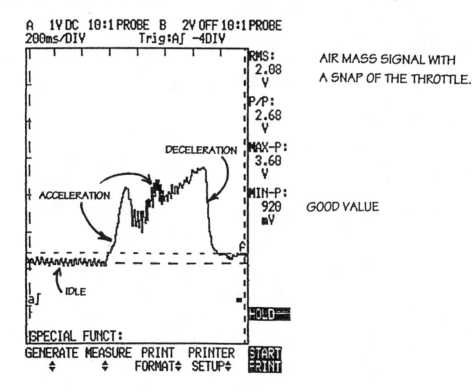

Figure 18-17 All the detail necessary to catch intermittent glitches and fast-occurring faults is present and controllable with a modern lab scope. *(Courtesy Mercedes-Benz, North America, Inc.*

all manufacturers and have reviewed how car-makers have similar emissions control strategies. Finally, we looked at the main tools for OBD II analysis: the scan tool and the oscilloscope.

▲ DIAGNOSTIC EXERCISE

1. An OBD II-equipped car has a burned

valve. How will the system prevent fuel from passing through the cylinder and overheating the catalytic converter?

2. What advantage comes from reading sensors' information and testing actuators through the OBD II connector rather than directly testing them on the car with the harness disconnected?

MIL CODE	GENERIC SCAN TOOL CODE	HEX CODE	DRB SCAN TOOL DISPLAY	DESCRIPTION OF DIAGNOSTIC TROUBLE CODE
11	P1391**	9D	Intermittent Loss of CMP or CKP	Intermittent loss of either camshaft or crankshaft position sensor
	or			
		28	No Crank Reference Signal at PCM	No crank reference signal detected during engine cranking.
	or			
	P1398**	BA	Misfire Adaptive Numerator at Limit	CKP sensor target windows have too much variation
12*			Battery Disconnect	Direct battery input to PCM was disconnected within the last 50 Key-on cycles.
13**	P1297	27	No Change in MAP From Start to Run	No difference recognized between the engine MAP reading and the barometric (atmospheric) pressure reading from start-up.
14**	P0107	24	MAP Sensor Voltage Too Low	MAP sensor input below minimum acceptable voltage.
	or			
	P0108	25	MAP Sensor Voltage Too High	MAP sensor input above maximum acceptable voltage.
	or			
15**	P0500	23	No Vehicle Speed Sensor Signal	No vehicle speed sensor signal detected during road load conditions.

Figure 18-18 A comparison of Chrysler two digit flash out codes versus OBD generic five digit codes. *(Courtesy of Chrysler Corp.)*

MIL CODE	GENERIC SCAN TOOL CODE	HEX CODE	DRB SCAN TOOL DISPLAY	DESCRIPTION OF DIAGNOSTIC TROUBLE CODE
17**	P0125	80	Closed Loop Temp Not Reached	Engine does not reach 50°F within 5 minutes with a vehicle speed signal.
	or			
17		21	Engine to Cold Too Long	Engine did not reach operating temperature within acceptable limits.
21**	P0131	9B	Upstream O2s Voltage Shorted to Ground	Tested after key off and at start to run.
	or			
	P0132	3E	Upstream O2 Sensor Shorted to Voltage	Upstream oxygen sensor input voltage maintained above the normal operating range.
	or			
	P0133	66	Upstream O2 Sensor Slow Response	Upstream oxygen sensor response slower than minimum required switching frequency or value does not go above .65 volts.
	or			
	P0134	20	Upstream O2 Sensor Stays at Center	Neither rich or lean condition detected from the upstream oxygen sensor input.
	or			
	P0135	67	Upstream O2 Sensor Heater Failure	Upstream oxygen sensor heating element circuit malfunction
	or			
	P0137	9C	Downstream O2s Voltage Shorted to Ground	Tested after key off and at start to run.
	or			
	P0138	7E	Downstream O2 Sensor Shorted to Voltage	Downstream oxygen sensor input voltage maintained above the normal operating range.
	or			
	P0140	81	Downstream O2 Sensor Stays at Center	Neither rich or lean condition detected from the downstream oxygen sensor.
	or			
	P0141	69	Downstream O2 Sensor Heater Failure	Downstream oxygen sensor heating element circuit malfunction
22**	P0117	1E	ECT Sensor Voltage Too Low	Engine coolant temperature sensor input below minimum acceptable voltage.
	or			
	P0118	1F	ECT Sensor Voltage Too High	Engine coolant temperature sensor input above maximum acceptable voltage.
23**	P0112	39	Intake Air Temp Sensor Voltage Low	Intake air temperature sensor input below the maximum acceptable voltage (2.4L only).
	or			
	P0113	3A	Intake Air Temp Sensor Voltage High	Intake air temperature sensor input above the minimum acceptable voltage (2.4L only).

Figure 18-18 A comparison of Chrysler two digit flash out codes versus OBD generic five digit codes, continued. *(Courtesy of Chrysler Corp.)*

MIL CODE	GENERIC SCAN TOOL CODE	HEX CODE	DRB SCAN TOOL DISPLAY	DESCRIPTION OF DIAGNOSTIC TROUBLE CODE
24**	P0121	84	TPS Voltage Does Not Agree With MAP	TPS signal does not correlate to MAP sensor
	or			
	P0122	1A	Throttle Position Sensor Voltage Low	Throttle position sensor input below the minimum acceptable voltage
	or			
	P0123	1B	Throttle Position Sensor Voltage High	Throttle position sensor input above the maximum acceptable voltage.
25**	P0505	19	Idle Air Control Motor Circuits	A shorted or open condition detected in one or more of the idle air control motor circuits.
	or			
	P1294	8A	Target Idle Not Reached	Actual idle speed does not equal target idle speed.
27**	P0201	15	Injector #1 Control Circuit	Injector #1 output driver does not respond properly to the control signal.
	or			
	P0202	14	Injector #2 Control Circuit	Injector #2 output driver does not respond properly to the control signal.
	or			
	P0203	13	Injector #3 Control Circuit	Injector #3 output driver does not respond properly to the control signal.
	or			
	P0204	3D	Injector #4 Control Circuit	Injector #4 output driver does not respond properly to the control signal.
	or			
	P0205	45	Injector #5 Control Circuit	Injector #5 output driver does not respond properly to the control signal.
	or			
	P0206	46	Injector #6 Control Circuit	Injector #6 output driver does not respond properly to the control signal.
31**	P0441	71	Evap Purge Flow Monitor Failure	Insufficient or excessive vapor flow detected during evaporative emission system operation.
	or			
	P0442	AO	Evap System Small Leak	A small leak has been detected by the leak detection monitor.
	or			
	P0443	12	EVAP Purge Solenoid Circuit	An open or shorted condition detected in the duty cycle purge solenoid circuit.
	or			
	P0455	A1	EVAP System Gross Leak	The leak detection monitor is unable to pressurize Evap system, indicating a large leak.
	or			

Figure 18-18 A comparison of Chrysler two digit flash out codes versus OBD generic five digit codes, continued. *(Courtesy of Chrysler Corp.)*

MIL CODE	GENERIC SCAN TOOL CODE	HEX CODE	DRB SCAN TOOL DISPLAY	DESCRIPTION OF DIAGNOSTIC TROUBLE CODE
	P1486	BB	EVAP System Obstruction	Plug or pinch detected between purge soleniod and fuel tank.
	or			
	P1494	B8	Leak Detect Pump Pressure Switch	Leak detection pump switch does not respond to input.
31**	P1495	B7	Leak Detection Pump Solenoid Circuit	Leak detection pump solenoid circuit fault (open or short).
32**	P0401	2E	EGR System Failure	Required change in air/fuel ratio not detected during diagnostic test.
	or			
	P0403	11	EGR Solenoid Circuit	An open or shorted condition detected in the EGR transducer solenoid circuit.
33*		10	A/C Clutch Relay Circuit	An open or shorted condition detected in the A/C clutch relay circuit.
	or			
		5A	A/C Pressure Sensor Sensor Volts Too High	Sensor input voltage is above 4.9 volts.
	or			
		5B	A/C Pressure Sensor Sensor Volts Too Low	Sensor input voltage is below .098 volts.
34*		0F	Speed Control Solenoid Circuits	An open or shorted condition detected in the Speed Control vacuum or vent solenoid circuits.
35**	P1491	0E	Rad Fan Control Relay Circuit	An open or shorted condition detected in the low speed radiator fan relay control circuit.
37**	P0740	94	Torque Converter Clutch No RPM Drop At Lockup	Relationship between engine speed and vehicle speed indicates no torque converter clutch engagement (2.4L w/31TH and 3.0L only).
	or			
	P0743	0C	Torque Converter Clutch Soleniod CKT	An open or shorted condition detected in the torque converter part throttle unlock solenoid control circuit (2.4L w/31TH and 3.0L only).
41***		0B	Generator Field Not Switching Properly	An open or shorted condition detected in the generator field control circuit.
42*		65	Fuel Pump Relay Control Circuit	An open or shorted condition detected in the fuel pump relay control circuit.
	or			
		0A	Auto Shutdown Relay Control Circuit	An open or shorted condition detected in the auto shutdown relay circuit.
	or			
		2C	No ASD Relay Output Voltage at PCM	An Open condition Detected In The ASD Relay Output Circuit.
	or			

Figure 18-18 A comparison of Chrysler two digit flash out codes versus OBD generic five digit codes, continued. *(Courtesy of Chrysler Corp.)*

MIL CODE	GENERIC SCAN TOOL CODE	HEX CODE	DRB SCAN TOOL DISPLAY	DESCRIPTION OF DIAGNOSTIC TROUBLE CODE
		95	Fuel Level Sending Unit Volts Too Low	Open circuit between BCM and fuel gauge sending unit.
	or			
		96	Fuel Level Sending Unit Volts Too High	Circuit shorted to voltage between BCM and fuel gauge sending unit.
	or			
		97	Fuel Level Unit No Change Over Miles	No movement of fuel level sender detected.
43**	P0300	6A	Multiple Cylinder Misfire	Misfire detected in multiple cylinders.
	or			
	P0301	6B	Cylinder #1 Misfire	Misfire detected in cylinder #1.
	or			
	P0302	6C	Cylinder #2 Misfire	Misfire detected in cylinder #2.
	or			
	P0303	6D	Cylinder #3 Misfire	Misfire detected in cylinder #3.
	or			
	P0304	6E	Cylinder #4 Misfire	Misfire detected in cylinder #4.
	or			
	P0305	AE	Cylinder #5 Misfire	Misfire detected in cylinder #5.
	or			
	P0306	AF	Cylinder #6 Misfire	Misfire detected in cylinder #6.
	or			
	P0351	2B	Ignition Coil #1 Primary Circuit	Peak primary circuit current not achieved with maximum dwell time.
	or			
	P0352	2A	Ignition Coil #2 Primary Circuit	Peak primary circuit current not achieved with maximum dwell time.
	or			
	P0353	29	Ignition Coil #3 Primary Circuit	Peak primary circuit current not achieved with maximum dwell time.
	or			
44**	P1492	9A	Battery Temp Sensor Voltage Too High	Battery temperature sensor input voltage above an acceptable range.
	or			
	P1493	99	Battery Temp Sensor Voltage Too Low	Battery temperature sensor input voltage below an acceptable range.
45**	P0700	89	EATX Controller DTC Present	An automatic transmission input DTC has been set in the transmission controller.
	or			
	P1899	72	Park/Neutral Switch Failure	Incorrect input state detected for the Park/Neutral switch.

Figure 18-18 A comparison of Chrysler two digit flash out codes versus OBD generic five digit codes, continued. *(Courtesy of Chrysler Corp.)*

MIL CODE	GENERIC SCAN TOOL CODE	HEX CODE	DRB SCAN TOOL DISPLAY	DESCRIPTION OF DIAGNOSTIC TROUBLE CODE
46***		06	Charging System Voltage Too High	Battery voltage sense input above target charging voltage during engine operation.
47***		05	Charging System Voltage Too Low	Battery voltage sense input below target charging during engine operation. Also, no significant change detected in battery voltage during active test of generator output circuit.
51**	P0171	77	Fuel System Lean	A lean air/fuel mixture has been indicated by an abnormally rich correction factor.
52**	P0172	76	Fuel System Rich	A rich air/fuel mixture has been indicated by an abnormally lean correction factor.
53**	P0601	02	Internal Controller Failure	PCM Internal fault condition detected.
	or			
	P0600	44	PCM Failure SPI Communications	PCM Internal fault condition detected.
54**	P0340	01	No Cam Signal at PCM	No camshaft signal detected during engine cranking.
55*				Completion of fault code display on Check Engine lamp.
61**	P0106	3C	Barometric Pressure Out of Range Key On	MAP sensor has a baro reading below an acceptable value.
62**	P!697	30	PCM Failure SRI Mile Not Stored	Unsuccessful attempt to update EMR mileage in the PCM EEPROM.
63**	P1698	31	PCM Failure EEPROM Write Denied	Unsuccessful attempt to write to an EEPROM location by the PCM.
64**	P0420	70	Catalytic Converter Efficiency Failure	Catalyst efficiency below required level.
65**	P0703	98	Brake Switch Sense Circuit	No release of brake switch seen after too many accelerations or no brake switch activation seen on several DECELS to RESET (OMPH).
66		61	No CCD Messages From Body Control Module	No messages received from BCM.
	or			
66**	P1698	60	No CCD Messages From TCM	No messages received from Transmission Control Module.
71	P1496	92	5 volt supply output too low	5 volt output from regulator does not meet minimum requirement

Figure 18-18 A comparison of Chrysler two digit flash out codes versus OBD generic five digit codes, continued. *(Courtesy of Chrysler Corp.)*

REVIEW QUESTIONS

1. What are the advantages for the independent shop and independent technician of the OBD II program?

2. What vehicles are covered by the requirement that they have OBD II self-diagnostics?

3. True or False: OBD II represents a revolutionary change in the way computers control engines to maximize performance, economy and emissions.

4. What is meant by the "snapshot" feature of the OBD II self-diagnostics program?

5. How does the misfire detection feature of the OBD II program work? What is the purpose of detecting these misfires?

6. OBD II-compliant vehicles use either tandem oxygen sensors, one before and one after the catalytic converter; or on V-form engines, three or four sensors, one on each bank and one behind the catalytic converter(s). How do these additional oxygen sensors affect the way the engine runs?

7. Give two examples of how an OBD II-compliant computer can run tests of its sensors and actuators to see whether they work properly.

8. What should be done first with an OBD II-compliant vehicle in which the check engine light is on?

9. If one or more of the terminals of the DLC-3 connector are not connected to any wires, is this reason to suspect emissions tampering?

10. True or False: an OBD II-compliant vehicle with the check engine light on can be driven without damage to the vehicle as long as no driveability symptoms are noticeable.

ASE-type Questions (Actual ASE tests are rarely so product specific.)

11. Technician A says that the reason OBD II was adopted was that it will make driveability complaint diagnosis more uniform. Technician B says that the main reason OBD II was adopted was concern about automotive emissions and air pollution. Who is correct?
   a. A only.
   b. B only.
   c. both A and B.
   d. neither A nor B.

12. Technician A says that "freeze frame" is the ability of the system to record data from all sensors and actuators at the instant the MIL is turned on. Technician B says that GM's "failure records" does the same thing as "freeze frame," but adds any fault stored in memory, not just those related to emissions. Who is correct?
   a. A only.
   b. B only.
   c. both A and B.
   d. neither A nor B.

13. Technician A says a DTC that begins with P1 is for the powertrain and is specific to the car maker. Technician B says a DTC that begins with P1 is a standard, universal SAE code. Who is correct?
   a. A only.
   b. B only.
   c. both A and B.
   d. neither A nor B.

14. Technician A says that a specific drive cycle is required in order for all OBD II monitoring sequences to be performed. Technician B says that all OBD II tests are performed instantaneously when the engine is started. Who is correct?
   a. A only.
   b. B only.
   c. both A and B.
   d. neither A nor B.

15. Technician A says that in every case the first time an emissions-related fault is detected, a "pending," "maturing," or "failed last time" code is set, but that the MIL will not be turned on unless the same failure occurs during the next drive cycle. Technician B says the misfire monitor will turn on the MIL in response to its first detection of a type A misfire. Who is correct?

a. A only.
b. B only.
c. both A and B.
d. neither A nor B.

16. Technician A says that OBD II checks catalytic converter effectiveness by comparing the signals of the upstream and downstream oxygen sensors. Technician B says that if the downstream oxygen sensor signal begins to fluctuate in harmony with that of the upstream sensor, the catalyst is bad. Who is correct?
a. A only.
b. B only.
c. both A and B.
d. neither A nor B.

17. Technician A says that the Bosch Motronic computerized engine management system used on many European cars has OBD II capability built into the main PCM. Technician B says that European cars are not required to meet OBD II requirements because their engine management systems are so sophisticated. Who is correct?
a. A only.
b. B only.
c. both A and B.
d. neither A nor B.

18. Technician A says that OBD II is a step backwards because the information that has been available for years from each car maker's data stream with its own dedicated scan tool is more comprehensive. Technician B says that both OBD II and specific manufacturer diagnostic information co-exist. Who is correct?
a. A only.
b. B only.
c. both A and B.
d. neither A nor B.

19. Technician A says that from 1996 on all cars must have just one generic DLC. Technician B says that new cars may have a make-specific diagnostic connector along with the generic DLC. Who is correct?
a. A only.
b. B only.
c. both A and B.
d. neither A nor B.

20. Technician A says the digital lab scope is ideal for troubleshooting complex computerized engine management systems. Technician B says you can do the same thing with the traditional live analog ignition scope. Who is correct?
a. A only.
b. B only.
c. both A and B.
d. neither A nor B.

Glossary

A4LD transmission (automatic four-speed light-duty) – An overdrive transmission featuring an ECA-controlled torque converter clutch for rear-wheel drive vehicles.

AC – A parts manufacturing and distribution division of General Motors.

Accumulator – A reservoir located between the evaporator (under dash unit) and the compressor of an air conditioning system. It allows refrigerant and any remaining liquid to separate. Also used to describe a component in any hydraulic circuit (like an automatic transmission) that moderates the application of some clutch or band.

Actuator – Any mechanism, such as a solenoid, relay, or motor, that when activated causes a change in the performance of a given system or circuit.

A/D – A converter circuit that translates an analog signal into digital on/off code.

A/D Interface – Analog/digital interface, a connection between an analog sensor and the computer's digital system. On many such connections, the sensor includes a small microprocessor that makes the actual conversion.

Adaptive memory – The ability of an engine control computer to assess the success of its actuator and sensor signals, and modify its internal calculations to correct them if the desired result is not achieved.

Adaptive strategy/calibration modification – The internal programs by which the Ford ECA learns from the results of its monitoring of sensors and activation of actuators, and can then modify its programming to better achieve driveability, emissions, and fuel economy objectives.

Alternate fuel compatibility – Probably the next big complication for engine designers will be to allow the engines to run on various fuels. The first step is the use of methanol- and ethanol-compatible seals and other parts in the fuel delivery system. The use of entirely different fuels, such as natural gas, is predicted to be the next step towards fuel economy.

Ampule – A strong glass tube that forms the body of an electrical component such as a vacuum tube.

Analog – A signal that continuously varies within a given range and with time used for the change to occur.

Aneroid – An accordionlike capsule that expands or contracts in response to external pressure changes.

Annulus ring – On the 6.5-liter GM diesel and others with similar injection pumps, a perforated ring in the injection pump that delivers fuel to the injection pump.

Anode – A conductor with positive voltage potential applied.

Aspirator – A device using airflow through a restriction in a tube to create a vacuum in another tube.

Astroroof – A term used by Cadillac to identify a power window in the roof of a vehicle.

Aurora/Northstar – A new engine model for Oldsmobile and Cadillac, incorporating virtually all features on earlier computer control systems.

AXOD (automatic transaxle overdrive) – A four-speed automatic transmission featuring an ECA-controlled torque converter clutch used in some front-wheel-drive vehicles.

Background noise – Modern knock sensors are effectively microphones informing the computer of all the engine's internal sounds, which the computer then sorts out for frequencies characteristic of knock. By tracking the increase in background noise with higher engine speed and load, the computer can test the effectiveness of the newer type of knock sensor.

Barometric pressure – The actual ambient pressure of the air at the engine. Since this pressure varies somewhat with the weather, the information is needed to correctly determine air/fuel mixture.

Baud – A variable unit of data transmission speed (as one bit per second). In a stream of binary signals, one baud is one bit per second. Baud refers to the speed at which information can be sent and received over a data bus. The limitation comes from the sending and receiving units, not from the wire itself. Many modern automotive computers can transmit and receive at a baud in excess of 10,000. The term *baud* is in honor of the Frenchman J. M. E. Baudot, who is responsible for much of the early development in unit data transmission.

Binary code – This is the form of all the signals transferred on a data bus. It consists of a series of on/off signals corresponding to the digits of the binary numbers, each with a designated meaning. Binary code transfer is very fast.

Bit – A single piece of information in a binary code system; somewhat comparable to a single letter within a word or a single digit within a number.

Bleed – A controlled leak.

Block learn and Integrator – The adaptive memory systems used on General Motors cars. As the vehicle ages and driving conditions change (like seasonal weather or altitude), the computer can make immediate and semipermanent changes to the fuel mixture and spark maps in its keep-alive memory to optimize driveability, emissions, and fuel economy.

Breakout-box – A diagnostic tool using special connectors and a terminal box that enables the technician to check voltages on all circuits of a harness while the system is in operation.

Byte – Eight pieces of binary code information strung together to make a complete unit of information.

C3I – A name used by General Motors to describe their three-coil ignition system with a module instead of a distributor.

Calpak – A removable part of the ECM that plugs in like the PROM and is primarily responsible for storing information regarding fuel injection calibration. Used on the GM 2.0-liter engine.

Capacitor – An electric storage device, used in most pressure sensors by Ford EEC IV systems. The unit's capacitance is transformed into a frequency signal internally and sent to the ECA.

Catalyst – An agent that causes a chemical reaction between other agents without being changed itself.

Cathode – A conductor with negative voltage potential applied.

Ceramic zirconia oxygen sensor – One of the two general types of oxygen sensor, it uses ceramic zirconia as a solid electrolyte in what is essentially a low current, low voltage cell (as in a battery) to generate a voltage signal corresponding to the amount of oxygen remaining in the exhaust. The oxygen sensor must become hot before it starts to work.

Clamping diode – A diode placed across a coil or winding to trap the voltage spikes produced when the device is turned on or off.

Closed loop – Describes an intimate, signal response, triangular relationship between the O_2 sensor, the ECM, and the M/C solenoid.

Closed loop/open loop – Closed loop is the mode in which the computer adjusts the air/fuel mixture in response to signals from the oxygen sensor, which in turn responds to the changed mixture ratio. Open loop refers to control of the mixture based on other information inputs, but not the oxygen sensor signal.

Compression ignition (CI) – As opposed to spark ignition, a compression-ignition engine is a diesel engine. It ignites the air/fuel mixture entirely by the heat generated by the compression stroke. Since the typical diesel engine has a compression ratio of 17- to 23-to-1, the temperature generated is usually between 300 and 450 degrees Fahrenheit. Fuel is injected into the compression-ignition combustion chamber at the point flame propagation should begin.

Continuous injection – One of the two kinds of fuel injector fuel spray techniques, the other being pulsed injection. A continuous injector (basically limited to the Bosch K-Jetronics systems), sprays fuel whenever the engine is running. The quantity of fuel injected is determined, not by the injectors, but by the fuel distributor, a separate component.

Crowd – A steady-state, heavy-throttle condition that is less than full throttle.

D/A Converter – A converter circuit that translates digital signals to analog voltage values.

Data bus network – The data bus network is the information transfer harness between the elements on the multiplexing system. In most vehicles, this consists of a special pair of twisted wires going to each element. The wires are twisted to prevent introduction of spurious signals from electrical fields generated by speed sensors, horn and headlight circuits, radios (particularly high power or CB), and spark plug cables. If the computer industry is a dependable example (and it has been for the last ten years), the number of wires in the data bus can be expected to increase in the future, allowing parallel transmission of data. The data bus wires must remain connected and shielded to work properly.

Default mode – A limp-in mode, in which the EEC system sets all the controls to deactivated. The vehicle can still run, but with reduced performance and economy, and adverse emissions.

Detonation – Uncontrolled, rapid burning or explosion of the air/fuel charge that results in cylinder knock.

Diagnostic Routine – A section in the service manual that lists possible causes of a driveability complaint. It is the beginning point of the Diagnostic Procedure once the complaint has been verified.

Diagnostic trouble code (DTC) – A set of alphanumeric sequences the computer stores in its memory when it senses failures in certain sensor or actuator circuits. DTC is the term used in OBD II literature; trouble codes or fault codes were used before.

Digital – Using values (numbers) expressed as digits with specific meanings to represent the variables in a set of operations.

Direct ignition/waste spark ignition – A direct ignition system eliminates the ignition distributor and fires spark plugs directly from the coil. If the direct ignition system is of the waste spark type (most are), each coil fires two spark plugs simultaneously. One spark occurs at the end of its cylinder's compression stroke just before the power stroke; the other spark occurs at the cylinder 180 degrees apart in firing order from the first, during the end of that cylinder's exhaust stroke. Each spark is of opposite polarity. Nonwaste spark direct ignition systems use one coil per spark plug.

Direct injection – A type of diesel engine in which the fuel is injected directly into the center of the combustion chamber rather than into a prechamber or Ricardo chamber. Direct injection diesels are, because of the reduced area for heat to transfer into the cooling system, more fuel efficient than other types of diesel.

Divert mode – When the ECM directs the air management system to dump air into the air cleaner or to the atmosphere rather than to the exhaust system.

Driveability – Those factors, including ease of

starting, idle quality, and acceleration without hesitation, that affect the ease of driving and reliability.

Dualjet – A two-venturi carburetor consisting basically of the primary bores and circuits of a Quadrajet, usually referred to by General Motors as E2ME. The last "E" means it is equipped for CCC systems.

Duraspark – An electronic ignition system developed by Ford. Three generations of the system have evolved: Duraspark I, II, and III.

Duty cycle – The portion of time during each cycle when an electrical device is turned on, when the device is cycled at a fixed number of cycles per unit of time.

E4OD (electronic four-speed overdrive) – An automatic transmission with some valve body functions performed by electronically controlled solenoids. It features an ECA-controlled torque converter clutch and is used primarily in trucks and vans.

E-cell – A module containing a cathode and an anode, each of which is gradually sacrificed as current passes through them, eventually causing the circuit to open.

EEPROM – Electronically-erasable programmable read-only memory, an improved and updated kind of PROM. Like the PROM, it can retain tabular information for the PCM even if power is withdrawn from the computer. Unlike the PROM this information can be changed by the computer. Also, unlike the PROM, the EEPROM is a nonreplaceable part of the computer.

EGR temperature sensor – On late model cars, particularly those that are OBD II-compliant, the EGR temperature sensor serves to signal the computer that recirculated exhaust gas flows when it sends the command for it to do so. This sensor makes use of the fact that the exhaust gas is very hot, so the temperature of the passage provides a reliable indicator of exhaust flow.

Engine calibration – Adjustments or settings that produce desired results such as spark timing, air/fuel mixture, and EGR control.

Engine calibration assembly (ECA) – The name Ford gives to the microprocessor or computer that manages all the engine and combustion control systems. On some models it performs other control functions, too.

Engine metal overtemp mode – A unique feature of the Aurora/Northstar engine, whereby under circumstances of very high engine temperature (if for example the coolant has leaked out), the computer selectively disables four of the fuel injectors to prevent a temperature that would cause permanent mechanical damage to engine components.

EVAP system pressure sensor – Used on many OBD II-compliant vehicles to test the fuel vapor canister system for leaks. On many vehicles, special plumbing allows the MAP sensor to provide this function during the period between ignition on and engine running.

Fail-safe – A design feature of a system to insure continued operation at reduced effectiveness or system capacity in the event of certain component or system failures, or to prevent damage to components by disabling or deactivating certain other system components in the event of certain system failures.

"Flagged" malfunction codes – On some OBD II systems, the computer will detect and set a code, but will not turn on the check engine light unless the code is repeated. For those that "flag" malfunction codes, even if the problem does not repeat, the malfunction is retained in memory in a flagged condition, indicating that the problem did once occur. Some will erase flagged codes after a relatively large number of trips without repetition of the problem. On others, the only way the code is erased is by a scan tool or a very extended disconnection of the battery.

Flow bench – A specialty tool, chiefly used at carmakers' factories, to properly set feedback carburetors to deliver proper air/fuel mixtures. The settings determined on a flow bench are normally marked with paint to indicate that they are not to be changed.

Freeze-frame and Snapshot – Another feature of the new systems, these are computer capacities that make fault diagnosis clearer. Whenever a diagnostic trouble code is set, the computer records the information from all the other sensors and actuators that might be relevant to the fault. When this information is called from memory, there is often a useful clue what went wrong.

Front and rear oxygen sensors – To ensure that the engine control system can go into closed loop quickly, both the front and rear oxygen sensors include heater circuits to get them to operating temperature quickly as well.

Fuel temperature sensor – On some cars, a sensor mounted in or near the fuel tank to inform the computer of the fuel temperature. This information is used to calculate fuel viscosity and susceptibility to vapor lock.

Fuel trim compensation factor – A term describing the changes made by the computer to its fuel delivery memory in response to what it has learned from the previous driving operations. Also called self-learn, it is similar to, but simpler than, the GM block learn and integrator.

Gallium arsenate crystal – A semiconductor material that changes its conductivity when exposed to a magnetic field.

Glow plugs – Starting aids for diesel engines, these are electrical resistance heaters extending into the combustion chamber. By heating the glow plugs, the high temperature required for compression ignition can be achieved even when the engine is cold.

Hall-effect switch – A magnetic switching device often used to signal crankshaft position and speed to the computer.

High-pressure cutout switch – A pressure-operated switch in the high-pressure side of the air conditioning system that opens when freon pressure goes too high. When the switch opens, it disables the A/C compressor clutch circuit.

Hot-film air mass sensor – A type of air mass sensor using changes in the temperature of the heated element to determine the mass of the air passing into the engine.

Hot wire air flow sensor – An air flow sensor that keeps a special wire at a fixed number of degrees above ambient temperature in the intake air flow. The amount of current required to maintain the heat is converted into a frequency signal that conveys information about the mass of the intake air.

Hz (hertz) – Cycles per second; a measure of frequency.

Idle tracking switch – An on/off switch indicating to the computer that it is in charge of the engine for idle speed.

Impedance – The total circuit resistance including resistance and reactance.

Inductance – The property of an electric circuit by which an alternating current through it produces a varying magnetic field that indices voltages in the same circuit or in a nearby circuit. In the reactive sense, the capacity of an electric circuit to produce a counterelectromotive force when a current through it changes.

Inductive pickup/Hall-effect pickup – Two types of rotation sensor are used on many vehicles, the inductive pickup and the Hall-effect. Inductive pickups consist of a wire coil wound around a permanent magnet. When a metal tooth passes the tip of the magnet, a small, polarity-reversing alternating current is generated (this current can grow much larger with higher speed). The Hall-effect pickup receives a reference voltage from the computer and converts it into an on/off signal. Hall-effect sensors do not vary the signal voltage output with speed.

Inertia switch – A Ford trademark safety feature and occasional no-start puzzle. To protect against fuel spray after an accident, the inertia switch disables the fuel pump circuit after a vehicle jolt. Sometimes the jolt happens at other times.

Initialization mode – The beginning state of an engine control computer when the switch has been turned to on, but the engine not yet started. The computer performs a variety of internal tests and sensor/actuator checks.

Injection pressure – The pressure of the fuel at the injector's delivery end. On gasoline engines, this is controlled ordinarily by a pressure regulator to under 100 psi. On diesel engines the pressure is much higher, since the fuel must flow into the combustion chamber against the power stroke pressures. Diesel injection pressures can reach almost 30,000 psi.

Injection timing stepper motor – On certain electronically-controlled diesel systems, the electric motor used by the computer to set the injection timing. See also **stepper motor.**

Interface – Circuitry that converts both input and output information to analog or digital signals as needed, and filters external circuit voltage to protect computer circuits.

KAM – Keep-alive memory, an electronic memory in the computer that must have a constant source of voltage to continue to exist.

Lambda – A Greek letter (λ) used to indicate how far the actual air/fuel mixture deviates from the ideal air/fuel mixture. Lambda equals actual inducted air quantity divided by the theoretical air required (14.7). The actual ideal mixture will depend on the characteristics of the fuel and of the air drawn in through the intake system.

Lift pump – An electric fuel pump in or near the fuel tank that pressurizes the fuel slightly (about 15 psi) and delivers it to the main fuel pump. Found on both gasoline and diesel engines. Sometimes called a priming pump.

Light-emitting diode (LED) – A semiconductor device that generates a small amount of light when forward biased (when voltage is applied with the polarity such that current flows across the PN junction).

Limp-in mode – A state of a computer engine management system in which one or more major component circuits have failed, and a substitute value is used to retain driveability until the vehicle can be repaired.

Logic module and power module – Early to mid-1980s Chrysler products separated the computer functions into two components, the power module, which carried the higher current and worked the actuators, and the logic module, usually in a more protected place, usually more electrically protected. The logic module assesses sensor inputs and instructs the power module when to inject fuel, fire plugs and work actuators.

Longitudinal – Parallel to the length of the car (opposed to transverse).

Manifold absolute pressure (MAP)/intake manifold vacuum – Manifold absolute pressure (MAP) is the actual pressure in the intake manifold, regardless of the difference between it and ambient air pressure. Intake manifold vacuum, the traditional concept, refers to the relative difference in pressure between the two. Intake manifold vacuum is always a relative measure. MAP measurements are not relative.

Maximum authority – A term describing the maximum idle speed the ISC motor can provide.

Memory – A computer's information storage capability.

Micron – A unit of measurement that describes the size of a particle one-millionth of an inch in diameter.

Microprocessor – A processor contained on an integrated circuit; processor (central processing unit) that makes the arithmetic and logic decisions in a microcomputer.

Milliamp – One thousandth of an amp, more correctly milliampere.

Millivolt – One thousandth of a volt.

Minimum authority – A term describing the minimum idle speed achievable by retracting the ISC plunger until the throttle lever rests on the idle stop screw.

Misfire detection – Part of the required OBD II program, misfire detection closely monitors engine rpm for the slight variations when combustion misfire occurs. In most cases, the computer will disable the injector for a misfiring cylinder (to protect the catalytic converter from overheating) and will set an appropriate DTC.

Modulate – To find a position between two extremes; to alternate between on and off.

Modulated displacement – A system in which a series of ECM-controlled electromechanical devices can selectively disable the rocker arms of particular cylinders, preventing the valves from opening. The engine then operates on eight, six or four cylinders, depending on power requirements.

Motronic – The latest of the Bosch engine management systems, Motronic is their first full engine control system, regulating fuel injection and ignition timing and dwell in one unit.

Multiplexing – One of the most significant extensions of computer control technology in vehicles, multiplexing connects many or all the control units in a vehicle together over a single wire network, the data bus. Information from all components is sent over the data bus and received by all.

Noise-vibration-harshness (NVH) – A special acoustic diagnosis tool that can identify the frequencies of various noises and vibrations to compare with known frequencies. Useful to determine whether a noise or vibration is engine-speed-relative, drivetrain-speed-relative, wheel-speed-relative or random.

Normally aspirated – An engine that uses ambient atmospheric pressure to force air into the engine (opposed to turbocharged and supercharged).

OBD II – A standardized computer control self-diagnostic program, mandated for most vehicles as of the 1996 model year. Nomenclature, trouble codes and even the size, shape, and many circuits of the diagnostic connector are standardized across the industry.

Open loop – An operational mode in which the air/fuel mixture is calculated based on coolant temperature, engine speed, and engine load without benefit of the oxygen sensor. See also **closed loop.**

Palladium (Pd) – A silver-white, ductile, metallic element; number 46 on the Periodic Table.

Parameter – One of the factors defining operating conditions for a system, such as coolant temperature, engine load, and air/fuel mixture ratio.

Pickup coil – A small coil of wire in which the signal of a pulse magnetic generator is produced, such as in many electronic distributors.

Piezoelectric – A crystal that produces a voltage signal when subjected to physical stress such as pressure or vibration.

Piezoresistive – A semiconductor material whose electrical resistance changes in response to changes in physical stress.

Piezoresistor – A special kind of resistor that generates a characteristic voltage signal if it experiences vibration of a specific frequency. On engines, this frequency is chosen so as to detect detonation or knock. The piezoresistor is the basic element of the knock sensor.

Pinpoint Procedure – A service manual section that gives specific directions to find the solution of identified faults. This section should not be used until either the Diagnostic Routine or the Self-Test has identified it as the next step.

PIP signal – The profile ignition pickup signal (PIP) comes from a crankshaft- or distributor-mounted sensor indicating cylinder position to the computer.

Platinum (Pt) – A heavy, grayish white, ductile, metallic element; number 78 on the Periodic Table.

Polarity—magnetic and electrical – Both magnetism and electricity are so-called bipolar phenomena, that is, there is a north and a south to every magnetic field, just as there is a positive and a negative to every (direct current) electric circuit. Magnetic polarity is used in some Chrysler camshaft position sensors to identify whether a piston is approaching its power or its intake stroke.

Potentiometer – A three-terminal variable resistor that acts as a voltage divider. Frequently used as a position sensor for throttle angle and so on. The potentiometer can be linear or rotary.

Power module – Early to mid-1980s Chrysler products separated the computer functions into two components, the power module, which carried the higher current and worked the actuators, and the logic module, usually in a more protected place, usually more electrically protected. The logic module assesses sensor inputs and instructs the power module when to inject fuel, fire plugs and work actuators.

Powertrain control module (PCM) – The term now used (since OBD II) for the main computer in charge of engine management.

Prechamber – A prechamber is a small cavity in the cylinder head of a diesel engine into which most of the air is squeezed during the compression stroke and into which the fuel is injected. The glow plug, if present, is also in the chamber. A prechamber will often have a small diffuser pin in it to turbulate the air and fuel still more, for better combustion. Opposed to direct injection. See also **"Ricardo chamber."**

Pressure cycling switch – A pressure-operated switch in the low-pressure side of the air conditioning system that opens as freon pressure drops below a specific pressure. When the switch opens, the A/C compressor clutch is disabled. The purpose of this switch is to protect the compressor against operation when the system has leaked empty.

Primary and secondary oxygen sensors – The oxygen sensors are normal types, heated for rapid warm-up and closed-loop acquisition. What is new is the use of tandem sensors (also called primary and secondary oxygen sensors), one before and one after the catalytic converter. This combination serves as a test of the converter's effectiveness. So long as the downstream, secondary sensor shows a relatively flat signal, whatever its output voltage, the catalyst is working. When the secondary oxygen sensor starts the characteristic mixture dithering like the primary one (still used for fuel mixture trim), the catalyst is out of business.

PROM – Programmable read-only memory.

Propagation – As used in automotive technology, the spread of the flame front across the combustion chamber of an engine.

Pulsair – The name General Motors applies to the pulse-type air injection system.

Pulsed injection – One of the two kinds of fuel injector fuel spray techniques, the other being continuous injection. In pulsed injection the injectors are turned on and off, either in sequence with the engine cycles on port injection systems, or at a fixed rate with throttle body systems. The fuel quantity delivered depends on the duty cycle, the ratio of on to off, and whether or not the injector is pulsed.

Pulse width and duty cycle – These concepts are related, but not identical. Pulse width refers to the time period, usually in milliseconds, the injector is open and fuel flows. Duty cycle refers to the percentage of time the injector is on rather than off. Single-point, throttle body injectors usually cycle at a fixed rate regardless of engine speed, so when they vary their pulse width they also vary the duty cycle in direct proportion. Multipoint fuel injection systems typically inject fuel once per engine cycle per cylinder (sometimes twice when the engine is cold). Pulse width is the only important measure of their injection volume and the mixture because engine speed changes the scale of the duty cycle time.

Pumping losses – The load on the engine caused by the work needed to pump the air through the system. On gasoline engines, because of the vacuum developed behind the throttle, pumping losses are greatest at the most closed throttle positions, idle and deceleration. The aerodynamic friction of air through the intake system also contributes to a lesser extent to the pumping losses, though at high speeds and open throttle, that becomes the principal source. Diesel engines, because they are unthrottled, have very low pumping losses compared to gasoline engines.

Purge valve – A vacuum-controlled valve controlling the removal of stored HC (fuel) vapors from the charcoal canister.

Quad driver/output driver – A power transistor in the PCM, capable of working either four or seven different actuators (such as injectors or ignition coils). Each quad driver or output driver works when the PCM uses it to ground a component's circuit, completing it.

Quick Test – The whole set of procedures that make up the Ford diagnostic procedure.

RAM – Random access memory.

Range-switching sensors – Sensors, particularly temperature sensors, with the capacity to "telescope" their information within the range of greatest interest. Thus Chrysler range-switching temperature sensors can, in effect, spread out the temperature scale from the lower reaches of normal warm operating conditions to the upper. Range-switching also allows for the nonlinear response of sensors. Their purpose is to provide much more detailed information about a particular area of their monitoring responsibility.

Reactance – Resistance, expressed in ohms, within an AC circuit that is a result of inductance or capacitance.

REF pulse – The basic timing signal sent to the computer, indicating crankshaft position. This signal is modified by the computer to set the actual timing.

Rheostat – A variable resistor that usually has two leads and that serves as a voltage or current limiter in series with a load. Used for controlling current in an electrical circuit.

Rhodium (Rh) – A white, hard, ductile, metallic element; a member of the noble metals family; number 45 on the Periodic Table.

Ricardo chamber – Similar to a prechamber, a Ricardo chamber is a small cavity in the cylinder head of a diesel engine into which most of the air is squeezed during the compression stroke and into which the fuel is injected. The glow plug, if present, is also in the chamber. Opposed to direct injection.

ROM – Read-only memory.

Scan tool – A diagnostic tool used to retrieve from a vehicle's computer memory any malfunction codes it has retained. On earlier systems, each manufacturer used a scan tool connector and setup unique to that manufacturer (though there are aftermarket scan tools with adapters that can read many of them). With OBD II-compliant vehicles, the scan tool is a standardized tool that can be used for the same purpose on any vehicle. Many scan tools can also be used to get direct measurements of various sensors and to exercise various actuators to check on their effectiveness.

Schmitt trigger – A device that trims an analog signal and converts it to a digital signal.

Secondary oxygen sensor – The oxygen sensors are normal types, heated for rapid warm-up and closed-loop acquisition. What is new is the use of tandem sensors (also called primary and secondary oxygen sensors), one before and one after the catalytic converter. This combination serves as a test of the converter's effectiveness. So long as the downstream, secondary sensor shows a relatively flat signal, whatever its output voltage, the catalyst is working. When the secondary oxygen sensor starts the characteristic mixture dithering like the primary one (still used for fuel mixture trim), the catalyst is out of business.

Self-diagnostic/on-board diagnostics – These terms mean essentially the same thing: the capacity of the computer to monitor its sensors' and actuators' circuits as well as its own internal functions. In a self-diagnostic system, the computer retains in electronic memory any malfunctions it may encounter, stored as codes the technician can recall with a scan tool.

Self Test – The self-diagnostic portion of the Ford quick-test in which the computer tests itself and its circuits for faults.

Sensor – All the inputs and outputs to the computer are either sensors that provide information or actuators that do work. Most sensors are sent a reference voltage which is reduced in a predictable way by whatever parameter they monitor. Some generate voltages of their own. Actuators (except for some midrange single-point injectors) are ordinarily externally powered, and the computer triggers them by providing a return ground path.

Sequential fuel injection – Sequential fuel injection systems inject fuel into the intake runner just upstream of each cylinder in the firing order of the engine to better vaporize the fuel and prevent fuel condensation on other intake manifold components.

Signal (or voltage signal) – The voltage values that a computer uses as communication signals.

Single-point/multipoint fuel injection – Single-point injection refers to fuel injection using the equivalent of an electric carburetor. On four-cylinder engines, this usually involves a single injector, but on some V-form engines two injectors are combined in one unit. Also called throttle-body fuel injection. Multipoint fuel injection employs one (or rarely two) injectors for each cylinder.

Snapshot and Freeze-frame – Another feature of the new systems, these are computer capacities that make fault diagnosis clearer. Whenever a diagnostic trouble code is set, the computer records the information from all the other sensors and actuators that might be relevant to the fault. When this information is called from memory, there is often a useful clue what went wrong.

Spark ignition (SI) – A spark-ignition engine depends on the heat generated by a spark across the electrodes of a spark plug to ignite the air/fuel mixture. Opposed to compression ignition.

Speed density formula – A term referring to a method of calculating the flow rate and density of the intake air. The result of the calculation indicates the air's ability to evaporate fuel.

Spout – The Ford Spark Output (SPOUT) signal comes from the ECA and indicates to the ignition module when to fire the next spark plug. The exact timing is calculated on the basis of inputs from all the computer's sensors.

Standardized diagnosis – The new feature of OBD II is the standardization of much of the engine management systems across manufacturers' car lines and between manufacturers. OBD II, if successful, will allow the use of common scan tools and other diagnostic aids for the solution of like problems, regardless of vehicle manufacturer.

Standardized DLC-3 16-PIN diagnostic socket – On OBD II-compliant vehicles, the diagnostic socket to which the scan tool connects during diagnosis. This socket is standardized for all cars, though there are a number of circuits that can be used by the manufacturer for special purposes.

Star Tester – A tester designed to test MCU and EEC IV systems that activates and deactivates the Self-Test procedure and displays a digital readout of the service codes.

Stepper motor – A type of direct-current electric motor with the unique capacity to move an exact number of turns in response to a computer's signal. Typically such a motor, used to adjust the idle speed bypass opening, for example, has a range of 256 different possible positions it can be set to. This allows the computer to open or close to an exact position. Such motors are also increasingly used in some heater door control systems.

Stoichiometric – In automotive terminology it refers to an air/fuel ratio in which all combustible materials are used with no deficiencies or excesses; 14.7 parts air to 1 part fuel, by weight, at sea level.

Thermac – The term used by General Motors to identify the heated air inlet system. The term is a shortened form of thermostatic air cleaner.

Thermactor – The name Ford Motor Company uses to identify its air injection pump for exhaust emissions reduction.

Thermistor – A temperature-sensing device containing a coil of wire whose resistance changes dramatically as the temperature changes. Most thermistors have a negative temperature coefficient (the resistance goes up as the temperature goes down).

Thermo-time switch – An electrical switch usually applied to a bimetallic strip that uses heat to open or close a set of electrical contacts. The amount of heat is controlled, so the time it takes to open or close the contacts is always the same.

Thick film integrated (TFI) – A term that identifies the type of chip used in the ignition module that is attached to the side of the distributor for EEC IV system applications.

Throttle body backup (TBB) – A backup circuit in the ECM of most EFI systems that operates the injector solenoid in the event of a partial system failure.

Throttle-body injection (TBI) – Throttle-body injection uses a single fuel injector in a throttle body (or a pair of two for V-form engines) that delivers the fuel to the intake manifold. The TBI unit is very similar to an electric carburetor.

Throttle kicker (TK) – A device operated either by vacuum or electrically that bumps the throttle open slightly more when activated during idle, ordinarily used to compensate for an added load, such as air conditioning compressor engagement.

Timer core – A star-shaped wheel attached near the top of the distributor shaft in an HEI distributor; also called the reluctor. It has one point per engine cylinder.

Torque converter lockup clutch – A computer-controlled, hydraulically actuated clutch in the automatic transmission's torque converter that locks the transmission's input shaft to the engine's crankshaft for fuel economy. The computer determines when to send this command on the basis of sensor inputs.

Torque management – A set of capacities the engine control system has, allowing it to reduce engine torque output under specific circumstances. Its first strategy is to retard spark; in more extreme circumstances it can selectively shut off fuel injectors. The traction control system employs part of the torque manage-

ment system to prevent front wheel spin under power when the pavement is slippery.

Transducer – A device that receives energy from one system and transfers it to another system, usually in a different energy form.

Transfer pump – On some diesel systems, a pump that receives the fuel either unpressurized from the tank or from the lift pump, and delivers it to the injection pump.

Transverse – Perpendicular to the centerline of the vehicle (opposed to longitudinal).

Two-trip malfunction detection – On some vehicles, the computer waits to record certain faults until they appear on two separate driving trips. Not all circuits fall under this provision; the most important will set codes after the first failure.

Unthrottled – Diesel engines are said to be unthrottled because they have no throttle and draw in as much air as possible for each cylinder cycle. Hence air/fuel ratio is unimportant for the diesel, except that more fuel means more power up to the limits of the available oxygen.

Vacuum control valve (VCV) – A switch opening or closing its ports to supply vacuum in response to coolant temperature change. It is sometimes referred to as a *ported vacuum switch* or a *thermal vacuum switch*. Thermal vacuum switch, however is more accurately used to refer to a switch located in the air cleaner, opening or closing a port in response to air temperature.

Vacuum fluorescent (VF) – A glass tube containing anodes and a cathode. The anodes are coated with a fluorescent material and are positioned so they form all of the segments of alphanumerical characters. As current passes through the cathode, it gives off electrons. If the anodes have positive voltage applied, the electrons given of by the cathode are attracted to them and strike the fluorescent material, causing it to glow.

Vacuum pump – On automotive and light truck diesels, a mechanically driven pump to develop vacuum. It is used for various purposes like operating heater doors, and is necessary on a diesel engine because there is no effective vacuum developed by the engine itself.

Varajet – A two-venturi carburetor consisting basically of one primary and one secondary bore and related circuitry from a Quadrajet; usually referred to by General Motors as E2SE, the last "E" meaning it is equipped for CCC systems.

Voltage drop test – A more precise substitute for ground resistance tests, the voltage drop test checks through each connection from power to ground to see whether there is a change in voltage through the connection. This test capitalizes on the fact that voltmeters are very accurate at the bottom of their scale, while ohmmeters are less accurate near zero.

Voltage spike – A sudden increase in the voltage in a particular circuit. Virtually any time an electric power consumer turns on or off, it generates some sort of voltage spike, but these are dangerous only if the circuit includes more delicate components, especially computers. Cars are designed with separate circuits to protect spike-sensitive elements during normal operation, but care must be taken in diagnosis not to circumvent these protections. If in doubt, measure a circuit separately with a high-impedance DVOM.

Volumetric efficiency (VE) – A measure of cylinder-filling efficiency. VE is expressed as the percentage of atmospheric pressure to which the cylinder is filled at the completion of the compression stroke. This efficiency varies with engine speed; the point of highest volumetric efficiency (which can rise above 100% in some cases) is the point of highest engine output torque.

Vortec injection system – A new type of fuel injection system by GM in which there is a central unit with the metering functions and separate, passive injectors on tubes for each cylinder.

Wastegate – A valve or door that when opened allows the exhaust gas to bypass the exhaust turbine of a turbocharger. The wastegate is used to limit turbocharger boost pressure.

Waste spark – An ignition system using one coil between every complementary pair of cylinders, firing both plugs simultaneously, one at the end of the compression stroke and the other at the end of the exhaust stroke. A waste spark system has a coil for every pair of cylinders.

WOT – Wide-open throttle

Zirconium dioxide (ZrO_2) – A white crystalline compound that becomes oxygen-ion conductive at about 315 degrees Centigrade (600 degrees Fahrenheit).

Index